Applied Hydrology

Applied Hydrology

Editor: Allison Sergeant

STATES
ACADEMIC PRESS
www.statesacademicpress.com

States Academic Press,
109 South 5th Street,
Brooklyn, NY 11249, USA

Visit us on the World Wide Web at:
www.statesacademicpress.com

ISBN: 978-1-63989-052-1 (Hardback)

Cataloging-in-Publication Data

Applied hydrology / edited by Allison Sergeant.
p. cm.
Includes bibliographical references and index.
ISBN 978-1-63989-052-1
1. Hydrology. 2. Water. I. Sergeant, Allison.
GB661.2 .A66 2022
551.48--dc23

Table of Contents

Preface

Hydrology is the discipline that focuses on the scientific study of water present on Earth or other planets. It includes the movement, quality and distribution of water on the planets including water resources, water cycle and environmental watershed sustainability. It focuses on analyzing water related problems such as water management, natural disasters, environmental preservation and provide their solutions. Hydrology is sub-divided into groundwater hydrology, surface water hydrology and marine water hydrology. Surface hydrology, hydrometeorology, hydrogeology, drainage basin management and water quality are some of the other domains of hydrology. Water circulation or water cycle is the central aspect of hydrology. It is concerned with how water circulates across the Earth through various pathways. This book unravels the recent studies in the field of hydrology. Different approaches, evaluations, methodologies and advanced studies have been included herein. Those in search of information to further their knowledge will be assisted by this book.

The information shared in this book is based on empirical researches made by veterans in this field of study. The elaborative information provided in this book will help the readers further their scope of knowledge leading to advancements in this field.

Finally, I would like to thank my fellow researchers who gave constructive feedback and my family members who supported me at every step of my research.

Editor

Preface

1

The integrated water balance and soil data set of the Rollesbroich hydrological observatory

Wei Qu, Heye R. Bogena, Johan A. Huisman, Marius Schmidt, Ralf Kunkel, Ansgar Weuthen, Henning Schiedung, Bernd Schilling, Jürgen Sorg, and Harry Vereecken

Agrosphere Institute (IBG-3), Forschungszentrum Jülich GmbH, 52425 Jülich, Germany

Correspondence to: Heye R. Bogena (h.bogena@fz-juelich.de)

Abstract. The Rollesbroich headwater catchment located in western Germany is a densely instrumented hydrological observatory and part of the TERENO (Terrestrial Environmental Observatories) initiative. The measurements acquired in this observatory present a comprehensive data set that contains key hydrological fluxes in addition to important hydrological states and properties. Meteorological data (i.e., precipitation, air temperature, air humidity, radiation components, and wind speed) are continuously recorded and actual evapotranspiration is measured using the eddy covariance technique. Runoff is measured at the catchment outlet with a gauging station. In addition, spatiotemporal variations in soil water content and temperature are measured at high resolution with a wireless sensor network (SoilNet). Soil physical properties were determined using standard laboratory procedures from samples taken at a large number of locations in the catchment. This comprehensive data set can be used to validate remote sensing retrievals and hydrological models, to improve the understanding of spatial temporal dynamics of soil water content, to optimize data assimilation and inverse techniques for hydrological models, and to develop upscaling and downscaling procedures of soil water content information.

1 Introduction

Climate and land use changes are taking place on different spatial and temporal scales, affecting all environmental compartments. Soil water content is known to be a major control for evapotranspiration, precipitation–runoff response, and heat transfer between soil and atmosphere, and it plays an essential role for climate projections, weather and flood forecasting, water and soil resources management, agriculture, and water quality control (Albertson and Kiely, 2001; Betts et al., 1996; Crow et al., 2005; Robinson et al., 2008; Vereecken et al., 2008; Western et al., 2002). However, the highly heterogeneous pattern of soil water content leading to complex and scale-dependent patterns of water, energy, and matter fluxes makes it challenging to predict terrestrial system responses for both scientists and policymakers (Jaeger and Seneviratne, 2011; Seneviratne et al., 2010). Therefore, integrated observations of soil water content and the exchange of water and heat between the soil, vegetation, and atmosphere are critical to improving our understanding of the terrestrial system response to changes in climatic conditions and land management (Dirnbock et al., 2003; Foley et al., 1998; Hinzman et al., 2005; Refsgaard, 1997; Seneviratne et al., 2010; Guo and Lin, 2016) and serve as key data in validating remote sensing data products (e.g., Rötzer et al., 2014; Cosh et al., 2016).

To this end, a network of integrated observation platforms has been established in the framework of the Terrestrial Environmental Observatories (TERENO) initiative (funded by Helmholtz Association of German Research Centers) to investigate the consequences of global change on terrestrial ecosystems (Bogena et al., 2012; Zacharias et al., 2011). TERENO aims to collect long-term time series of system states and fluxes using state-of-the-art monitoring technologies, e.g., to close the local water balance (Graf et al., 2014) and to investigate of the effects of deforestation on water, energy, and matter fluxes in an integrative manner (Bogena et al., 2015). The Rollesbroich grassland experimental catchment is part of the TERENO observatory Eifel/Lower Rhine Valley. All components of the water balance (e.g., precipitation, evapotranspiration, runoff, soil water content) are continuously monitored using state-of-the-art instrumentation, providing detailed information about the spatial and temporal variation of the local water cycle for the evaluation of hydrological models (Bloschl and Sivapalan, 1995; Thompson et al., 2011). In addition, using water balance data within the context of hydrological modeling helps to determine measurement errors, to diagnose such errors, and to avoid misattribution of water balance components (Evett et al., 2012; Kampf and Burges, 2010; Vasilenko, 2004). Finally, quantification of water balance components is helpful for understanding the availability of water resources, the potential of hydrologic extremes such as floods and droughts, and the interactions between the land surface and the atmosphere (Flerchinger and Cooley, 2000; Huntington, 2006).

Here, we present a comprehensive hydrological data set recorded in the Rollesbroich catchment from 1 May 2011 to 31 December 2013. The hydrological data set includes time series of meteorological forcing (i.e., precipitation, air temperature, air humidity, radiation components, and wind speed), actual evapotranspiration, runoff, as well as soil water content data from a wireless sensor network (SoilNet). In addition, information on soil physical properties and vegetation (i.e., leaf area index, LAI) useful for the parameterization of hydrological models is presented.

2 Catchment description

The Rollesbroich catchment (50°37′27″ N, 6°18′17″ E) is located in the Eifel mountain range and covers an area of about 40 ha with altitudes ranging from 474 to 518 m a.s.l. The catchment mean annual air temperature and precipitation are 7.7 °C and 103.3 cm, respectively, for the period from 1981 to 2001. These data are recorded by a meteorological station operated by the North Rhine-Westphalian State Environment Agency with a distance of 4 km from the Rollesbroich catchment. The dominant soils are Cambisols in the southern part and Stagnosols in the northern part of the catchment. The grassland vegetation is dominated by perennial ryegrass (Lolium perenne) and smooth meadow grass (Poa pratensis). The average slope within the hydrological observatory is 1.63° (min.: 0.35°, max.: 3.12°).

3 Methods

3.1 Meteorological data

Meteorological data, i.e., precipitation, air temperature, air humidity, radiation components, and wind speed, were recorded at a micrometeorological tower located in the center of the almost flat terrain in the southern part of the Rollesbroich catchment (see Fig. 1). Wind speed was obtained with a sonic anemometer at 2.6 m above surface (CSAT3, Campbell Scientific, Inc., Logan, USA). The H_2O concentration was measured using an open-path infrared gas analyzer (LI7500, LI-COR Inc., Lincoln, NE, USA) at the same height. Air temperature and humidity (HMP45C, Vaisala Inc., Helsinki, Finland) were measured at 2.6 m height above the ground surface. Incoming short- and long-wave radiation were determined using a NR01 four-component net radiometer (Hukseflux Thermal Sensors, Delft, Netherlands). Data of all instruments including diagnostic data were recorded with a logger (CR3000, Campbell Scientific, Logan, UT, USA) at 20 Hz. Precipitation was recorded by a heated Hellmann-type tipping bucket rain gauge (eco-Tech GmbH, Bonn, Germany). In July 2013, a weight-based precipitation gauge (Pluvio^2, OTT Hydromet GmbH, Kempten, Germany) was added to the nearby backup climate station, providing more accurate measurements of all precipitation types. Both precipitation gauges were installed at a height of 1 m above ground surface as recommended by the German Weather Service for sites at an elevation above 500 m a.s.l. with occasional heavy snowfall. All meteorological measurements were stored at 10 min intervals.

3.2 Actual evapotranspiration

Latent heat flux was obtained by the eddy covariance (EC) technique. The EC post-processing software TK3.1 (Mauder and Foken, 2011) was used to calculate latent heat flux from the vertical wind velocity obtained by the sonic anemometer (CSAT3, Campbell Scientific, Inc., Logan, USA) and water vapor density obtained by an infrared gas analyzer (LI7500, LI-COR Inc., Lincoln, NE, USA). The processing and quality assurance of the EC data followed the corresponding TERENO strategy presented in Mauder et al. (2013). Actual evapotranspiration was calculated from the latent heat flux using

$$\mathrm{ET_a} = \frac{\mathrm{LH}}{\rho_\mathrm{w} \cdot L_\mathrm{water}} \tag{1}$$

$$L_\mathrm{water} = 10^{-3} \cdot (2500.8 - 2.36 \cdot T + 0.0016 \cdot T^2 - 0.00006 \cdot T^3), \tag{2}$$

where $\mathrm{ET_a}$ is actual evapotranspiration ($\mathrm{m\,s^{-1}}$), LH is latent heat flux ($\mathrm{W\,m^{-2}}$), ρ_w is water density ($\mathrm{kg\,m^{-3}}$), L_water is latent heat of condensation of water in the temperature range from −25 to 40 °C ($\mathrm{J\,kg^{-1}}$), and T is air temperature (°C).

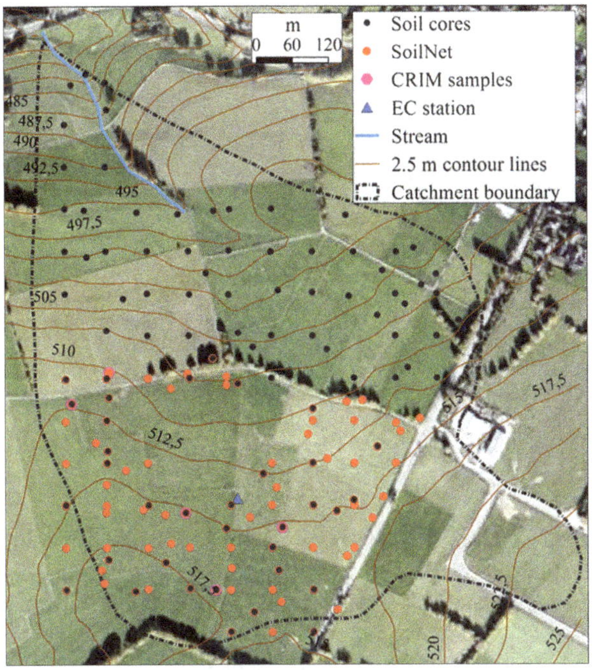

Figure 1. Map of the Rollesbroich catchment showing locations of the SoilNet sensor network, locations of the soil samples to determine soil physical and chemical properties, locations of the soil samples for site-specific calibration of the CRIM model, the location of the eddy covariance (EC) station, 2.5 m contour lines, and catchment boundary.

3.3 Runoff

Runoff was measured at the catchment outlet using a gauging station equipped with a combination of a V-notch weir for low flow measurements and a Parshall flume to measure normal to high flows. Runoff data of the two weir types were combined by using V-notch values for water levels below 5 cm, Parshall flume values for water levels greater than 10 cm and the weighted mean of V-notch and Parshall flume values for water levels between 5 and 10 cm, where the water levels refer to those of the V-notch weir.

3.4 Soil water content

Soil water content was measured at 87 SoilNet locations within the southern part of catchment (Fig. 1) using SPADE soil moisture sensors (Qu et al., 2013; Hübner et al., 2009). The SPADE sensors were installed at 5, 20, and 50 cm depth. Two SPADE sensors were installed in parallel at each depth with a distance of ~ 10 cm to increase the sensing volume and to allow examination of inconsistencies in sensor reading. The measurement frequency was 15 min.

The SPADE sensor is a ring oscillator. The oscillator frequency is a function of the dielectric permittivity of the surrounding medium (Qu et al., 2013), which strongly depends on the water content of the soil because of the high permittivity of water ($\varepsilon_w \approx 80$), as compared to mineral soil solids ($\varepsilon_s \approx 2 \sim 9$), and air ($\varepsilon_a \approx 1$). The temperature of the soil was also determined by the SPADE sensor using a digital thermometer (DS18B20) with an accuracy of ± 0.5 °C in the range from -10 to 85 °C. The two-step calibration procedure suggested by Jones et al. (2005) was used to relate sensor reading to soil moisture. In a first step, reference liquids with a known dielectric permittivity were used to calibrate the following empirical model (Eq. 3) that relates sensor reading to apparent dielectric permittivity:

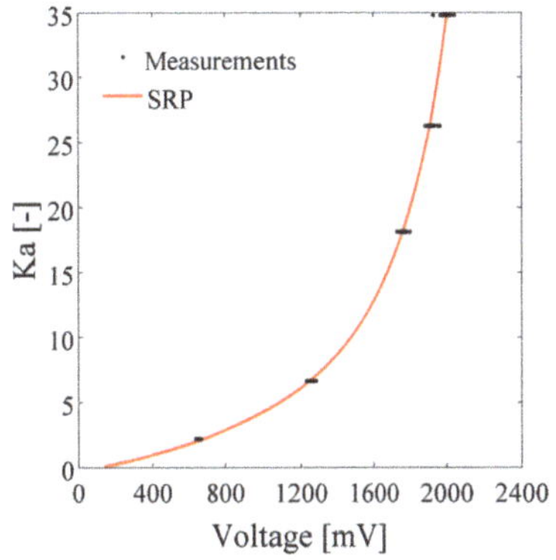

Figure 2. Sensor output of 60 SPADE sensors in five reference liquids (K_a refers to dielectric permittivity). The fitted "universal" calibration relationship (Eq. 3) is also presented.

$$K_a = \gamma + \frac{1}{\alpha + \beta / \nu}, \tag{3}$$

where K_a is the dielectric permittivity, α, β, and γ are the fitting parameters, and ν is sensor output (unit, V). Prior to installation, 60 SPADE sensors were calibrated in five reference liquids that covered a permittivity range from 2.2 to 34.8. The outputs for the 60 sensors as well as the fitted model are shown in Fig. 2. The root mean square error (RMSE) between known and predicted dielectric permittivity was 0.0188, and the best fitting parameters of α, β, and γ were -0.1502, 0.3612, and -0.1599, respectively.

In a second step, a site-specific calibration between dielectric permittivity and soil water content was obtained using a complex refraction index model (CRIM, Eq. 4) as proposed by Birchak et al. (1974):

$$\theta = \frac{K_a^{0.5} - (1-\eta) \cdot K_{\mathrm{solid}}^{0.5} - \eta \cdot K_{\mathrm{air}}^{0.5}}{K_{\mathrm{water}}^{0.5} - K_{\mathrm{air}}^{0.5}}, \tag{4}$$

where η is the porosity of the soil, $1-\eta$ is the solid fraction, K_a is the permittivity of soil, and K_{water}, K_{solid}, and K_{air} are the permittivity of water, solids, and air component of soil, respectively. In order to estimate the unknown value of K_{solid} and to assess the accuracy of this relationship, 15 undisturbed samples (length = 7.7 cm, diameter = 5 cm) were taken from the two main soil types in 5, 20, and 50 cm

Table 1. Parameters and the RMSE of the CRIM model for 5, 20, and 50 cm depth.

	5 cm	20 cm	50 cm
K_{water}	78.54	78.54	78.54
K_{solid}	2.08	3.78	4.40
K_{air}	1.00	1.00	1.00
η	0.59	0.49	0.41
RMSE	0.028	0.025	0.022

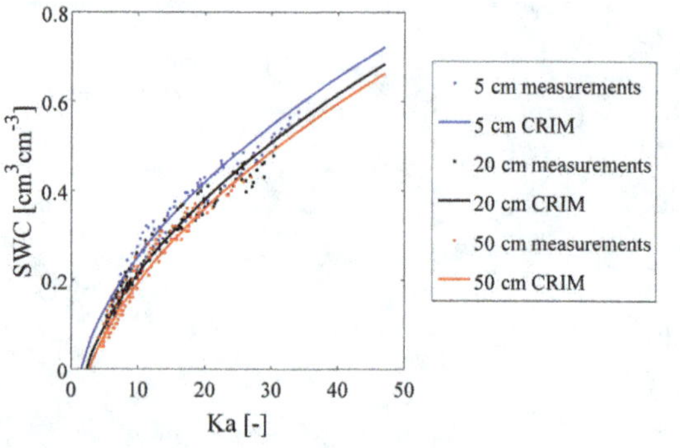

Figure 3. Relationship between dielectric permittivity and soil water content for the Rollesbroich test site and the derived K_a-θ models (CRIM).

depth. These samples were first saturated with deionized water and then CS 640-L three-rod TDR probes with a length of 7.5 cm were inserted in the middle of the sample. These probes were connected to a TDR100 system (Campbell Scientific, Inc., Logan, USA) to determine the dielectric permittivity of the soil samples using a custom-made Matlab algorithm based on the travel time analysis algorithm of Heimovaara and Bouten (1990). Subsequently, the samples were dried and both weight and dielectric permittivity of each sample were determined in regular time intervals. After drying at room temperature, the remaining water was removed by oven-drying at 105 °C for 24 h so that the dry bulk density, porosity, and the volumetric soil water content could be determined from the recorded weights.

The measured dielectric permittivity and soil water content and the fitted relationships are shown in Fig. 3. Because of the large difference in porosity at the three depths, the mean porosity at each depth was determined from the calibration samples and used to parameterize three different relationships. After fitting the solid permittivity for each depth, the performance of these calibration relationships was judged by the RMSE (Table 1). It was found that the three relationships performed well with a RMSE ranging from 0.022 to 0.028 $cm^3 cm^{-3}$.

As already briefly discussed in Qu et al. (2014), we found that sensor output showed artificially pronounced diurnal variations after the deployment of the sensor network. This behavior was attributed to a charging capacitor that affected the first reading of the SPADE sensor. If multiple sensor readings were made sequentially without turning off the sensor, the stability of the measurement considerably improved and the temperature dependence of the measurements disappeared (Qu et al., 2014). To correct these temperature-dependent oscillations in sensor reading, the acquisition software of SoilNet was changed temporarily so that two measurements were saved from 5 September 2012 to 3 March 2013. After 3 March 2013, the software was updated again so that only the second more accurate measurement was saved. Figure 4 shows an example of the measured voltage for the first and second measurements and the soil temperature for a selected sensor (location 053, 5 cm depth). It can be seen that the difference between the two measurements (Δv) is strongly correlated with soil temperature and could be fitted with an empirical second-order polynomial function with a RMSE of 5.18 mV (Fig. 5). Such second-order polynomial functions were obtained for all sensors individually and subsequently used to correct measurements made between April 2011 and September 2012. After the correction, measurements from the closely spaced sensors at a single measurement location agreed well with each other with a RMSE that varied from 0.010 to 0.035 $cm^3 cm^{-3}$. The uncorrected and corrected voltage as well as the associated soil water content of one representative sensor is plotted in Fig. 6. It is clear that the corrected measurements before September 2012 better match the expected soil water content after September 2012. On average, the corrected soil water content was 0.07 $cm^3 cm^{-3}$ lower than the uncorrected values. The plausibility of the corrected soil water content values is further supported by the fact that the increase in soil water content in the winter of 2012 disappeared after the temperature correction. The corrected soil water content is now relatively constant in winter and the maximum of the soil water content corresponds well with the porosity determined from the soil samples (both 0.59 $cm^3 cm^{-3}$).

3.5 Soil physical properties

Soil cores were taken at all locations where soil water content sensors were installed (length of 100 cm, diameter of 8 cm; Carl Hamm GmbH, Essen, Germany). Then, soil samples were taken from three pedological horizons (0–10, 10–20, and 20–40 cm) within the soil cores (see Fig. 1), which resulted in a total of 282 soil samples. Sand, silt, and clay fractions were determined using a combination of wet sieving (sand fractions) and sedimentation (silt and clay fraction) following ISO-11277 (2009). Organic carbon (C) and total nitrogen (N) content were determined by a combination of standard laboratory analyses (ISO-10694, 1995) and mid-infrared spectroscopy (MIRS) using partial least square regression (PLSR) (Bornemann et al., 2008). Mid-infrared spectra were recorded in diffuse reflectance mode by a Ten-

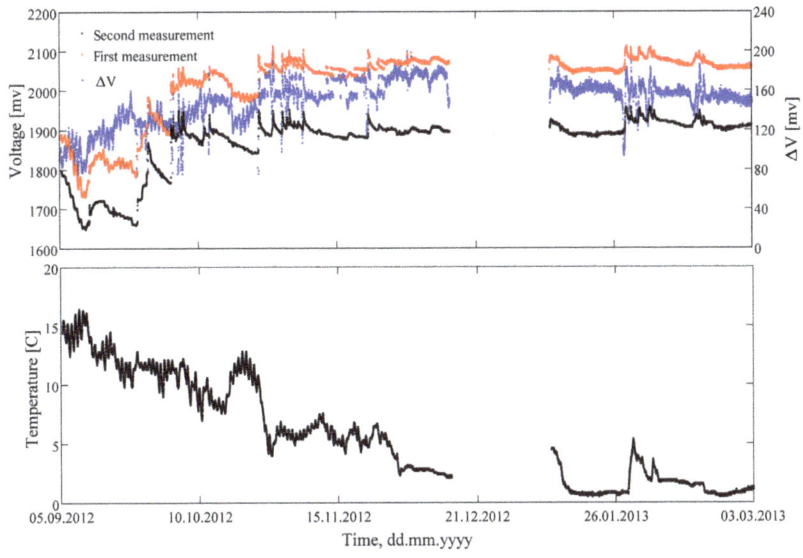

Figure 4. Time series of the first and second voltage measured after software update as well as the respective voltage differences (ΔV) and the associated temperatures.

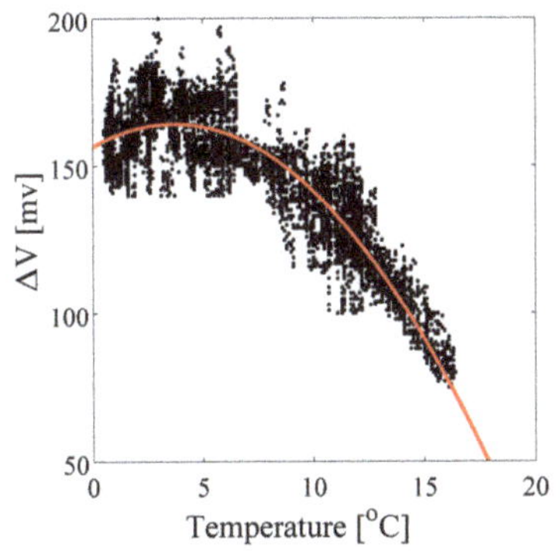

Figure 5. Second-order polynomial function fitted to the relationship between soil temperature and the difference between the first and second voltage measurements.

sor 27 (Bruker, Billerica, United States), extended with a high throughput device (HTS-XT) equipped with a liquid N_2-cooled mercury–cadmium–telluride detector. The OPUS 6.0 software (Bruker) was used for spectrum analyses and evolution of prediction models for C and total N. Models were validated through cross- (leave-one-out) and test-set validation (10 %). Cross validation yielded coefficients of determination of 0.99 (for both C and total N).

3.6 Leaf area index

The agricultural management of the different fields in the Rollesbroich catchment is very similar. Heterogeneity of the grass cover is mainly caused by different mowing times, which typically vary only by a few days. Therefore, we assume that the grass cover is homogeneous on the long-term in the Rollesbroich catchment. The effective leaf area index ($\mathrm{LAI_{eff}}$) that contributes to actual evapotranspiration was computed from grass height, h, using the following equations (Allen et al., 2006; Rochette et al., 1991):

$$\mathrm{LAI} = 24 \cdot h, \tag{5}$$

$$\mathrm{LAI_{eff}} = \frac{\mathrm{LAI}}{0.3 \cdot \mathrm{LAI} + 1.2}. \tag{6}$$

Average grass height in the Rollesbroich catchment was determined weekly by measuring grass height at five representative locations in the catchment.

Time series of LAI were also derived from RapidEye images using the normalized difference vegetation index (NDVI) approach (Myneni et al., 1997). The NDVI is calculated from infrared (NIR) and red-edge spectral bands. The NDVI of a plant with high LAI has a high ratio between NIR and red reflectance that can be detected by RapidEye. The LAI was computed from the NDVI values using a radiative transfer model (Myneni et al., 1997). Detailed information about the procedure can be found in Ali et al. (2013).

4 Data management and quality control

The distributed spatial data infrastructure TEODOOR (http://www.tereno.net) was developed to handle, describe, exchange, and publish all monitored environmental data of the TERENO project (Kunkel et al., 2013). Each institution hosting an observatory maintains its local data infrastructure. The observatories are connected via OGC-compliant web services, while the TERENO Data Discovery Portal (DDP) as a central application enables data searching, visualization and

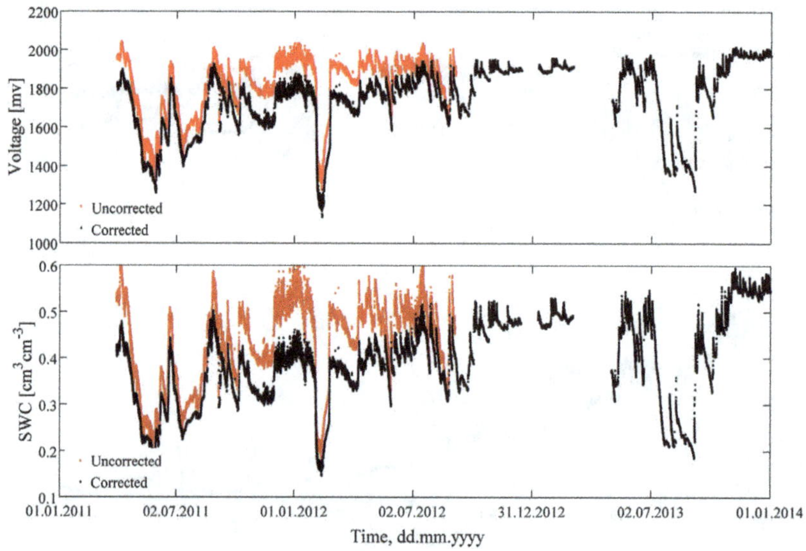

Figure 6. Time series of the uncorrected (red) and corrected (black) voltage and soil water content (SWC) measurements of one representative sensor.

download. According to the TERENO data policy it is obligate that each data set is described by standardized metadata elements, like ISO19115, OGC SensorML, or NetCDF's CF Conventions.

The observation data sets are processed and assessed within TEODOOR in three different ways (for details see Devaraju et al., 2015). In the first data processing type, the imported data undergo automated quality checks (e.g., minimum/maximum thresholds) and subsequently are published after visual inspection by experts. For example, data from weather stations and river gauges are processed in this way. Secondly, more complex data can be externally processed and assessed by the principal investigators and subsequently imported into TEODOOR, e.g., EC data. The third data processing type also involves automatic data import, but in this case the data quality assessment is executed using an external evaluation method developed by the responsible scientist, e.g., wireless sensor network data. Subsequently, the flagging information is updated by TEODOOR after the quality assessment has been completed.

Characterization of data quality is done by three descriptors, which are stored together with each observation: data uncertainty, data processing levels, and data quality flags. The observed data values remain unchanged in any case. Data uncertainty arises from the observation process itself and is mainly determined by the accuracy of the sensors used. Data processing levels indicate the status of data handling. For instance, (unpublished) raw data are termed "level 1" and "level 2" refers to data subjected to quality control, whereas the higher levels refer to derived data products. The flagging scheme consists of two tiers: the first tier includes generic flags, e.g., "good", "unevaluated", "suspicious", or "bad". The second tier is use-case-specific and indicates either the result of individual quality tests, e.g., failed gradient checks, or background events affecting data values, e.g., icing events. In the following section, we present in more detail how data were checked for plausibility to derive the data sets used in this study.

4.1 Meteorological data and latent heat flux

Meteorological data were checked for quality by a multi-step quality control including the use of diagnostic information provided by the instruments, the application of site specific plausibility limits, visual inspections of the data series, and cross-checks with data from the nearby backup weather station. The quality control of the latent heat flux was in accordance with the standardized method for the processing and quality assessment of EC data as suggested by Mauder et al. (2013). This scheme includes site-specific plausibility limits and the application of a spike removal algorithm based on median absolute deviation of raw measurements. Processed half-hourly fluxes and statistics were checked using three different flags (high, moderate, and low) based on tests of integral turbulence and stationarity (Foken and Wichura, 1996). As a result of this quality check, 48.5 % of all possible 30 min observations of latent heat flux were marked as good-quality data, 12.2 % were marked as moderate quality, 35.4 % were marked as low-quality data (35.4 %), while 3.9 % were marked as missing data because of power failure and sensor maintenance. In this study, only data of high and moderate quality were used. A more detailed description of the treatment of EC data can be found elsewhere (Gebler et al., 2015; Post et al., 2015).

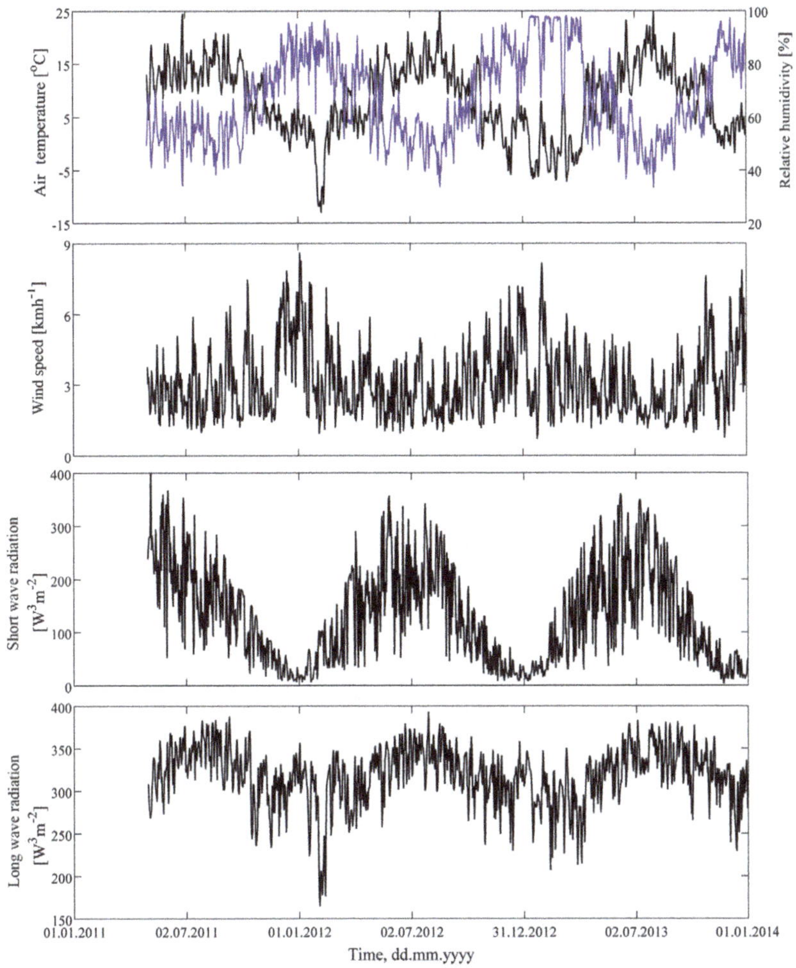

Figure 7. Daily averages of air temperature, relative humidity, wind speed, and incoming short- and long-wave radiation measured at the eddy covariance station from 1 May 2011 to 31 December 2013.

4.2 Runoff

As a first quality check, time series of the V-notch weir and a Parshall flume were compared for consistency. In addition, both runoff time series were visually inspected for inexplicable outliers (e.g., runoff peak without preceding rainfall event) and sensor failures. As outlined above, unreliable data were identified and appropriate flags were set.

4.3 Soil water content

Measurements of soil moisture outside the physical plausibility range (0.05 to 0.85 $cm^3\,cm^{-3}$) were identified and flagged (please note that we used a conservative upper limit that can also be used for organic soils). Subsequently, unreliable measurements were identified by analyzing the first derivative of the soil water content time series. In case an increase larger than 2 times the standard deviations of the preceding 24 h was observed, this measurement was flagged as an unreliable measurement. In addition, the whole data set was visually inspected to verify the results of the automatic flagging procedures.

5 Data sets

5.1 Hydrometeorological data

Temporal dynamics of the most important meteorological data (i.e., air temperature, air humidity, radiation components, wind speed, precipitation, actual evapotranspiration, and runoff) and water balance components (i.e., precipitation, actual evapotranspiration, runoff, and soil water content) from 1 May 2011 to 31 December 2013 are plotted in Figs. 7 and 8. The air temperature, relative humidity, shortwave radiation, and evapotranspiration showed a clear annual pattern. The highest runoff amounts occurred during the winter seasons due to high precipitation amounts and low evapotranspiration rates, as well as overland flow due to saturation excess (Fig. 8). Generally, soil water content showed a strong dependence on precipitation events especially at 5 and 20 cm

Table 2. Descriptive statistics of soil properties determined from 273 soil samples taken in the Rollesbroich catchment.

		Clay %	Sand %	Silt %	Bulk density ($g\,cm^{-3}$)	Carbon content ($g\,kg^{-1}$)	Porosity ($cm^3\,cm^{-3}$)
5 cm	mean	18.99	19.90	61.10	0.94	54.47	0.65
	SD	2.00	3.82	3.79	0.12	15.82	0.05
20 cm	mean	18.03	20.76	61.20	1.28	34.08	0.52
	SD	1.99	4.03	3.46	0.15	16.84	0.05
50 cm	mean	16.50	22.00	61.50	1.52	11.22	0.43
	SD	2.40	5.68	4.53	0.16	6.01	0.06

Table 3. ANOVA results of LAI determination with grass height and RapidEye from May to December 2011.

Source of variation	SS	df	MS	F	P value	F critical
Variability between group	0.311	1	0.311	0.717	0.411	2.145
Variability within group	6.074	14	0.434			
Total	6.385	15				

depth. Quick increases in soil water content can be observed after rainfall events, which were followed by a slow recession during periods without precipitation. A detailed analysis of the spatial variability of the soil moisture has been presented by Qu et al. (2014).

In addition, there seems to be a rising trend in the soil moisture signal in the first year, which might be due to slow soil recompaction after installation (see, e.g., Poltoradnev et al., 2014) or a temporal drift in the electronics of the SPADE sensors. Unfortunately, the exact reasons are unclear. The trend in the average soil moisture content could be corrected using independent cosmic-ray soil moisture measurements (see Baatz et al., 2014), but this is not possible for the individual sensors. Since any correction of individual sensors is going to be subjective because of a lack of appropriate reference measurements, we have decided to refrain from doing further data corrections at this stage. We are convinced that the data are still useful despite possible (minor) sensor drifts, as also demonstrated by a range of recent publications relying on this data (e.g., Qu et al., 2014, 2015; Baatz et al., 2014, 2015).

5.2 Water balance closure

The water balance can be written as

$$P = R + \mathrm{ET_a} + \Delta S, \tag{7}$$

where P is precipitation, R is runoff, $\mathrm{ET_a}$ is actual evapotranspiration, and ΔS is the storage term. Because of the relatively low hydraulic conductivity (10^{-9} to $10^{-7}\,\mathrm{m\,s^{-1}}$) of the aquifer bedrock (HK100, 2009), we neglected deep percolation. Mean average annual precipitation was partitioned into about 57 % actual evapotranspiration and 50 % runoff. The residual of the balance was within 7 % of precipitation for the whole time period as shown in Fig. 8. This residual is related to measurement uncertainty and soil water storage depletion. As presented by Graf et al. (2014), soil water storage can be derived from the measured soil water content, θ, as follows:

$$S(t) = \int \theta(x, y, z, t)\,\mathrm{d}x\mathrm{d}y\mathrm{d}z\mathrm{d}t \approx \sum_{i=1}^{N} c_i \theta(i, t) + \varepsilon, \tag{8}$$

where t is time, z is depth, the integral refers to the 3-D domain as defined by the catchment boundary. The discrete version of storage is expressed by the right-hand side of the equation, where i is the number of sensors, and c_i is the empirical coefficient representing the 3-D domain which is well represented by the sensor. The residual of ε is the storage affected by vegetation or groundwater and not represented well by the sensor. The changes in ΔS are expected to correspond well with the residual of the major water budget component, P-$\mathrm{ET_a}$-R (Eq. 7). Thus, the time derivative of soil water storage should be linearly related to this residual, although measurement errors within the other water balance components and the unaccounted storage terms (e.g., vegetation storage) will deteriorate this relation. Our results show that there is a linear relationship between the derivative of soil water content and the water balance residual (Fig. 9). The R^2 is 0.60, which means that 60 % of the residual of the water balance is explained by soil water storage changes in the catchment.

5.3 Soil physical properties

The mean and standard deviation of sand, silt, and clay percentages, organic carbon content, and bulk density are shown

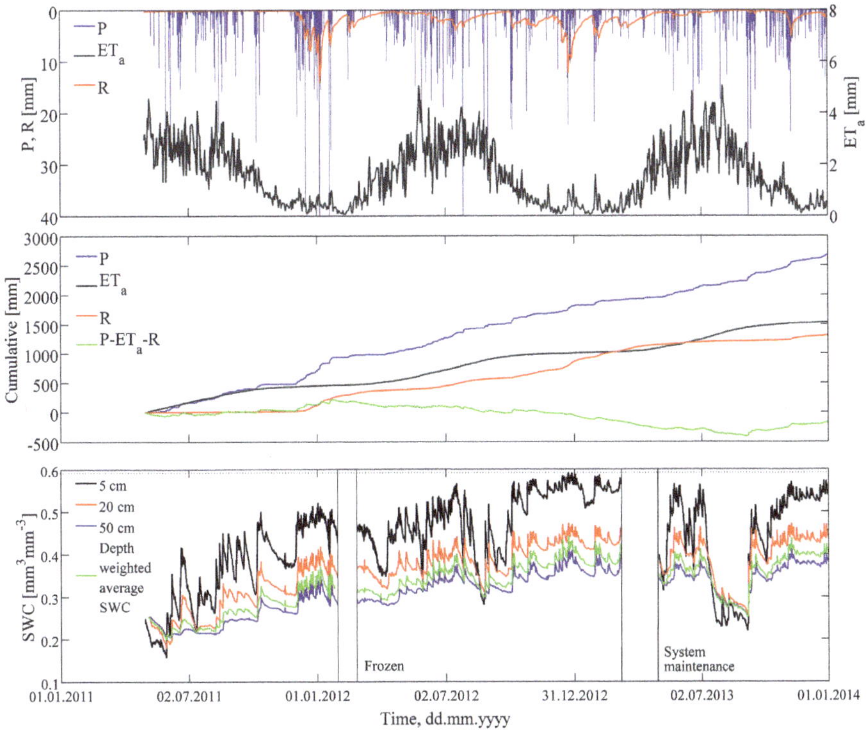

Figure 8. Daily and cumulative time series (1 May 2011 to 31 December 2013) of precipitation (P), runoff (R), actual evapotranspiration (ET_a), and spatial mean soil water content (SWC) at three depths. The SWC data contains two major gaps due to frozen soil conditions and maintenance of the SoilNet system. The dashed line indicates the soil porosity determined from the soil samples ($0.59\,cm^3\,cm^{-3}$).

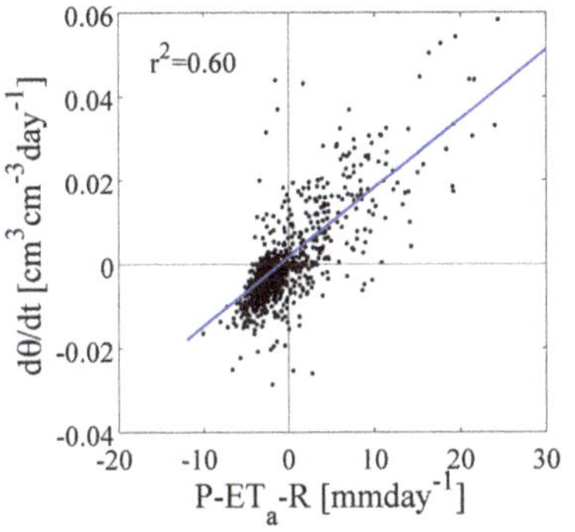

Figure 9. Time derivative of volumetric soil water content (average of all depth) vs. the water balance residual.

in Table 2. The bulk density ranged from 0.94 to $1.52\,g\,cm^{-3}$ and generally increased with depth. Porosity ranged from 0.43 to $0.65\,cm^3\,cm^{-3}$ and decreased with depth. The higher spatial variability of porosity of the subsoil is caused by the higher and more variable stone content at this depth. In addition, former agriculture land management activities reduced the spatial variability of porosity of the top soil (plough layer, 20 to 30 cm thick). The measured soil texture fractions and soil density were used to estimate the spatial distribution of soil hydraulic properties with the pedotransfer function Rosetta (Schaap et al., 2001). Figure 10 shows the spatial distribution of Mualem–van Genuchten (MvG) soil hydraulic parameters (van Genuchten, 1980) derived with Rosetta at 5 cm depth. Some soil hydraulic parameters show a distinct pattern. For instance, the MvG parameter α is generally larger in the northern part than in the southern part of the catchment. Such information is important for the investigation of controlling factors of spatial patterns of soil water content.

5.4 Leaf area index

The LAI derived from measured grass height agrees well with the LAI obtained from RapidEye images (Fig. 11). Both LAI time series showed a distinct annual pattern with the highest values during the summer time. We averaged the monthly LAI derived from measured grass height and an ANOVA was conducted to test whether there was a significant difference between the LAI obtained from grass height and RapidEye in the time period of May to December in 2011. The results of this ANOVA (Table 3) confirmed that there is no significant difference between the two methods to determine LAI.

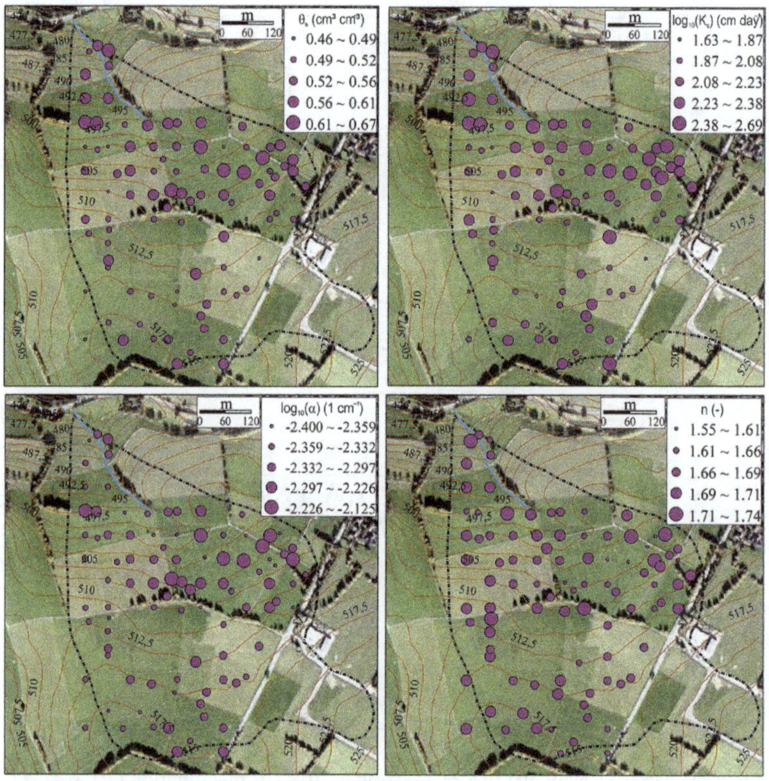

Figure 10. Spatial distribution of soil hydraulic properties (θ_s, $\log_{10}(K_s)$, $\log_{10}(\alpha)$, and n) at 5 cm depth derived from soil information obtained from the soil cores taken in the Rollesbroich catchment.

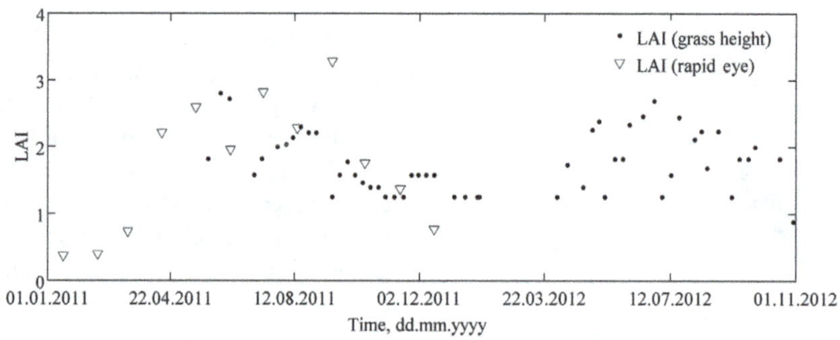

Figure 11. Time series of leaf area index (LAI) computed from measured grass height and extracted from the RapidEye images of the Rollesbroich catchment.

6 Data availability

We presented data from the intensively instrumented hydrological observatory Rollesbroich providing long-term hydrometeorological data with high spatial and temporal resolution. Our results showed that the catchment water balance is reasonably closed by the provided measurements and that 60 % of the water balance residual could be related to soil water storage changes within the Rollesbroich catchment. In addition, important soil physical and chemical properties (e.g., hydraulic properties) have been reported in addition to catchment-scale information on vegetation. This comprehensive hydrological data set can be used for the calibration, validation and improvement of hydrological models, e.g., in hydrological model intercomparison projects (Breuer et al., 2009; Maxwell et al., 2014; Refsgaard, 1997; Smith et al., 2004) and for the calibration and validation of remote sensing data products (Bastiaanssen et al., 1998; Jackson et al., 2010; Le Hegarat-Mascle et al., 2002; Njoku et al., 2003). All the presented data and data from further monitoring stations located in the Rollesbroich catchment are freely available from the TERENO data portal (http://www.tereno.net). In addition, three persistent identifiers are associated with the data set described here:

- climate/runoff/water quality station: doi:10.5880/TERENO.2016.001 (Bogena, 2016a)
- EC/climate station Rollesbroich: doi:10.5880/TERENO.2016.004 (Schmidt, 2016)
- SoilNet Rollesbroich: doi:10.5880/TERENO.2016.003 (Bogena, 2016b)

Please note that the NetCDF format is used, which is a standard format for data exchange. There is free conversion software available to convert the NetCDF data format in other data formats, e.g., at http://www.giss.nasa.gov/tools/panoply/. In addition, an Excel file containing the basic soil information is available at the "SoilNet Rollesbroich" download site and two Excel files containing the LAI information presented in Fig. 11 are available at the "EC/Climate station Rollesbroich" download site.

Acknowledgements. We gratefully acknowledge the support of the SFB-TR32 "Pattern in Soil–Vegetation–Atmosphere Systems: Monitoring, Modeling and Data Assimilation" funded by the Deutsche Forschungsgemeinschaft (DFG), and TERENO funded by the Helmholtz Association. Martina Kettler and Daniel Dolfus are thanked for supporting the acquisition of vegetation information.

Edited by: N. Verhoest

References

Ali, M., Montzka, C., Stadler, A., Menz, G., and Vereecken, H.: Estimation and validation of leaf area index time series for crops on 5 m scale from space, International Geosciences and Remote Sensing Symposium (IGARSS), Melbourne, Australia, 21–26 July 2013.

Albertson, J. D. and Kiely, G.: On the structure of soil moisture time series in the context of land surface models, J. Hydrol., 243, 101–119, 2001.

Allen, R. G., Pruitt, W. O., Wright, J. L., Howell, T. A., Ventura, F., Snyder, R., Itenfisu, D., Steduto, P., Berengena, J., Yrisarry, J. B., Smith, M., Pereira, L. S., Raes, D., Perrier, A., Alves, I., Walter, I., and Elliott, R.: A recommendation on standardized surface resistance for hourly calculation of reference eto by the fao56 penman-monteith method, Agr. Water Manage., 81, 1–22, 2006.

Baatz, R., Bogena, H., Hendricks Franssen, H.-J., Huisman, J. A., Wei, Q., Montzka, C., and Vereecken, H.: Calibration of a catchment scale cosmic-ray soil moisture network: A comparison of three different methods, J. Hydrol., 516, 231–244, 2014.

Baatz, R., Bogena, H., Hendricks Franssen, H.-J., Huisman, J. A., Montzka, C., and Vereecken, H.: An empirical vegetation correction for soil water content quantification using cosmic ray probes, Water Resour. Res., 51, 2030–2046, 2015.

Bastiaanssen, W. G. M., Pelgrum, H., Wang, J., Ma, Y., Moreno, J. F., Roerink, G. J., and van der Wal, T.: A remote sensing surface energy balance algorithm for land (sebal) – 2. Validation, J. Hydrol., 212, 213–229, 1998.

Betts, A. K., Ball, J. H., Beljaars, A. C. M., Miller, M. J., and Viterbo, P. A.: The land surface-atmosphere interaction: A review based on observational and global modeling perspectives, J. Geophys. Res.-Atmos., 101, 7209–7225, 1996.

Birchak, J. R., Gardner, C. G., Hipp, J. E., and Victor, J. M.: High dielectric constant microwave probes for sensing soil moisture, P. IEEE, 62, 93–98, 1974.

Bloschl, G. and Sivapalan, M.: Scale issues in hydrological modeling – a review, Hydrol. Process., 9, 251–290, 1995.

Bogena, H.: TERENO (Eifel-Rur), Climate/Runoff/Water Quality station Rollesbroich, Germany, GFZ Data Services, doi:10.5880/TERENO.2016.001, 2016a.

Bogena, H.: TERENO (Eifel-Rur), SoilNet Rollesbroich, Germany, GFZ Data Services, doi:10.5880/TERENO.2016.003, 2016b.

Bogena, H., Kunkel, R., Pütz, T., Vereecken, H., Krüger, E., Zacharias, S., Dietrich, P., Wollschläger, U., Kunstmann, H., Papen, H., Schmid, H. P., and Munch, J. C.: Tereno – long-term monitoring network for terrestrial research, Hydrol. Wasserbewirts., 56, 138–143, 2012.

Bogena, H. R., Bol, R., Borchard, N., Brueggemann, N., Diekkrueger, B., Druee, C., Groh, J., Gottselig, N., Huisman, J. A., Luecke, A., Missong, A., Neuwirth, B., Puetz, T., Schmidt, M., Stockinger, M., Tappe, W., Weihermueller, L., Wiekenkamp, I., and Vereecken, H.: A terrestrial observatory approach to the integrated investigation of the effects of deforestation on water, energy, and matter fluxes, Science China-Earth Sciences, 58, 61–75, 2015.

Bornemann, L., Welp, G., Brodowski, S., Rodionov, A., and Amelung, W.: Rapid assessment of black carbon in soil organic matter using mid-infrared spectroscopy, Org. Geochem., 39, 1537–1544, 2008.

Breuer, L., Huisman, J. A., Willems, P., Bormann, H., Bronstert, A., Croke, B. F. W., Frede, H. G., Graeff, T., Hubrechts, L., Jakeman, A. J., Kite, G., Lanini, J., Leavesley, G., Lettenmaier, D. P., Lindstroem, G., Seibert, J., Sivapalan, M., and Viney, N. R.: Assessing the impact of land use change on hydrology by ensemble modeling (luchem). I: Model intercomparison with current land use, Adv. Water Resour., 32, 129–146, 2009.

Cosh, M. H., Ochsner, T. E., McKee, L., Dong, J., Basara, J. B., Evett, S. R., Hatch, C. E., Small, E. E., Steele-Dunne, S. C., Zreda, M., and Sayde, C.: The Soil Moisture Active Passive Marena, Oklahoma, In Situ Sensor Testbed (SMAP-MOISST): Testbed Design and Evaluation of In Situ Sensors, Vadose Zone J., 15, doi:10.2136/vzj2015.09.0122, 2016.

Crow, W. T., Ryu, D., and Famiglietti, J. S.: Upscaling of field-scale soil moisture measurements using distributed land surface modeling, Adv. Water Resour., 28, 1–14, 2005.

Devaraju, A., Jirka, S., Kunkel, R., and Sorg, J.: Q-SOS – A Sensor Observation Service for Accessing Quality Descriptions of Environmental Data, ISPRS Int. J. Geo-Inf., 4, 1346–1365, 2015.

Dirnbock, T., Dullinger, S., and Grabherr, G.: A regional impact assessment of climate and land-use change on alpine vegetation, J. Biogeogr., 30, 401–417, 2003.

Evett, S. R., Schwartz, R. C., Casanova, J. J., and Heng, L. K.: Soil water sensing for water balance, Agr. Water Manage., 104, 1–9, 2012.

Flerchinger, G. N. and Cooley, K. R.: A ten-year water balance of a mountainous semi-arid watershed, J. Hydrol., 237, 86–99, 2000.

Foken, T. and Wichura, B.: Tools for quality assessment of surface-based flux measurements, Agr. Forest Meteorol., 78, 83–105, 1996.

Foley, J. A., Levis, S., Prentice, I. C., Pollard, D., and Thompson, S. L.: Coupling dynamic models of climate and vegetation, Glob. Change Biol., 4, 561–579, 1998.

Gebler, S., Hendricks Franssen, H.-J., Pütz, T., Post, H., Schmidt, M., and Vereecken, H.: Actual evapotranspiration and precipitation measured by lysimeters: a comparison with eddy covariance and tipping bucket, Hydrol. Earth Syst. Sci., 19, 2145–2161, doi:10.5194/hess-19-2145-2015, 2015.

Graf, A., Bogena, H. R., Druee, C., Hardelauf, H., Puetz, T., Heinemann, G., and Vereecken, H.: Spatiotemporal relations between water budget components and soil water content in a forested tributary catchment, Water Resour. Res., 50, 4837–4857, 2014.

Guo, L. and Lin, H.: Critical Zone Research and Observatories: Current Status and Future Perspectives, Vadose Zone J., 15, doi:10.2136/vzj2016.06.0050, 2016.

Heimovaara, T. J. and Bouten, W.: A computer-controlled 36-channel time domain reflectometry system for monitoring soil-water contents, Water Resour. Res., 26, 2311–2316, 1990.

Hinzman, L. D., Bettez, N. D., Bolton, W. R., Chapin, F. S., Dyurgerov, M. B., Fastie, C. L., Griffith, B., Hollister, R. D., Hope, A., Huntington, H. P., Jensen, A. M., Jia, G. J., Jorgenson, T., Kane, D. L., Klein, D. R., Kofinas, G., Lynch, A. H., Lloyd, A. H., McGuire, A. D., Nelson, F. E., Oechel, W. C., Osterkamp, T. E., Racine, C. H., Romanovsky, V. E., Stone, R. S., Stow, D. A., Sturm, M., Tweedie, C. E., Vourlitis, G. L., Walker, M. D., Walker, D. A., Webber, P. J., Welker, J. M., Winker, K., and Yoshikawa, K.: Evidence and implications of recent climate change in northern alaska and other arctic regions, Climatic Change, 72, 251–298, 2005.

HK100: Hydrogeological map of north Rhine-Westphalia (1 : 100 000), Geological Survey, North Rhine-Westphalia, Krefeld, Germany, 2009.

Hübner, C., Cardell-Oliver, R., Becker, R., Spohrer, K., Jotter, K., and Wagenknecht, T.: Wireless soil moisture sensor networks for environmental monitoring and vineyard irrigation, Helsinki University of Technology, No. 1, 408–415, 2009.

Huntington, T. G.: Evidence for intensification of the global water cycle: Review and synthesis, J. Hydrol., 319, 83–95, 2006.

ISO-10694: Soil quality – determination of organic and total carbon after dry combustion (elemental analysis), Beuth, Berlin, 1995.

ISO-11277: Soil quality – determination of particle size distribution in mineral soil material – method by sieving and sedimentation, Beuth, Berlin, 2009.

Jackson, T. J., Cosh, M. H., Bindlish, R., Starks, P. J., Bosch, D. D., Seyfried, M., Goodrich, D. C., Moran, M. S., and Du, J.: Validation of advanced microwave scanning radiometer soil moisture products, IEEE T. Geosci. Remote Sens., 48, 4256–4272, 2010.

Jaeger, E. B. and Seneviratne, S. I.: Impact of soil moisture-atmosphere coupling on european climate extremes and trends in a regional climate model, Clim. Dynam., 36, 1919–1939, 2011.

Jones, S. B., Blonquist, J. M., Robinson, D. A., Rasmussen, V. P., and Or, D.: Standardizing characterization of electromagnetic water content sensors: Part 1. Methodology, Vadose Zone J., 4, 1048–1058, 2005.

Kampf, S. K. and Burges, S. J.: Quantifying the water balance in a planar hillslope plot: Effects of measurement errors on flow prediction, J. Hydrol., 380, 191–202, 2010.

Kunkel, R., Sorg, J., Eckardt, R., Kolditz, O., Rink, K., and Vereecken, H.: Teodoor: A distributed geodata infrastructure for terrestrial observation data, Environmental Earth Sciences, 69, 507–521, 2013.

Le Hegarat-Mascle, S., Zribi, M., Alem, F., Weisse, A., and Loumagne, C.: Soil moisture estimation from ers/sar data: Toward an operational methodology, IEEE T. Geosci. Remote Sens., 40, 2647–2658, 2002.

Mauder, M. and Foken, Th.: Documentation and Instruction Manual of the Eddy-Covariance Software Package TK3. Arbeitsergebnisse, Universität Bayreuth, Abteilung Mikrometeorologie 46, Bayreuth, available at: https://epub.uni-bayreuth.de/id/eprint/342 (last access: 14 October 2016), 2011.

Mauder, M., Cuntz, M., Druee, C., Graf, A., Rebmann, C., Schmid, H. P., Schmidt, M., and Steinbrecher, R.: A strategy for quality and uncertainty assessment of long-term eddy-covariance measurements, Agr. Forest Meteorol., 169, 122–135, 2013.

Maxwell, R. M., Putti, M., Meyerhoff, S., Delfs, J.-O., Ferguson, I. M., Ivanov, V., Kim, J., Kolditz, O., Kollet, S. J., Kumar, M., Lopez, S., Niu, J., Paniconi, C., Park, Y.-J., Phanikumar, M. S., Shen, C., Sudicky, E. A., and Sulis, M., 2014, Surface-subsurface model intercomparison: A first set of benchmark results to diagnose integrated hydrology and feedbacks, Water Resour. Res., 50, 1531–1549, 2014.

Moncrieff, J. B., Massheder, J. M., deBruin, H., Elbers, J., Friborg, T., Heusinkveld, B., Kabat, P., Scott, S., Soegaard, H., and Verhoef, A.: A system to measure surface fluxes of momentum, sensible heat, water vapour and carbon dioxide, J. Hydrol., 189, 589–611, 1997.

Myneni, R. B., Nemani, R. R., and Running, S. W.: Estimation of global leaf area index and absorbed par using radiative transfer models, IEEE T. Geosci. Remote Sens., 35, 1380–1393, 1997.

Njoku, E. G., Jackson, T. J., Lakshmi, V., Chan, T. K., and Nghiem, S. V.: Soil moisture retrieval from amsr-e, IEEE T. Geosci. Remote Sens., 41, 215–229, 2003.

Poltoradnev, M., Ingwersen, J., and Streck, T.: Calibration and Application of Aquaflex TDT Soil Water Probes to Measure the Soil Water Dynamics of Agricultural Topsoil in Southwest Germany, J. Irrig. Drain. E.-ASCE, 141, doi:10.1061/(ASCE)IR.1943-4774.0000838, 2014.

Post, H., Hendricks Franssen, H. J., Graf, A., Schmidt, M., and Vereecken, H.: Uncertainty analysis of eddy covariance CO_2 flux measurements for different EC tower distances using an extended two-tower approach, Biogeosciences, 12, 1205–1221, doi:10.5194/bg-12-1205-2015, 2015.

Qu, W., Bogena, H. R., Huisman, J. A., and Vereecken, H.: Calibration of a novel low-cost soil water content sensor based on a ring oscillator, Vadose Zone J., 12, 2013.

Qu, W., Bogena, H. R., Huisman, J. A., Martinez, G., Pachepsky, Y., and Vereecken, H.: Effects of soil hydraulic properties on the spatial variability of soil water content: Evidence from sensor network data and inverse modeling, Vadose Zone J., 13, 2014.

Qu, W., Bogena, H. R., Huisman, J. A., Vanderborght, J., Schuh, M., Priesack, E., and Vereecken, H.: Predicting sub-grid variability of soil water content from basic soil information, Geophys. Res. Lett., 42, 789–796, 2015.

Refsgaard, J. C.: Parameterisation, calibration and validation of distributed hydrological models, J. Hydrol., 198, 69–97, 1997.

Robinson, D. A., Campbell, C. S., Hopmans, J. W., Hornbuckle, B. K., Jones, S. B., Knight, R., Ogden, F., Selker, J., and Wendroth,

O.: Soil moisture measurement for ecological and hydrological watershed-scale observatories: A review, Vadose Zone J., 7, 358–389, 2008.

Rochette, P., Pattey, E., Desjardins, R. L., Dwyer, L. M., Stewart, D. W., and Dube, P. A.: Estimation of maize (zea-mays-l) canopy conductance by scaling up leaf stomatal conductance, Agr. Forest Meteorol., 54, 241–261, 1991.

Rötzer, K., Montzka, C., Bogena, H., Wagner, W., Kidd, R., and Vereecken, H.: Catchment scale validation of SMOS and ASCAT soil moisture products using hydrological modelling and temporal stability analysis, J. Hydrol., 519, 934–946, 2014.

Schaap, M. G., Leij, F. J., and van Genuchten, M. T.: Rosetta: A computer program for estimating soil hydraulic parameters with hierarchical pedotransfer functions, J. Hydrol., 251, 163–176, 2001.

Schmidt, M.: TERENO (Eifel-Rur), Eddy Covariance/Climate station Rollesbroich 1, Germany, GFZ Data Services, doi:10.5880/TERENO.2016.004, 2016.

Seneviratne, S. I., Corti, T., Davin, E. L., Hirschi, M., Jaeger, E. B., Lehner, I., Orlowsky, B., and Teuling, A. J.: Investigating soil moisture-climate interactions in a changing climate: A review, Earth-Sci. Rev., 99, 125–161, 2010.

Smith, M. B., Seo, D. J., Koren, V. I., Reed, S. M., Zhang, Z., Duan, Q., Moreda, F., and Cong, S.: The distributed model intercomparison project (dmip): Motivation and experiment design, J. Hydrol., 298, 4–26, 2004.

Thompson, S. E., Harman, C. J., Troch, P. A., Brooks, P. D., and Sivapalan, M.: Spatial scale dependence of ecohydrologically mediated water balance partitioning: A synthesis framework for catchment ecohydrology, Water Resour. Res., 47, W00J03, doi:10.1029/2010WR009998, 2011.

van Genuchten, M. T.: A closed-form equation for predicting the hydraulic conductivity of unsaturated soils, Soil Sci. Soc. Am. J., 44, 892–898, 1980.

Vasilenko, N. G.: Water balance of small russian catchments in the southern mountainous taiga zone: "Mogot" case study, in: Northern research basins water balance, edited by: Kane, D. L. and Yang, D., Vol. 290, Wallingford, Int. Assoc. Hydrological Sciences, 65–77, 2004.

Vereecken, H., Huisman, J. A., Bogena, H., Vanderborght, J., Vrugt, J. A., and Hopmans, J. W.: On the value of soil moisture measurements in vadose zone hydrology: A review, Water Resour. Res., 44, 1–21, doi:10.1029/2008WR006829, 2008.

Western, A. W., Grayson, R. B., and Bloschl, G.: Scaling of soil moisture: A hydrologic perspective, Annu. Rev. Earth Planet. Sc., 30, 149–180, 2002.

Zacharias, S., Bogena, H., Samaniego, L., Mauder, M., Fuss, R., Puetz, T., Frenzel, M., Schwank, M., Baessler, C., Butterbach-Bahl, K., Bens, O., Borg, E., Brauer, A., Dietrich, P., Hajnsek, I., Helle, G., Kiese, R., Kunstmann, H., Klotz, S., Munch, J. C., Papen, H., Priesack, E., Schmid, H. P., Steinbrecher, R., Rosenbaum, U., Teutsch, G., and Vereecken, H.: A network of terrestrial environmental observatories in germany, Vadose Zone J., 10, 955–973, 2011.

2

Hydrometeorological data from Baker Creek Research Watershed, Northwest Territories, Canada

Christopher Spence and Newell Hedstrom

Environment and Climate Change Canada, Saskatoon, SK, Canada

Correspondence: Christopher Spence (chris.spence@canada.ca)

Abstract. It is uncommon to collect long-term coordinated hydrometeorological and hydrological data in northern circumpolar regions. However, such datasets can be very valuable for engineering design, improving environmental prediction tools or detecting change. This dataset documents physiographic, hydrometeorological and hydrological conditions in the Baker Creek Research Watershed from 2003 to 2016. Baker Creek drains water from 155 km^2 of subarctic Canadian Shield terrain in Canada's Northwest Territories. half-hourly hydrometeorological data were collected each year, at least from April to October, from representative locations, including exposed Precambrian bedrock ridges, peatlands, open black spruce forest and lakes. Hydrometeorological data include radiation fluxes, rainfall, temperature, humidity, winds, barometric pressure and turbulent energy fluxes. Terrestrial sites were monitored for ground temperature and soil moisture. Spring maximum snowpack water equivalent, depth and density data are included. Daily streamflow data are available for a series of nested watersheds ranging in size from 9 to 128 km^2. These data are unique in this remote region and provide scientific and engineering communities with an opportunity to advance understanding of geophysical processes and improve infrastructure resiliency.

1 Introduction

The subarctic Canadian Shield is typified by a landscape dominated by exposed Precambrian bedrock, numerous lakes, open forest and wetlands. It is a large region, occupying 1.26 million km^2 and 13 % of Canada. It extends from the Northwest Territories in the west and to the east through the Nunavut territory, the provinces of Saskatchewan, Manitoba, Québec, and Newfoundland and Labrador. By its sheer size, it represents an important region for Canada. It is rich in mineral deposits, including gold, diamonds, rare earth elements, nickel and copper. There are several active mines across the region, but there are also many abandoned mines that represent a significant environmental liability. Hydroelectric power generation is a significant economic activity. How the hydrometeorology of this region may or may not be changing has consequences beyond its borders. This includes, but is not limited to, implications of changes to the timing and magnitude of freshwater flux into the Arctic Ocean. The importance of freshwater content in the Arctic Ocean to its circulation (Carmack et al., 2016) may have cascading impacts on storm tracks to more southern latitudes (Vihma et al., 2016). The region produces an estimated 10 % of the freshwater yield that flows towards the Arctic Ocean (Prowse and Flegg, 2000), so changes here may not be inconsequential elsewhere.

The ability to make informed decisions about engineering design and make short- and long-term predictions about environmental conditions in this region can be hindered by a lack of data. The large areas and limited transportation infrastructure make intensive data collection expensive. There are no long-term climate stations in the region for which there are up-to-date precipitation data adequate for long-term change detection (Laudon et al., 2017). It is rare to have multiple gauges on streams that can provide data to help understand how water is transferred through watersheds and across landscapes. Such information is also important for developing and testing water and energy cycling algorithms used in both weather and climate models.

To help fill this gap, the Baker Creek Research Watershed was instituted in the early 2000s. The hydrometeorological data produced from activities in this watershed describe atmospheric conditions, soil climate and streamflow among a series of nested watersheds. These data constitute the only integrated hydrological and hydrometeorological small basin dataset in the western subarctic Canadian Shield. The data described here represent a valuable resource of hydrometeorological conditions that can be used for engineering design, improving environmental prediction tools or detecting change. While these data are not representative of the conditions in locations further afield in the subarctic Canadian Shield, they can be used to understand hydrological processes and inform an environmental model structure that can be used across the region. The cautious use of some of the data with longer periods of record could be used in change detection studies. As shown below, the period of record in 2005–2016 represents an almost complete wet–dry cycle useful for understanding extremes in the area. The years 2008 to 2011 were among the wettest on record compared to nearby long-term meteorological data. This was followed by a drought from 2012 and 2016. This drought was most severe in 2015 and was a major contributor to forest fires in 2014 that were the most extensive in 2 decades (Walker et al., 2018).

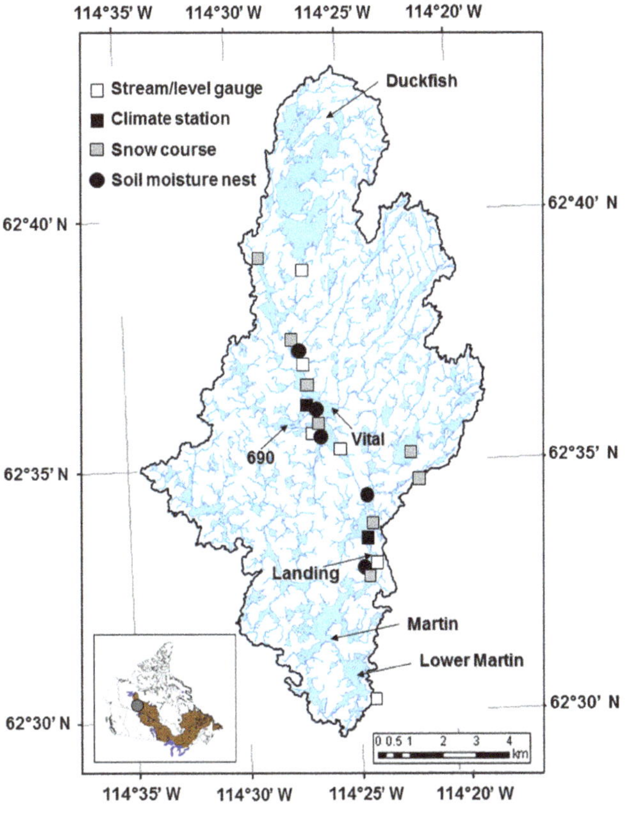

Figure 1. The Baker Creek watershed illustrating the drainage network (derived from 1 : 50 000 scale National Topographic System maps) and observation sites. Lakes that are monitored are identified. The outlet of the watershed defined by the map is at Lower Martin Lake. The inset map of the Northwest Territories shows the Canadian Shield ecozones and Baker Creek's location in the grey circle.

2 Site description

The Baker Creek Research Watershed is located in the subarctic Canadian Shield landscape of Canada's Northwest Territories near the city of Yellowknife. Baker Creek is a stream characterized by lakes connected by short channels that drains water from $\sim 165\,\text{km}^2$ at its outlet into Great Slave Lake (Fig. 1). The observation focus is in the upper reaches of the stream, specifically that area of the watershed upstream of the Water Survey of Canada (WSC) hydrometric gauge Baker Creek at the outlet of Lower Martin Lake (07SB013), which has a topographic drainage area of $155\,\text{km}^2$. There are 349 perennial lakes in the basin, with many being small, and the median and mean lake areas are 5400 and $88\,800\,\text{m}^2$, respectively. The basin is in the zone of discontinuous permafrost. Large changes in topography, vegetation, winter snow accumulation, local hydrology and surficial geology over short distances result in abrupt transitions from permafrost to non-permafrost conditions. Glaciolacustrine clays, outwash and organic deposits are typically underlain by permafrost, whereas bedrock and well-drained glaciofluvial sands are typically unfrozen (Morse et al., 2016). The thickness of organic soils or fine-grained materials derived from glaciolacustrine sediments, glaciofluvial sands or outwash range from less than 1 m to more than 10 m, depending on underlying bedrock topography. Well- and ground-thermistor string installations associated with this dataset rarely involve drilling greater than 10 m before encountering bedrock. An organic soil layer of about 0.25 m is ubiquitous over fine-grained soils. Any forest canopy that grows in these areas is typically quite open. Predominant vegetation includes black spruce (*Picea mariana*), jack pine (*Pinus banksiana*), paper birch (*Betula papyrifera*), Labrador tea (*Ledum groenlandicum*), moss (*Sphagnum* spp.) and lichen (*Cladonia* spp.).

Data from the nearby Meteorological Service of Canada (MSC) climate station Yellowknife A characterize a regional climate that has short cool summers (July daily average temperature of 17 °C) and long cold winters (January daily average temperature of −27 °C). Annual unadjusted precipitation averages at 289 mm, with 40 % of that falling as snow. Annual snow cover begins in October and lasts until the end of April and beginning of May (Spence et al., 2010). In most years, the largest input of water to the basin is during the spring freshet (Spence et al., 2011) and the hydrological regime of the basin is described best as subarctic nival. The average annual streamflow at the outlet of Lower Martin Lake is $0.24\,\text{m}^3\,\text{s}^{-1}$ or $47\,\text{mm}\,\text{a}^{-1}$, providing an annual

Table 1. Land cover percentages from SPOT-5 classification exercise.

Cover	km^2	%
Coniferous forest hillslopes	31.88	20.78
Deciduous forest hillslopes	1.15	0.75
Exposed bedrock	61.24	39.92
Peatland	15.45	10.07
Water	34.72	22.63
Wetland	8.98	5.85

runoff ratio of 0.17. This was calculated using published precipitation data that were unadjusted for undercatch, which implies that actual runoff ratios may be lower. The runoff regime exhibits remarkable variation for a basin with almost 350 lakes, with a standard deviation in annual streamflow of $0.18\,m^3\,s^{-1}$ or $37\,mm\,a^{-1}$ and annual runoff ratios that range from 0.005 (2015) to 0.34 (2001). A maximum daily streamflow of $8.7\,m^3\,s^{-1}$ has been observed, but there are common prolonged periods of zero flow below the outlet of Lower Martin Lake.

3 Physiographic data

Physiographic data available for the Baker Creek Research Watershed include a satellite-derived land cover classification at a 10 m resolution (Fig. 2). Land cover was mapped using a maximum likelihood supervised classification of a composite image of two SPOT5 MS satellite images collected on 24 May 2008 and 20 June 2009. Four multispectral bands and the normalized difference vegetation index of both images were used as input information. Using two images, one prior to and one after leaf-out, assists in differentiating vegetation. Following classification, a mode-based filter with a 3 cell by 3 cell mask was passed over the image to filter out errors commonly experienced at edges due to mixed pixel signatures. The accuracy of the land cover classification was evaluated using a random sample of 314 points recorded during field surveys and marked with a handheld global positioning system (GPS) accurate to within ±8 m. The overall accuracy of the land cover map was 86 %, and the kappa coefficient was 0.82. Land cover was classified into six primary types (Table 1), of which exposed bedrock and lakes and ponds comprise 61 % (Phillips et al., 2011). Variation in soil properties among the different soil-covered land cover types is listed in Table 2. Soil samples were collected during a study summarized in Guan et al. (2010a, b) and properties calculated using the methods and equations found in Dingman (2015). Saturated hydraulic conductivity was measured in piezometers installed at representative locations and calculated with the Luthin approach (1966).

Elevation data (Fig. 3) were collected during a Light Detection and Ranging (LiDAR) survey on 21 August 2007.

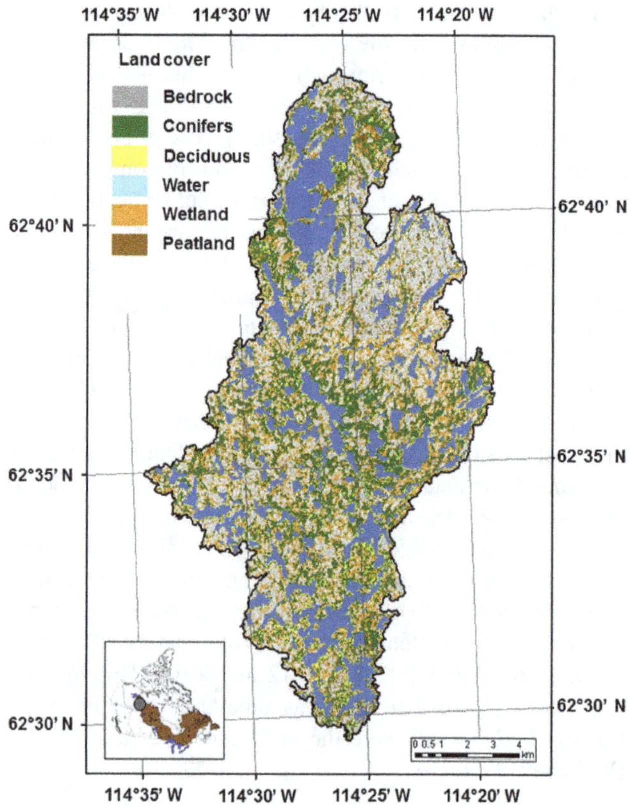

Figure 2. Baker Creek land cover derived from SPOT satellite imagery.

During the survey, 1 s interval GPS base-station data were collected from the Natural Resources Canada's Canadian Active Control Site GPS site in nearby Yellowknife. Laser-point positions were computed by differentially correcting all base-station data, integrating these with the airborne GPS and an inertial measurement unit (IMU) to differentially correct the airborne trajectory, calibrating survey data over known targets. Data post-processing included examining for mismatches between flight lines and isolating high and low laser-pulse returns that were well above the canopy or below the ground surface. Ground returns were classified from the point cloud within TerraScan assuming at least one ground point was within a $400\,m^2$ area, the terrain angle was less than 88°, the iteration angle was less than 6° and the iteration distance was less than 1.4 m to the plane. Datasets were adjusted to account for local geoidal undulation. Orthometric heights and geoid separation were determined using the HTv2.0 model (CGG2000 Scientific model + HRG01 Corrector Surface, allowing the direct transformation of NAD83 or International Terrestrial Reference Frame (ITRF) ellipsoidal heights to CGVD28 orthometric heights). Ground-only points were gridded to 1 m node spacing using 2nd power inverse distance to power functions within a 15 m search radius. This 1 m dataset was degraded to 10 m to align with the land cover data described above. Both the eleva-

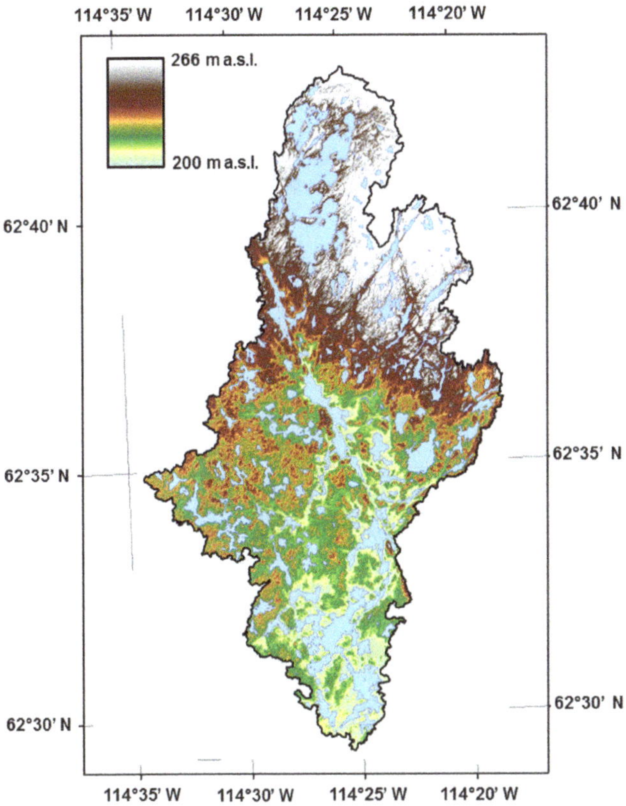

Figure 3. Baker Creek elevation (m a.s.l.) derived from the LiDAR survey.

tion and land cover datasets are projected to NAD1983 UTM Zone 11N (Spence et al., 2010).

4 Hydrometeorological data and conditions

There are two long-term primary climate towers operating in the catchment (Table 3). These occupy locations above an exposed bedrock ridge and a lake (Figs. 1 and 4). As noted above, these two cover types represent the majority of the catchment area (61 %). They are also arguably the most difficult surfaces over which to estimate turbulent energy fluxes, so both stations are equipped with eddy covariance instrumentation. Meteorological variables measured at each tower include air temperature, relative humidity or vapour pressure, net radiation and its components, wind speed, barometric pressure, and rainfall. The tower specifications, including variables and units, sensor types and heights, and surface characteristics are summarized in Table 4. Radiometers (Kipp and Zonen NR-Lite and CNR-1s and Li-Cor LI-200SAs) are mounted to face south to reduce shadows on the sensors. Temperature and relative humidity sensors (Vaisala HMP45Cs) are deployed within 10 plate shields to mitigate the influence of wind and sunlight on measurements. Wind speed is measured with three-cup anemometers (Met One 013As) at both towers, while wind direction is measured with a wind vane (NRG) at Landing Lake. Surface-water temperature is measured with an infrared thermometer (Apogee SI-111) mounted four metres above the water and pointed towards the water. There is a Texas Instruments TE-525M tipping bucket rain gauge mounted at the top of the bedrock tower. This type of gauge is not designed to measure solid precipitation. Hence, the dataset includes only rainfall data and not precipitation data. The eddy covariance systems at both sites employ CSAT-3 sonic anemometers. Water vapour content is measured at the bedrock tower with a Li-Cor Li-7500 and at the lake with a Campbell Scientific KH-20. Both systems measure wind speed and water vapour content at 10 Hz and fluxes are calculated over a half-hour period. Corrections to the eddy covariance measurements include coordinate rotation (Kaimal and Finnigan, 1994), the Webb–Pearman–Leuning (WPL) adjustment (Webb et al., 1980), sonic path length, high frequency attenuation and sensor separation (Massman, 2000; Horst, 1997), and oxygen extinction (Blanken et al., 2011). Barometric pressure is measured with different sensors at each tower. At the bedrock tower, the Li-Cor Li-7500 from the eddy covariance system is used, and at the lake tower, an RM Young 612 is deployed. All meteorological data are observed every 5 s, and half-hourly averages (or totals) are logged on Campbell Scientific 23X or CR3000 data loggers. The dataset includes these half-hourly values. The period of record of the bedrock ridge tower extends from 2005 to 2016. The lake tower's period of record is from 2008 to 2016. There are numerous gaps in the data due to sensor malfunction, none of which were filled. The stations typically operate from early April to early October each year and are not operated through winter because data integrity cannot be guaranteed. This is primarily due to frost being allowed to collect on the towers because of infrequent site visits. This frost either prevents batteries from being charged or coats sensors that prevent accurate readings.

The strong seasonality of radiation and energy exchanges in this landscape are evident from the climate tower on the exposed bedrock outcrop near Vital Lake (Fig. 5). Incoming long-wave radiation remains relatively consistent month to month compared to incoming short-wave radiation, the latter peaking near $250\,\mathrm{W\,m^{-2}}$ in June and decreasing to $< 100\,\mathrm{W\,m^{-2}}$ by September. Net long-wave radiation is almost always negative from exposed bedrock surfaces during the snow-free season. short-wave albedos are typically low (0.12), except when snow is present (~ 0.7). Net radiation rises from monthly averages of $75\,\mathrm{W\,m^{-2}}$ in April to $140\,\mathrm{W\,m^{-2}}$ in June and then steadily decreases to $50\,\mathrm{W\,m^{-2}}$ by September. Bowen ratios are high in this dry location, almost always exceeding unity, with monthly averages commonly near 3.0. Only during exceptionally wet autumn periods (2008) were monthly Bowen ratios less than 1. At the tower located on Landing Lake, the seasonal cycle begins with predominantly negative Bowen ratios at the end of winter, as sensible heat fluxes are directed to the snow- and ice-

Table 2. Soil characteristics in each typical soil-covered landscape type present in the Baker Creek Research Watershed. K denotes hydraulic conductivity.

Trait	Peatland	Forested Hillslope	Wetland
Porosity	0.85	0.83	0.8
Bulk density ($kg\,m^{-3}$)	78	113	104
Particle density ($kg\,m^{-3}$)	574	644	567
Specific yield	0.15	0.19	0.25
K (0–0.5 m) ($m\,s^{-1}$)	10^{-6}	10^{-5} to 10^{-7}	10^{-6} to 10^{-7}
K (> 0.5 m) ($m\,s^{-1}$)	10^{-7} to 10^{-8}	10^{-8} to 10^{-9}	10^{-6} to 10^{-9}

Table 3. List of monitoring locations in the Baker Creek Research Watershed.

Site	Land cover	Latitude	Longitude
Snow courses			
Vital Lake	Water	62.6087°	−114.4489°
Ryan Lake	Water	62.5868°	−114.3709°
Northern wetland	Wetland	62.6529°	−114.4800°
Landing portage	Coniferous forest hillslope	62.5483°	−114.4051°
Jack pine above Vital	Coniferous forest hillslope	62.6242°	−114.4534°
Outcrop above camp	Exposed bedrock	62.5937°	−114.4387°
Outcrop near Ryan Lake road	Exposed bedrock	62.5689°	−114.3639°
Landing Lake eastern shore	Deciduous forest hillslope	62.5631°	−114.4001°
Climate towers			
Vital upland	Exposed bedrock	62.6042°	−114.4475°
Landing Lake	Water	62.5593°	−114.4117°
Hydrometric gauges			
Baker Creek at outlet of Landing Lake	–	62.5499°	−114.4005°
Baker Creek at Vital Narrows	–	62.5792°	−114.4159°
Outlet of Lake 690	–	62.5946°	−114.4436°
Baker Creek above Vital Lake	–	62.6190°	−114.4545°
Baker Creek at outlet of Duckfish Lake	–	62.6476°	−114.4477°
Ground temperature and moisture nests			
Landing portage	Coniferous forest hillslope	62.5483°	−114.4051°
Wetboot peatland	Peatland	62.5728°	−114.4071°
Camp valley	Coniferous forest hillslope	62.5938°	−114.4375°
Tower Peatland	Peatland	62.5998°	−114.4425°
Wetland above Vital	Wetland	62.6190°	−114.4545°
Vital upland	Exposed bedrock	62.6042°	−114.4476°

covered surface. Monthly average latent heat fluxes are always higher than sensible heat fluxes. Sensible heat increases relative to latent heat at the very end of the open-water season, as conditions get cool and damp (Fig. 6). Short-wave albedos of the lake surface stay high (0.5) until lake ice break-up in early to mid-June, after which they average 0.15.

The meteorological data presented in Figs. 7 and 8 do not always fully demonstrate the strong annual cycles of temperature that occur in this region, but monthly average air temperatures usually rise above 0 °C in May, peak just below 20 °C in July, average 8 °C in September and drop below freezing again in October. Mean annual air temperature measured by the Meteorological Service of Canada at the nearby Yellowknife A station from 2005 to 2016 was −3.4°. Relative humidity rises through the warm season, averaging 58 % and 74 % in July and September, respectively. Average wind speed for the period of record is $3\,m\,s^{-1}$, with a maximum recorded half-hourly average of $12\,m\,s^{-1}$. Wind speeds over $10\,m\,s^{-1}$ occur less than one percent of the time. The period in 2005–2016 represents an almost complete wet–dry cycle. Data from Yellowknife A beginning in the 1940s provide excellent context for the data collected at Baker Creek,

Table 4. Sensor suites at the primary climate stations at the bedrock ridge and Landing Lake. Vegetation height is the maximum observed at the bedrock ridge, with displacement and roughness height calculated with equations found in Oke (1987). NA – not available

	Bedrock ridge		Landing Lake	
Variable	Sensor	Height (m)	Sensor	Height (m)
Air temperature (°C) and Relative humidity (RH) or vapour pressure (kPa)	Vaisala HMP45C	2.8 and 4.4	Vaisala HMP45C	1.4
Surface temperature (°C)	NA		Apogee SI-111 infrared thermometer	4.0
Turbulent fluxes ($W m^{-2}$)	Campbell Scientific CSAT-3 and LiCor LI-7500	4.7	Campbell Scientific CSAT-3 and KH20	3.2
Incoming short-wave radiation ($W m^{-2}$)	Kipp and Zonen CNR1	4.2	Li-Cor LI-200SA	3.5
Incoming long-wave radiation ($W m^{-2}$)	Kipp and Zonen CNR1	4.2	NA	
Outgoing short-wave radiation ($W m^{-2}$)	Kipp and Zonen CNR1	4.2	Li-Cor LI-200SA	0.95
Outgoing long-wave radiation ($W m^{-2}$)	Kipp and Zonen CNR1	4.2	NA	
Net radiation ($W m^{-2}$)			Kipp and Zonen NRLite	1.65
Wind speed ($m s^{-1}$)	Met One 013A	4.4	Met One 013A	1.1
Barometric pressure (kPa)	Li-Cor Li-7500	4.7	RM Young 612	1.5
Rainfall (mm)	Texas Instruments TE525M	5.3	NA	
Wind direction (°)	NA		NRG systems wind vane	4.3
Vegetation height (m)		4.0		NA
Displacement height (m)		2.6		NA
Roughness height (m)		0.5		NA

Figure 4. Photos of **(a)** the climate tower at Landing Lake and **(b)** the climate tower on the bedrock ridge above Vital Lake.

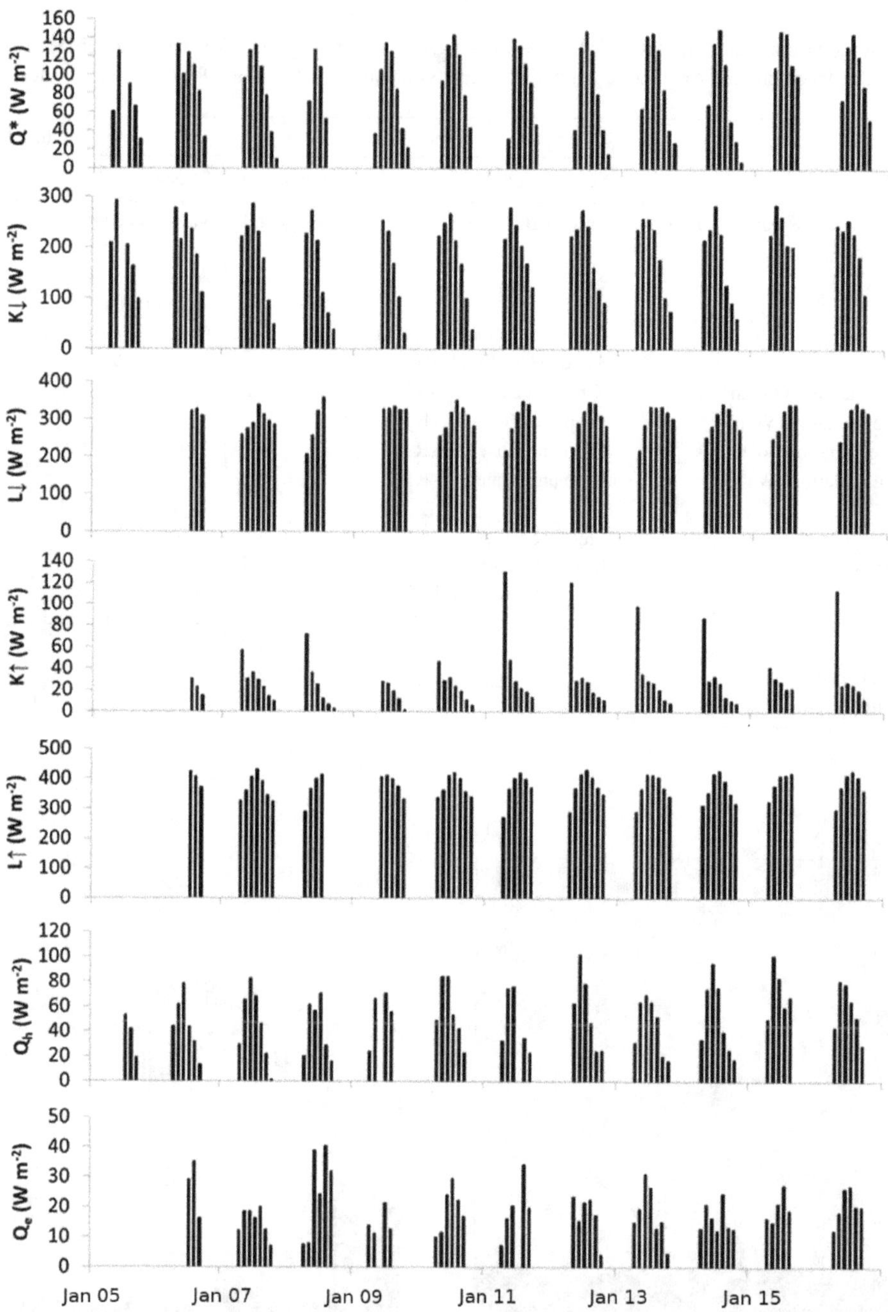

Figure 5. Monthly average radiation and energy fluxes at the bedrock ridge climate tower in 2005–2016. The radiation terms Q^*, $K\downarrow$, $K\uparrow$, $L\downarrow$ and $L\uparrow$ are net radiation, downwelling short-wave radiation, upwelling short-wave radiation, downwelling long-wave radiation and upwelling long-wave radiation, respectively. The turbulent energy fluxes are sensible heat, Q_h, and latent heat, Q_e.

with an average April–September precipitation of 180 mm. The gauges in Baker Creek averaged a total of 147 mm over these months from 2005 to 2011, but this decreased by 27 % to 107 mm between 2012 and 2016. This drought was most severe in 2015, with only 70 mm of rainfall.

5 Spring snowpacks

Each year since 2003 (except 2006), a land cover stratified spring maximum snow pack survey has been conducted within the Baker Creek basin at the locations listed in Table 3. These are typically done in 1 day within the first 2 weeks of April. Within each of the five land cover types de-

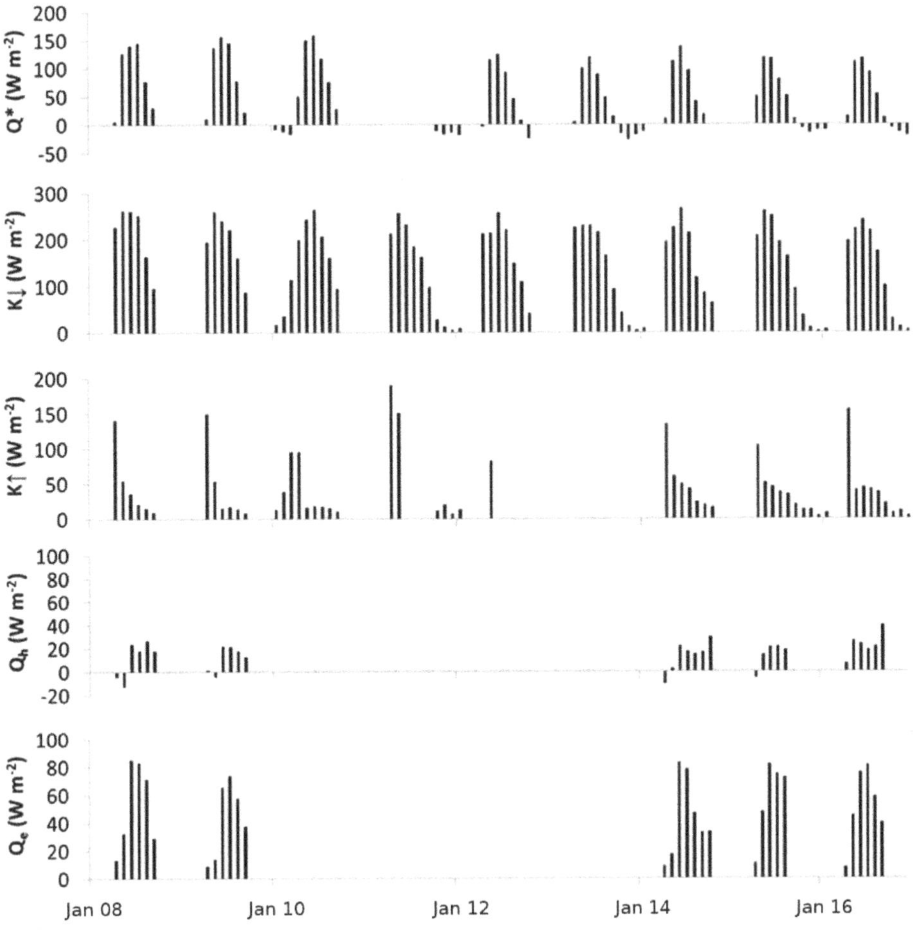

Figure 6. Monthly average radiation and energy fluxes at the Landing Lake climate tower in 2008–2016. Nomenclature is identical to that in Fig. 5.

scribed above – coniferous forest, deciduous forest, lakes, exposed bedrock and wetlands – there is at least one designated 25-point snow course (Fig. 9). Wetlands and peatlands are combined into one cover type for the purposes of the spring snow survey. At each course 25 depths are recorded at least 10 m apart using an aluminum rod marked every 1 cm. At each fifth depth measurement, a density measurement is taken using an ESC-30 snow corer. From these measurements, spring maximum snowpack depth (cm) and snow density (kg m^{-3}) are estimated for each land cover type. The snow water equivalent (SWE) (mm) was calculated as the product of depth and density, with a conversion to change the units to millimetres. Spring snowpacks are rarely over 60 cm deep, and as Pomeroy and Gray (1995) suggest, these shallow snowpacks do not exhibit covariance between depth and density. A basin-wide average is calculated by pro-rating these estimates by land cover fractions provided in Table 1. The accuracy of such a snow survey is expected to be within 15 % (Pomeroy and Gray, 1995).

Results indicate that the basin-wide spring maximum snow pack averaged 80 ± 17 mm SWE in 2003–2016 (Fig. 10). Snow depths are shallow, reflecting the dry climate, averaging 38 ± 8 cm. The deepest snowpacks are typically in the wetlands, perhaps as the shrubs that proliferate there experience less interception than the forest and better trap the snow (Sturm et al., 2001). The thinnest and most homogenous snowpack are over lakes, as these tend to be quite open and wind-blown (Rees et al., 2006; Derksen et al., 2006). The exposed bedrock almost always has the most variable snowpack, as the snow is transported from open areas into depressions across the uneven topographic surface (Fig. 9).

6 Streamflow data

The nested hydrometric network in the Baker Creek basin (Table 3) includes one WSC gauge and five others operated by the Science and Technology Branch of Environment and Climate Change Canada. Baker Creek at the outlet of Lower Martin Lake (07SB013) has been operated by

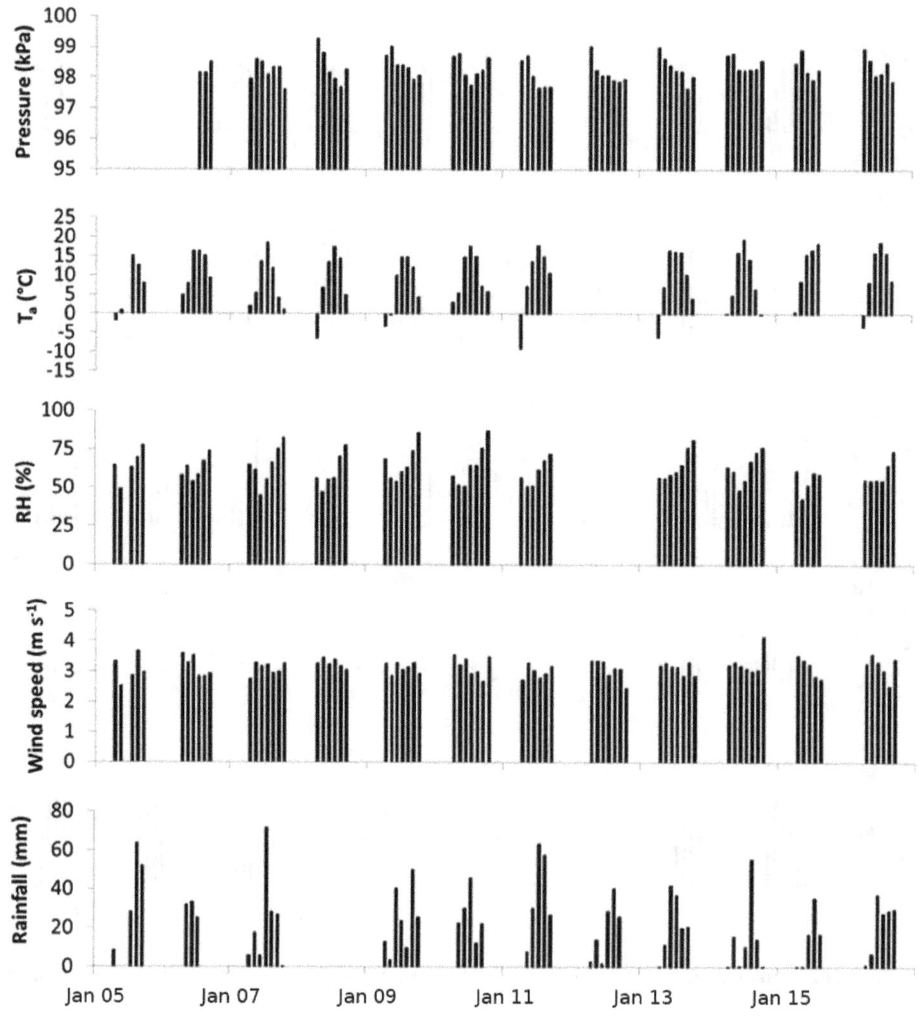

Figure 7. Monthly average meteorological conditions at the bedrock ridge climate tower in 2005–2016. T_a refers to air temperature, RH denotes relative humidity, and pressure refers to barometric pressure.

the WSC since 1983. These streamflow data are available from https://wateroffice.ec.gc.ca/search/historicale.html (last access: 20 June 2018). The outlet of Lower Martin Lake often remains open, and the stream cross section is bedrock and extremely stable. This results in a nearly complete (95 %) record of high-quality streamflow data that is a foundational component of research catchment operations.

The next oldest hydrometric gauge is at the outlet of Landing Lake (Fig. 1), which was installed in 2003 for the Mackenzie Global Energy and Water Cycle Experiment (GEWEX) Study (Woo, 2008). This was followed by Baker Creek at Vital Narrows in 2005, Baker Creek above Vital Lake and Moss Creek at the outlet of Lake 690 in 2008 and Baker Creek at the outlet of Duckfish Lake in 2009. These last three gauges were initially installed to support Canada's contribution to the International Polar Year. Permanent benchmarks were installed in 2017 that allow stages at all the gauges (save Baker Creek above Vital) to be referenced to elevations in metres above sea level. These benchmark locations are referenced to NAD83, but unlike the elevation dataset, they use orthometric heights in CGVD2013. Any future exercise that merges the water and topographic elevations must take this into account.

A combination of lower flows, beaver activity and icing-susceptible cross sections at these gauge sites higher in the basin make it more difficult to obtain continuous accurate stage measurements, stable rating curves and streamflow estimates. Of the non-WSC gauges, the Landing and Vital Narrows stations are meant to be capable of continuous stage measurements, while the others are seasonal, as stage recorders are removed in late fall to avoid the freezing of the transducers. This precludes many instances of known zero-flow periods from appearing in the record. The seasonal stage recorders at Baker Creek above Vital, at Duckfish Lake and

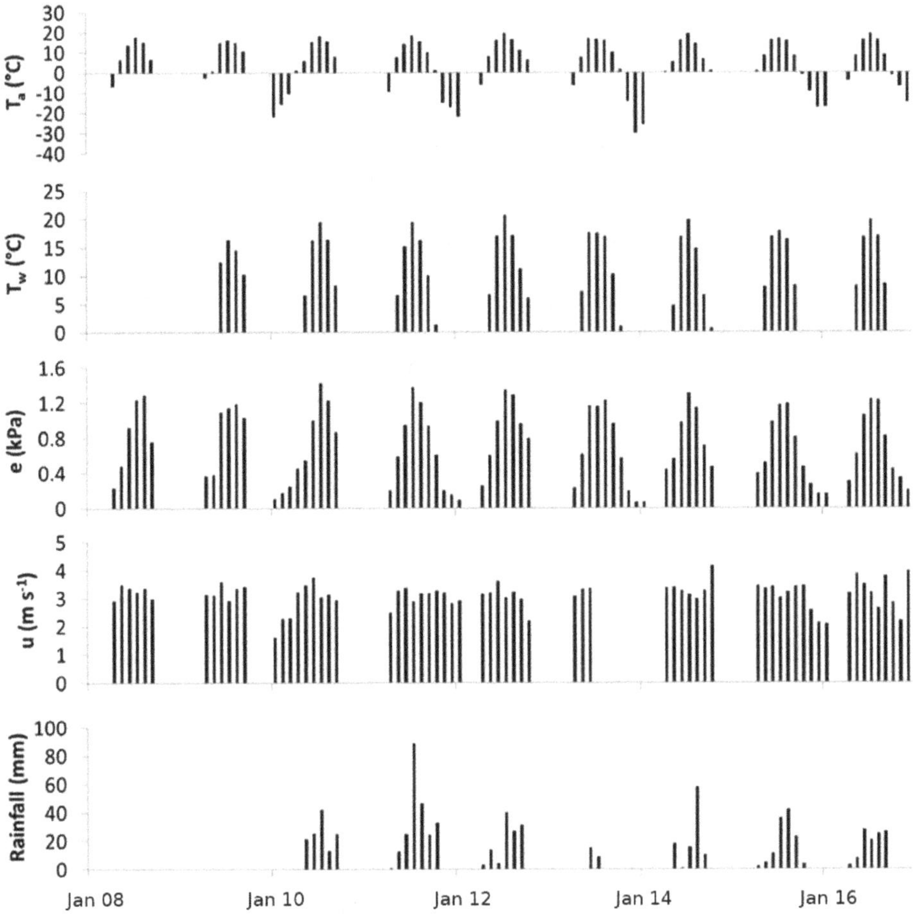

Figure 8. Monthly average meteorological conditions over Landing Lake in 2008–2016. T_a and T_w denote air and surface-water temperature, respectively. Vapour pressure is e, and u is wind speed.

at Moss Creek at the outlet of Lake 690 have recently been replaced with instrumentation capable of continuous stage measurements. These non-continuous records make it difficult to calculate and compare annual or multi-year basin yields, but the data are useful for hydrological model testing and process studies (Spence, 2006; Spence et al., 2010, 2014, 2018; Phillips et al., 2011).

The long WSC dataset from Baker Creek at the outlet of Lower Martin Lake provides context for the shorter records of the upstream gauges. The period beginning in 2003 was typically characterized by average to above average streamflow in the watershed, reflective of precipitation inputs (Figs. 7 and 10). The high rainfall was sometimes heavy enough to generate late warm-season runoff events (Spence et al., 2011, 2014) (e.g. 2008, 2011) that exceeded spring freshet events. These late fall and early winter events could not be captured at all gauges in all years because of icing conditions and are more evident at the Vital Narrows and Landing Lake gauges, which stay ice-free longer under high water conditions. The meteorological drought noted earlier (Fig. 7) that began in 2013 also manifested into a hydrological drought (Fig. 11). This effect is most pronounced at the highest gauge, Duckfish Lake, which has not experienced outflow since 2014 up until the time of writing (2018), as water levels remain below the lake outlet elevation. This kind of hydrologic disconnectivity is common across the watershed during dry conditions, and this can reduce streamflow so much that lake evaporation along the watercourse can prevent water from areas actively producing runoff from proceeding downstream (Spence, 2006; Spence et al., 2018). This was particularly evident in 2015 and 2016 with the differences in streamflow at the Vital Narrows, Landing Lake and Lower Martin gauges. During these types of periods, the sub-catchments of the Baker Creek watershed behave as individuals, rather than as part of a cohesive watershed.

7 Ground temperature and soil moisture data

Shallow ground temperatures and soil moisture have been measured since 2007 at locations listed in Table 3. Decagon

Figure 9. The beginning of one of the exposed bedrock snow courses. Note the sparse jack pine individuals and exposed rock behind Ross Phillips.

Figure 10. Basin average spring maximum snow water equivalent in 2004–2017.

EM50 ECH$_2$0 data loggers with Decagon 5TM capacitance soil moisture and temperature sensors at the ground surface (i.e. 1 cm depth) and 25 cm depth are deployed at five locations throughout the basin. These sites represent the different typologies and topologies of areas with overburden; these are peat-filled bedrock depressions, forested hillslopes below bedrock ridges and wetlands between lakes (Table 2). These data have been used to estimate storage changes in these cover types across the basin, assuming that variability among such different land cover types is larger than that within the individual land cover types (Spence et al., 2010). Half-hourly data are typically available from April to December, capturing both the spring thaw and fall freeze-back (Fig. 12). The sensors have been prone to animal disturbance, so data completeness is uncommon. The sensors do not measure total water content, but total unfrozen water content, so values are unavailable through winter, and the soil moisture time series often captures an increase in unfrozen water during the spring thaw and a decrease during the freeze-back. This should not be interpreted as a change in storage but as a change in phase. The active layer depth varies among the sites, often shallowest (0.3 m) in wetlands and deepest in the peatlands ($\sim$ 1.5 m) (Guan et al., 2010a). Soils near the surface often reach saturation during spring snowmelt but then quickly dry (Fig. 13). The dry climate means the elevation of the water table tends to follow the descent of the frost table in the drier hillslopes, with commensurate decreases in unsaturated soil moisture. Soil moisture at a depth in wetlands between lakes is much more stable than in the hillslopes, likely because of moisture distributed from runoff travelling through the wetland. Moisture distribution in these wetlands is known to be uneven (Spence et al., 2011), and edges of these wetlands may actually be better represented

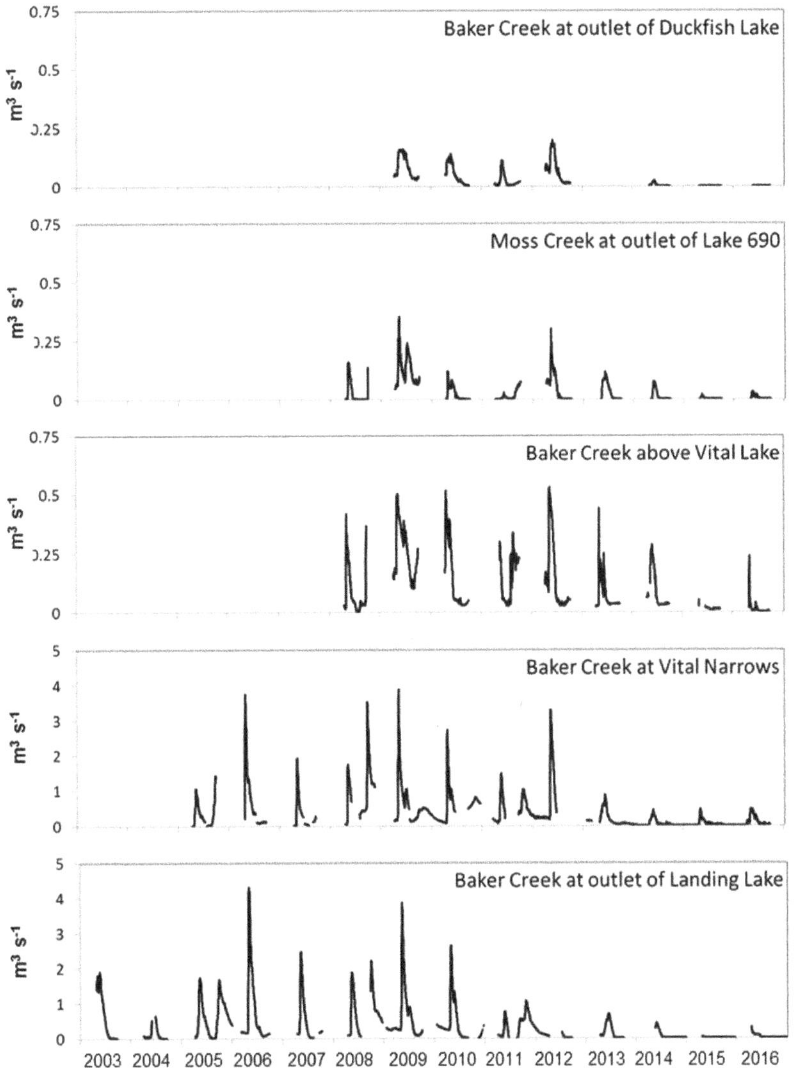

Figure 11. Daily streamflow (2003–2016) from the series of nested hydrometric gauges within the Baker Creek Research Watershed.

by measurements taken in forested hillslopes. The water table in peatlands is often near the surface, and the soil column is often near saturation through the entire warm season.

To calculate ground heat fluxes within the footprint of the bedrock ridge tower, a shallow 0.5 m borehole was drilled in 2007 and instrumented with HOBO pendant thermistors initially at the surface (20 cm, 30 cm, and 46 cm) and only from April to September each year. In 2010, the thermistors were changed to HOBO Pro v2 soil/water probes, and one at 10 cm was added. These sensors provide half-hourly data to an external logger that permits year-round measurements. The high thermal conductivity of the granite in this location results in much warmer ground temperatures than in areas of overburden. The temperature of the shallow bedrock often exceeds 25 °C in July, while unfrozen soil columns remain closer to 12 °C (Fig. 14). Conversely, in the middle of winter, these data imply that shallow depths in bedrock outcrops can be as cold as −17 °C, while soil columns do not get colder than −10 °C.

8 Data availability

In partnership with Portage, Compute Canada, and the Canadian Association of Research Libraries, the University of Saskatchewan launched the Federated Research Data Repository (FRDR), which is a single online location from which research data can be shared, preserved, discovered, curated and cited. The data described here are available at: https://www.frdr.ca/repo/handle/doi:10.20383/101.026 (Spence and Hedstrom, 2018).

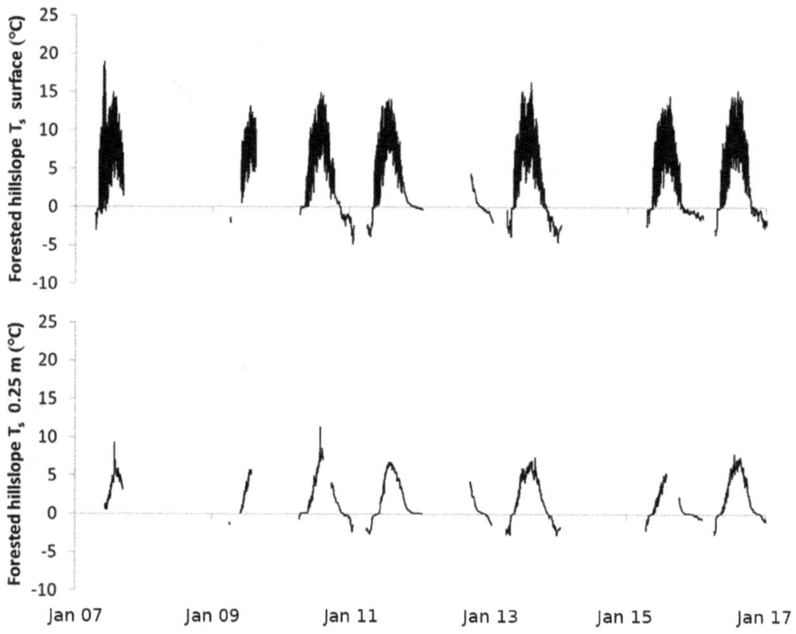

Figure 12. Half-hourly surface and 0.25 m soil temperatures from one of the forested hillslope soil moisture nests from 2007 to 2016.

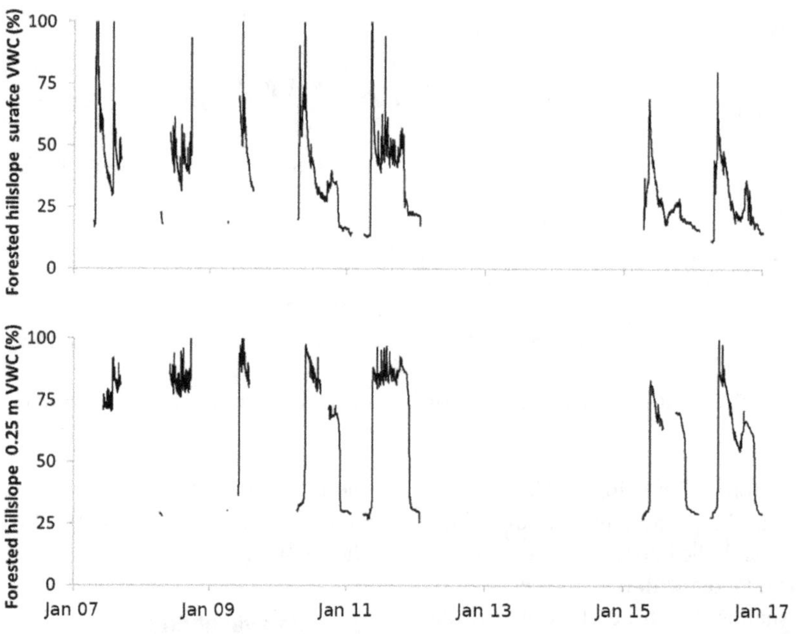

Figure 13. Half-hourly soil surface and 0.25 m volumetric water content at the same forested hillslope soil moisture nest illustrated in Fig. 9.

9 Summary

These data represent an effort to measure all components of the energy and water cycle in a catchment in the subarctic Canadian Shield landscape. The period of record for some components extends from 2003. The goal of making the data available to the research and applied hydrology communities is twofold. First, it is meant to support and inform water resource management decision-making. The record is an important source of baseline data that can be used to assess the effect of disturbance, such as fire or resource development. Second, these data are provided to allow others to also investigate hydrological processes, medium-term patterns and potential changes in this extensive landscape. Measurements have employed consistent methods to ensure comparability within the research catchment and perhaps eventually permit

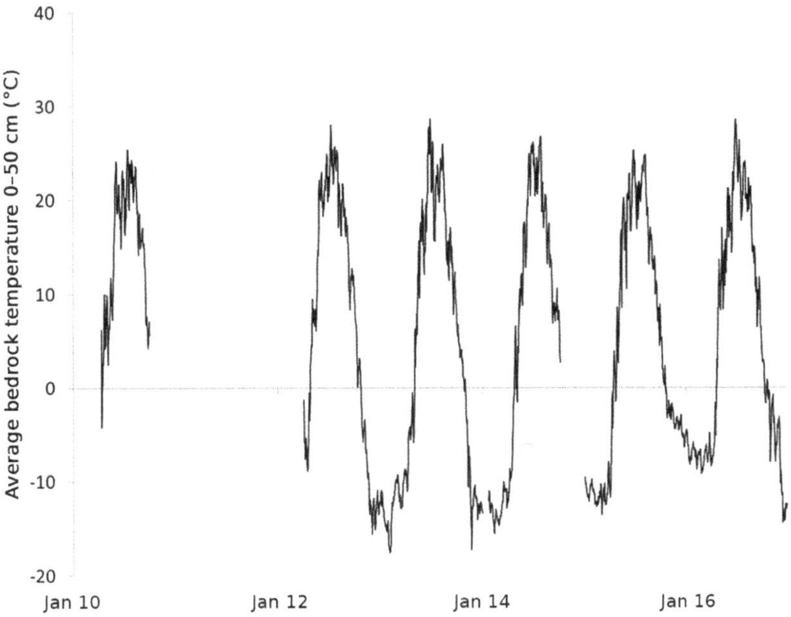

Figure 14. Daily average 0–0.5 m bedrock temperatures near the exposed bedrock ridge climate tower in 2010–2016.

time series analysis. The data have proven fit for the purpose of supporting hydrological and hydrometeorological process research and this can ensure the proper interpretation of spatial and temporal patterns and causal attribution.

Author contributions. Both authors were involved in field work to collect the data discussed here, including processing, quality assurance and quality control. Both authors contributed to writing the manuscript.

Competing interests. The authors declare that they have no conflict of interest.

Acknowledgements. There have been several people over the years that have contributed to data collection and deserve recognition. These include Shawne and Steve Kokelj, Meg McCluskie, Stefan Goodman, Scott Dowler and Ryan Gregory from the government of the Northwest Territories; May Guan, Ross Phillips, and Kirby Ebel from the University of Saskatchewan; and Cuyler Onclin, Mark Russell and Kerry Pippy from the Environment and Climate Change Canada Water Science and Technology Directorate. The Yellowknife office of the Water Survey of Canada has provided outstanding logistical support for the research catchment; thank you to Jason Friesen, Bob Reid, Derek Forsbloom, Rick Klakowich and Dale Ross. Chris Hopkinson and the Canadian Consortium for Lidar Environmental Applications Research (C-CLEAR) collected the LiDAR data.

Edited by: Chris DeBeer

References

Blanken, P. D., Spence, C., Hedstrom, N., and Lenters, J. D.: Evaporation from Lake Superior: 1. Physical Controls and Processes, J. Great Lakes Res., 37, 707–716, 2011.

Carmack, E., Yamamoto-Kawai, M., Haine, T. W. N., Bacon, S., Bluhm, B. A., Lique, C., Melling, H., Polyakov, L. V., Staneo, F., Timmermans, M. L., and Williams, W. J.: Fresh water and its role in the Arctic Marine System: Sources, disposition, storage, export, and physical and biogeochemical consequences in the Arctic and global oceans, J. Geophys. Res.-Biogeo., 121, 675–171, https://doi.org/10.1002/2015JG003140, 2016.

Derksen, C., Strum, M., Liston, G. E., Holmgren, J., Huntington, H., Silis, A., and Solie, D.: Northwest Territories and Nunavut Snow Characteristics from a Subarctic Traverse: Implications for Passive Remote Sensing, J. Hydrometeorol., 10, 448–463, 2006.

Dingman, S. L.: Physical Hydrology, Waveland Press, Long Grove, 643 pp., 2015.

Guan, X. J., Spence, C., and Westbrook, C. J.: Shallow soil moisture – ground thaw interactions and controls – Part 2: Influences of water and energy fluxes, Hydrol. Earth Syst. Sci., 14, 1387-1400, https://doi.org/10.5194/hess-14-1387-2010, 2010a.

Guan, X. J., Westbrook, C. J., and Spence, C.: Shallow soil moisture – ground thaw interactions and controls – Part 1: Spatiotemporal patterns and correlations over a subarctic landscape, Hydrol. Earth Syst. Sci., 14, 1375–1386, https://doi.org/10.5194/hess-14-1375-2010, 2010b.

Horst, T. W.: A simple formula for attenuation of eddy fluxes measured with first order response scalar sensors, Bound.-Lay. Meteorol., 94, 517–520, 1997.

Kaimal, J. C. and Finnigan, J. J.: Atmospheric Boundary Layer Flows – Their Structure and Measurement, Oxford University Press, New York, 1994.

Laudon, H., Spence, C., Buttle, J., Carey, S. K., McDonnell, J. J., McNamara, J., Soulsby, C., and Tetzlaff, D.: Save Northern High-Latitude Catchments, Nat. Geosci., 10, 324–325, 2017.

Luthin, J. N.: Drainage Engineering, Wiley, New York, 1966.

Massman, W. J.: A simple method for estimating frequency response corrections for eddy covariance systems, Agr. Forest Meteorol., 104, 185–198, 2000.

Morse, P. D., Wolfe, S. A., Kokelj, S. V., and Gaanderse, A. J. R.: The occurrence and thermal disequilibrium state of permafrost in forest ecotopes of the Great Slave Region, Northwest Territories, Canada, Permafrost Periglac., 27, 145–162, 2016.

Oke, T. R.: Boundary Layer Climates, Routledge, New York, 435 pp., 1987.

Phillips, R. W., Spence, C., and Pomeroy, J. W.: Connectivity and Runoff Dynamics in Heterogeneous Basins, Hydrol. Process., 25, 3061–3075, 2011.

Pomeroy, J. W. and Gray, D. M.: Snowcover: Accumultaion, Relocation and Management, National Hydrology Research Institute Science Report #7, 144 pp., 1995.

Prowse, T. D. and Flegg, P. O.: The magnitude of river flow to the Arctic Ocean: dependence on contributing area, Hydrol. Process., 14, 3185–3188, 2000.

Rees, A., Derksen, C., English, M., Walker, A., and Duguay, C.: Uncertainty in snow mass retrievals from satellite passive microwave data in lake-rich high-latitude environments, Hydrol. Process., 20, 1019–1022, 2006.

Spence, C.: Hydrological processes and streamflow in a lake dominated water course, Hydrol. Process., 20, 3665–3681, 2006.

Spence, C. and Hedstrom, N.: Baker Creek Research Catchment Hydrometeorological and Hydrological Data, https://doi.org/10.20383/101.026, 2018.

Spence, C., Guan, X. J., Phillips, R. W., Hedstrom, N., Granger, R., and Reid, R.: Storage Dynamics and Streamflow in a Catchment with a Variable Contributing Area, Hydrol. Process., 24, 2209–2221, 2010.

Spence, C., Guan, X. J., and Phillips, R. W.: The Hydrological Functions of a Boreal Wetland, Wetlands, 31, 75–85, 2011.

Spence, C., Kokelj, S. A., Kokelj, S. V., and Hedstrom, N.: The Process of Winter Streamflow Generation in a Subarctic Precambrian Shield Catchment, Hydrol. Process., 28, 2179–2190, 2014.

Spence, C., Ali, G., Oswald, C. J., and Wellen, C.: An application of the T-TEL assessment method to evaluate connectivity in a lake-dominated watershed after drought, J. Am. Water Resour. As., in review, 2018.

Sturm, M., McFadden, J. P., Liston, G. E., Chapin, F. S., Racine, C. H., and Holmgren, J.: Snow-shrub interactions in arctic tundra: a hypothesis with climate implications, J. Climate, 14, 336–344, 2001.

Vihma, T., Screen, J., Tjernström, M., Newton, B., Zhang, X., Popova, V., Deser, C., Hollan, M., and Prowse, T. D.: The atmospheric role in the Arctic water cycle: A review on processes, past and future changes, and their impacts, J. Geophys. Res.-Biogeo., 121, 586–620, 2016.

Walker, X. J., Rogers, B. M., Baltzer, J. L., Cumming, S. G., Day, N. J., Goetz, S. J., Johnstone, J. F., Schuur, W. A. G., Turetsky, M. R., and Mack, M. C.: Cross-scale controls on carbon emissions from boreal forest megafires, Glob. Change Biol., 24, 4251–4265, https://doi.org/10.1111/gcb.14287, 2018.

Webb, E. K., Pearman, G. I., and Leuning, R.: Correction of flux measurements for density effects due to heat and water vapour transfer, Q. J. Roy. Meteor. Soc., 106, 85–100, 1980.

Woo, M. K.: Cold Region Atmospheric and Hydrologic Studies The Mackenzie GEWEX Study Experience, Volume 2: Hydrologic Processes, Springer, Berlin, 2008.

3

Spatially distributed water-balance and meteorological data from the Wolverton catchment, Sequoia National Park, California

Roger C. Bales[1], **Erin M. Stacy**[1], **Xiande Meng**[1], **Martha H. Conklin**[1], **Peter B. Kirchner**[2,3], **and Zeshi Zheng**[4]

[1]Sierra Nevada Research Institute, University of California, Merced, CA 95343, USA
[2]Southwest Alaska Network, Inventory & Monitoring, National Park Service, Anchorage, AK 99501, USA
[3]W.A. Franke College of Forestry and Conservation, University of Montana, Missoula, MT 59812, USA
[4]Department of Civil and Environmental Engineering, University of California, Berkeley, CA 94720, USA

Correspondence: Roger C. Bales (rbales@ucmerced.edu)

Abstract. Accurate water-balance measurements in the seasonal, snow-dominated Sierra Nevada are important for forest and downstream water management. However, few sites in the southern Sierra offer detailed records of the spatial and temporal patterns of snowpack and soil-water storage and the fluxes affecting them, i.e., precipitation as rain and snow, snowmelt, evapotranspiration, and runoff. To explore these stores and fluxes we instrumented the Wolverton basin (2180–2750 m) in Sequoia National Park with distributed, continuous sensors. This 2006–2016 record of snow depth, soil moisture and soil temperature, and meteorological data quantifies the hydrologic inputs and storage in a mostly undeveloped catchment. Clustered sensors record lateral differences with regards to aspect and canopy cover at approximately 2250 and 2625 m in elevation, where two meteorological stations are installed. Meteorological stations record air temperature, relative humidity, radiation, precipitation, wind speed and direction, and snow depth.

1 Introduction

The western slope of the Sierra Nevada, California, has an elevation-driven climate gradient where wintertime snowpack accumulation provides for a large part of California's annual water needs. Small temperature increases will raise the snowline elevation and alter snowmelt, and subsequent water use by vegetation, streamflow, and forest dryness (Goulden and Bales, 2014). The implications of such a shift affect forest management, downstream water management, and ultimately, regional water policy. This research program in the Wolverton basin (1720–3500 m elevation) of Sequoia National Park is part of a larger effort to quantify the two main near-surface stores of water, snowpack, and soil-water storage, across the current Sierra Nevada rain–snow transition.

As part of this research, we used distributed snow and soil sensors, as well as remotely sensed data, to characterize snow accumulation, snowmelt, and other components of the distributed water balance. The data set covers late 2006 through the end of water year (WY) 2016. This period includes the drought years of 2012–2015 as well as the high snow year in WY 2011.

Kirchner et al. (2014) found that snowmelt at the upper elevations occurs later in the season and at faster rates compared to lower elevations. Historically, the Wolverton basin has been in the seasonal snow zone of the southern Sierra Nevada, characterized by winter snow accumulation and spring ablation. At the lower elevations, winter snow cover is ephemeral in some years, with 70 %–95 % of the snow-covered season having slow melt rates (Kirchner et al., 2014). At upper elevations, melt occurs during less than 65 %

of the season. Based on a 2010 set of lidar flights (Anderson et al., 2012), snow depth near the date of accumulation is greatest at an elevation of 3300 m, which is 200 m below the highest ridge. Between 1850 and 3300 m, snow depth increases at a rate of approximately 15 cm per 100 m; at higher elevations, snow depth decreases at a rate of 48 cm per 100 m (Kirchner et al., 2014). Elevation is one of the most important factors for determining snow depth across the catchment, though snow depth is also mediated by canopy penetration fraction northness and vegetation height (Zheng et al., 2016; Tennant et al., 2017). The difference between open canopy and under-canopy areas increases through the rain–snow transition up to 25–45 cm at high elevations (Fig. 1).

Meteorological factors, as well as the interference of canopy coverage, impact snow accumulation and melt at these sites. Sub-canopy direct-beam irradiance and a sky-view factor explain the most variation in snow ablation rates, especially at finer timescales (Musselman et al., 2012). A time-varying canopy parameter in snow modeling reduces errors by 7 days in the simulated snow disappearance date, and errors in the timing of soil-water fluxes by 11 days, on average, compared to a bulk parameterization of radiation transfer through the canopy (Kirchner et al., 2014).

Studies of this basin indicate that it is sensitive to a changing climate. The timing and rate of snowmelt indicate that this elevation range is sensitive to seasonal meteorology, especially where upper elevations may begin to experience snow melt during more of the snow-covered season (Kirchner et al., 2014). At both the upper and lower sites, peak soil moisture precedes the average date of snow disappearance, meaning that soil moisture declines even while snowmelt is infiltrating into the soil system (Harpold et al., 2015). With peak snow depth around 3300 m (Kirchner et al., 2014), such changes to the hydrological system could have major implications for snowpack water storage and runoff.

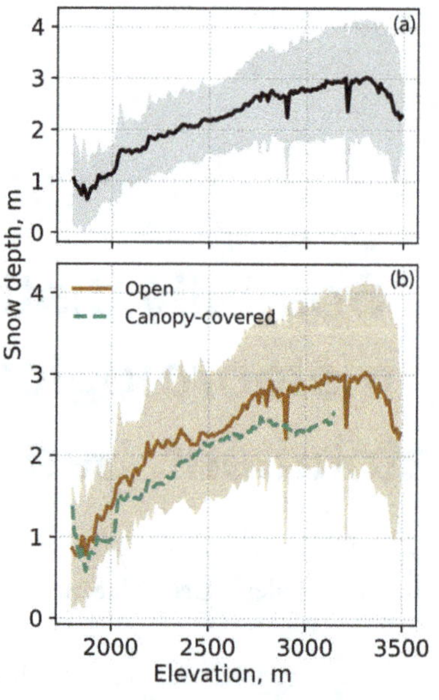

Figure 1. Snow-depth changes over elevation observed from the lidar data in the Wolverton basin **(a)** for all snow-depth pixels and **(b)** for snow-depth pixels in the open area and under canopy. On average, snow depth increases with elevation nonlinearly with elevation below 3300 m and decreases when elevation is above 3300 m. Lidar data from Anderson et al. (2012).

2 Site Description

The Wolverton basin is a montane, forested catchment northeast of Giant Forest in Sequoia National Park (Fig. 2). In the 5.4 km^2 catchment, the dominant aspect is 214° from the north, with an average slope of 21.1°, determined by a Lidar flight (Harpold et al., 2014). Basin coverage is 45 %–55 % forest (Harpold et al., 2014). The forest transitions from mixed conifer at the lower elevations to subalpine forests at the higher elevations. Dominant species are red fir (*Abies manifica*), lodgepole pine (*Pinus contorta*), western white pine (*Pinus monticola*), incense cedar (*Calocedrus decurrens*), and Jeffrey pine (*Pinus jeffreyi*; Kirchner, 2014). A long, narrow meadow occupies the lower reaches of the western fork, and small meadows are present in the eastern fork.

Most of the catchment is undeveloped forest with no history of thinning, though there was a small downhill ski area in the basin for about 50 years until 1994. There is some recreational infrastructure at the base of the catchment, including trailhead parking lots, a water treatment plant, and recreational buildings. In addition, there are several recreational hiking trails around the catchment.

During the WY 2007–2016 period, mean annual precipitation was 728 mm at the nearby cooperative Lodgepole temperature and weather station operated by Sequoia and Kings Canyon National Parks. The mean annual temperature was 6.0 °C. The 2008–2010 water years in the Wolverton basin represent dry, normal, and wet years, where peak mean snow depths were 73 %–165 % of the 30-year average (Harpold et al., 2015). Soil texture in the upper 30 cm of soil was 6 % clay, 17 %–23 % silt, and 71 %–77 % sand (Harpold et al., 2015). Parts of the basin were previously glaciated. The geology of the basin is granodiorite, granite, and a mix of glacial and alluvial till in the lower reaches (U.S. Geological Survey, map GQ-1636).

3 Meteorological data

We installed and instrumented two meteorological stations (met) in 2008. The Wolverton meteorological station is at 2206 m, on a 7 m steel triangular-frame tower. The Panther meteorological station, at 2618 m, is a 6.1 m steel pole with cross-arms near the apex. Sensor instruments were installed for air temperature, relative humidity, wind speed, wind direction, net radiation, solar radiation, snow depth, and soil moisture and soil temperature. Sensor manufacturers and in-

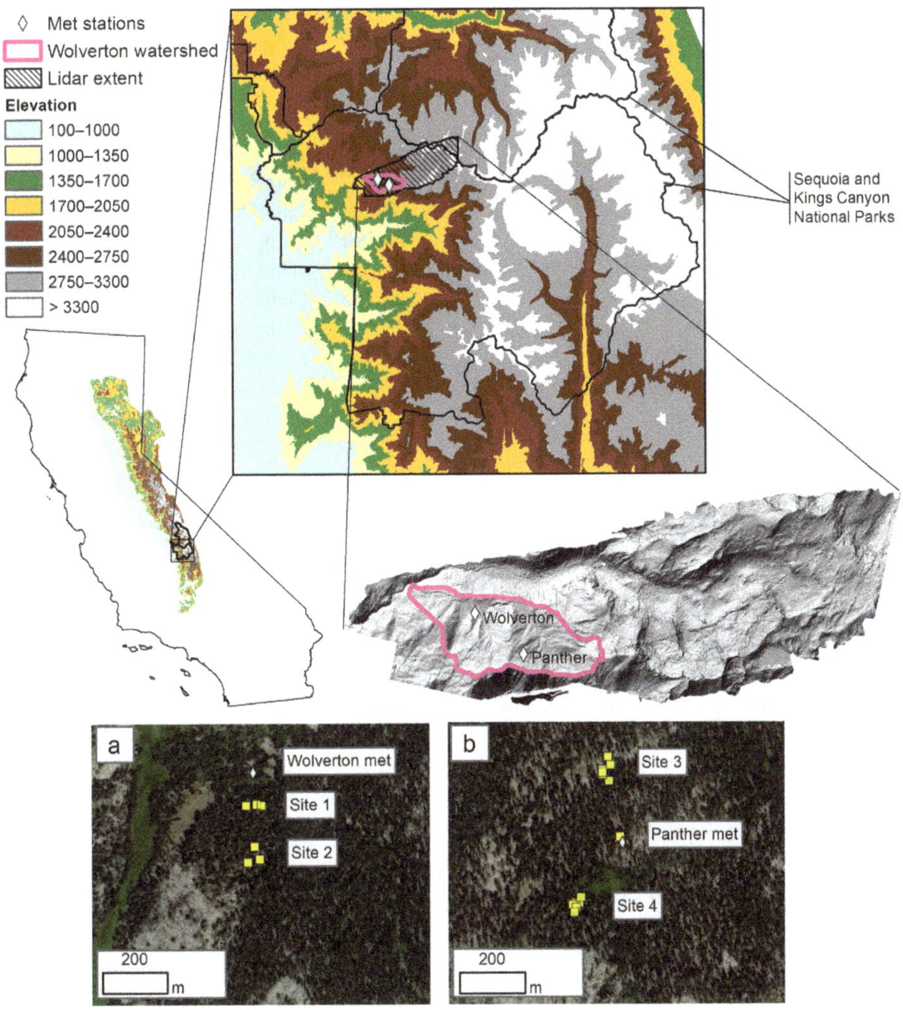

Figure 2. Wolverton catchment in the Marble Fork of the Kaweah River, Sequoia and Kings Canyon National Parks. Distributed instruments are clustered around the **(a)** Wolverton and **(b)** Panther meteorological stations (topographic data: EDNA-filled DEM grid, U.S. Geological Survey, 2003. Lidar data: Anderson et al. (2012), snow-off surface raster. Imagery sources: Esri, DigitalGlobe, Earthstar Geographics, CNES/Airbus DS, GeoEye, USDA FSA, USGS, Aerogrid, IGN, IGP, and the GIS User Community, 20 July 2017.)

stallation heights are listed in Table 1. Data are recorded at 60 min intervals on a on Campbell Scientific CR1000 data logger. Data acquisition programs are located on the UC Merced-SNRI digital library with the data.

4 Distributed instrument clusters

The distributed sensor nodes were installed in four clusters at the upper and lower elevations in roughly north and south aspects (Fig. 2). The lower-elevation sites, 1 and 2, are south of Wolverton meteorological station and face north and southeast, respectively, at elevations of around 2226 and 2260 m. The upper-elevation sites, 3 and 4, are near the Panther meteorological station and face southeast and north, respectively, at elevations of around 2596 and 2636 m. Soil moisture and temperature, snow depth, and meteorological data (air temperature, relative humidity, and solar radiation) are measured at these clusters across the network at hourly intervals. Snow depth is stratified by canopy coverage, with 15 measurements at open canopy locations, six at the canopy drip edge, and five under the tree canopy (canopy classification was verified on-site). Distance to snow and soil surface and air temperature are measured with Judd Communications ultrasonic depth sensors, using analog control. Total solar radiation is measured with the LI-COR Environmental LI-200R Pyranometer, like the met stations. Soil volumetric water content (VWC) and soil temperature are measured using the Campbell Scientific CS 616 Water Content Reflectometer and 107 temperature probes at depths of 10, 30, and 60 cm below the mineral soil surface. The standard calibration for the CS-616 sensors was used. The fact that the sensors were not individually

Table 1. Meteorological station instrumentation.

Variable	Unit	Wolverton installation height (m)	Panther installation height (m)	Instrument
Air temperature	°C	6	5.2	Campbell Scientific HMP45C
Relative humidity	%	6	5.2	Campbell Scientific HMP45C
Wind speed, direction	$m\,s^{-1}$, °	6.8	5.1	Campbell Scientific R. M. Young 05103 Wind Speed & Direction Sensor
Rain	mm	5.1	4.5	Campbell Scientific TE525MM/TE525M Rain Gauge
Net radiation	$W\,m^{-2}$	5.7	–	Kipp & Zonen NR-LITE
Solar radiation	$W\,m^{-2}$	4.5	5	LI-COR LI200R Pyranometer
Snow depth, air temperature	cm, °C	4.4	4.4	Judd Sonic Snow Depth Sensor

calibrated to soils at each site is one of the flaws in the original experimental design, but the standard calibration does have an accuracy of ±2.5 % VWC. Soil matric potential is measured in one location at each site using the Decagon Devices MPS-1 dielectric water potential sensor at depths of 10, 30, 60, and 90 cm below the mineral soil surface (http://www.decagon.com/products/discontinued-products/mps-1-dielectric-water-potential-sensor/, last access: 8 June 2018).

Data are recorded at 60 min scan intervals, with control and storage on Campbell Scientific CR1000 data logger, using an AM16/32B multiplexer. Program for data acquisition are located on the UC Merced-SNRI digital library with the data.

5 Example data

In water years 2007–2016, snowpack was deeper and more persistent at the higher-elevation sites (2640 m; Fig. 3a). Warmer temperatures at the lower-elevation sites (2245 m; Fig. 3b) result in earlier soil wet-up, higher mid-winter soil moisture storage, and earlier peak soil moisture storage in wet and dry years. Long data gaps at Site 2 in 2013 and at Site 1 in 2016 are due to battery and logger issues. Comparing average (WY 2010), wet (WY 2011), and dry (WY 2012) years, snowpack is persistent at both the upper- and lower-elevation sites (Fig. 4). However, in WY 2014 and 2015, snowpack receded multiple times during each winter. Soil moisture peaks earlier at the two lower-elevation clusters. The Wolverton met station experienced a wider swing in temperatures than the Panther met station. Mean annual temperature was 6.3 °C at the upper met station Panther, and 5.7 °C at the lower-elevation Wolverton station, with both warmer maximum and cooler minimum air temperatures at the Wolverton site. Water-level data are available for two locations in the meadow.

Some of the sensors have inherent possible biases. The precipitation sensors are unshielded and unheated tipping buckets. Precipitation measurements in each data set will underreport precipitation through undercatch, freezing, and coverage by snow. Considering this bias, the precipitation records shown for comparison with the other meteorological data (in Fig. 3a and b). That meteorological station has a weighing gauge with a long-term record, though the sensor is also unshielded.

6 Data availability

Data are available at hourly intervals by water year (WY; 1 October–30 September), from WY 2007 through WY 2016. Data are available through online data repositories. Distributed snow depth, air temperature, and soil moisture and temperature are available through https://doi.org/10.6071/M3S94T (Bales et al., 2017), which links to data storage on a server hosted at University of California, Merced, or are available directly from our digital library at https://eng.ucmerced.edu/snsjho/files/MHWG/Field/SEKI/Wolverton (last access: 13 November 2018). These data from the distributed networks have been processed to level 1 (QA/QC) and level 2 (gap-filled, derived) data. Raw data for matric potential, sap flow, and solar radiation at each site are also available. Gaps are filled through regression with a nearby sensor node, which is selected based on the best correlation. Multiple neighboring nodes may be selected if needed, and different neighboring nodes may be used to fill each measurement. Short gaps, or gaps in soil temperature, may be filled through linear interpolation. Process notes (metadata) are available from the data repositories. Data in at each site of clusters (e.g., Site 1 or Site 3) are organized by and named by installation location (P1, P2, etc., with canopy coverage indicated as "DE" for drip edge, "UC" for under canopy, or "O" for open).

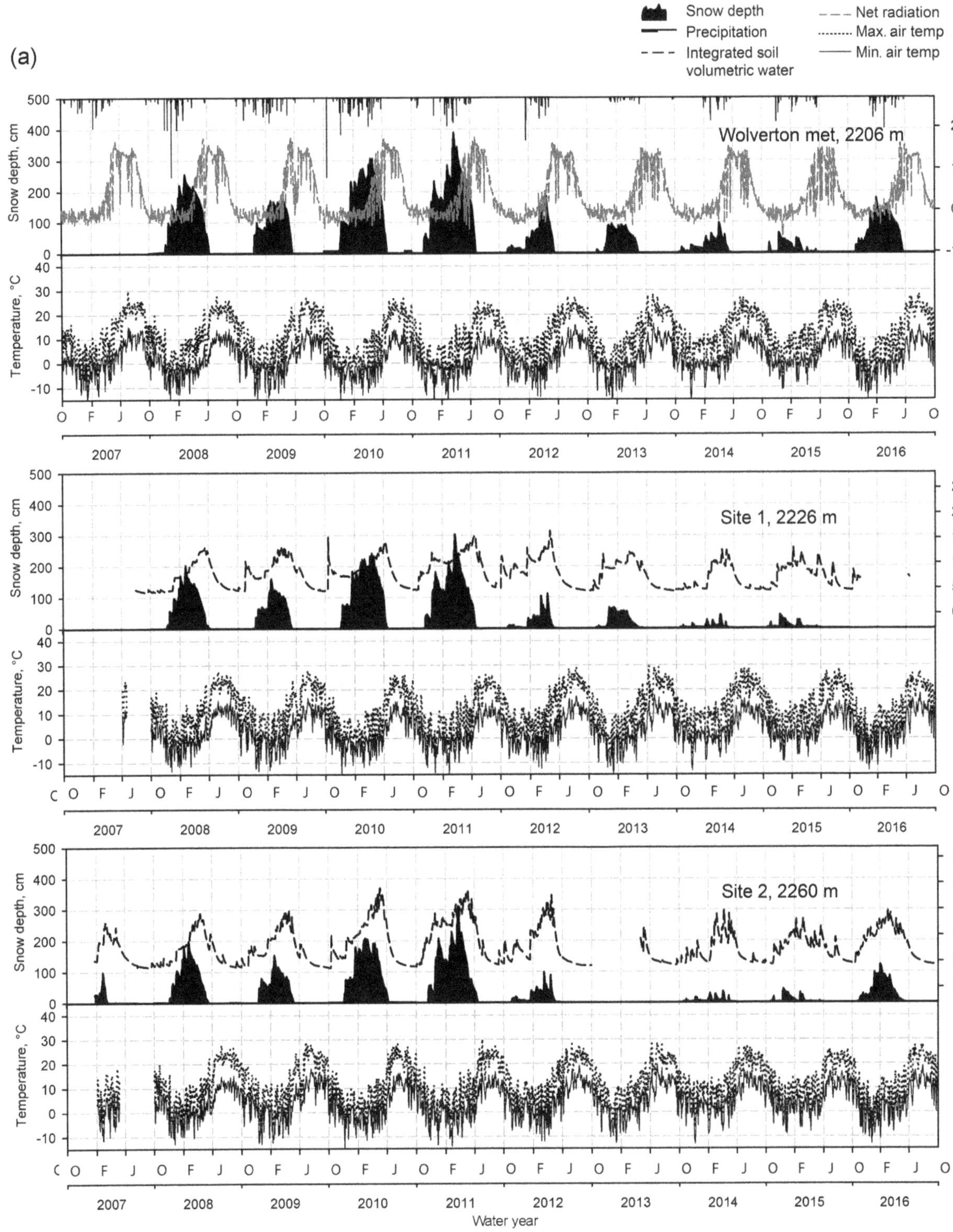
Snow depth
Precipitation
Integrated soil volumetric water
Net radiation
Max. air temp
Min. air temp
(a)
Wolverton met, 2206 m
Site 1, 2226 m
Site 2, 2260 m
Snow depth, cm
Temperature, °C
Net radiation, w m⁻²
Precipitation, cm
Water storage, cm in top 1 m
Water year

Figure 3.

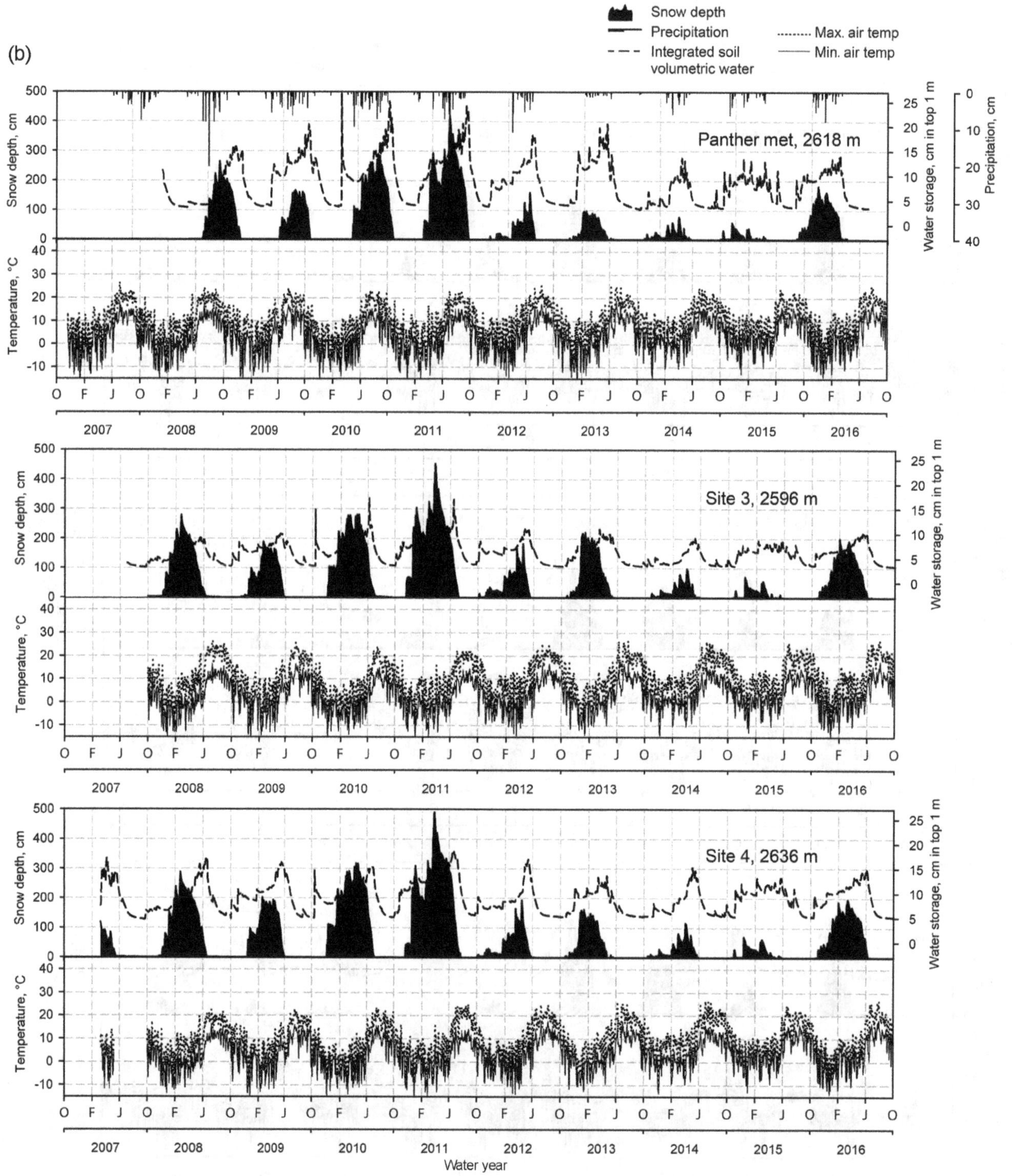

Figure 3. **(a)** record of soil moisture, snow depth, and air temperature (max. and min.) at the lower-elevation instrument clusters. Near Wolverton meteorological station, Site 1 has a north aspect, and Site 2 has a southeast aspect. Precipitation from nearby meteorological station at Lodgepole is shown for comparison. **(b)** Record of soil moisture, snow depth, and air temperature (max. and min.) at the higher elevation instrument clusters. Near Panther meteorological station, Site 3 has a southeast aspect, and Site 4 has a north aspect. Precipitation from nearby meteorological station at Lodgepole is shown for comparison.

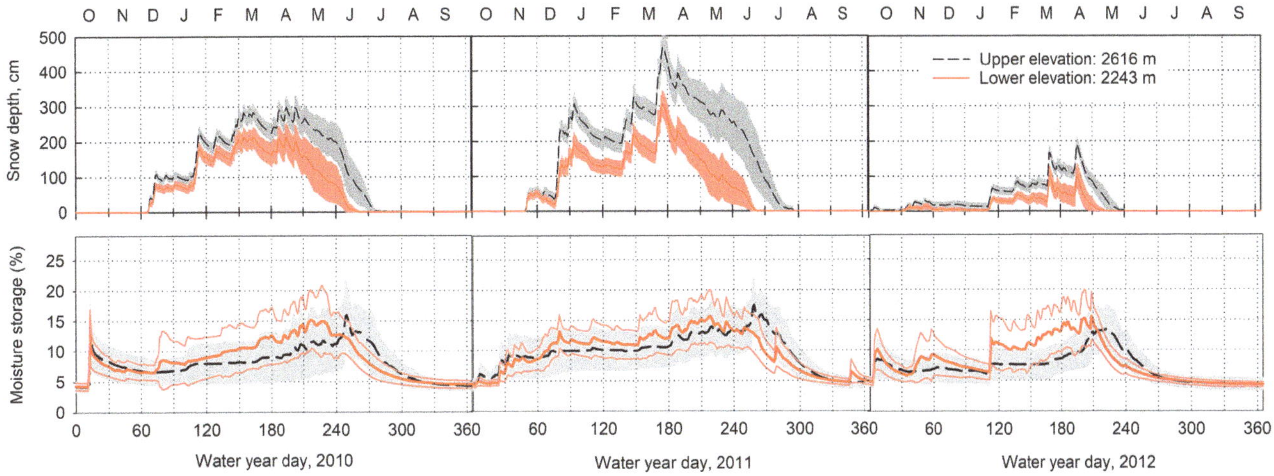

Figure 4. The 24 sensor nodes highlight the spatial variability of soil moisture and snow depth through water years 2010 (average), 2011 (record-setting wet year), and 2012 (extremely dry). Soil moisture can be higher at the lower elevation than the higher elevation, and peak soil moisture may predate the end of snowmelt.

7 Summary

A 10-year meteorological and hydrologic data record is presented for a catchment in Sequoia National Park, in the southern Sierra Nevada. Distributed snow depth and soil temperature and moisture combined with two meteorological stations provide a means of testing and evaluating hydrologic processes in a productive montane forest. The Wolverton basin research site serves as a southern comparison point with installations in the Kings River (SSCZO; 2008–present; Bales et al., 2018), Merced River (MRB; 2006–present; Roche, 2017), Stanislaus River (2013–present, Pickard, 2015), American River (2014–present; Zhang et al., 2016, 2017a, b), and Feather River (2016–present; Avanzi et al., 2018) basins. Studies compiled to date indicate that hydrological variables, including snow depth, the timing and rate of snow melt, and soil moisture, are susceptible to changing climate patterns like warmer temperature or increased vegetation water demands in this part of the Sierra Nevada. We invite others to use this data for their own studies of the basin or as a comparison point to further the discussion.

Author contributions. RCB, MHC, and PBK designed the sensor networks. PBK, EMS, XM, and ZZ installed and maintained the sensor networks and processed the sensor-network data. RCB, EMS, and ZZ prepared the paper with contributions from all authors.

Competing interests. The authors declare that they have no conflict of interest.

Acknowledgements. We thank the Sequoia and Kings Canyon National Parks research and permitting staff and the staff and research team at the Southern Sierra Critical Zone Observatory. This research was funded in part by NSF EAR-0725097, EAR-1239521, and EAR-1331939 for the Southern Sierra Critical Zone Observatory, and the University of California, Merced.

Edited by: Danny Marks

References

Anderson, S. P., Qinghua, G., and Parrish, E. G.: Snow-on and snow-off Lidar point cloud data and digital elevation models for study of topography, snow, ecosystems and environmental change at Southern Sierra Critical Zone Observatory, California, Univ. Cal. Merced, digital media, https://doi.org/10.5069/G9BP00QB, 2012.

Avanzi, F., Maurer, T. P., Malek, S. A., Glaser, S. D., Bales, R. C., and Conklin, M. H.: Feather River Hydrologic Observatory: Improving hydrological snow pack forecasting for hydropower generation using intelligent information systems, State of California Energy Commission, Report CCCA4-CEC-2018-001, Sacramento, CA, 2018.

Bales, R. C., Stacy, E. M., Meadows, M., Kirchner, P. B., Conklin, M. H., and Meng, X.: Southern Sierra Critical Zone Observatory (SSCZO), Wolverton Creek meteorological data, soil moisture and temperature, v3, UC Merced Dash, Dataset, https://doi.org/10.6071/M3S94T, 2017.

Bales, R., Stacy, E., Safeeq, M., Meng, X., Meadows, M., Oroza, C., Conklin, M., Glaser, S., and Wagenbrenner, J.: Spatially distributed water-balance and meteorological data from the rain–snow transition, southern Sierra Nevada, California, Earth Syst. Sci. Data, 10, 1795–1805, https://doi.org/10.5194/essd-10-1795-2018, 2018.

Goulden, M. L. and Bales, R. C.: Mountain runoff vulnerability to increased evapotranspiration with vegetation expansion, P. Natl. Acad. Sci. USA, 111, 14071–14075, https://doi.org/10.1073/pnas.1319316111, 2014.

Harpold, A. A., Guo, Q., Molotch, N., Brooks, P. D., Bales, R., Fernandez-Diaz, J. C., Musselman, K. N., Swetnam, T. L., Kirchner, P., Meadows, M., Flanagan, J., and Lucas, R.: LiDAR-derived snowpack data sets from mixed conifer forests across the Western United States, Water Resour. Res., 50, 2749–2755, https://doi.org/10.1002/2013WR013935, 2014.

Harpold, A. A., Molotch, N. P., Musselman, K. N., Bales, R. C., Kirchner, P. B., Litvak, M., and Brooks, P. D.: Soil moisture response to snowmelt timing in mixed-conifer subalpine forests, Hydrol. Process., 29, 2782–2798, https://doi.org/10.1002/hyp.10400, 2015.

Kirchner, P.: Snow distribution over an elevation gradient and forest snow hydrology of the southern Sierra Nevada, California, Environmental Systems Graduate Group, Univ. Cal. Merced, PhD Dissertation, 192 pp., Merced, CA, 2014.

Kirchner, P. B., Bales, R. C., Molotch, N. P., Flanagan, J., and Guo, Q.: LiDAR measurement of seasonal snow accumulation along an elevation gradient in the southern Sierra Nevada, California, Hydrol. Earth Syst. Sci., 18, 4261–4275, https://doi.org/10.5194/hess-18-4261-2014, 2014.

Musselman, K. N., Molotch, N. P., Marguilis, S. A., Kirchner, P. B., and Bales, R. C.: Influence of canopy structure and direct beam solar irradiance on snowmelt rates in a mixed conifer forest, Agr. Forest. Meteorol., 161, 46–56, https://doi.org/10.1016/j.agrformet.2012.03.011, 2012.

Pickard, M. R.: Influence of within-stand tree spatial arrangement on snowpack distribution and ablation in the Sierra Nevada, CA, Environmental Systems Graduate Group, Univ. Cal. Merced, MS Thesis, Merced, CA, 82 pp., 2015.

Roche, J. W.: Evaluating water balance components in the Sierra Nevada: Snowpack sensitivity to climate warming and forest evapotranspiration reduction potential, Environmental Systems Graduate Group, Univ. Cal. Merced, Merced, CA, PhD Dissertation, 152 pp., 2017.

Tennant, C. J., Harpold, A. A., Lohse, K. A., Godsey, S. E., Crosby, B. T., Larsen, L. G., Brooks, P. D., Van Kirk, R. W., and Glenn, N. F.: Regional sensitivities of seasonal snowpack to elevation, aspect, and vegetation cover in western North America, Water Resour. Res., 53, 6908–6926, https://doi.org/10.1002/2016WR019374, 2017.

Zhang, Z., Glaser, S. D., Watteyne, T., and Malek, S.: Long-term monitoring of the Sierra Nevada snowpack using wireless sensor networks, IEEE Internet Things, 12, 1–9, 2016.

Zhang, Z., Glaser, S. D., Bales, R. C., Conklin, M. H., Rice, R., and Marks, D. G.: Insights into mountain precipitation and snowpack from a basin-scale wireless sensor network, Water Resour. Res., 53, 6626–6641, 2017a.

Zhang, Z., Glaser, S. D., Bales, R. C., Conklin, M. H., Rice, R., and Marks, D. G.: Technical report: The design and evaluation of a basin-scale wireless sensor network for mountain hydrology, Water Resour. Res., 53, 4487–4498, https://doi.org/10.1002/2016WR019619, 2017b.

Zheng, Z., Kirchner, P. B., and Bales, R. C.: Topographic and vegetation effects on snow accumulation in the southern Sierra Nevada: a statistical summary from lidar data, The Cryosphere, 10, 257–269, https://doi.org/10.5194/tc-10-257-2016, 2016.

4

Estimating the thickness of unconsolidated coastal aquifers along the global coastline

Daniel Zamrsky[1], **Gualbert H. P. Oude Essink**[1,2], **and Marc F. P. Bierkens**[1,2]

[1]Department of Physical geography, Utrecht University, Utrecht, the Netherlands
[2]Deltares, Utrecht, the Netherlands

Correspondence: Daniel Zamrsky (d.zamrsky@uu.nl)

Abstract. Knowledge of aquifer thickness is crucial for setting up numerical groundwater flow models to support groundwater resource management and control. Fresh groundwater reserves in coastal aquifers are particularly under threat of salinization and depletion as a result of climate change, sea-level rise, and excessive groundwater withdrawal under urbanization. To correctly assess the possible impacts of these pressures we need better information about subsurface conditions in coastal zones. Here, we propose a method that combines available global datasets to estimate, along the global coastline, the aquifer thickness in areas formed by unconsolidated sediments. To validate our final estimation results, we collected both borehole and literature data. Additionally, we performed a numerical modelling study to evaluate the effects of varying aquifer thickness and geological complexity on simulated saltwater intrusion. The results show that our aquifer thickness estimates can indeed be used for regional-scale groundwater flow modelling but that for local assessments additional geological information should be included.

1 Introduction

Coastal aquifers provide fresh groundwater for more than 2 billion people worldwide (Ferguson and Gleeson, 2012). Multiple local and regional studies have shown that these fresh groundwater resources are threatened not only by natural disasters such as storm surges and tsunamis (Cardenas et al., 2015), but also increasingly by climate-induced sea-level rise (Carretero et al., 2013; Rasmussen et al., 2013; Sefelnasr and Sherif, 2014) and urbanization that lead to coastal aquifer over-exploitation combined with reduced groundwater recharge (Custodio, 2002).

Comparing coastal aquifer vulnerabilities worldwide in a consistent manner requires a global-scale study (Döll, 2009). However, many necessary input datasets, both physical and societal, are only available on regional or local scales and can therefore only be used in coastal aquifer vulnerability investigations on a regional scale. Notable work on global coastal aquifer vulnerability are studies by Ranjan et al. (2009) and Michael et al. (2013) looking at coastal aquifer vulnerability to seawater intrusion and by Nicholls and Cazenave (2010) taking social–economic factors into account. Related observation-based studies are performed by van Weert et al. (2008) on global saline groundwater occurrence assessment and by Post et al. (2013) on the existence of offshore fresh or brackish groundwater. Lacking global information, the few global studies that attempted a modelling approach (i.e. Ranjan et al., 2009, and Michael et al., 2013) used globally or regionally homogenous hydraulic parameters, including aquifer thickness. Indeed, recent reviews concluded that most of the past modelling studies until the present day (on both local and global scales) considered a homogeneous aquifer system (Werner et al., 2013; Ketabchi et al., 2016). This pinpoints that there is still a large gap in our knowledge about coastal aquifer hydrogeological settings in many parts of the world. Since the local and regional hydrogeological conditions largely determine the coastal aquifer vulnerability to sea-level rise (Michael et al., 2013) and groundwater pumping (Ferguson and Gleeson, 2012), it is important to im-

Table 1. Statistics for individual GLIM classes in the coastal ribbon (200 km and less from the coastline). The population numbers are based on the 2015 global population count (CIESIN, 2017).

GLIM class (xx)	GLIM class (name)	Total % in coastal ribbon	Population sum	Population % in coastal ribbon
ev	Evaporites	0.27	4 612 359	0.12
ig	Ice and Glaciers	0.22	3480	0.00
mt	Metamorphic rocks	20.59	395 567 421	10.52
nd	No Data	0.03	1 211 402	0.03
pa	Acid Plutonic rocks	5.66	248 153 378	6.60
pb	Basic Plutonic rocks	0.74	19 480 337	0.52
pi	Intermediate Plutonic rocks	0.52	13 959 834	0.37
py	Pyroclastics	1.00	39 688 219	1.06
sc	Carbonate Sedimentary rocks	8.96	268 875 153	7.15
sm	Mixed Sedimentary rocks	13.69	350 423 700	9.32
ss	Siliclastic Sedimentary rocks	14.23	487 261 398	12.95
su	Unconsolidated Sediments	25.78	1 562 019 536	41.53
va	Acid Volcanic rocks	1.22	60 196 503	1.60
vb	Basic Volcanic rocks	4.39	165 959 005	4.41
vi	Intermediate Volcanic rocks	2.29	128 173 527	3.41
wb	Water Bodies	0.43	15 979 647	0.42
Total		100.00	3 761 564 898	100

Table 2. Summary of the global datasets used for aquifer thickness estimation.

Dataset name	Description	Resolution	Reference
GEBCO 2014	Global topography and bathymetry	30 arcsec	Weatherall et al. (2015)
Average soil and sedimentary deposit thickness	A gridded global dataset of soil, intact regolith, and sedimentary deposit thicknesses for regional and global land surface modelling; max. estimated depth is 50 m.	30 arcsec	Pelletier et al. (2016)
PCR-GLOBWB	Thickness of the groundwater layer from the global model (5 arcmin)	5 arcmin	de Graaf et al. (2015)
GLIM	Global Lithological Map – Rock types of the Earth surface (16 basic classes), more than 1 200 000 polygons	vector	Hartmann and Moosdorf (2012)
Natural Earth coastline	Global coastline	vector	Natural Earth (2017)

prove our insight into the local and regional coastal aquifer hydrogeology worldwide.

The goal of this study is to estimate the unconsolidated aquifer system thickness along the global coastline. This constitutes a first step towards a more complete hydrogeological characterization of coastal aquifers. Our focus is limited to aquifer systems formed by unconsolidated sediments (Table 1) that constitute around 25 % of the coastal ribbon (200 km far or less from the coastline) based on the GLIM (Global lithological map) dataset (Hartmann and Moosdorf, 2012). In contrast, more than 36 % is shaped by different types of sedimentary rocks where aquifers can also be expected. These sedimentary rock formations most probably form the coastal aquifer systems that are missed in this study. However, more than 40 % of people living in the coastal ribbon (CIESIN, 2017) are located on top of unconsolidated sediment aquifer systems (Table 1), while less than 30 % live in areas with sedimentary rock aquifers. This means that there is potentially more pressure on freshwater availability in areas with unconsolidated sediment aquifer systems.

To be globally applicable and comparable, our method of aquifer thickness estimation makes use of already available open-source global datasets (see Table 2). These datasets contain information on elevation, surficial lithology, regolith thickness and overall sedimentary thickness. What motivated this study is that none of the globally available thickness datasets are individually suited to representing coastal aquifer thickness. Two of these datasets only provide estimated regolith (surficial sediment layer) or soil thickness (Pelletier et al., 2016; Shangguan et al., 2017). The soil or regolith layer is only part of the aquifer system formed by unconsolidated sediments and therefore unfit to use in building a hydrogeological model representing the flow in the whole aquifer system. Conversely, the other two datasets (Whittaker et al., 2013; de Graaf et al., 2015) estimate the total porous media thickness without making a distinction between unconsolidated and consolidated sediments (rocks), and therefore tend to overestimate the unconsolidated aquifer system thickness.

The resulting dataset consists of 26 968 cross sections perpendicular to the global coastline with unconsolidated aquifer thickness estimated along each cross section. Additionally, the uncertainty ranges in aquifer thicknesses are provided for each cross section. In order to illustrate how to use the new dataset in a regional groundwater modelling setting, we will show the results of variable-density groundwater flow and coupled salt transport models for three distinctly different coastal cross sections. We also show the sensitivity of modelling results to varying the aquifer thickness and geological complexity.

2 Materials and methods

2.1 Coastal aquifer unconsolidated sediment thickness estimation

We collected state-of-the-art open-source global datasets (Table 2) that provide information on topography and bathymetry (Weatherall et al., 2015), regolith thickness estimation (Pelletier et al., 2016), global-scale aquifer thickness estimated by de Graaf et al. (2015), lithology (Hartmann and Moosdorf, 2012) and coastline position (Natural Earth, 2017). The core of our aquifer thickness estimation (ATE) method is to combine topographical and lithological information. This enables us to find the topographical slope of outcropping bedrock formations and to determine the coastal plain extent. The latter is defined by a low topographical slope (Weatherall et al., 2015), a lithology consisting of unconsolidated sediments (Hartmann and Moosdorf, 2012) and a regolith thickness thicker than 50 m (Pelletier et al., 2016). This is the first study that directly combines lithology and topographic information to estimate the coastal unconsolidated sediment aquifer systems thickness at global scale.

Given the large variety of coastal environments, ranging from steep cliffs to extensive deltaic flat areas, it is important to develop a robust method that distinguishes between these different coastal types and also takes into account variations of inland bedrock formations. To achieve this, the coastal zones are represented as perpendicular cross sections to the coastline and are placed equidistantly (5 km) along the coastline. The intersections between the cross sections and the coastline are called coastal points. Along the cross section, a set of equidistant points (0.5 km) is positioned (cross-section points) and marks the locations where values from the datasets listed above are extracted (Fig. 1a). The cross sections span 200 km both inland and offshore from the coastal point to capture the bathymetrical and topographical profile. This distance was chosen to safely cover the necessary stretch both landward and offshore for groundwater flow and coupled salt transport modelling. Recent studies dealing with the latter set the landward boundary less than 200 km from the coastline even in deltaic areas (Delsman et al., 2014; Larsen et al., 2017; Nofal et al., 2016). Similarly, previous studies showed that submarine groundwater discharge can occur more than 100 km offshore (Kooi and Groen, 2001; Post et al., 2013).

Figure 1b shows an example of a cross section running through a coastal point. All the necessary values from the individual datasets are aggregated and used to determine the coastal plain extent and the anchor point position. The inland boundary point of the coastal plain is defined as a cross-section point that has a lithological class different than a water body (to take into account e.g. lagoons and bays) or unconsolidated sediments based on the GLIM dataset. Hartmann and Moosdorf (2012) state that uncertainty in the GLIM dataset is still significant based on the amount of mixed sediment class ($\sim 15\,\%$ of the world area), so it is likely that some unconsolidated sediment coastal areas have been missed in our study.

Once the coastal plain extent is known, the next step is to define the anchor point using the regolith thickness dataset (Pelletier et al., 2016). Taking note of the fact that the Pelletier et al. (2016) dataset generally has increasing thickness values towards the coast in case of unconsolidated sediments and that a thickness larger than 50 m is not mapped, we define the anchor point as the last cross-section point (moving from land to coast) with soil and sedimentary deposit thickness smaller than 50 m. Pelletier et al. (2016) state that areas with low relief, such as coastal plains, generally have a thicker sedimentary layer (> 50 m) than hillslopes, so the transition zone between these two relief types is modelled with acceptable accuracy on global scale. The anchor point represents the last point where soil and sedimentary deposit thickness is known and is located below ground at the indicated depth by this dataset. A histogram of anchor point distances to coastline and of total coastal plain extent values is shown in Fig. 2. The ATE is then performed for all cross-section points located between the anchor point and the coastline.

Due to a large variety in the coastal plain extent, topography and geological diversity of the coastal cross sections worldwide, four different estimation techniques are proposed to increase the overall estimation method robustness. The differences between these techniques are in the topographical points' selection; these points are used to simulate the bedrock slope (Fig. 1c). The anchor point is added to the set of topographical points in every estimation technique.

The first technique selects all cross-section point elevation values of the first peak located prior to the coastal plain, regardless of lithological class. The second technique selects all the cross-section point elevation values of the highest peak located in a bedrock formation (any other class than unconsolidated sediments). The third selection technique consists of selecting all cross-section point elevations located between the coastal plain end and the bedrock formation

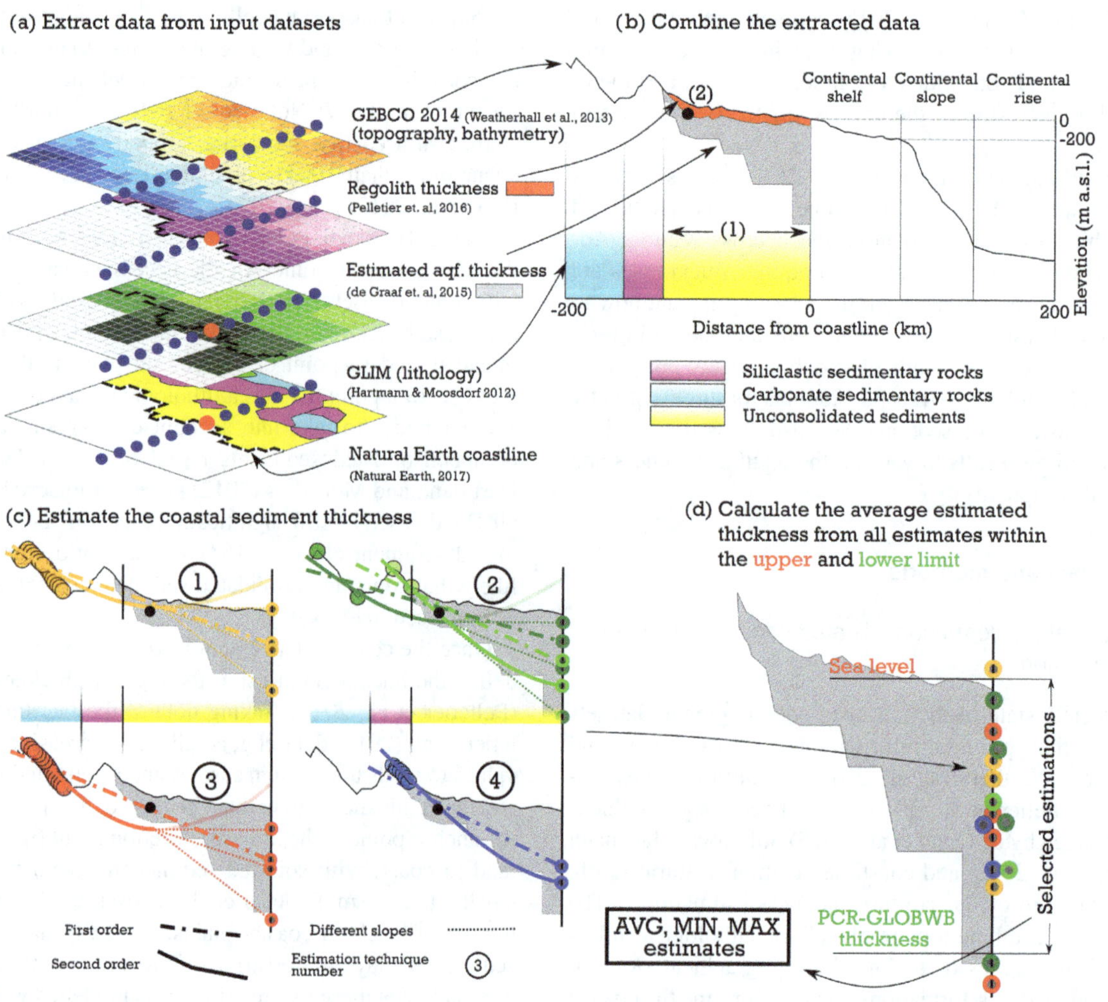

Figure 1. Schematization of the ATE method using available open-source global datasets. **(a)** Combining input datasets and extracting the values at cross-section points along a perpendicular cross section to the coastline running through a coastal point (red dot), only a few are schematized in the figure (in reality 800 per cross section). **(b)** Determine the extent of the coastal plain (1) and position of the anchor point (2). Extent of the cross section is set to 200 km landward and offshore; **(c)** the estimation is performed via topographical points selected based on the coastal plain extent, the position of the anchor points and the lithological classes from the GLIM dataset. The second-order estimation line is not used for estimation in case its minimum is reached before the coastline (transparent). **(d)** Final step of calculating the average, minimum and maximum estimated values.

end. The last technique selects only the local minimum and maximum points of every peak located behind the coastal plain. This diversity in selecting cross-section points based on lithological and topographical information combinations allows for a more robust method that is fit for various coastal environments.

For each selection technique described above, a first- and second-order curve fitting is performed to simulate the bedrock formation slope (Fig. 1c). If the minimum point of the second-order curve is situated before the coastline, we use three different linear function types to extend the bedrock slope simulation and estimate the sediment thickness by extending it beyond the coastline. All three lines start at the minimum point of the second-order estimation and run through the coastline. The first line is a constant horizontal line, the second simulates the average continental shelf slope (defined as shallower than −200 m b.s.l.) and the last line simulates the average slope of the whole 200 km cross-section offshore segment.

The global-scale aquifer thickness estimated by de Graaf et al. (2015) is chosen as the lower boundary since it tends to overestimate the coastal aquifer thickness because its underlying method is more fit for the inland areas and uses river networks and basins as a basis for thickness estimates (de Graaf et al., 2015). Finally, all the points are used to estimate the mean, minimum and maximum aquifer thicknesses at the coastline and the mean coastal profile for the unconsolidated sediment extent. The dataset that is stored contains

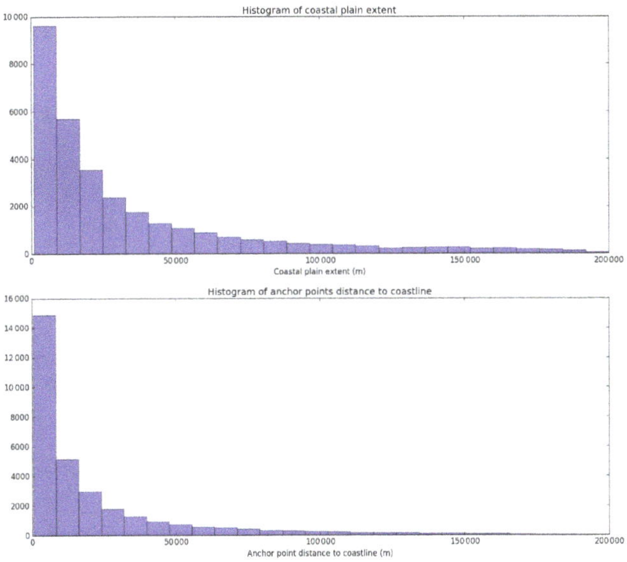

Figure 2. Histogram of coastal plain extents and anchor point distance to coastline values.

per coastal cross section the mean profile as well as the maximum and minimum depth and the depth at the coastline standard deviation. For each coastal cross section, the anchor point position and depth are also included.

2.2 Validation methods

Two different validation approaches are applied to test the fit of our estimated aquifer thicknesses with measured values. First, the results are compared with information from available open-source geological borehole datasets. The second validation method consists of comparing the average estimated aquifer thickness with measured values gathered via a literature review.

A dataset incorporating 168 geological borehole descriptions was collected and sorted out from open-source datasets and web services, mostly located in the Netherlands, the USA, Brazil and Australia. After digitizing the borehole reports, we translated the geological information into overall unconsolidated sediment thickness to compare it to our final thickness estimates. This means that all the unconsolidated sediment types such as sand, clay or silt were merged into the same stratigraphic unit and their overall thickness is taken as the final sediment thickness. Figure 3 shows the collected borehole location, and the data sources are presented in Table S1 in the Supplement. Since some boreholes are not located in direct proximity to the coastline, we chose to extrapolate the estimated sediment thickness by calculating the estimated sediment thickness for each cross-section point. This was done by creating a line between the anchor point depth and the estimated sediment depth at the coastline (Fig. 4). Next, the average cross-section point thickness in a circle with radius of 2.5 km around the borehole is compared to the thickness in the borehole.

The final literature validation set is composed of maximum, minimum and/or average aquifer thickness values (unconsolidated sediment) for 64 coastal areas worldwide. However, not all the literature sources provide the average unconsolidated sediment thickness. In the cases where it does not, it is calculated as half the maximum indicated thickness in case only the maximum value is provided. If both maximum and minimum thicknesses are given, the average thickness is set to be halfway between these two values. The table with literature source references and the sediment thickness values provided by these sources are listed in Tables S2 and S3. The final estimated average sediment thickness values were compared with the literature dataset and evaluated based on the relative error and relative improvement compared to the overall average thickness value from all literature sources. The relative error is based on the following equation:

$$\mathrm{RE} = \frac{Z_{\mathrm{est}} - Z_{\mathrm{lit}}}{Z_{\mathrm{lit}}}, \tag{1}$$

where Z_{est} is the ATE by our method and Z_{lit} is the average thickness given by the literature. The RE can be either positive or negative, which implies that the ATE over- or under-estimates the aquifer thickness respectively (compared to values indicated by the literature). The percentage relative error is calculated as

$$\mathrm{PRE} = \mathrm{RE} \cdot 100. \tag{2}$$

The average global aquifer thickness value based on all literature sources was calculated using the equation below:

$$Z_{\mathrm{lit_{avg}}} = \frac{1}{N}\sum\nolimits_{i=1}^{n} Z_i, \tag{3}$$

where $Z_{\mathrm{lit_{avg}}}$ is the overall average value of all literature values Z_i.

The mean absolute error was then calculated for both the overall average value and the estimated average thickness values suggested by our method; see the equations below:

$$\mathrm{MAE_{lit}} = \frac{1}{N}\sum\nolimits_{i=1}^{n} |Z_{\mathrm{lit_{avg}}} - Z_i|, \tag{4}$$

$$\mathrm{MAE_{est}} = \frac{1}{N}\sum\nolimits_{i=1}^{n} |Z_{\mathrm{est}} - Z_i|. \tag{5}$$

Subsequently, the relative improvement rate and percentage relative improvement are calculated as follows:

$$\mathrm{RI} = \frac{\mathrm{MAE_{lit}} - \mathrm{MAE_{est}}}{\mathrm{MAE_{lit}}}, \tag{6}$$

$$\mathrm{PRI} = \mathrm{RI} \cdot 100. \tag{7}$$

The same validation criteria are calculated using the borehole data.

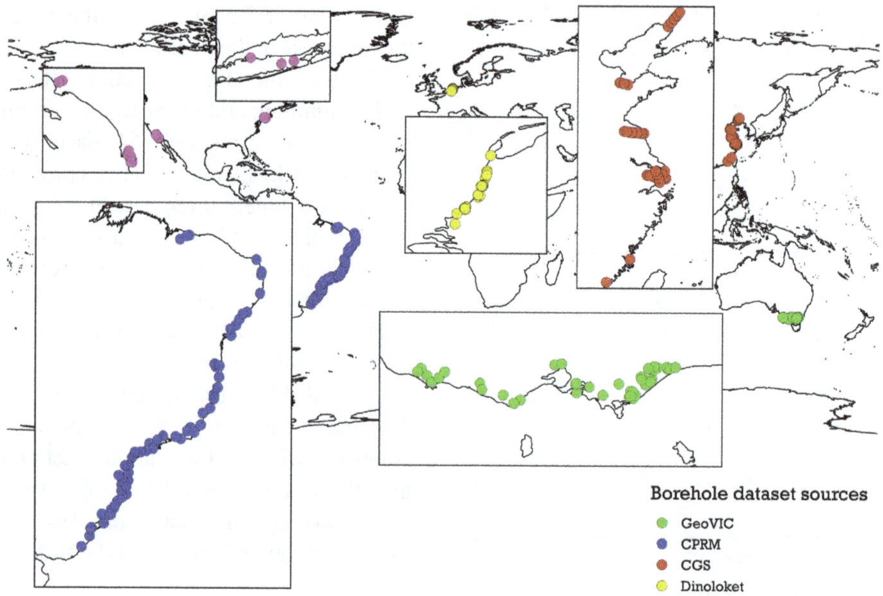

Figure 3. Location of the borehole data used as a validation dataset; sources are listed in Table S3. The borehole information in Brazil and Australia was manually digitized, while the subsurface information in China was gathered by interpreting the cross section provided in the hydrogeological maps.

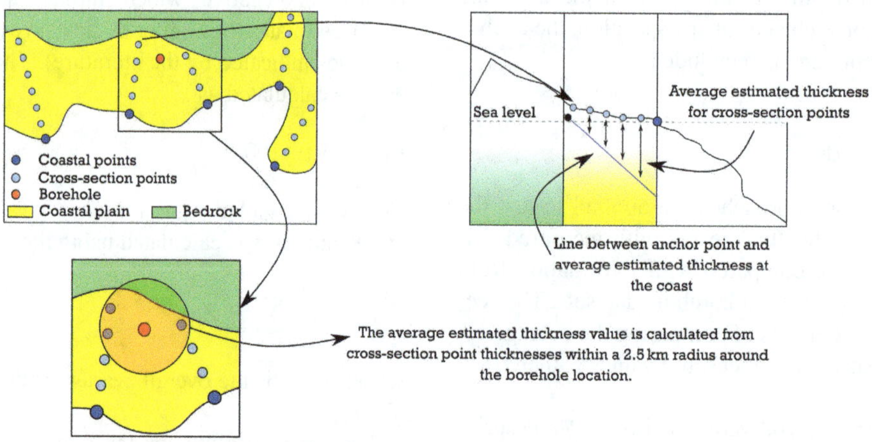

Figure 4. Schematization of the borehole validation process. A set of points lying within a given distance is selected for each borehole and their estimated sediment thickness is averaged. The final comparison between these average values and measured values from the boreholes is shown in Fig. 3 in the main article.

2.3 Groundwater flow and salt transport modelling

The main motivation behind building numerical models simulating the groundwater flow and salt transport as part of this study is to examine the effects of varying aquifer thickness and its geological complexity (absence or presence of low permeable aquitard layers) on simulated saltwater intrusion. Better understanding of these sensitivities will help create improved large-scale hydrogeological models in coastal areas, which in turn will lead to more accurate present and future fresh groundwater volume predictions. To achieve that, a set of variable-density groundwater flow models with varying aquifer thickness and geological complexity (heterogeneous versus homogeneous systems) is created. To evaluate the sensitivity of saltwater intrusion to aquifer thickness and geological complexity, we compare, at a fixed time, the salinity profiles of all simulations as well as the freshwater cell percentage in the coastal zone.

The models with different parameter settings were set up for three cross sections located in Italy in the Versilia plain (Pranzini, 2002), the coast of Virginia in the USA (Trapp Jr. and Horn, 1997) and in the Mediterranean aquifer in Is-

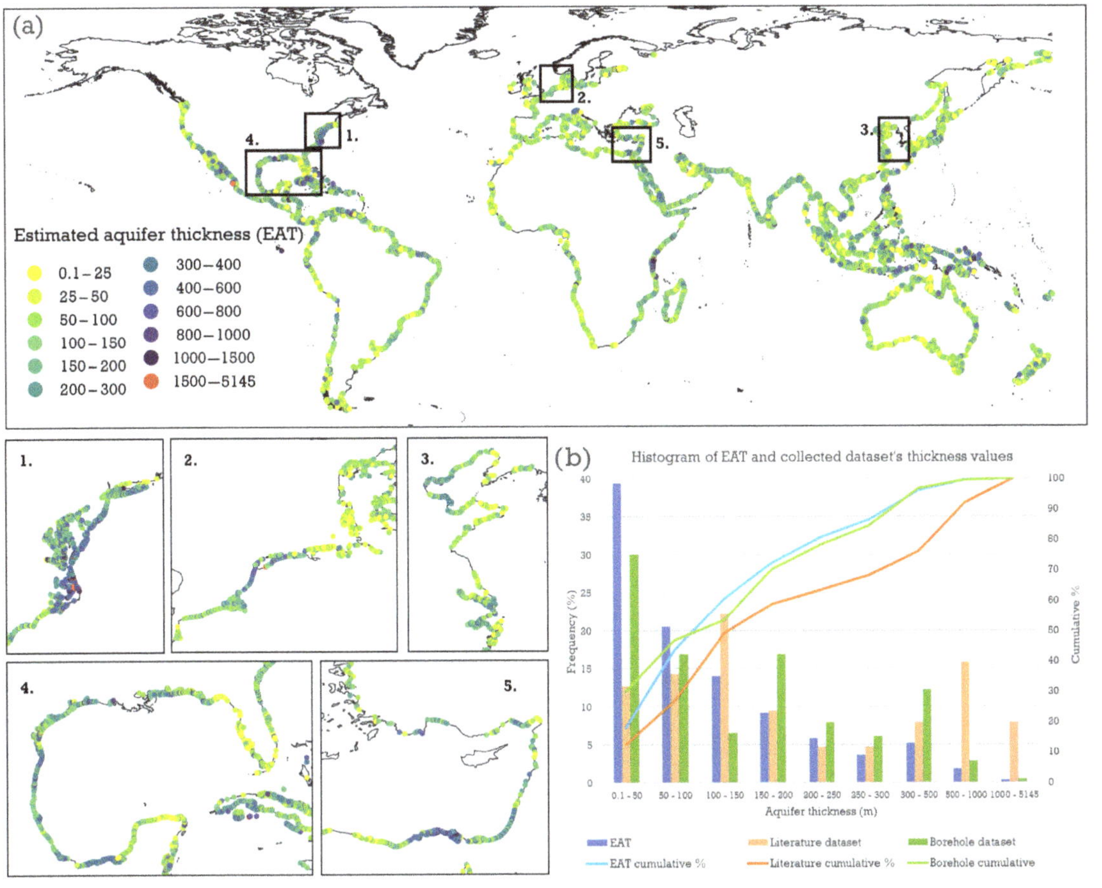

Figure 5. (a) Global map of EAT at the coastline and zoomed areas (1–5) showing regional variations of estimated thickness in various coastal zones around the world. The coastal points are magnified, giving the impression that more than the stated 20 % of the global coastline is covered, which is not the case (see the plain black line). **(b)** Histogram of EAT values with cumulative frequency in %.

rael (Yechieli et al., 2010). We use these studies to build the heterogeneous geological scenarios based on provided cross sections indicating the exact position of low permeable aquitard layers. This was done to evaluate the relative importance of aquifer thickness to the effect of geological complexity. Since the main motivation of this numerical modelling study is to investigate the sensitivity to aquifer thickness and geological complexity, we kept both the aquifer and aquitard layer hydraulic conductivities constant for all simulations (see Table S4). The hydraulic conductivity values were based on the GLHYMPS dataset by taking the highest value of the unconsolidated sediment class as the hydraulic conductivity of the aquifer and the lowest value (fine grained) as the aquitard hydraulic conductivity (Gleeson et al., 2014). To build these models we use the SEAWAT code (Guo and Langevin, 2002) and the Python Flopy library (Bakker et al., 2016). The model schematizations and input parameter values list are presented in Fig. S3 and Table S4.

3 Results

3.1 Sediment thickness estimation

The aquifer thickness is estimated for 26 968 coastal points around the globe, which cover roughly one-fifth of the global coastline. The rest of the global coastline is covered by other lithological types than unconsolidated sediments and is not taken into account by the ATE method. The overall estimated aquifer thickness (EAT) results are presented in Fig. 5a. It shows that the aquifer thickness estimates range between 0.1 and 5145 m, with a mean value close to 170 m. In total 87 % out of all the EAT values predict a thickness lower than or equal to 300 m (see Fig. 5b). A slightly different result is observed in the literature source analysis, where 69 % of the studied areas have an aquifer thickness lower than or equal to 300 m. This difference is explained by the fact that a disproportionally large number of deltaic areas with thick sediment layers is included in the literature validation dataset. Figure 6 shows the areas where there are no EAT results, largely due to the absence of unconsolidated sediments.

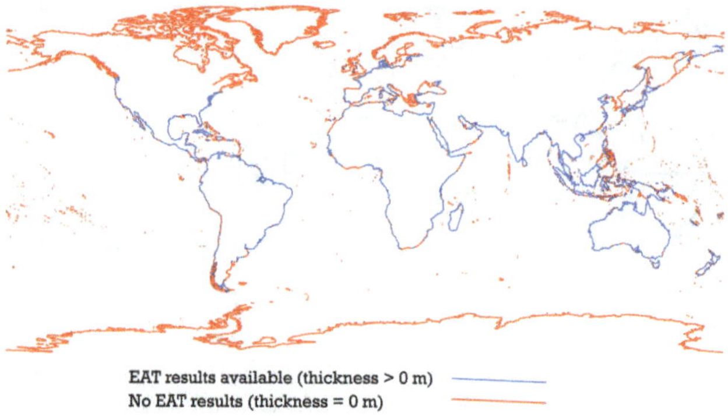

Figure 6. Map showing the spatial distribution of EAT values (unconsolidated sediment aquifer thickness > 0 m) and areas where the unconsolidated sediment aquifer thickness is 0 m.

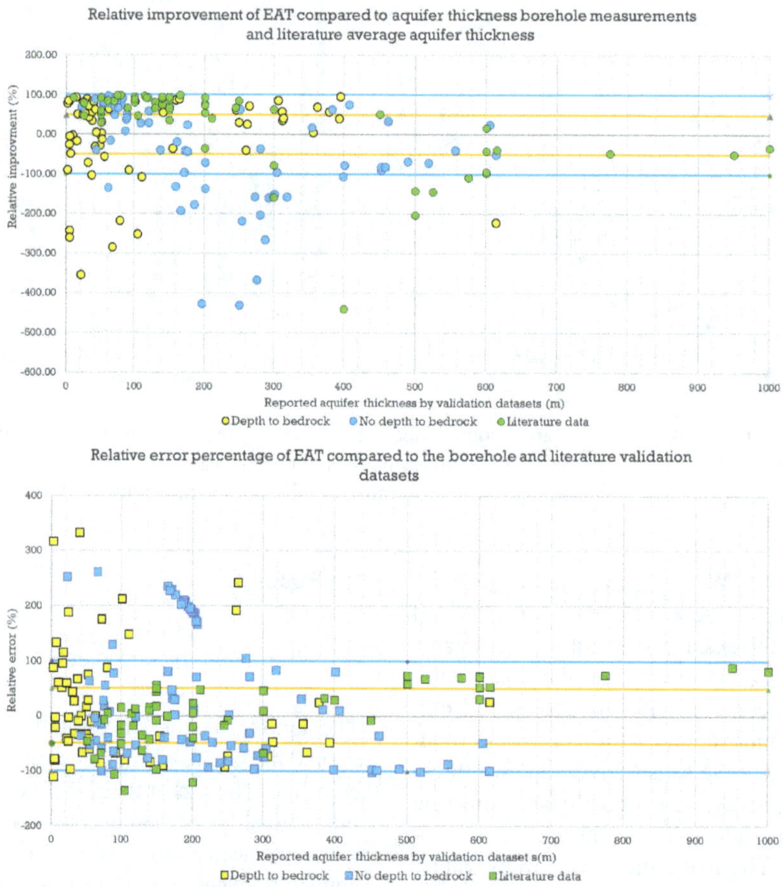

Figure 7. Overall borehole and literature validation results of the EAT results.

3.2 Validation of EAT results

When comparing our EAT with the information collected from the borehole dataset, it is clear that our ATE method provides estimates in the right order of magnitude, but it cannot capture local variations of aquifer thickness. Figure 7a shows that the majority of EAT have relative error values (Eq. 1) lower than 100 %, meaning that our results are of the same order of magnitude as observed values from the borehole dataset. However, the relative improvement of the EAT, as compared with using the average of the borehole thicknesses as an estimate (Eq. 6), is inconclusive as the number of positive values is nearly equal to the total negative values (Fig. 7b).

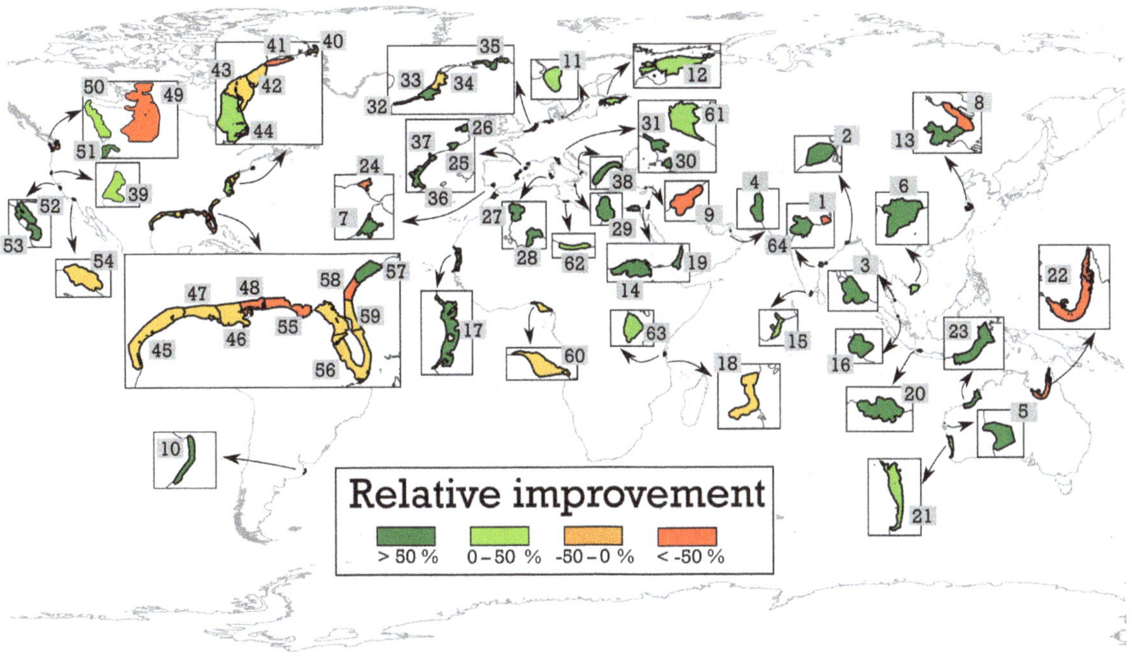

Figure 8. Relative improvement of our estimated sediment thickness compared to using the overall average thickness value from all literature sources.

The results of the validation with the coastal sediment thickness values gathered via a literature review show a more positive result compared to the borehole validation. The overall average thickness of the literature dataset is 353 m, with 69 % of all studied areas having a sediment thickness of 300 m or lower. The relative improvement of sediment thickness estimates using our method is about 22 % compared to the overall thickness value average as indicated by the literature (see Table S3). The relative improvement for individual literature validation areas is shown in Fig. 8. The majority of the areas show an improvement, while estimates for the large coastal plains of the eastern and southern coasts of the USA suggest the opposite. This will be discussed further in Sect. 4. However, in coastal zones that have an average sediment thickness of 300 m or less, the relative improvement of our method is around 59 %. Since our results suggest that 87 % of the global coastline that is composed of unconsolidated sediments has an average thickness of 300 m or lower, the higher relative improvement achieved by our method gains extra importance.

Overall, about 48 % of the validation areas have the absolute relative error percentage below 50 %, while 35 % of validation areas have the absolute relative error percentage between 50 % and 100 % (Fig. 7b). Still, 17 % of the validation areas show an absolute relative error percentage higher than 100 %. A closer look at Fig. 7b reveals that the majority of these validation areas have the average thickness (based on the literature) lower than 100 m. However, the overall results for validation areas with average thickness lower than 300 m show that 59 % have a relative error percentage lower than 50 %: this is an 11 % increase compared to the overall validation dataset.

3.3 Groundwater flow and salt transport modelling

Figure 9 presents a sample of simulated salinity profiles for selected aquifer thickness values for the three test cases. The complete set of the simulated salinity profiles together with the model conceptualization and model parameters and variables is given in Figs. S3 to S6 and Table S5. While comparing the salinity profiles for different aquifer thicknesses, it is apparent that aquifer thickness variations for homogenous geological conditions (figures on the right) do not have large effects on the fresh–saline distribution, except for the lowest aquifer thickness value (Fig. 9a, c). Figure 9b shows that the thicker the aquifer at the coastline, the more saline water intrudes inland and, in some cases, upconing under low-lying areas can be observed (Fig. 9c).

The implementation of complex geological conditions based on the literature description that existed about these sites consisted mainly of introducing low conducting layers (aquitards). As Fig. 9 shows, an aquitard has a substantial effect on the final salinity profile when compared to the salinity profile for homogenous geological conditions with the same aquifer thickness. The aquitard position combined with varying aquifer thicknesses has a large effect on the salinity profile and potential fresh (or brackish) groundwater offshore reserves (Fig. 9b, left column). In particular, the simulations with larger aquifer thickness values show fresh (or brackish)

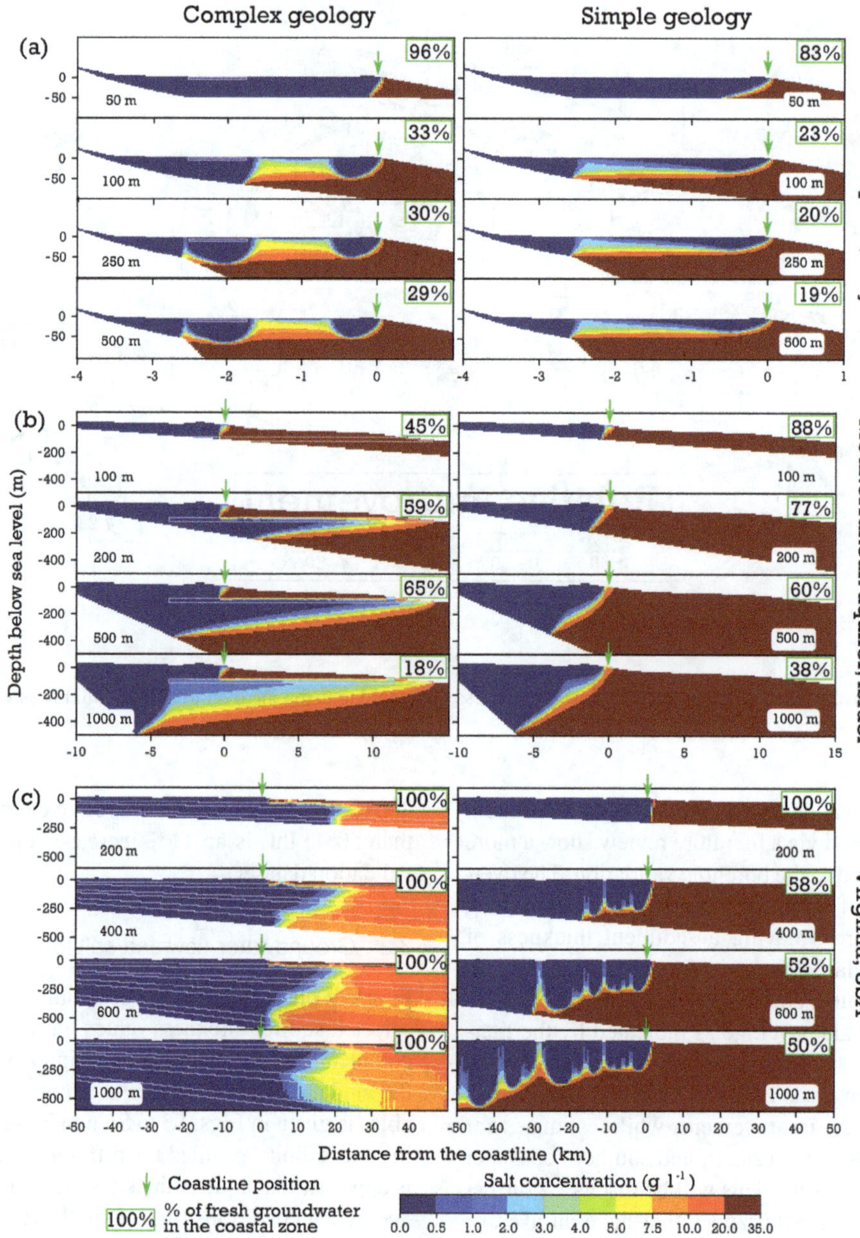

Figure 9. Simulation results as salinity concentration profiles for the cross sections located in the **(a)** Versilia plain, Italy, **(b)** Mediterranean aquifer, Israel, and **(c)** Virginia, USA, with varying aquifer thicknesses and two geology scenarios. The local geological information for each area, **(a)** Pranzini (2002), **(b)** Yechieli et al. (2010), and **(c)** Trapp Jr. and Horn (1997), was implemented (left column) together with a homogeneous aquifer system (right column) to investigate the effects of geological complexity and aquifer thickness on simulated salinity profiles.

offshore groundwater below the aquitard layer. Similar patterns can be observed in the last test case (Fig. 9c), where the aquitard layers prevent saline water from intruding inland and show large offshore brackish water volumes.

Comparison of fresh groundwater cell percentage within the coastal zone of all three test cases (Fig. 9) shows a trend where the geological scenario (homogenous versus complex) has a larger effect on the amount of estimated fresh groundwater reserves than varying aquifer thickness. For the same geological scenario, the largest differences are observed between the aquifer thickness extreme values (thinnest versus thickest).

4 Data availability

The final output data provide both the EAT at the coastline and the location and depth of the corresponding anchor points. These data are given as shapefile and comma separated value files. The data can be downloaded via https://doi.pangaea.de/10.1594/PANGAEA.880771.

5 Discussion and conclusion

Although in the right order of magnitude (Fig. 7a), the ATE validation with borehole measurements is worse than literature dataset validation. The large scale discrepancy between our global estimated aquifer thickness (EAT) dataset and boreholes is the most obvious cause of this. It shows that our approach is not detailed enough to estimate very local variations in aquifer thickness as picked up by boreholes. Boreholes will generally lie between profile locations, which means that local variation also results in spatial dislocation errors, even though spatial averaging is used to bridge the scale gap (Fig. 4). Still, even when compared to boreholes, we observe an overall ATE method performance improvement for coastal areas with measured thicknesses between 100 and 300 m. The comparison between literature values comes out more favourably, because the data synthesis in the form of spatial statistics and geological profiles is a spatial aggregation form that better matches the ATE method scale. We have used the validation data that could be collected during the course of this study, but the validation set is far from exhaustive. The validation dataset should be expanded and continuously improved to achieve better EAT along the global coastline.

Our method tends to underestimate the aquifer thickness in deeper systems, such as large complex deltaic sedimentary structures with measured average aquifer thicknesses larger than 500 m (Fig. 7). This could be due to the limited cross-section length that spans at most 200 km inland and offshore from the coastline depending on the coastal plain extent. If the latter exceeds this maximum length, then no bedrock formation is found and thus no aquifer thickness is estimated. In case the bedrock formation is only partially taken into account (e.g. only the foothills of a mountain range), its topographical slope will be lower, which leads to lower EAT values at the coastline. The opposite happens for coastal areas with measured average thicknesses lower than 100 m. In these cases, our average EAT values tend to be overestimated (Fig. 7). This could again be caused by the input datasets' resolution (see Table 2) which creates larger errors on local scale and for shallow systems which by themselves have a smaller size than more extensive coastal plains. Compared to the other two datasets providing thickness estimates (de Graaf et al., 2015; Pelletier et al., 2016) the lowest EAT values correspond to the range of values provided by Pelletier et al. (2016). The histogram in Fig. 5b suggests that nearly 20 % of coastal areas covered by our study have EAT between 0 and 50 m. On the other side of the spectrum, our maximum EAT value is 5145 m, which is of the order of magnitude of the de Graaf et al. (2015) dataset.

The numerical modelling results show that only the simulations with extreme EAT values give substantially different results from the simulations with average or close to average EAT values. More variation in the fresh groundwater cells fraction in the coastal zone can be observed in the test case with intermediate aquifer thickness (Fig. 9b). In the other two test cases (Fig. 9a and c) the variation in the fresh groundwater cells fraction is very low for both geological scenarios. On the other hand, the model results also show that geological complexity (multi-layering) has a big impact on the results. Thus, for locally meaningful results, the aquifer thickness is but a first result, and a global estimate of multi-layering (aquifers and aquitards) is a necessary next step. Werner et al. (2013) stress that accounting for geological heterogeneities is important to accurately simulate the saline groundwater distribution in coastal areas. Previous regional- to global-scale studies (e.g. Michael et al., 2013; Solórzano-Rivas and Werner, 2018; Knight et al., 2018) considered the geological conditions (permeability and aquifer thickness) to be homogeneous and our EAT dataset could provide a first constraint on unconsolidated sediment thickness for these types of studies.

When comparing our numerical modelling output (with the complex geology incorporated) with the salinity profiles reported from the individual studies (Pranzini, 2002; Yechieli et al., 2010; Trapp Jr. and Horn, 1997) we find that differences for the cases (a) and (b) are small and a 2-D schematization suffices. However, for cross section (c) the differences are considerable. This is most likely due to the presence of strong alongshore flows in the area, a more complex upper hydrological system and the groundwater withdrawal distribution in the area. This shows that the 2-D modelling approach does not always suffice to estimate coastal groundwater flow.

In conclusion, we showed that it is possible to obtain, at first order, coastal aquifer thickness estimates by using available global datasets and a simple methodology consisting of simulating the bedrock slope from the geological outcrops. Our dataset complements the existing datasets listed in Table 2 by providing an estimate of the complete unconsolidated part of coastal aquifer systems. In such a way it is now possible to build more detailed and vertically stratified regional- and global-scale hydrogeological models based on the herein provided dataset. By combining our dataset with existing sedimentary thickness estimates by e.g. de Graaf et al. (2015) we can distinguish the unconsolidated aquifer system (our dataset) overlaying the sedimentary rocks. However, our dataset is not suitable for building detailed local hydrogeological models, as in such a case additional local geological information should be included. Furthermore, the local-scale geological complexity seems to play a larger role in simulated salinity concentration profiles than aquifer thickness (except for extreme values). Thus, our EAT dataset provides a satisfactory first step towards a global coastal aquifer characterization that should be followed by the assessment of the coastal aquifers' geological complexity for local application.

Author contributions. DZ designed the concept of the methodology and computer models with the help of GOE and MB. The manuscript was written by DZ and commented on and revised by GOE and MB. The final dataset was created by DZ.

Competing interests. The authors declare that they have no conflict of interest.

Acknowledgements. This research was partially funded by the Netherlands Organisation for Scientific Research (NWO) and the Ministry of Infrastructure and Water Management under TTW Perspectives program Water Nexus (project no. 14298). We thank two anonymous reviewers for their thorough reading and comments that have considerably improved the quality of this paper.

Edited by: David Carlson

References

Bakker, M., Post, V., Langevin, C. D., Hughes, J. D., White, J. T., Starn, J. J., and Fienen, M. N.: Scripting MODFLOW Model Development Using Python and FloPy, Groundwater, 54, 733–739, https://doi.org/10.1111/gwat.12413, 2016.

Cardenas, M. B., Bennett, P. C., Zamora, P. B., Befus, K. M., Rodolfo, R. S., Cabria, H. B., and Lapus, M. R.: Devastation of aquifers from tsunami-like storm surge by Supertyphoon Haiyan, Geophys. Res. Lett., 42, 2844–2851, https://doi.org/10.1002/2015GL063418, 2015.

Carretero, S., Rapaglia, J., Bokuniewicz, H., and Kruse, E.: Impact of sea-level rise on saltwater intrusion length into the coastal aquifer, Partido de La Costa, Argentina, Cont. Shelf Res., 61–62, 62–70, https://doi.org/10.1016/j.csr.2013.04.029, 2013.

CIESIN – Center for International Earth Science Information Network, Colombia University: Gridded Population of the World, Version 4 (GPWv4): Population count, Revision 10, NASA Socioeconomic Data and Application Center (SEDAC), Palisades, NY, https://doi.org/10.7927/H4PG1PPM (last access: 30 May 2018), 2017.

Custodio, E.: Aquifer overexploitation: what does it mean?, Hydrogeol. J., 10, 254–277, https://doi.org/10.1007/s10040-002-0188-6, 2002.

Delsman, J. R., Hu-a-ng, K. R. M., Vos, P. C., de Louw, P. G. B., Oude Essink, G. H. P., Stuyfzand, P. J., and Bierkens, M. F. P.: Paleo-modeling of coastal saltwater intrusion during the Holocene: an application to the Netherlands, Hydrol. Earth Syst. Sci., 18, 3891–3905, https://doi.org/10.5194/hess-18-3891-2014, 2014.

Döll, P.: Vulnerability to the impact of climate change on renewable groundwater resources: a global-scale assessment, Environ. Res. Lett., 4, 35006, https://doi.org/10.1088/1748-9326/4/3/035006, 2009.

Ferguson, G. and Gleeson, T.: Vulnerability of coastal aquifers to groundwater use and climate change, Nat. Clim. Change, 2, 342–345, https://doi.org/10.1038/nclimate1413, 2012.

Gleeson, T., Moosdorf, N., Hartmann, J., and Van Beek, L. P. H.: A glimpse beneath earth's surface: GLobal HYdrogeology MaPS (GLHYMPS) of permeability and porosity, Geophys. Res. Lett., 41, 3891–3898, https://doi.org/10.1002/2014GL059856, 2014.

de Graaf, I. E. M., Sutanudjaja, E. H., van Beek, L. P. H., and Bierkens, M. F. P.: A high-resolution global-scale groundwater model, Hydrol. Earth Syst. Sci., 19, 823–837, https://doi.org/10.5194/hess-19-823-2015, 2015.

Guo, W. and Langevin, C. D.: User's Guide to SEAWAT: A Computer Program For Simulation of Three-Dimensional Variable-Density Ground-Water Flow, Techniques of Water-Resources Investigations, 06-A7, 77 pp., 2002.

Hartmann, J. and Moosdorf, N.: The new global lithological map database GLiM: A representation of rock properties at the Earth surface, Geochem. Geophy. Geosy., 13, 1–37, https://doi.org/10.1029/2012GC004370, 2012.

Ketabchi, H., Mahmoodzadeh, D., Ataie-Ashtiani, B., and Simmons, C. T.: Sea-level rise impacts on seawater intrusion in coastal aquifers: Review and integration, J. Hydrol., 535, 235–255, https://doi.org/10.1016/j.jhydrol.2016.01.083, 2016.

Knight, A. C., Werner, A. D., and Morgan, L. K.: The onshore influence of offshore fresh groundwater, J. Hydrol., 561, 724–736, https://doi.org/10.1016/j.jhydrol.2018.03.028, 2018.

Kooi, H. and Groen, J.: Offshore continuation of coastal groundwater systems; predictions using sharp-interface approximations and variable-density flow modelling, J. Hydrol., 246, 19–35, https://doi.org/10.1016/S0022-1694(01)00354-7, 2001.

Larsen, F., Tran, L. V., Van Hoang, H., Tran, L. T., Christiansen, A. V., and Pham, N. Q.: Supp Material: Groundwater salinity influenced by Holocene seawater trapped in incised valleys in the Red River delta plain, Nat. Geosci., 10, 376–381, https://doi.org/10.1038/ngeo2938, 2017.

Michael, H. A., Russoniello, C. J., and Byron, L. A.: Global assessment of vulnerability to sea-level rise in topography-limited and recharge-limited coastal groundwater systems, Water Resour. Res., 49, 2228–2240, https://doi.org/10.1002/wrcr.20213, 2013.

Natural Earth, http://www.naturalearthdata.com/downloads/10m-physical-vectors, last access: 21 August 2017.

Nicholls, R. J. and Cazenave, A.: Sea-level rise and its impact on coastal zones, Science, 328, 1517–1520, https://doi.org/10.1126/science.1185782, 2010.

Nofal, E. R., Amer, M. A., El-didy, S. M., and Fekry, A. M.: Delineation and modeling of seawater intrusion into the Nile Delta Aquifer?: A new perspective, Water Sci., 29, 156–166, https://doi.org/10.1016/j.wsj.2015.11.003, 2016.

Pelletier, J. D., Broxton, P. D., Hazenberg, P., Zeng, X., Troch, P. A., Niu, G.-Y., Williams, Z., Brunke, M. A., and Gochis, D.: A gridded global data set of soil, intact regolith, and sedimentary deposit thicknesses for regional and global land surface modeling, J. Adv. Model. Earth Syst., 8, 41–65, https://doi.org/10.1002/2015MS000526, 2016.

Post, V. E. A., Groen, J., Kooi, H., Person, M., Ge, S., and Edmunds, W. M.: Offshore fresh groundwater reserves as a global phenomenon, Nature, 504, 71–78, https://doi.org/10.1038/nature12858, 2013.

Pranzini, G.: Groundwater salinization in Versilia (Italy), in: 17th Salt Water Intrusion Meeting, Delft, The Netherlands, 6–10 May 2002.

Ranjan, P., Kazama, S., Sawamoto, M., and Sana, A.: Global scale evaluation of coastal fresh groundwater resources, Ocean Coast. Manage., 52, 197–206, https://doi.org/10.1016/j.ocecoaman.2008.09.006, 2009.

Rasmussen, P., Sonnenborg, T. O., Goncear, G., and Hinsby, K.: Assessing impacts of climate change, sea level rise, and drainage canals on saltwater intrusion to coastal aquifer, Hydrol. Earth Syst. Sci., 17, 421–443, https://doi.org/10.5194/hess-17-421-2013, 2013.

Sefelnasr, A. and Sherif, M.: Impacts of Seawater Rise on Seawater Intrusion in the Nile Delta Aquifer, Egypt, Groundwater, 52, 264–276, https://doi.org/10.1111/gwat.12058, 2014.

Shangguan, W., Hengl, T., Mendes de Jesus, J., Yuan, H., and Dai, Y.: Mapping the global depth to bedrock for land surface modeling, J. Adv. Model. Earth Syst., 9, 65–88, https://doi.org/10.1002/2016MS000686, 2017.

Solórzano-Rivas, S. C. and Werner, A. D.: On the representation of subsea aquitards in models of offshore fresh groundwater, Adv. Water Resour., 112, 283–294, https://doi.org/10.1016/j.advwatres.2017.11.025, 2018.

Trapp Jr., H. and Horn, M. A.: Ground Water Atlas of the United States: Segment 11, Delaware, Maryland, New Jersey, North Carolina, Pennsylvania, Virginia, West Virginia, available at: http://pubs.er.usgs.gov/publication/ha730L (last access: 23 August 2018), 1997.

Van Weert, F., J., Gun, Van der, J., and Reckman, J.: World-wide overview of saline and brackish groundwater at shallow and intermediate depths, International Groundwater Resources Assessment Centre (IGRAC), Rep. GP 20, 87 pp., 2008.

Weatherall, P., Marks, K. M., Jakobsson, M., Schmitt, T., Tani, S., Arndt, J. E., Rovere, M., Chayes, D., Ferrini, V., and Wigley, R.: A new digital bathymetric model of the world's oceans, Earth and Space Science, 2, 331–345, https://doi.org/10.1002/2015EA000107, 2015.

Werner, A. D., Bakker, M., Post, V. E. A., Vandenbohede, A., Lu, C., Ataie-Ashtiani, B., Simmons, C. T., and Barry, D. A.: Seawater intrusion processes, investigation and management: Recent advances and future challenges, Adv. Water Resour., 51, 3–26, https://doi.org/10.1016/j.advwatres.2012.03.004, 2013.

Whittaker, J. M., Goncharov, A., Williams, S. E., Müller, R. D., and Leitchenkov, G.: Global sediment thickness data set updated for the Australian-Antarctic Southern Ocean, Geochem. Geophy. Geosy., 14, 3297–3305, https://doi.org/10.1002/ggge.20181, 2013.

Yechieli, Y., Shalev, E., Wollman, S., Kiro, Y., and Kafri, U.: Response of the Mediterranean and Dead Sea coastal aquifers to sea level variations, Water Resour. Res., 46, W12550, https://doi.org/10.1029/2009WR008708, 2010.

Zamrsky, D., Oude Essink, G. H. P., and Bierkens, M. F. P.: Aquifer thickness along the global coastline: link to shape files, PANGAEA, https://doi.org/10.1594/PANGAEA.880771 (last access: 23 August 2018), 2017.

5

The Cariboo Alpine Mesonet: sub-hourly hydrometeorological observations of British Columbia's Cariboo Mountains and surrounding area since 2006

Marco A. Hernández-Henríquez[1], **Aseem R. Sharma**[2], **Mark Taylor**[1], **Hadleigh D. Thompson**[2], **and Stephen J. Déry**[1]

[1]Environmental Science and Engineering Program, University of Northern British Columbia, 3333 University Way, Prince George, British Columbia, V2N 4Z9, Canada
[2]Natural Resources and Environmental Studies Program, University of Northern British Columbia, 3333 University Way, Prince George, British Columbia, V2N 4Z9, Canada

Correspondence: Stephen J. Déry (sdery@unbc.ca)

Abstract. This article presents the development of a sub-hourly database of hydrometeorological conditions collected in British Columbia's (BC's) Cariboo Mountains and surrounding area extending from 2006 to present. The Cariboo Alpine Mesonet (CAMnet) forms a network of 11 active hydrometeorological stations positioned at strategic locations across mid- to high elevations of the Cariboo Mountains. This mountain region spans 44 150 km^2, forming the northern extension of the Columbia Mountains. Deep fjord lakes along with old-growth western redcedar and hemlock forests reside in the lower valleys, montane forests of Engelmann spruce, lodgepole pine and subalpine fir permeate the mid-elevations, while alpine tundra, glaciers and several large ice fields cover the higher elevations. The automatic weather stations typically measure air and soil temperature, relative humidity, atmospheric pressure, wind speed and direction, rainfall and snow depth at 15 min intervals. Additional measurements at some stations include shortwave and longwave radiation, near-surface air, skin, snow, or water temperature, and soil moisture, among others. Details on deployment sites, the instrumentation used and its precision, the collection and quality control process are provided. Instructions on how to access the database at Zenodo, an online public data repository, are also furnished (https://doi.org/10.5281/zenodo.1195043). Information on some of the challenges and opportunities encountered in maintaining continuous and homogeneous time series of hydrometeorological variables and remote field sites is provided. The paper also summarizes ongoing plans to expand CAMnet to better monitor atmospheric conditions in BC's mountainous terrain, efforts to push data online in (near-)real time, availability of ancillary data and lessons learned thus far in developing this mesoscale network of hydrometeorological stations in the data-sparse Cariboo Mountains.

1 Introduction

As in many regions worldwide, amplified climate change is altering the hydrometeorology of mountainous basins in northern British Columbia (BC). Climate change in north-central BC's mountains is accelerating permafrost degradation and thawing (Hasler et al., 2015), snowpack thinning (Kang et al., 2016) and glacier retreat (Beedle et al., 2015) and decreasing the proportion of snowfall to total precipitation (Kang et al., 2016). The Cariboo Alpine Mesonet (CAMnet; MacLeod and Déry, 2007) is a mesoscale network ($\sim$ 100s km^2) initiated in 2006 by the Northern Hydrometeorology Group (NHG) at the University of Northern British Columbia (UNBC) to collect long-term hydroclimatological data for the Cariboo Mountains and surrounding regions. Given the remote and difficult terrain in mountain environments where atmospheric and snowpack conditions vary substantially over small spatial scales, paucity of

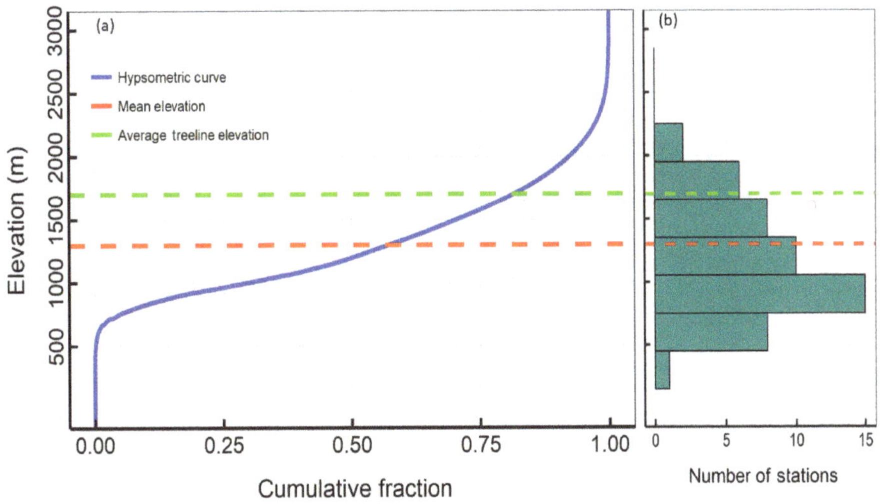

Figure 1. **(a)** Hypsometric curve showing the cumulative fraction of the elevation ranges in the Cariboo Mountains region. **(b)** The number of active weather stations at different elevations within the Cariboo Mountains region.

observational climate data exists in the world's alpine regions (Beniston, 2006; Bradley et al., 2004). This was indeed the case for the Cariboo Mountains where there is relatively good coverage of meteorological stations below the treeline elevation of 1700 m a.s.l.; however, coverage is particularly sparse in the higher elevations (Fig. 1). Indeed, there are only 49 active weather stations in the Cariboo Mountains region that spans 44 150 km^2, equivalent to a density of 1.1 station per 1000 km^2, which is well below the range of one station per 100–250 km^2 recommended by the World Meteorological Organization (WMO) (Miles et al., 2003) and found in other regions of Canada (e.g., Mekis et al., 2018) or in mountainous environments such as the European Alps (Gubler et al., 2017). This issue is especially acute at higher elevations as there are currently only six active meteorological stations above the treeline in the Cariboo Mountains, three of which are part of CAMnet. Thus, CAMnet fills an important observational gap to detect climate change at high elevations and its impacts to snowpacks, glaciers and water resources.

Since the summer of 2006, we have established 11 active and 7 deactivated automatic weather stations and 2 radio repeater stations in valley and mountain settings, at elevations ranging from 683 to 2105 m a.s.l. (Déry et al., 2010). Data from these weather stations have supported modelling studies of seasonal snowpack evolution (Younas et al., 2017), blowing snow fluxes (Déry et al., 2010), turbulent fluxes on a mountain glacier (Radić et al., 2017), glacial retreat (Beedle et al., 2009, 2015) and pro-glacial sediment transport dynamics (Leggat et al., 2015; Stott et al., 2016). These data have also been used to validate remote sensing products of snow (Tong et al., 2009a, b, 2010), gridded meteorological datasets (Sharma and Déry, 2016) and output from numerical weather prediction models over complex terrain (Schirmer and Jamieson, 2015). More recently, CAMnet stations have provided meteorological data to researchers investigating the long-term physical, chemical and biological effects of the Mount Polley mine tailings impoundment breach in August 2014 (see Sect. 3.2) (Petticrew et al., 2015). The CAMnet meteorological datasets thus have the potential to serve multiple users including the operational forecasting and climate modelling communities, the developers of gridded climate datasets such as ANUSPLIN (McKenney et al., 2011) and ClimateWNA (Wang et al., 2012), water resource managers, natural hazard mitigators (e.g., for the control of wildfire activity and avalanche potential), and the tourism and recreation industries (e.g., heli-skiing operations).

The purpose of this article is to document the development and current status of CAMnet, including the setting in which the meteorological stations are deployed; the hydrometeorological variables being monitored and collected; efforts to quality control the data; and data processing, archiving and availability. The paper also describes some of the challenges and opportunities incurred during the data collection process in the remote terrain of the Cariboo Mountains, the availability of ancillary data and prospects for future expansion of the network in the Cariboo Mountains and beyond.

2 Study area

The Cariboo Mountains region spans ~44 150 km^2 in the central interior of BC, forming the northernmost range of the Columbia Mountains between the Interior Plateau and Rocky Mountain Trench (Fig. 2). Elevations in the Cariboo Mountains range from 330 m a.s.l. in the valley bottoms to 3520 m a.s.l. at its highest peak, Mount Sir Wilfrid Laurier (Sharma, 2014). The climate in the Cariboo Mountains remains drier than the Coast Mountains to the west, but wetter than the Interior Plateau to the west and Rocky Mountains

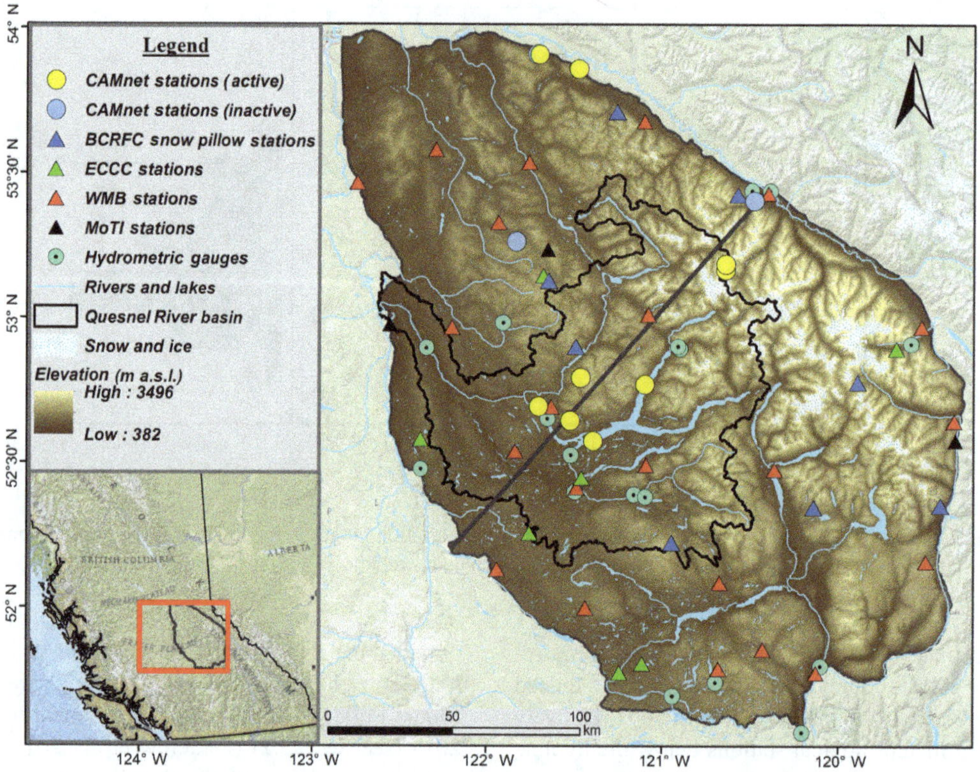

Figure 2. Topographic map of the Cariboo Mountains region including the Quesnel River basin. Dots and triangles show the location of active weather, snow pillow and hydrometric stations operated by different agencies in this region including the BC River Forecast Centre (BCRFC), Environment and Climate Change Canada (ECCC), BC Wildfire Management Branch (WMB), and BC Ministry of Transportation and Infrastructure (MoTI). The grey line depicts a transect between Williams Lake and McBride, BC, for which climate data are extracted (see Fig. 6).

to the east (Beedle et al., 2015). On the windward side of the Cariboo Mountains in Quesnel, BC, the total annual precipitation averages ≈ 500 mm but increases to > 2000 mm 150 km further east (Burford et al., 2009). Annual air temperature averages $1.8\,°C \pm 1.0\,°C$ over the Cariboo Mountains region, with January being the coldest month (average air temperature of $-10.1\,°C \pm 1.1\,°C$) and July the warmest (average air temperature of $12.9\,°C \pm 1.8\,°C$). Average annual total precipitation in this region reaches $739\,mm \pm 140\,mm$ with nearly half falling as snow (Sharma, 2014). The relative amount of snow accumulation increases with altitude (Burford et al., 2009). Peak snow accumulation commonly surpasses 1000 mm water equivalent near the treeline (Déry et al., 2014) and snow cover often persists from October through June (Tong et al., 2009a). Annual peak flows in the creeks and rivers draining the Cariboo Mountains follow snowmelt with general declines in summer although moderated through glacier melt into fall prior to entering the low-flow season during winter when precipitation accumulates in the seasonal snowpack (Burford et al., 2009).

Within the North Cariboo region of BC lies the $\sim$ 12 000 km^2 Quesnel River basin, covering six biogeoclimatic zones (Boreal Altai Fescue Alpine, Interior Mountain-heather Alpine, Engelmann Spruce-Subalpine Fir, Interior Cedar-Hemlock, Sub-boreal Spruce, and Sub-boreal Pine and Spruce). The terrain varies from relatively flat to mountainous with a mean elevation of 1375 m a.s.l, rising from $\sim$ 500 m a.s.l. in the west to $\sim$ 3000 m a.s.l. in the northeast. Glaciers cover $\sim 2\,\%$ of the watershed while deep fjord lakes (e.g., Quesnel Lake with depths surpassing 500 m) reside in steep valleys. The mountainous terrain and snowy climate provide a suitable environment for the world's largest inland temperate rainforest (Stevenson et al., 2011; Farley, 1979). Situated along the Quesnel River sits UNBC's Quesnel River Research Centre (QRRC; http://www.unbc.ca/qrrc, last access: 3 July 2018) that includes a large workshop, laboratory, office space and housing residence. This unique combination of landscape and research infrastructure makes the watershed an ideal location for collecting high-elevation hydrometeorological data.

Table 1. Details for each CAMnet station.

Station	Name	Latitude (° N)	Longitude (° W)	Elevation (m a.s.l.)	Terrain	Date of deployment	Date of decommission
1	QRRC[a]	52°37′06″	121°35′24″	743	Flat short grass	11 Aug 2006	
2	Spanish Mountain	52°33′48″	121°24′35″	1509	Sloped forest regrowth	30 Jun 2006	
3	Browntop Mountain	52°42′28″	121°20′02″	2030	Alpine ridge	23 Aug 2006	
4	Blackbear Mountain radio repeater[b]	52°36′53″	121°26′12″	1590	Sloped forest regrowth	24 Aug 2006	
5	Upper Castle Creek Glacier	53°02′34″	120°26′15″	2105	Bedrock ridge	28 Aug 2007	
6	Mount Tom	53°11′32″	121°39′49″	1490	Sloped cutblock	20 Sep 2007	15 Sep 2009
7	Lower Castle Creek Glacier	53°03′45″	120°26′04″	1803	Flat moraine	13 Aug 2008	
8	Castle Creek Glacier radio repeater	53°02′03″	120°25′57″	2219	Bedrock ridge	1 Aug 2009	
9	Ancient Forest	53°46′21″	121°13′44″	774	Old-growth forest	7 Oct 2009	
10	Lunate Creek	53°49′56″	121°27′32″	953	Regenerating clear cut	17 May 2010	
11	Lucille Mountain	53°16′22″	120°14′18″	1587	Planted cutblock	7 Jul 2010	10 Jul 2013
12	Upper Lunate Creek	53°50′02″	121°28′21″	1134	Regenerating clear cut	23 Aug 2011	29 Jul 2012
13	Ness Lake	54°01′27″	123°05′35″	763	Flat short grass	30 Jul 2012	
14	Coles Lake	59°47′22″	122°36′43″	479	Flat – bulldozed	11 Jun 2013	29 Jul 2015
15	Kiskatinaw	55°45′04″	120°31′27″	726	Flat long grass	20 Aug 2015	15 Jun 2017
16	Tatuk Lake	53°32′41″	124°16′36″	938	Flat short grass	29 Sep 2015	
17	Plato Point	52°29′15″	121°17′03″	728	Sandy beach	17 Aug 2016	
18	Long Creek	52°40′06″	120°57′34″	728	Gravel beach	26 Jun 2017	

[a] Quesnel River Research Centre. [b] The Blackbear Mountain weather station was dismantled in July 2007 and became a radio repeater station to facilitate communication and data download at the QRRC.

3 Hydrometeorological stations

3.1 Overview

As of 31 March 2018, CAMnet comprises 11 active meteorological stations and two radio repeaters positioned at strategic locations, predominantly at mid- to high elevations within the Quesnel River basin and the surrounding Cariboo Mountains of north-central BC (Figs. 3 and 4, Table 1). Specifically, the Quesnel watershed hosts five active CAMnet stations, the Upper and Lower Castle Creek Glacier stations reside on its eastern divide, and four others reside near Prince George and its vicinity. At each station, instruments measure atmospheric pressure, 2 m air temperature and relative humidity (with respect to water), wind speed and direction, liquid precipitation and snow depth among other variables that are only measured at specific stations (e.g., soil moisture, solar radiation, and near-surface air, snow, soil and water temperature) (Supplement Table S1). All the sensors are powered by 12 V batteries charged during the day by solar panels; however, both the QRRC and Ness Lake weather stations run on alternating current (AC) supplemented by batteries in case of electrical interruptions. The QRRC station records incoming and outgoing shortwave and longwave radiation in addition to the parameters measured by the remote stations. Table 2 lists the specifications of each instrument, including their accuracy and precision. Data loggers record the meteorological parameters and store data every 15 min. Stations located within the QRRC's proximity periodically send data via spread-spectrum radio to the operational base computer located in the housing residence (see Sect. 4.2). Despite remote data access for several stations, site visits for data downloads and routine maintenance occur at a minimum once a year at all stations (see Sect. 4.1).

3.2 Chronological development

Having identified a significant observational gap in the Cariboo Mountains and addressing the needs of the QRRC's expanding research operations, UNBC's NHG established its first four CAMnet weather stations during the summer of 2006 in the vicinity of Likely, BC. Given the QRRC's availability to other UNBC researchers and those at other institutions, a dedicated effort was made to establish a standard meteorological station with an extended suite of instrumentation on the research centre's property. This station became operational on 11 August 2006 and is located 2 km southwest of Likely, BC, in an incised valley ∼ 100 m south of the Quesnel River at an elevation of 743 m a.s.l. This is the only CAMnet weather station with a permanent 10 m tower with a concrete foundation for the tower footing and guy wires for anchoring. There is 50 % sky visibility at the site and, given its location, accessibility remains easy. Vegetation consists of grasses, western redcedar, hemlock and some Douglas fir. A CR23X data logger controls the entire weather station and accommodates the additional programming capacity and power supply needed to run a full radiometer and associated heater, installed 5 m above bare ground. Winds are measured at 10 m above the surface while air temperature and relative humidity are gauged 2 m above bare ground. This site also benefits from access to AC power to operate a heated precipitation gauge surrounded by an Alter shield. Data are periodically sent to the QRRC operational base computer via spread-spectrum radio.

Table 2. Specifications for sensors used at each CAMnet station. NA – not available.

Sensor	Model	Measurements	Units	Sensitivity	Accuracy	Operating range
Spread-spectrum radio	RF401	NA	NA	−109 dBm	NA	−25 to 50 °C
Barometric pressure sensor	61205V 61205V-10	Atmospheric pressure	hPa	0.1 hPa	±0.5 hPa	600–1100 hPa
	61302V	Atmospheric pressure	hPa	0.1 hPa	±0.2 hPa (25 °C) ±0.3 hPa (−50 to +60 °C)	500–1100 hPa
	CS106	Atmospheric pressure	hPa	NA	±0.3 hPa (20 °C) ±0.6 hPa (0 to 40 °C) ±1.0 hPa (−20 to +45 °C) ±1.5 hPa (−40 to +60 °C)	−40 to 60 °C 500–1100 hPa
Relative humidity and air temperature probe	HMP35C HMP45C HMP45C212	Air temperature	°C	NA	±0.1 °C	−40 to 60 °C
		Relative humidity	%	NA	±2 % (0–90 % RH) at 20 °C ±3 % (90–100 % RH) at 20 °C	
	HMP60	Air temperature	°C	NA	±0.6 °C (−40 to +60 °C)	
		Relative humidity	%	NA	±3 % RH (0–90 % RH) ±5 % RH (90–100 % RH)	
	HC-S3	Air temperature	°C	NA	±0.2 °C (−30 to 60 °C)	−40 to 60 °C
		Relative humidity	%	NA	±1.5 % at 23 °C	
Temperature probe	107B	Soil temperature	°C	NA	±0.4 °C (−24 to 48 °C) ±0.9 °C (−35 to 50 °C)	−50 to 100 °C
	109 109B	Soil temperature	°C	NA	±0.60 °C (−50 to 70 °C) ±0.25 °C (−10 to 70 °C)	
RM Young wind monitor	05103-10 05103-45	Wind speed	$m\,s^{-1}$	NA	$\pm0.3\,m\,s^{-1}$	−50 to 50 °C
		Wind direction	degrees	NA	±5°	
Tipping bucket rain gauge	34-HT-P	Liquid/solid precipitation	mm	0.25 mm per tip	Precipitation rate: ±0.25 mm up to 20 $mm\,h^{-1}$ ±3 % over 20 $mm\,h^{-1}$	−20 to 50 °C
	TE525WS	Liquid precipitation	mm	0.254 mm per tip	Precipitation rate: ±1 % up to 25.4 $mm\,h^{-1}$ +0, −2.5 % from 25.4 to 50.8 $mm\,h^{-1}$ +0, −3.5 % from 50.8 to 76.2 $mm\,h^{-1}$	0 to 50 °C
	TE525M	Liquid precipitation	mm	0.1 mm per tip	Precipitation rate: ±1 % up to 25.4 $mm\,h^{-1}$ +0, −2.5 % from 25.4 to 50.8 $mm\,h^{-1}$ +0, −3.5 % from 50.8 to 76.2 $mm\,h^{-1}$	
	TR-525USW	Liquid precipitation	mm	0.254 mm per tip	Precipitation rate: 1 % at 25.4 $mm\,h^{-1}$ or less	
Ultrasonic distance measurement	SR50	Snow depth	cm	0.1 mm	±1 cm or 0.4 % of distance to target (whichever is greater)	
	SR50A	Snow depth	cm	0.25 mm	±1 cm or 0.4 % of distance to target (whichever is greater)	−45 to 50 °C
Water content reflectometer	CS616	Soil moisture	% volumetric water content (VWC)	< 0.1 % VWC	±2.5 % VWC using standard calibration with bulk electrical conductivity (EC) ≤ 0.5 $dS\,m^{-1}$ and bulk density ≤ 1.55 $g\,cm^{-3}$ in measurement range 0 to 50 % VWC	0 to 70 °C
	CS650	Soil moisture	% VWC	< 0.05 % VWC	±3 % (typical with factory VWC model) where solution EC < 3 $dS\,m^{-1}$	−50 to 70 °C

Table 2. Continued.

Sensor	Model	Measurements	Units	Sensitivity	Accuracy	Operating range
Radiometer/ pyranometer	CNR1	Shortwave up/down, Longwave up/down	W m^{-2}	6.18 µV (W m^{-2})$^{-1}$	±10 % of daily totals	−40 to 70 °C
	CMP3	Shortwave up/down	W m^{-2}	5 to 20 µV (W m^{-2})$^{-1}$	±10 % of daily totals	−40 to 80 °C
	SP LITE2	Shortwave up/down	W m^{-2}	81 µV (W m^{-2})$^{-1}$ < ±2 % shift per year	±5 % of daily totals	−40 to 80 °C
	LI 200R	Shortwave up/down	W m^{-2}	7 to 15 µV (W m^{-2})$^{-1}$	Absolute error in natural daylight is ±5 % maximum; ±3 % typical	−40 to 65 °C
	LI 190SB			5 µA per 1000 µmoles s^{-1} m^{-2}	Maximum deviation of 1 % up to 10 000 µmoles s^{-1} m^{-2}	
Precision infrared radiometer	SI-111 (IRR-P)	Surface temperature	°C	±0.1 °C @ −10 to 65 °C ±0.3 °C @ −40 to 70 °C	±0.2 °C @ −10 to 65 °C (where target temperature is within 20 °C of sensor body temperature) ±0.5 °C @ −40 to 70 °C (where target temperature is > 20 °C of sensor body temperature)	−55 to 80 °C 0 to 100 % RH

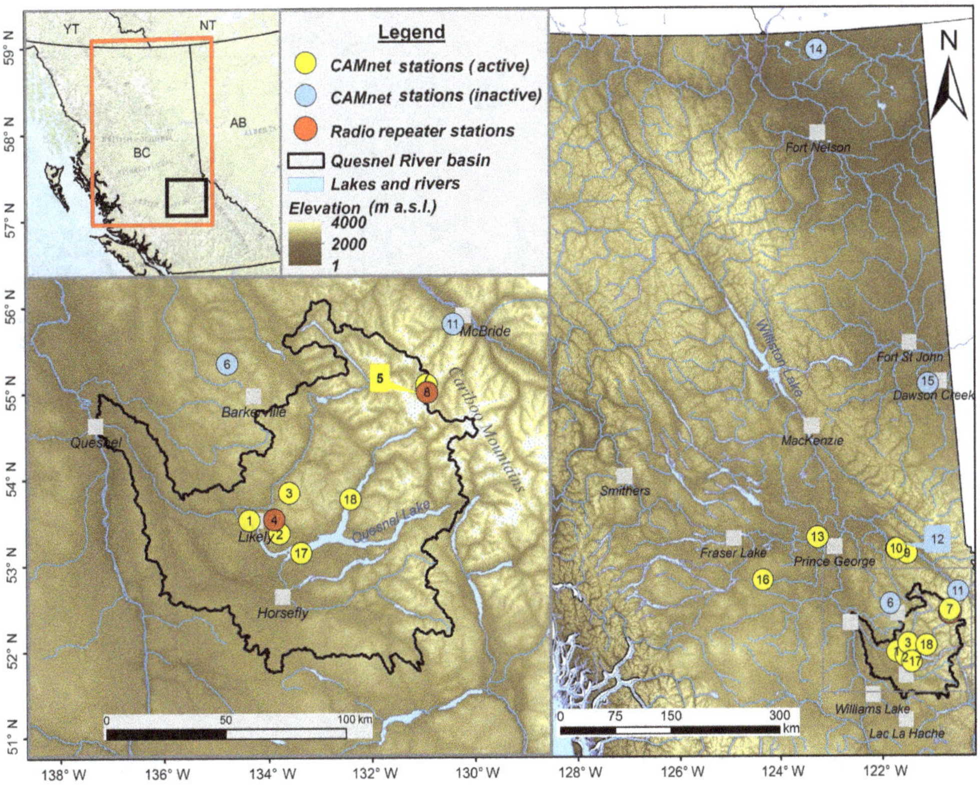

Figure 3. Map of north-central BC showing the location of active and inactive CAMnet weather and radio repeater stations, numbered as in Table 1.

Three additional weather stations mounted on 3 m tripods were also deployed during summer 2006 near Likely, BC, in an attempt to capture the effects of elevation gradients on atmospheric conditions in the Cariboo Mountains. Factors leading to the selection of Spanish, Browntop and Blackbear mountains as deployment sites included their range in elevations (1500–2000 m a.s.l.) and accessibility (deactivated forestry roads and/or hiking trails) and were within line of sight of each other, with one station being within line of sight of the QRRC. All were equipped with identical instrumentation to measure 2 m air temperature and relative humidity, 3 m wind speed and direction, atmospheric pressure, snow depth, rainfall and soil temperature at 10–15 cm in depth. Stations at Spanish and Blackbear mountains were installed

Figure 4.

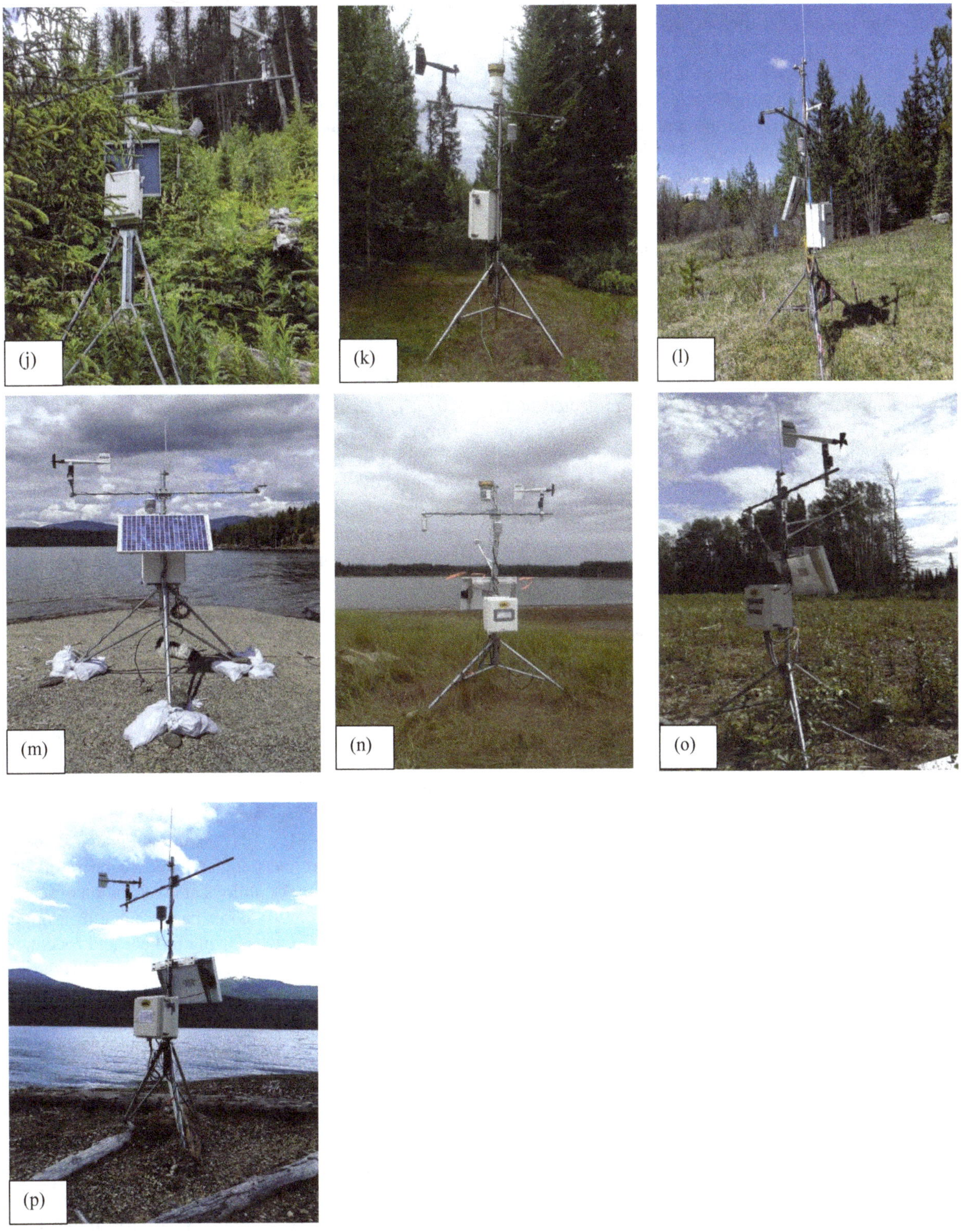

Figure 4. Photos of CAMnet weather stations: **(a)** QRRC in May 2017 (el. 743 m), **(b)** Spanish Mountain in June 2017 (el. 1509 m), **(c)** Blackbear Mountain radio repeater in July 2015 (el. 1509 m), **(d)** Browntop Mountain in June 2017 (el. 2030 m), **(e)** Upper Castle Creek Glacier in September 2015 (el. 2105 m), **(f)** Mount Tom in September 2008 (el. 1490 m), **(g)** Lower Castle Creek Glacier in September 2015 (el. 1803 m), **(h)** Castle Creek Glacier radio repeater in August 2009 (el. 2219 m), **(i)** Ancient Forest in May 2017 (el. 774 m), **(j)** Lunate Creek in June 2017 (el. 953 m), **(k)** Ness Lake in June 2015 (el. 763 m), **(l)** Tatuk Lake in May 2017 (el. 938 m), **(m)** Plato Point in May 2017 (el. 728), **(n)** Kiskatinaw in August 2015 (el. 726 m), **(o)** Coles Lake in July 2015 (el. 479 m) and **(p)** Long Creek in June 2017 (el. 728 m).

in similar terrain (regenerating cutblocks of lodgepole pine stands) and elevations such that the meteorological conditions between the two exhibited little differences. As such, the meteorological instrumentation at Blackbear Mountain was decommissioned during the summer of 2007 but the radio repeater was retained to act as the communication hub between the remote Spanish and Browntop Mountain stations and the operational base computer. The Browntop Mountain weather station lies at 2030 m a.s.l. in the alpine terrain on an exposed ridge and endures excessive winds approaching $40\,m\,s^{-1}$ in some storms. These high winds have been quite challenging in keeping the entire station operational, with the continuity of readings frequently disrupted through instrument damage (including the loss of multiple anemometers), icing and power malfunctions. A tipping bucket rain gauge was eventually dismantled on 11 July 2012 due to its low capacity to capture precipitation even when protected with an Alter shield. While a SR50 remains in operation at this site it rarely measures > 10 cm of snow as winds continually erode it from the ridge top towards the lee side of Browntop Mountain where massive snowdrifts form and persist during summer.

Equipment previously installed at Blackbear Mountain was moved during August 2007 to a bedrock ridge at 2105 m a.s.l. near Castle Creek Glacier, 40 km west of McBride, BC, to support research of the Western Canadian Cryospheric Network. Despite its exposed and remote location, the station has experienced only a few minor issues mainly from strong winds affecting precipitation measurements, with the loss on several occasions of the tipping bucket rain gauge's funnel and/or filter. A second weather station was installed at 1803 m a.s.l. in 2008, ~ 700 m in the Castle Creek Glacier forefield to complement the data being collected at higher elevation. The Lower Castle Creek Glacier weather station is unique in the CAMnet network in measuring air temperature, relative humidity, wind speed and direction at two heights (4.1 and 2.7 m) above the surface on a 6 m mast, allowing estimates of turbulent sensible and latent heat fluxes. This site also contains an array of sensors to measure near-surface air or snow, skin surface and soil temperatures. A tipping bucket rain gauge protected by an Alter shield was added to the site on 21 August 2012. Both Castle Creek Glacier weather stations are within line of sight of a radio repeater mounted in August 2009 on a bedrock ridge at 2219 m a.s.l. facing eastward towards Castle Creek Valley. Combined, the two weather stations at Castle Creek Glacier provide a nearly continuous record of atmospheric and snow conditions exceeding a decade that complement well ongoing efforts to monitor the glacier's mass balance using push moraines, geodetic and remote sensing techniques.

A basic weather station was installed in 2007 for a period of 2 years at Mt. Tom, in the Cariboo Highlands near Wells and Barkerville, BC, to support the research projects of two UNBC graduate students. This research was in collaboration with the BC Ministry of Forests, Lands, Natural Resource Operations and Rural Development's (FLNRORD) Mt. Tom adaptive management trial. The spatial distribution of snow cover in forested and clear-cut areas of different sizes as well as soil moisture memory was the focus of these two projects. Weekly snow survey data during one ablation season in addition to snow/surface temperatures were collected during this experiment.

Following the completion of the two graduate student projects at Mt. Tom, the equipment was transferred in October 2009 to a site at the Ancient Forest ~ 100 km east of Prince George at an elevation of 775 m a.s.l. The station was installed a few hundred metres from the Yellowhead Highway in a small clearing in the inland temperate rainforest situated on the northwestern border of the Cariboo Mountains (Stevenson et al., 2011), now within the boundary of the Ancient Forest/Chun T'oh Whudujut Provincial Park created in 2016. The damp, low-lying characteristics of this old-growth forest limit sky visibility and expose the site to abundant precipitation in all seasons with heavy snowfall during winter. Ground cover vegetation consists of large ferns, tall grasses and dense Devil's club. Measurements at this site include soil moisture at three depths down to 65 cm. The CR10X data logger at this site was replaced with a CR1000 in August 2013, while the tripod and a cross arm were replaced in August 2015 following a near-direct hit by a felled old-growth western redcedar. This station provides critical information on BC's great "snowforest" in a changing climate including support for ongoing ecosystem, ecological and outdoor recreation and tourism research at UNBC. It also serves routinely as a field site for a snow survey conducted by UNBC students enrolled in a course titled "Snow and Ice".

In proximity to the Ancient Forest, a basic meteorological station at Lunate Creek was acquired from the BC Ministry of FLNRORD in May 2010. The equipment at the station was fully updated and deployed in a regenerating western redcedar and hemlock cutblock at an elevation of 953 m a.s.l. The tower sits on a 20° north-facing slope overlooking the Rocky Mountain Trench. Owing to bear and rodent activity along with the minimal incoming solar radiation during winter this site has experienced periodic power interruptions and thus suffers extended data gaps. A second weather station was installed on a neighbouring cutblock at 1134 m a.s.l. to monitor upslope conditions during a period of 1 year in support of a graduate student project that quantified the contribution of snowmelt to soil moisture replenishment on the area's steep inclines. The toe slopes of the Cariboo Mountains are well known for their abundant soil moisture and springs even during dry summer periods, with these water resources vital in sustaining the Ancient Forest's old-growth forest stands.

Similarly to the Lunate Creek site, another weather station was acquired from BC Ministry of FLNRORD in July 2010 at Lucille Mountain, west of McBride, BC. The weather station is deployed in a regenerating cutblock (planted) at an elevation of 1587 m a.s.l. and faces the Rocky Mountain

Trench. Due to its difficult access and animal activity, the station has not been visited nor maintained since 2013.

Following the conclusion of the intensive field campaign at Lunate Creek, its upper station was transferred in July 2012 to Ness Lake, ~ 35 km northwest of UNBC on the property of the senior and corresponding author. The weather station lies ~ 25 m from the eastern shore of Ness Lake at an elevation of 763 m a.s.l. The tower has only ~ 75 % sky visibility since it is deployed in an area with large Douglas firs and two nearby residences. Data collected contribute to studying the effects that public recreational use has on Ness Lake wave activity and shoreline erosion. This site also serves to train students and research staff on the use of the meteorological equipment; testing new, damaged or repaired equipment; and data logger programming and has also provided meteorological resources for UNBC courses. This site benefits from access to AC power and the internet, facilitating automatic data transfers to a computer server at UNBC.

To support a graduate student's project investigating the water balance of a small boreal lake and its watershed, a weather station was deployed at Coles Lake, ~ 100 km north of Fort Nelson, BC, near the border to the Northwest Territories in June 2013. The weather station resided ~ 500 m north of Coles Lake on the property of Quicksilver Inc., an oil and gas extraction company that, along with the BC Ministry of FLNRORD, supported the research. Apart from the meteorological data collected at the weather station, ancillary data included rainfall at three additional sites under various vegetation canopies, snow depth and snow water equivalent through biweekly snow surveys at three sites during one ablation season, Coles Lake inflows and outflows, shallow groundwater levels with three piezometer nests along the lakeshore and lake skin surface water temperature.

Completion of the field campaign at Coles Lake allowed its station to be moved to the Kiskatinaw watershed in the summer of 2015. The site was located 300 m south of Arras Road and the John Hart Highway, ~ 19 km west of Dawson Creek, BC. The tower was installed in a ~ 4000 m^2 opening atop a small hill ~ 60 m north of the Hansen Reservoir. Ground cover consisted of unmaintained grass and the surrounding area is composed of mixed deciduous–coniferous trees and agricultural plots. Data for this station supported a graduate student's study on the prediction of the spring freshet in the Kiskatinaw River that supplies domestic, commercial and industrial water for Dawson Creek and surrounding communities.

In a continuing effort to improve and integrate knowledge on the Nechako River (a major tributary to the Fraser River) and its watershed, a weather station was deployed at Tatuk Lake, BC, in September 2015. The weather station stands ~ 250 m from the northern shore of Tatuk Lake in the headwaters of the Chilako River, ~ 75 km south of Vanderhoof, BC. It resides on the Tatuk Lake Resort's private property, which provides safe and secure year-round access. The tower sits at an elevation of 938 m a.s.l. in an open section (~ 1000 m^2) where there is ~ 85 % sky visibility. Tall grasses, shrubs and rocky soil characterize the site. This station also serves as a reliable source of meteorological information for local communities of the Nechako watershed and as a demonstration site for local elementary and middle school students.

The catastrophic failure of a tailings pond impoundment at Mount Polley mine near Likely, BC, in August 2014 (Petticrew et al., 2015) renewed interest in monitoring closely atmospheric conditions at Quesnel Lake and the surrounding area. While CAMnet included three weather stations near Likely, none was located on the shores of Quesnel Lake. To fill this observational gap, two lake-level sites were chosen for weather station deployment in August 2016 and 2017. The Plato Point weather station is located 22 km southeast of Likely on the sandy shoreline of Quesnel Lake just east of the Plato Island Resort. The tower sits at an elevation of 728 m a.s.l. and has ~ 90 % sky visibility. The station has excellent exposure in all cardinal directions except towards the south, allowing measurement of long-fetch winds aligned along the principal axis of Quesnel Lake's main basin. During spring snowmelt, water levels rise, partially submerging the station. Two additional probes measure water and near-surface air temperatures. In the summer of 2017, the equipment from the Kiskatinaw watershed weather station was moved to Long Creek, on a gravel beach along the shoreline of Quesnel Lake approximately halfway up its north arm. The tower is deployed at an elevation of 728 m a.s.l. and has ~ 75 % sky visibility. Small shrubs and mature conifers flank the western side of the station that otherwise benefits from excellent exposure in all other cardinal directions. Both the Plato Point and Long Creek weather stations allow detection of wind storms that generate basin-scale hydrodynamic processes in Quesnel Lake, leading to possible resuspension of mine tailings sediments deposited on the lake bottom during the 2014 spill.

3.3 Equipment used at each site

The majority of the meteorological equipment was purchased from Campbell Scientific Canada (CSC) and its suppliers; however, this report should not be construed as an endorsement of their products. Sole use of CSC equipment at nearly all CAMnet sites facilitates data cross comparisons, equipment substitutions, student and research staff training, and ensures the homogeneity of the hydrometeorological records (see Sect. 5.2). Most remote stations are assembled on 3 m tripods (i.e., CM110 or UT10) anchored to the ground to maintain the orientation of the tower, ensure proper positioning of sensors and prevent electrical damage from lightning strikes. Most of the sensors are installed and configured as recommended in the manuals (e.g., anemometers ~ 3 m above the ground surface and the air temperature and humidity sensors ~ 2 m above the ground surface). However, the precipitation gauges were an exception since the construction of concrete footings was not feasible at the remote sites. Instead, the precipitation gauges were typically installed on 1.2 m × 1.2 m platforms anchored to the ground with rebar to support and level the rain gauges as well as the Alter wind screens that reduce the effects of wind and turbulence on precipitation measurements. Most of the remote stations employ CR1000 data loggers from CSC to power and operate the sensors. A CR23X data logger is used at the QRRC station to run the sensors as it can provide the necessary power to engage the CNR1 net radiometer heater. Supplement Table S1 provides a detailed summary of the equipment and sensors used at each CAMnet station.

3.4 Precision and accuracy of the instrumentation

Table 2 provides the specifications, including precision and sensitivity, for sensors used at each CAMnet station. These sensors are manufactured to withstand strong variations in environmental conditions and often remain unattended for lengthy periods of time. Most of the instruments, including the tipping bucket rain gauges, barometers and wind monitors, are initially calibrated and come with a certificate of accuracy. Depending on the amount of usage and the presence of adverse atmospheric conditions, some of the CAMnet instrumentation undergoes periodic recalibration, as maintaining the precision and quality of data remains of utmost importance to the development of the database. Recalibration and regular maintenance are completed if the quality of the data being sampled starts to become affected or if there is obvious damage to the instrumentation. Common errors in sampling precision and accuracy stem from worn-out equipment such as the depth-to-target data recorded by the SR50 ultrasonic depth sensor that can contain outlying values when the transducer wears out. This causes peaks in the depth-to-target data that are carried over to the snow depth data through calculations. These peaks can be filtered out relatively easily so they do not significantly affect data quality.

All instrumentation used at CAMnet weather stations includes a recommended temperature operating range where data collected outside these parameters are subject to larger errors. However, these operating ranges are quite vast (typically −40 to +70 °C), thus minimizing potential errors. Note that data loggers with extended temperature operating ranges are employed at all CAMnet sites given the frigid conditions often encountered during winter in the Cariboo Mountains and surrounding area. Snow and ice, like temperature, have the ability to affect the precision of the data and accuracy of the equipment. Rime build-up can cause the anemometer blades to freeze up, resulting in extended periods of data with wind speeds of 0 m s^{-1}. Snow accumulation on pyranometers and the radiation shields of temperature and relative humidity probes can also affect the precision of the sampled data. Regular site visits mitigate these issues and allow for periodic instrument check-ups to maintain the quality of the data.

4 Data collection

4.1 Frequency

Data loggers were programmed using the Loggernet SCWin, CRBasic and Edlog software from CSC. All of the station programs employ scan intervals of 1 min and data intervals of 15 min. Depending on the sensors and the specific variable being measured, sampling either occurs once every 15 min or is measured for the entire data interval and averaged. Some of the measurements are also simply summed over the 15 min period, such as precipitation and incoming solar radiation. Regardless of the sampling interval, all collected data are stored in 15 min timestamps in the data logger memory and the database spreadsheets.

All CAMnet weather stations are visited at least once annually for data downloads and maintenance, although repeat visits are not feasible at the most remote sites (i.e., Upper and Lower Castle Creek Glacier weather stations) given access is often by helicopter. All other sites, however, are usually visited two to four times a year, while others such as the Quesnel River Research Centre (QRRC) and Ness Lake are easily accessible and thus serviced much more regularly. The frequency of visits escalates during intensive field campaigns in association with a student's research project or accelerated to address promptly a known technical or instrumental issue.

4.2 Automated data transfers

Spread-spectrum radios facilitate communication and automated data transfers between the weather stations at Spanish Mountain, Browntop Mountain and Blackbear Mountain (no longer equipped with meteorological equipment) with the operational base computer. Spread-spectrum radios have a range of ∼ 20 km within line of sight (Campbell Scientific, 2017); however, obstructions such as hills and vegetation attenuate significantly the radio signals while the specific type of antenna used modulates signal strength. Thus, a radio signal range of ∼ 13 km was chosen during the site selection process for stations 1–4 (Table 1). The Blackbear Mountain radio repeater routes radio communication from the Spanish Mountain and Browntop Mountain stations once daily to the QRRC operational base computer that acts as a gateway to online data retrieval. Radio communication from the QRRC weather station to this computer occurs on an hourly basis, allowing online data transfers to a computer server on the UNBC campus in Prince George, BC, for web-based visualization and long-term archiving.

A second radio repeater station installed on a bedrock ridge at Castle Creek Glacier within line of sight of the two nearby weather stations and the Castle Creek Valley allowed remote data downloads at the end of a deactivated forestry road accessed by a 4 × 4 vehicle. A major landslide in 2012, however, has since blocked road access to the upper reaches of the Castle Creek Valley, rendering this radio repeater obsolete. Thus, the radio repeater station at Castle Creek Glacier may soon be relocated to a site near Quesnel Lake to extend remote communication at sites there.

4.3 Gaps and infilling

Some time series within the database have gaps (Fig. 5) either for entire stations or individual instruments and for a variety of reasons. Stations based in wilderness settings have experienced interruptions due to environmental factors such as falling trees, intense storms and wintertime icing (see ECCC, 2015). Less common, but potentially more destructive to the equipment and instrumentation, are the wildlife interactions, particularly with large mammals such as black and grizzly bears. Since the NHG's research requires long-term historical climate records, it is necessary to find other resources that may fill some of the data gaps. A few meteorological stations exist in the vicinity of the Quesnel River basin that may be able to provide additional data to complete this record (Fig. 2). Specifically, the BC Ministry of FLNRORD Likely Aerodrome weather station (52°36′05″ N, 121°30′48″ W; elevation 1046 m a.s.l.), the BC Ministry of Environment snow pillows at Yanks Peak East (52°49′ N, 121°21′ W; elevation 1683 m a.s.l.) and Barkerville (53°03′ N, 121°29′ W; elevation 1520 m a.s.l.), and the Environment and Climate Change Canada (ECCC) Barkerville weather station (53°04′09″ N, 121°30′53″ W; elevation 1283 m a.s.l.) have reliable and relatively long-term meteorological measurements for the Cariboo Mountains and in the vicinity of CAMnet weather stations. Note that the ECCC Barkerville record begins in 1888, making it BC's third longest observation-based meteorological dataset; however, this remains a weather station operated by volunteers, rendering data quality and homogeneity suspect at times. Nonetheless, these proximal weather and snow pillow stations provide valuable data on hydrometeorological conditions when CAMnet stations experienced data gaps. CAMnet stations are notably deficient in measuring cold season precipitation and snow water equivalence, variables that may be acquired from these additional weather stations.

Temporal gaps at one CAMnet station may also be infilled with data from another proximal one in the network. For instance, daily air temperatures for the QRRC, Spanish Mountain and Browntop Mountain weather stations are highly correlated ($r \geq 0.92$; Sharma, 2014). The local surface air temperature lapse rates vary from month to month with the steepest rates (6.6 °C km^{-1}) during summer and with more modest levels during winter (2.7 °C km^{-1}) (Sharma, 2014). Indeed, potent air temperature inversions often arise in winter such that care must be exercised in reconstructing this variable. Vigorous orographic enhancement of precipitation occurs in the Cariboo Mountains inducing strong precipitation gradients across their steep elevation gradients. Thus, site-specific conditions such as altitude, exposure, and land and vegetation cover induce considerable spatial variations in all quantities measured at CAMnet weather stations that require special attention in reconstructing missing data.

4.4 Data quality assessment and control

For CAMnet observed data, quality assessment is carried out for individual parameters at each station through statistical quality control procedures, analysis of frequency intervals, values, manual observations, etc. The CAMnet database consists of the original raw data allowing users to manipulate them according to their needs and desires. Missing timestamps are identified and filled with "NA" using codes developed in R (R Core Team, 2014) to maintain a complete yearly data record for each station. All data values assessed as erroneous, extreme or outside a predetermined frequency interval are simply flagged (highlighted in yellow) and recorded in the metadata documentation. Data from several sampled variables are also assessed for their quality through visual inspection and scatter plots such as snow depth and soil temperature. An outline of this quality assessment and control techniques can be found in a document located in the online database in each of the station's respective folders.

A particular concern with data quality is precipitation measurements in cold, windy environments such as the Cariboo Mountains. Apart from the freezing of tipping bucket rain gauges, high winds, evaporative losses and trace events are factors that often induce errors in precipitation measurements, especially for snowfall (Cherry et al., 2005). CAMnet data are not immune to inaccuracies due to precipitation measurement errors; however, the installation of single Alter shields at most of the CAMnet weather stations with tipping bucket rain gauges reduces the effects of wind undercatch of precipitation. Similarly to other CAMnet measurements, no modifications are performed on the precipitation data to correct for possible wind undercatch given the various strategies reported in the literature to address this (e.g., Sevruk, 1982). A few of the tipping bucket precipitation gauges at other CAMnet weather stations (most notably the Ancient Forest and Spanish Mountain sites) are equipped seasonally with CS705 snowfall adapters that are filled with antifreeze to melt and record solid precipitation. However, the antifreeze remains prone to dilution from precipitation and hence often freezes up, particularly during midwinter when air temperatures may reach ≤ -30 °C. Despite their known limitations, the tipping bucket rain gauges provide reasonable (liquid) precipitation records for the Cariboo Mountains and surrounding areas (see Sect. 4.6). Indeed, the accuracy of the CAMnet precipitation measurements remains comparable to that measured at nearby weather stations that meet WMO standards (Hay, 2018).

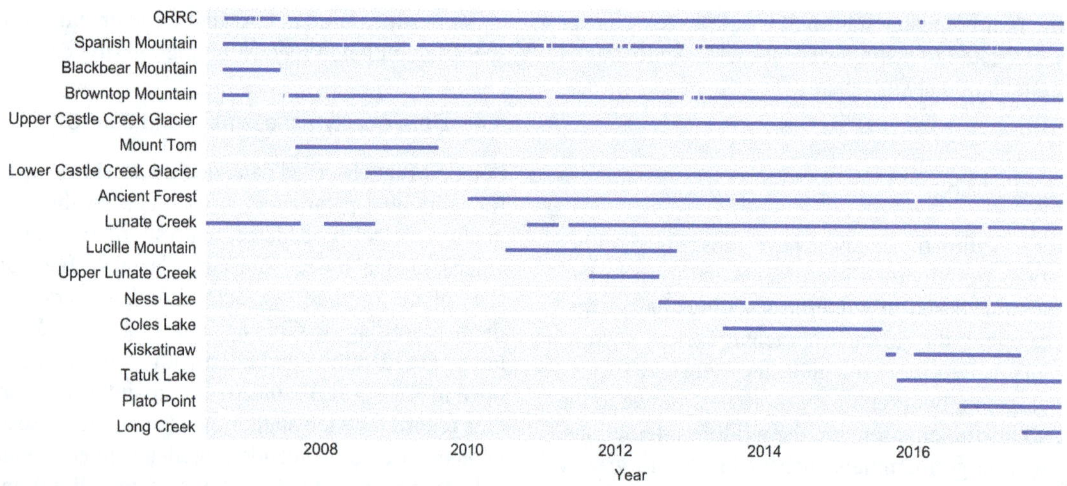

Figure 5. Operating periods for active and inactive CAMnet weather stations (1 January 2006–31 December 2017), deduced by analyzing the battery voltage required to power each station's data logger and instruments. Gaps in datasets occur due to a variety of reasons outlined in Sect. 5.1. Note that data may not be available for all instruments even when a station is operational (e.g., the icing of anemometers during winter).

4.5 Metadata

Metadata for each climate station in the CAMnet network are important to take into consideration when utilizing the associated data and are therefore summarized in a detailed .docx file. This document, a copy of which is available with the data for each station, outlines site location (coordinates, elevation, etc.) and access and provides a brief overview of the landscape, station equipment specifics (e.g., instrument heights) and data collection parameters.

Field notes and site logs are documented using a mediawiki.org web page allowing the ability to check on historical station visits dating back to the original deployment. These logs remain particularly valuable due to the high turnover of students and research staff in the NHG (e.g., summer field assistants) and serve as a complete and detailed resource. Weather conditions and equipment status, as well as work carried out, comprise important factors noted during each station visit and are therefore documented in these logs. Moreover, instrument serial numbers, recalibration and servicing schedules as well as an ongoing to-do list of field work activities are systematically updated on the CAMnet wiki.

4.6 Example data

CAMnet data compare favourably with measurements from independent meteorological networks that meet WMO standards such as those collected by Environment and Climate Change Canada and the BC Ministry of FLNRORD or with gridded climate products such as ANUSPLIN (McKenney et al., 2011). For example, monthly air temperature and precipitation collected at the BC Ministry of FLNRORD Likely Aerodrome weather station (52°36′54″ N, 121°30′48″ W; elevation 1046 m a.s.l.) correspond well with data from the CAMnet QRRC weather station 5.2 km away with differences of only 0.45 °C and 22 % (117 mm) during 2014, respectively (Figs. S1 and S2). Furthermore, profiles of air temperature and precipitation along a transect crossing the Cariboo Mountains between Williams Lake and McBride, BC, based on 1950–2010 ANUSPLIN data, reveal the expected cooler, wetter conditions at higher elevations (Fig. 6). Climatological data from proximal CAMnet stations to this transect match well the ANUSPLIN data, with differences explained partly by the altitudes of the in situ data relative to the smoothed topography of the gridded climate product. The QRRC, Spanish Mountain, Browntop Mountain and Upper Castle Creek Glacier stations exhibit strong elevational dependence in rainfall amounts, albeit with an appreciable precipitation undercatch relative to ANUSPLIN (Fig. 6a). Air temperatures at CAMnet stations cool from ~ 4 °C at the QRRC's valley bottom site to near -1 °C in the alpine settings of Browntop Mountain and the Upper and Lower Castle Creek stations, in close agreement with the gridded climate data (Fig. 6b).

CAMnet data remain central to the NHG's goal of investigating the meteorological drivers influencing the hydrology of north-central BC. Indeed, CAMnet allows monitoring the antecedent and real-time conditions of hydrological extremes including droughts and floods, as experienced in the region during 2010 and 2011, respectively (Fig. 7). In late winter–early spring 2010, shallow snow depths (maximum of 45 cm) followed by a hot, dry summer during an El Niño event led to scant total annual precipitation (346 mm), 36 % lower than the decadal mean observed at the QRRC weather station. This yielded some of the lowest annual discharge rates ever recorded for the Quesnel River (Déry et al., 2012). These dry conditions were followed by a much deeper snowpack (peak

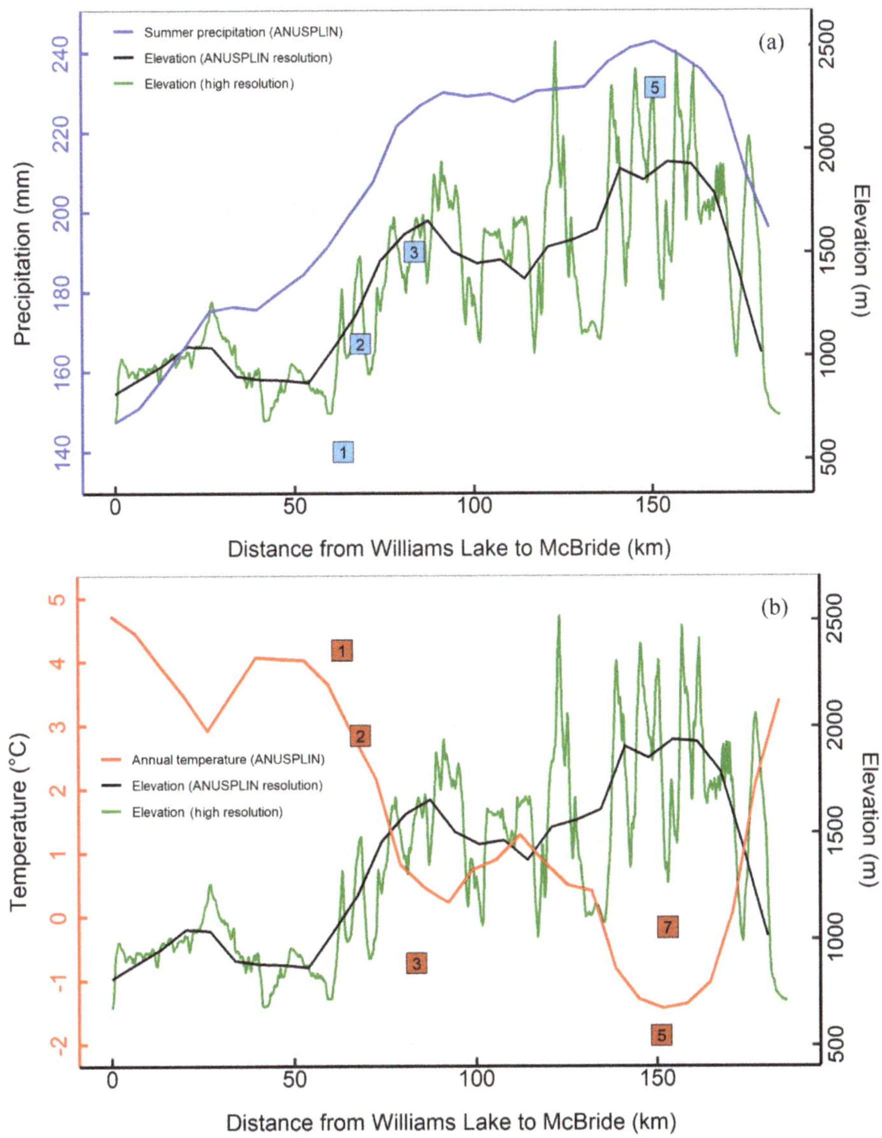

Figure 6. Profiles of average **(a)** total summer precipitation and **(b)** annual air temperature along an elevational transect (as shown in Fig. 2) from Williams Lake to McBride, BC, based on the 1950–2010 ANUSPLIN gridded product along with CAMnet weather station data. Elevations at high spatial resolution (~ 90 m) and of the ANUSPLIN smoothed topography (at a spatial resolution of ~ 10 km) are shown. Squares depict climatological results for CAMnet stations in proximity to the transect and are numbered as in Table 1. Results for CAMnet stations depend on data availability as follows: (1) QRRC (2007–2017), (2) Spanish Mountain (2007–2017), (3) Browntop Mountain (2007–2017), (5) Upper Castle Creek Glacier (2008–2017) and (7) Lower Castle Creek Glacier (2009–2016).

snow depth of 114 cm) in winter–spring of 2011 with precipitation during the first half of 2011 (322 mm) nearly equating the total for 2010. These meteorological conditions led to saturated soils, increased overland flow and elevated river runoff throughout 2011.

5 Challenges and opportunities

5.1 Physical and environmental challenges

The construction and maintenance of a highly sensitive, long-term database such as CAMnet generates distinct challenges on a variety of scales. Power outages due to battery failure or solar panel issues are a major contributor towards data loss and interruptions in sampling (see Sect. 4.3), a risk that is elevated during winter when snow and ice may cover solar panels and immerse instrumentation while cold tempera-

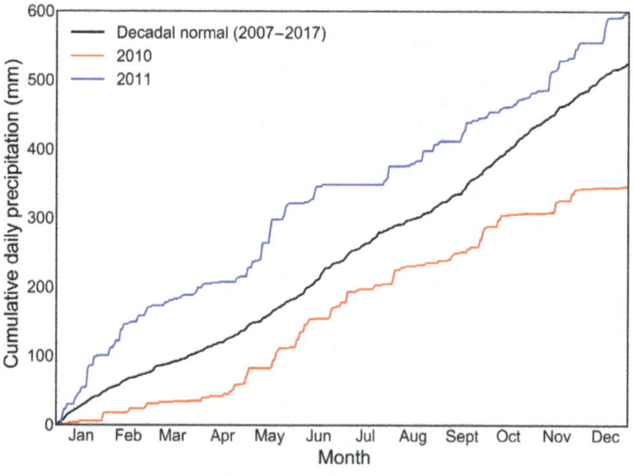

Figure 7. Cumulative daily precipitation recorded at the QRRC CAMnet station (using a heated Ogawa Seiki tipping bucket rain gauge) for 2010, 2011 and the overall mean for 2007–2017.

tures may accelerate the depletion of batteries. CAMnet stations have also been plagued with several animal incidents whereby the equipment has been displaced, damaged or rendered inoperable. Additionally, wildlife also has the ability to influence data solely by its presence, with snowpack measurements that rely on an undisturbed section of snow beneath the sensor being especially vulnerable. Other environmental considerations include lightning strikes that are mitigated by robustly grounding each station and the risk of damage from wildfires, which were extensive throughout the Cariboo region during the summer of 2017 and led to ashfall in nearby precipitation gauges. Physical challenges are presented by the complex terrain and remote locations of many stations, which also preclude frequent site visits; some stations such as those situated near Castle Creek Glacier are only visited once per year. Methods to access station sites include the following: through the use of a 4 × 4 vehicle (often on deactivated logging roads), a helicopter, a snowmobile and on foot (hiking in summer, snowshoeing in winter). Since these remote locations and environments are fundamental for local and regional hydrology, the importance for hydrometeorological data collection and the CAMnet network is pronounced; however, difficulty associated with securing funds to sustain this extensive network of remote weather stations (e.g., costs associated with field work and travel, equipment repairs, recalibration and purchases, student and research staff wages) remains a distinct and ongoing challenge.

5.2 Data homogeneity challenges

To minimize possible spurious trends or step changes in the hydrometeorological observations, all CAMnet sites are inspected and maintained regularly with faulty instrumentation repaired or replaced as necessary. While dedicated efforts are made in substituting deficient equipment with the same make and model, this is not always feasible if, for instance, the supplier no longer fabricates a given piece of equipment or specific model. Meteorological towers may shift under the influence of winds, frost heaving and heavy snowpacks, altering the measurement heights and possibly rendering the station unlevelled. Modifications to data logger programs (e.g., to switch on or off the heater for the QRRC radiometer) or operating systems may induce additional discontinuities in the CAMnet data.

Despite efforts to maintain consistency in the CAMnet datasets by routine station inspection and maintenance, various factors have resulted in a small fraction of data being influenced by external factors. Examples of these include the encroaching vegetation around the station on Spanish Mountain, the accidental rotation of the Browntop wind vane (away from north), and the replacement of instruments and equipment after various wildlife incidents and severe storms. As with data gaps, no post-processing of the data has been performed to assess their homogeneity, to remove suspect trends or erroneous data points. Site photos, field notes and post-trip updates of metadata remain invaluable resources in resolving issues related to these external influences and are available for any parties utilizing CAMnet data.

5.3 Opportunistic measurements

A primary objective of CAMnet remains to develop long-term hydrometeorological records in the poorly sampled Cariboo Mountains to assess climate change impacts on regional snow, glacier and water resources. This requires long-term homogeneous records of hydrometeorological conditions in the Cariboo Mountains at strategic locations such as the QRRC, Browntop and Spanish Mountains, and Castle Creek Glacier. A broader role of this long-term monitoring is the ability to answer unforeseen questions or research needs such as wildfires, pest infestations, mine spills (such as the Mount Polley mine accident), future resource extraction and land use and cover change in the area in addition to climate change. Indeed, CAMnet may assist in answering questions that have not yet been asked or even contemplated while providing baseline data in the event of future disturbances. Pushing CAMnet data online in (near-)real time also allows monitoring current atmospheric conditions remotely to assist in mitigating extreme events such as atmospheric rivers, floods or intense convective storms that can ignite wildfires such as in July 2017 in the BC interior. Finally, the mobile nature of the 3 m masts facilitates the transport and rapid deployment of weather stations to various sites, often to support a graduate student's research project or an intensive field campaign at a designated site.

5.4 Research, training and educational opportunities

CAMnet now assembles over a decade of sub-hourly hydrometeorological observations in the Cariboo Mountains and the surrounding area, forming the core data for multiple undergraduate and graduate student projects at UNBC and other institutions. It allows for training opportunities for students and research staff to develop skills in deploying meteorological instruments, data logger programming, data quality control and interpretation, as well as field expertise in remote terrain. CAMnet data and site photos are routinely used in undergraduate and graduate level courses to demonstrate regional examples of phenomena studied in class. CAMnet weather stations also provide staging sites for educational videos (e.g., for the Vancouver Aquarium's Year of Science series), for media interviews and for outreach activities for students of all levels. Finally, CAMnet provides opportunities to undertake interdisciplinary research at UNBC and institutions across Canada and abroad.

6 Ancillary data

Other studies focused on the Cariboo Mountains and surrounding area have generated ample ancillary data to the CAMnet hydrometeorological records. This includes an ongoing effort to collect seasonal and annual mass balance measurements at Castle Creek Glacier initiated in 2008 by colleagues at UNBC. This site is of particular interest given its record of annual push moraines, allowing a reconstruction of the glacier's retreat since the mid-1940s (Beedle et al., 2009). Eddy covariance measurements of turbulent fluxes on Castle Creek Glacier were also conducted during two summer field campaigns (in 2010 and 2012), during which additional atmospheric measurements were collected on the glacier (Radić et al., 2017). Sediment fluxes in Castle Creek and its main tributaries were also sampled during two other summer field campaigns (in 2008 and 2011; Leggat et al., 2015; Stott et al., 2016). Meteorological observations along the shores of Quesnel Lake support ongoing efforts to understand the long-term impacts of the Mount Polley mine tailings pond spill on the water quality and ecology of the system (Petticrew et al., 2015). A dedicated effort, based at the QRRC, is being made to monitor lake conditions (e.g., vertical profiles of water temperature, electrical conductivity, turbidity), as well as downstream water quality, temperature, and sediment concentrations and quality in the Quesnel River. This ongoing project benefits from historical, background data collected on Quesnel Lake starting in the early 2000s (Laval et al., 2008). Finally, provincial and federal networks operate additional meteorological and snow pillow stations while the Water Survey of Canada manages several hydrometric gauging stations on the main waterways draining the Cariboo Mountains including the Fraser, North Thompson, Clearwater, Quesnel, Horsefly, Doré, Willow and Bowron rivers (see Fig. 2 and Déry et al., 2012). This assemblage of unique ecosystems in pristine landscapes, availability of existing hydrometeorological data and a reliable research station to base field activities make the Cariboo Mountains particularly attractive for ongoing environmental and ecological research.

7 Recent and future expansion and application of CAMnet

During the summer of 2018, three additional meteorological stations were deployed along the shores of Quesnel Lake in support of an ongoing project investigating its resiliency to the Mount Polley mine tailings impoundment breach. Quesnel Lake has a complex morphometry with the steep surrounding topography often channelling winds. The area is also susceptible to mountain-valley circulations during clear summer days, with katabatic winds descending the steep, glacierized and snow-covered slopes towards the lower reaches of the valleys. Concerns over resuspension of lake-bottom sediments deposited during the tailings pond breach associated with wind-driven seiches require improved monitoring of atmospheric conditions including winds along Quesnel Lake.

In support of a new project titled the Storms and Precipitation Across the continental Divide Experiment (SPADE), a CAMnet weather station will be deployed near or within Kootenay National Park in southeastern BC during the fall of 2018. This will complement an existing array of meteorological stations operated by the University of Saskatchewan's Coldwater Laboratory at Fortress Mountain and Marmot Creek, Alberta, on the eastern flanks of the Canadian Rockies (Rothwell et al., 2016) and by the University of Calgary at Haig Glacier on the North American continental divide (Shea et al., 2005). An intensive field campaign will be conducted during May and June 2019 to obtain high-temporal-resolution atmospheric conditions with an emphasis on precipitation on a longitudinal transect across the continental divide.

8 Data availability

Data described in this article may be downloaded from Zenodo (https://doi.org/10.5281/zenodo.1195043, Déry, 2017). CAMnet meteorological data are available as Excel spreadsheets in .xlsx format. The database comprises a specific directory for each station that includes (i) yearly .xlsx data files named according to the site location and year and (ii) a .docx file outlining metadata describing the location, quality control performed to evaluate the data, installed instruments and sampling parameters of each instrument per station. Data files are listed chronologically and follow the numerical order listed in Table 1.

9 Conclusions

CAMnet was initiated from an interest and need to collect high-temporal-resolution data in elevated, complex terrain in a mountainous region where hydrometeorological data were previously quite sparse. Although challenges abound in obtaining high-quality data in a wilderness setting, our ever-expanding hydrometeorological datasets provide insight to the weather and climate of the Cariboo Mountains and surrounding region. We have presented an overview of CAMnet station locations, the instruments used, parameters measured and the resulting datasets since 2006. The recent and planned addition of stations around Quesnel Lake and along the continental divide demonstrates the ongoing expansion of the network, while an extensive collection of metadata contributes to the understanding of data inconsistencies and possible errors. We encourage interested parties to contact the NHG for information on using CAMnet data or for further discussion on the establishment and maintenance of a mesoscale network of hydrometeorological stations.

Author contributions. SJD initiated the development of CAMnet and designed the present report. MAHH wrote the first draft of the manuscript with contributions from ARS, MT, HDT and SJD, and all contributed to the discussion of the database and manuscript refinement and revisions.

Competing interests. The authors declare that they have no conflict of interest.

Acknowledgements. Thanks to UNBC, Future Forest Ecosystems Scientific Council, Environment and Climate Change Canada, Natural Sciences and Engineering Research Council of Canada through the Canadian Sea Ice and Snow Evolution network and the Discovery Grant and Research Tools and Instrumentation programs, Nechako Environmental Enhancement Fund, Canada Foundation for Innovation, Canadian Foundation for Climate and Atmospheric Sciences through the Western Canadian Cryospheric Network, BC Knowledge Development Fund, Global Water Futures and the Canada Research Chair program of the Government of Canada for funding. Thanks to Sam Albers, Michael Allchin, Darwyn Coxson, Art Fredeen, Scott Green, Rachel Hay, Rick Holmes, Peter Jackson, Brian Menounos, Jeremy Morris, Ken Otter, Phil Owens, Margot Parkes, Ellen Petticrew and Roger Wheate (all previously or currently affiliated with UNBC); Bernard Laval (UBC); Vanessa Foord, Richard Kabzems and Pat Teti (FLNRORD); and many field assistants and volunteers for their contributions. Thank you to the staff at the QRRC (including past managers Rick Holmes and Sam Albers and the current manager Michael Allchin) and the people of Likely and the surrounding area for their continued support and interest in our research. Constructive and insightful comments from Phil Owens (UNBC), two anonymous referees and the Topical Editor (John Pomeroy, University of Saskatchewan) led to a much improved paper.

Edited by: John Pomeroy

References

Beedle, M. J., Menounos, B., and Wheate, R.: Push moraines as climate proxy, Geophys. Res. Lett., 36, L20501, https://doi.org/10.1029/2009GL039533, 2009.

Beedle, M. J., Menounos, B., and Wheate, R.: Glacier change in the Cariboo Mountains, British Columbia, Canada (1952–2005), The Cryosphere, 9, 65–80, https://doi.org/10.5194/tc-9-65-2015, 2015.

Beniston, M.: Mountain weather and climate: A general overview and a focus on climatic change in the Alps, Hydrobiologia, 562, 3–16, https://doi.org/10.1007/s10750-005-1802-0, 2006.

Bradley, R. S., Keimig, F. T., and Diaz, H. F.: Projected temperature changes along the American cordillera and the planned GCOS network, Geophys. Res. Lett., 31, L16210, https://doi.org/10.1029/2004GL020229, 2004.

Burford, J. E., Déry, S. J., and Holmes, R. D.: Some aspects of the hydroclimatology of the Quesnel River Basin, British Columbia, Canada, Hydrol. Process., 23, 1529–1536, https://doi.org/10.1002/hyp.7253, 2009.

Campbell Scientific (Canada) Corp: Instruction Manual, RF401A-Series Spread Spectrum Radios, 51 pp., available at: https://s.campbellsci.com/documents/ca/manuals/rf401a-seriesman.pdf (last access: 25 March 2018), 2017.

Cherry, J. E., Tremblay, L. B., Déry, S. J., and Stieglitz, M.: Reconstructing solid precipitation from snow depth measurements and a land surface model, Water Resour. Res., 41, W09401, https://doi.org/10.1029/2005WR003965, 2005.

Déry, S.: Cariboo Alpine Mesonet (CAMnet) Database, Data set, Zenodo, https://doi.org/10.5281/zenodo.1195043, 2017.

Déry, S. J., Clifton, A., MacLeod, S., and Beedle, M. J.: Blowing snow fluxes in the Cariboo Mountains of British Columbia, Canada, Arct. Antarct. Alp. Res., 42, 188–197, https://doi.org/10.1657/1938-4246-42.2.188, 2010.

Déry, S. J., Hernández-Henríquez, M. A., Owens, P. N., Parkes, M. W., and Petticrew, E. L.: A century of hydrological variability and trends in the Fraser River Basin, Environ. Res. Lett., 7, 024019, https://doi.org/10.1088/1748-9326/7/2/024019, 2012.

Déry, S. J., Knudsvig, H. K., Hernández-Henríquez, M. A., and Coxson, D. S.: Net snowpack accumulation and ablation characteristics in the Inland Temperate Rainforest of the Upper Fraser Basin, Canada, Hydrology, 1, 1–19, https://doi.org/10.3390/hydrology1010001, 2014.

Environment Climate Change Canada (ECCC): Manual of Surface Weather Observations, 7th Edn., 463 pp., 2015.

Farley, A. L.: Atlas of British Columbia, University of British Columbia Press, Vancouver, BC, 136 pp., 1979.

Gubler, S., Hunziker, S., Begert, M., Croci-Maspoli, M., Konzelmann, T., Brönnimann, S., Schwierz, C., Oria, C., and Rosas, G.: The influence of station density on climate data homogenization, Int. J. Climatol., 37, 4670–4683, https://doi.org/10.1002/joc.5114, 2017.

Hasler, A., Geertsema, M., Foord, V., Gruber, S., and Noetzli, J.: The influence of surface characteristics, topography and continentality on mountain permafrost in British Columbia, The Cryosphere, 9, 1025–1038, https://doi.org/10.5194/tc-9-1025-2015, 2015.

Hay, R. W.: Sensitivity testing for the effects of sub-grid scale topography on snow evolution in the Canadian Land Surface Scheme, MSc Thesis, University of Northern British Columbia, Prince George, BC, 2018.

Kang, D. H., Gao, H., Shi, X., Islam, S. U., and Déry, S. J.: Impacts of a rapidly declining mountain snowpack on streamflow timing in Canada's Fraser River Basin, Sci. Rep.-UK, 6, 19299, https://doi.org/10.1038/srep19299, 2016.

Laval, B. E., Morrison, J., Potts, D. J., Carmack, E. C., Vagle, S., James, C., McLaughlin, F. A., and Foreman, M.: Wind-driven summertime upwelling in a fjord-type lake and its impact on downstream river conditions: Quesnel Lake and River, British Columbia, Canada, J. Great Lakes Res., 34, 189–203, https://doi.org/10.3394/0380-1330(2008)34, 2008.

Leggat, M. S., Owens, P. N., Stott, T. A., Forrester, B. J., Déry, S. J., and Menounos, B.: Hydro-meteorological drivers and sources of suspended sediment flux in the proglacial zone of the retreating Castle Creek Glacier, Cariboo Mountains, British Columbia, Earth Surf. Proc. Land., 40, 1542–1559, https://doi.org/10.1002/esp.3755, 2015.

MacLeod, S. and Déry, S. J.: The Cariboo Alpine Mesonet, CMOS Bulletin SCMO, 35, 45–51, 2007.

McKenney, D. W., Hutchinson, M. F., Papadopol, P., Lawrence, K., Pedlar, J., Campbell, K., Milewska, E., Hopkinson, R. F., Price, D., and Owen, T.: Customized spatial climate models for North America, B. Am. Meteorol. Soc., 92, 1611–1622, https://doi.org/10.1175/2011BAMS3132.1, 2011.

Mekis, E., Donaldson, N., Reid, J., Zucconi, A., Hoover, J., Li, Q., Nitu, R., and Melo, S.: An overview of surface-based precipitation observations at Environment and Climate Change Canada, Atmos.-Ocean, 56, 71–95, 2018.

Miles, M. and Associates Ltd. for Ministry of Water Land and Air Protection: British Columbia's Climate-Related Observation Networks: An Adequacy Review, available at: http://www.for.gov.bc.ca/hfd/library/documents/bib94904a.pdf (5 June 2018), 2003.

Petticrew, E. L., Albers, S., Baldwin, S., Carmack, E. C., Déry, S. J., Gantner, N., Graves, K., Laval, B., Morrison, J., Owens, P. N., Selbie, D. T., and Vagle, S.: The impact of a catastrophic mine tailings impoundment spill into one of North America's largest fjord lake: Quesnel Lake, British Columbia, Canada, Geophys. Res. Lett., 42, 3347–3355, https://doi.org/10.1002/2015GL063345, 2015.

Radić, V., Menounos, B., Shea, J., Fitzpatrick, N., Tessema, M. A., and Déry, S. J.: Evaluation of different methods to model near-surface turbulent fluxes for a mountain glacier in the Cariboo Mountains, BC, Canada, The Cryosphere, 11, 2897–2918, https://doi.org/10.5194/tc-11-2897-2017, 2017.

R Core Team: R: A language and environment for statistical computing, R Foundation for Statistical Computing, Vienna, Austria, R Foundation for Statistical Computing, available at: http://www.r-project.org/, last access: 16 March 2014.

Rothwell, R., Hillman, G., and Pomeroy, J. W.: Marmot Creek experimental watershed study, Forestry Chron., 92, 32–36, 2016.

Schirmer, M. and Jamieson, B.: Verification of analysed and forecasted winter precipitation in complex terrain, The Cryosphere, 9, 587–601, https://doi.org/10.5194/tc-9-587-2015, 2015.

Sevruk, B.: Methods of correction for systematic error in point precipitation measurement for operational use, Oper. Hydrol. Rep. 21, 91 pp., World Meteorological Organization, Geneva, Switzerland, 1982.

Sharma, A. R.: Trends and elevational dependence of the hydroclimatology of the Cariboo Mountains, British Columbia, MSc Thesis, University of Northern British Columbia, Prince George, BC, https://doi.org/10.13140/RG.2.1.2121.9444, 2014.

Sharma, A. R. and Déry, S. J.: Elevational dependence of air temperature variability and trends in British Columbia's Cariboo Mountains, 1950–2010, Atmos.-Ocean, 54, 153–170, https://doi.org/10.1080/07055900.2016.1146571, 2016.

Shea, J. M., Anslow, F. S., and Marshall, S. J.: Hydrometeorological relationships on Haig Glacier, Alberta, Canada, Ann. Glaciol., 40, 52–60, https://doi.org/10.3189/172756405781813465, 2005.

Stevenson, S. K., Armleder, H. M., Arsenault, A., Coxson, D., Delong, S. C., and Jull, M.: British Columbia's Inland Rainforest, UBC Press, Vancouver, BC, Canada, 432 pp., 2011.

Stott, T. A., Leggat, M. S., Owens, P. N., Forrester, B. J., Déry, S. J., and Menounos, B.: Suspended sediment dynamics in the proglacial zone of the rapidly retreating Castle Creek Glacier, British Columbia, Canada, in: Sources-to-Sink Fluxes in Undisturbed Cold Environments, edited by: Beylich, A. A., Dixon, J. C., and Zwolinski, Z., Cambridge University Press, 313–325, 2016.

Tong, J., Déry, S. J., and Jackson, P. L.: Topographic control of snow distribution in an alpine watershed of western Canada inferred from spatially-filtered MODIS snow products, Hydrol. Earth Syst. Sci., 13, 319–326, https://doi.org/10.5194/hess-13-319-2009, 2009a.

Tong, J., Déry, S. J., and Jackson, P. L.: Interrelationships between MODIS/Terra remotely sensed snow cover and the hydrometeorology of the Quesnel River Basin, British Columbia, Canada, Hydrol. Earth Syst. Sci., 13, 1439–1452, https://doi.org/10.5194/hess-13-1439-2009, 2009b.

Tong, J., Déry, S. J., Jackson, P. L., and Derksen, C.: Testing snow water equivalent retrieval algorithms for passive microwave remote sensing in an alpine watershed of western Canada, Can. J. Remote Sens., 36, S74–S86, https://doi.org/10.5589/m10-009, 2010.

Wang, T., Hamann, A., Spittlehouse, D. L., and Murdock, T. Q.: ClimateWNA – High resolution spatial climate data for western North America, J. Appl. Meteorol. Clim., 51, 16–29, 2012.

Younas, W., Hay, R. W., MacDonald, M. K., Islam, S. U., and Déry, S. J.: A strategy to represent impacts of subgrid-scale topography on snow evolution in the Canadian Land Surface Scheme, Ann. Glaciol., 58, 1–10, 2017.

6

A global data set of soil hydraulic properties and sub-grid variability of soil water retention and hydraulic conductivity curves

Carsten Montzka[1], Michael Herbst[1], Lutz Weihermüller[1], Anne Verhoef[2], and Harry Vereecken[1,3]

[1]Forschungszentrum Jülich GmbH, Institute of Bio- and Geosciences: Agrosphere (IBG-3), Jülich, Germany
[2]University of Reading, Department of Geography and Environmental Science, Reading, UK
[3]International Soil Modeling Consortium, Jülich, Germany

Correspondence to: Carsten Montzka (c.montzka@fz-juelich.de)

Abstract. Agroecosystem models, regional and global climate models, and numerical weather prediction models require adequate parameterization of soil hydraulic properties. These properties are fundamental for describing and predicting water and energy exchange processes at the transition zone between solid earth and atmosphere, and regulate evapotranspiration, infiltration and runoff generation. Hydraulic parameters describing the soil water retention (WRC) and hydraulic conductivity (HCC) curves are typically derived from soil texture via pedotransfer functions (PTFs). Resampling of those parameters for specific model grids is typically performed by different aggregation approaches such a spatial averaging and the use of dominant textural properties or soil classes. These aggregation approaches introduce uncertainty, bias and parameter inconsistencies throughout spatial scales due to nonlinear relationships between hydraulic parameters and soil texture. Therefore, we present a method to scale hydraulic parameters to individual model grids and provide a global data set that overcomes the mentioned problems. The approach is based on Miller–Miller scaling in the relaxed form by Warrick, that fits the parameters of the WRC through all sub-grid WRCs to provide an effective parameterization for the grid cell at model resolution; at the same time it preserves the information of sub-grid variability of the water retention curve by deriving local scaling parameters. Based on the Mualem–van Genuchten approach we also derive the unsaturated hydraulic conductivity from the water retention functions, thereby assuming that the local parameters are also valid for this function. In addition, via the Warrick scaling parameter λ, information on global sub-grid scaling variance is given that enables modellers to improve dynamical downscaling of (regional) climate models or to perturb hydraulic parameters for model ensemble output generation. The present analysis is based on the ROSETTA PTF of Schaap et al. (2001) applied to the SoilGrids1km data set of Hengl et al. (2014). The example data set is provided at a global resolution of 0.25° at https://doi.org/10.1594/PANGAEA.870605.

1 Introduction

Hydraulic properties have fundamental importance in the description of water, energy and carbon exchange processes between the land surface and the atmosphere (e.g. Ek and Cuenca, 1994; Xue et al., 1996). Therefore, agroecosystem models (e.g. SWAP; van Dam et al., 2008) and land surface models (LSMs; see below) require adequate parameterization of soil hydraulic properties – i.e. more specifically, the water retention curve (WRC) and the hydraulic conductivity curve (HCC). These properties regulate the relative magnitude of water balance fluxes such as evapotranspiration, infiltration and surface and sub-surface runoff (Vereecken et al., 2016), and thus the amount of water held in the soil at any one time. As well as affecting the water balance components, they also play a role in the land surface energy balance – directly via their effect on latent heat flux (evapotranspiration, e.g. via available soil water influencing plant water stress; see Verhoef and Egea, 2014), and indirectly because the size of the evapotranspiration co-determines land surface temperature, that in turn affects net radiation, sensible heat flux and soil heat flux (the latter is also affected via soil moisture dependency of soil thermal properties). With regard to the carbon balance, photosynthesis and soil respiration both strongly depend on soil moisture content and hence implicitly on choice of soil hydraulic models and their parameters.

State-of-the-art LSMs, e.g. NOAH (Niu et al., 2011), CLM (Oleson et al., 2008), VIC (Liang et al., 1994), JULES (Best et al., 2011) and ORCHIDEE (Ngo-Duc et al., 2007), are key components of regional and global climate models (RCMs/GCMs) and numerical weather prediction models (NWPMs). They form important components of reanalyses (e.g. ERA-Interim/Land; Balsamo et al., 2015) and model–data assimilation systems, such as the NASA Land Information System (LIS) or the Global Land Data Assimilation System (GLDAS; Rodell et al., 2004). LSMs solve Richard's equation for the water flow in the saturated/unsaturated zone. A fundamental problem is the adequate parameterization of the water retention and hydraulic conductivity function to solve Richard's equation. At point scale, a broad suite of experimental methods are available that allow measuring the WRC and HCC. These measurements are, however, expensive and time consuming, and often comprise intensive field sampling campaigns. Alternatively, parameters of the WRC and HCC can be estimated from in situ or remotely sensed data in combination with parameter estimation techniques (Scharnagl et al., 2011; Bauer et al., 2012; Dimitrov et al., 2014; Jadoon et al., 2012; Montzka et al., 2011). At larger scales, such as those where RCMS and GCMs are employed, the WRC and HCC are impossible to estimate (because underlying soil properties vary widely within grid cells) and are unobtainable by direct measurements.

To overcome this problem, the estimation of the required soil hydraulic properties is usually based on pedotransfer functions (PTFs) that use simple soil properties such as texture, organic matter and bulk density to derive the parameters of mathematical equations that describe the HCC and the WRC (Vereecken et al., 2010). The idea behind PTFs is that more easily available soil data such as soil texture, soil organic carbon content or bulk density can be used to predict the hydraulic parameters for the WRC and HCC. In the last three decades, soil scientists have developed a broad suite of PTFs that differ with respect to the parameterizations of soil hydraulic properties for which they are used, the type of soil properties needed as inputs to derive the model-dependent parameters, and their spatial patterns. PTFs were developed for the prediction of parameters used in the Campbell (1974) family of hydraulic functions (e.g. Clapp and Hornberger, 1978), Brooks and Corey (1964) (e.g. Rawls and Brakensiek, 1985), and Mualem–van Genuchten equations (e.g. Rawls and Brakensiek, 1985; Vereecken et al., 1989; Scheinost et al., 1997; Wösten et al., 1999; Weynants et al., 2009; Toth et al., 2015). Bouma (1989) distinguished between two types of PTFs, namely continuous and class type PTFs. Continuous PTFs use information on textural properties, bulk density and soil organic matter amongst others, whereas class type PTFs do not estimate the parameters based on continuous textural and other soil properties but estimate the parameters for defined textural classes – e.g. 12 USDA textural classes (Clapp and Hornberger, 1978; Toth et al., 2013). The disadvantage of the latter is that only the class average can be predicted and inner-class variability is neglected.

Another issue with the soil hydraulic parametrization in LSMs is caused by the spatial resolution of the model application under consideration (e.g. GCM runs for reanalyses or NWP model runs for weather forecasts), which is currently several (tens of) kilometres. This means that the soil input parameters to be provided at this scale have to be derived from existing data sources. Unfortunately, intrinsic soil properties are highly variable in space; in most cases several soil types can be found within a single grid cell of GCMs, for example. These soil types often differ strongly in soil texture, soil organic carbon content and bulk density, and soil depth and layering. Consequently, the fine-scale soil information, available from state-of-the-art soil maps such as the European LUCAS (Land Use/Land Cover Area Frame Survey) (Toth et al., 2013; Ballabio et al., 2016) at 500 m resolution or the global SoilGrid database at 1 km resolution (Hengl et al., 2014), has to be up-scaled to the scale at which the LSMs are being employed. The general problem of up-scaling, or change in spatial resolution of the input data by aggregating small-scale input data, and the resulting output uncertainty for various model states was reported, for example, by Cale et al. (1983), Rastetter et al. (1992), Pierce and Running (1995), Hoffmann et al. (2016), and Kuhnert et al. (2016). A practical example is GLDAS2-NOAH, where the porosity and the percentages of sand, silt and clay at the original scale of the input data from Reynolds et al. (2000) were horizontally resampled, i.e. spatially averaged, to the 0.25° GLDAS grid (Rodell et al., 2004). Despite their importance, only a few studies investigating the implications of the above-mentioned issues (up-scaling, aggregation or resampling) on the model results have been conducted in the past.

The most straightforward method to aggregate input parameters from small-scale soil maps to larger-scale grid cells of a GCM would be spatial averaging. For some soil properties such as soil organic carbon, bulk density, or soil depth this kind of approach seems reasonable, whereas for soil texture averaging it is associated with considerable problems. For example, a GCM grid cell containing a pure sand soil for half of its area with the other half a pure clay would provide a sandy clay on average, which neither adequately reflects the sand nor the clay soil physical properties. Additionally, averaging percentages may cause artefacts in closing the mass balance. A second method would be the averaging of soil hydraulic parameters (e.g. the van Genuchten parameters), whereby Zhu and Mohanty (2002) clearly showed that averaging of especially the shape parameters (a and n) is associated with considerable uncertainty. A third and most widely used aggregation technique for soil inputs at coarse model resolution is the one based on dominant soil types, where the dominant soil type within a coarse grid cell is derived from the fine-scale soil map. However, in using this approach some information will get lost in the GCM outputs because non-dominant, but physically very differently behaving soils will not be taken into account during the model runs. In consequence, fluxes from non-dominant areas of the grid cell will not be reproduced at large scale.

The theory introduced by Miller and Miller (1956) provided a technique to scale the relationships of pressure head and hydraulic conductivity by considering microscopic laws for capillary pressure forces and viscous flow in porous media based on a similarity assumption of the pore space structure (Warrick et al., 1977). Similarity scaling allows converting hydraulic characteristics (e.g. pressure head or conductivity) of one system (e.g. a soil sample) or location (e.g. a point scale measurement of WRC at field scale) towards corresponding characteristics of another system or location (Tillotson and Nielsen, 1984) under the condition that the internal geometry of the system only differs by size. Miller–Miller scaling therefore allows capturing the spatial variability of soil hydraulic properties in one single scaling parameter rather than having to specify the statistics for each single hydraulic parameter (Warrick et al., 1977). The set of scaling factors (i.e. each location or sample has one scaling factor) follow approximately a log-normal distribution (Simmons et al., 1979). Tuli et al. (2001) analysed a physically based scaling approach of unsaturated hydraulic conductivity and soil water retention functions from pore size distribution. They assumed that the relationship between both characteristics is log-normally distributed and that pores are geometrically similar, and showed that in this case scaling factors computed from median pore size or capillary pressure head can be used to describe the variability of unsaturated hydraulic conductivity functions. Using a fractal model, Pachepsky et al. (1995) showed that the spatial variability of water retention functions could be described by the spatial variability of a single dimensionless parameter. Ahuja et al. (1984) found that scaling factors for different soil depths are also related, and Clausnitzer et al. (1992) investigated the potential to simultaneously scale the WRC and HCC and found evidence that the results do not necessarily require independent scaling. In more detail, Hendrayanto et al. (2000) showed that separate scaling resulted in large estimation errors in either effective saturation or hydraulic conductivity. Further scaling methods have been developed based on the fractal method, e.g. the piecewise fractal approach proposed by Millan and Gonzalez-Posada (2005) or the wavelet transform modulus maxima introduced by Zeleke and Si (2007). Wang et al. (2009), as well as Fallico et al. (2010), analysed the multifractal distribution of scaling parameters for soil water retention characteristics. Shu et al. (2008) stressed the need for location-dependent scale analyses to improve the performance for soil water retention characteristic predictions. Jana and Mohanty (2011) showed that a Bayesian neural network can be applied across spatial scales to approximate fine-scale soil hydraulic properties. With this approach ground-, air-, and space-based remotely sensed geophysical parameters directly contribute to a PTF in a single processing step instead of aggregating/scaling the estimated parameters to other scales in an independent second step. Recently, Fang et al. (2016) established an amplification factor for soil hydraulic conductivity to compensate for the resulting retardation of water flow due to the loss of information content as a consequence of spatial aggregation. Liao et al. (2014) pointed out that uncertainty in the soil water retention parameters mainly results from the limited number of samples used for deriving PTFs and the spatial interpolation of basic soil properties. However, the latter error contribution dominates the potential to correctly determine spatial parameters, which leads to the assumption for our study that existing PTFs provide adequate parameters for global model applications. Nevertheless, the scaling uncertainty still needs to be considered.

The objectives of this study are therefore as follows: (i) to apply the Miller–Miller scaling approach to the state-of-the-art soil data set SoilGrids1km to provide a global consistent soil hydraulic parameterization for GCMs based on first principles; these soil hydraulic parameterizations can be used in models of the terrestrial system to predict soil water fluxes based on solving Richard's equation from local to global scale; (ii) to present a method to identify the sub-grid variability of WRC and HCC with reference to the 1 km resolution SoilGrids1km soil texture database – this scaling information can be used to perturb hydraulic parameters to generate ensemble runs with GCMs or to improve GCM downscaling; (iii) to evaluate the performance of the scaling approach for the calculation of the WRCs and HCCs against standard aggregation procedures based on two exemplary grid cells with varying variability in textural properties; and finally (iv) to demonstrate the importance of the scaling variance at different spatial resolutions for three larger regions in North America, Africa and Asia. We provide the corresponding aggregated global data set for the ROSETTA PTF (Schaap et al., 2001) at 0.25° regular grid spacing at https://doi.org/10.1594/PANGAEA.870605.

2 Material and methods

In this section the database and the approach to generate the global data set of soil hydraulic parameters is presented (see also Fig. 1). In the following, all analyses are based on the SoilGrids1km database in combination with ROSETTA (Schaap et al., 2001). Other soil databases and PTFs can be used similarly. The PTF will be applied to the high-resolution soil texture maps to predict high-resolution Mualem–van Genuchten (MvG) parameters. These are then scaled to cal-

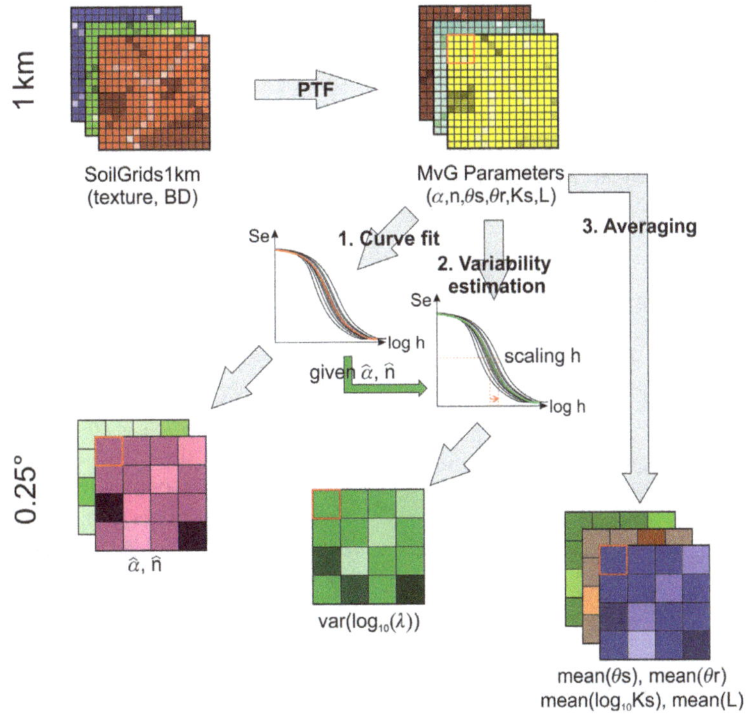

Figure 1. Proposed method to aggregate soil hydraulic properties and sub-grid variability of soil water retention and hydraulic conductivity curves.

culate soil hydraulic properties for the WRCs and HCCs employed at a coarser scale. In Sect. 2.3, the scaling approach is explained, and additionally the treatment of the sub-grid variance will be presented. In Sect. 2.4 details about the application to generate the global data set are given.

2.1 SoilGrids1km

The SoilGrids1km database (Hengl et al., 2014) is a consistent, coherent and global data set created by automated mapping (Vereecken et al., 2016). The main inputs are publicly available soil profile data, such as the USA National Cooperative Soil Survey Soil Characterization database (NCSS), the Land Use/Cover Area frame Statistical Survey LUCAS (Toth et al., 2013), and the Soil and Terrain Database (SOTER) (Van Engelen and Dijkshoorn, 2012). Moreover, additional information, derived from moderate-resolution imaging spectroradiometer (MODIS) satellite imagery and the Shuttle Radar Topography Mission (SRTM) digital elevation model, has been used. Artificial surfaces as well as bare rock areas, water bodies, shifting sands, permanent snow and ice were neglected. The resulting soil properties at seven predefined depths (0, 5, 15, 30, 60, 100 and 200 cm) are soil organic carbon ($\mathrm{g\,kg^{-1}}$), soil pH, sand, silt, and clay fractions (%), coarse fragments (gravel) (%), bulk density ($\mathrm{kg\,m^{-3}}$), cation-exchange capacity ($\mathrm{cmol^{+}\,kg^{-1}}$), soil organic carbon stock ($\mathrm{t\,ha^{-1}}$), depth to bedrock (cm), World Reference Base soil groups and USDA Soil Taxonomy suborders. SoilGrids1km implements model-based geostatistics and multiple linear regressions for predicting sand, silt and clay percentages, and bulk density, as well as general linear models with log-link function for predicting organic carbon content. Lower and upper confidence limits at 90 % probability of the predictions are also provided. In theory, the full prediction uncertainty could be used to estimate soil hydraulic property uncertainty but in our study we restricted the analysis on the mean predicted values. The findings of Hengl et al. (2014) indicate that the distribution of soil organic carbon content in horizontal direction is mainly controlled by climatic conditions (temperatures and precipitation), while the distribution of texture is mainly controlled by topography and lithology. The advantage of SoilGrids1km over other soil databases is that it provides pixel-based information rather than gridded vector information from class-based vector polygons. It should be noted that SoilGrids1km is stored in a World Geodetic System 84 (WGS84) regular grid with 1 km resolution at the equator. The resolution at other latitudes is therefore higher.

2.2 Pedotransfer functions, water retention and hydraulic conductivity functions

A global high-resolution hydraulic parameters data set is needed to infer the scaling characteristics for large-scale climate models. Based on the textural information stored in the SoilGrids1km database, PTFs can directly estimate the required hydraulic parameters (Fig. 1). Several PTFs have been developed; here, we focus on the widely used PTF ROSETTA model H3 by Schaap et al. (2001), which is based on neural network predictions for the estimation of the Mualem–van Genuchten (MvG) parameters θ_s, θ_r, α, n, K_s and L (van Genuchten, 1980), whereby the WRC to describe the effective volumetric saturation S_e is calculated according to

$$S_e(h) = \frac{\theta - \theta_r}{\theta_s - \theta_r} \left\{ \begin{array}{ll} 1 & h \geq 0 \\ \left[1 + (\alpha \left|h\right|)^n\right]^{-m} & h < 0,\ \alpha, m > 0,\ n > 1, \end{array} \right. \tag{1}$$

where θ_r ($cm^3\,cm^{-3}$) and θ_s ($cm^3\,cm^{-3}$) are the residual and saturated volumetric water content, respectively, and α (cm^{-1}), n (–) and m (–) ($m = n - \frac{1}{n}$) are shape parameters. Finally, the MvG approach to describe the HCC is given by

$$K(h) = K_s S_e^L \left[1 - \left(1 - S_e^{1/m}\right)^m\right]^2, \tag{2}$$

where K (–) is the unsaturated hydraulic conductivity, K_s ($cm\,d^{-1}$) is the saturated hydraulic conductivity and L is the pore connectivity parameter (–).

In a first step, ROSETTA was used to predict the MvG parameters based on the textural information of the SoilGrid map for each 1 km cell. In a next step, the water retention pairs (S_e versus h) were calculated for predefined pressure heads h (cm) using the pressure head vector $\boldsymbol{h}$:

$$\begin{aligned} \boldsymbol{h} = [&-1, -5, -10, -20, 30, -40, -50, -60, -70, -90, -110, \\ &-130, -150, -170, -210, -300, -345, -690, -1020, \\ &-5100, -15\,300, -20\,000, -100\,000, -1\,000\,000]. \end{aligned} \tag{3}$$

The pressure heads in $\boldsymbol{h}$ were chosen to reflect pressure steps commonly used in laboratory analysis. We assumed −300 cm to reflect field capacity whereas wilting point ($h \sim -1500$ cm) is generally found between −1020 and −5100. In pF terms ($\log_{10}(\boldsymbol{h})$) the $\boldsymbol{h}$ vector went up to 6.

2.3 Scaling approach and sub-grid variability estimation

In this study, the Warrick et al. (1977) scaling approach is applied to the parameters derived from the SoilGrids1km data for each soil depth separately. The procedure characterizes scaling factors to relate the hydraulic properties at a specific location to the mean hydraulic properties at a reference point or a point representative for a larger region.

In a first step we need to find adequate parameters for the retention function at the coarse scale (Fig. 1). This approach has been reported in Clausnitzer et al. (1992). For each subpixel i the relative saturation S_{e_i} is calculated by

$$S_{e_i}(\boldsymbol{h}) = f(\boldsymbol{h}, \alpha_i, n_i) = \left[1 + (\alpha_i \left|\boldsymbol{h}\right|)^{n_i}\right]^{-m_i}. \tag{4}$$

Next, the coarse-scale parameters $\hat{\alpha}$ and $\hat{n}$ of the water retention curve $f(h, \alpha_i n_i)$ need to be found that minimize the sum of squares of the deviations for all respective subpixels $i = 1 \ldots N$, with N being the number of subpixels within the coarse grid cell:

$$(\hat{\alpha}, \hat{n}) = \sum\nolimits_{i=1}^{N} \left[S_{e_i} - f(\boldsymbol{h}, \alpha_i, n_i)\right]^2. \tag{5}$$

The parameter fitting algorithm used in this study was the damped least-squares method of Levenberg–Marquardt (LM) (Marquardt, 1963) to find a global minimum. As initial values for LM fitting, the grid-specific spatial average of α ($\overline{\alpha}$) and n ($\overline{n}$) was used.

Russo and Bresler (1980), as well as Warrick et al. (1977), showed that scale factors for soil water retention and unsaturated hydraulic conductivity are not necessarily identical. However, Clausnitzer et al. (1992) reported that an independent fitting would lead to inconsistencies in the parameter space, and that a single scaling factor is well suited to describe the distribution and correlation structure of HCC and WRC. In this study we use the relationship between K_r and S_e in Eq. (2). Therefore, the scaling of the WRC can be directly transferred to the HCC, by using $\hat{\alpha}\hat{n}$ from Eq. (5), even though these scaling parameters might not be the optimum choice for the HCC. However, this approach was chosen for simplicity – to allow for easy handling of hydraulic parameters via a single scaling parameter for global Earth system model applications. For the coarse grid cell representative HCC the missing parameters K_s and L are spatially averaged from the sub-grid parameters – in the case of K_s in logarithmic space. Similarly, θ_s and θ_r were also spatially averaged, i.e. for the coarse resolution $\overline{\theta_s}$ and $\overline{\theta_r}$ were calculated.

In a second step, the sub-grid variability is estimated by introducing the scale parameter λ to the hydraulic head vector to simplify the description of the statistical variation of soil properties (Fig. 1). This is done by

$$\boldsymbol{h}^* = \frac{\boldsymbol{h}}{\lambda}. \tag{6}$$

After substituting $\boldsymbol{h}$ by $\boldsymbol{h}^*$ in the van Genuchten (1980) water retention function (Eq. 4), while using previously estimated $\hat{\alpha}$ and $\hat{n}$, only the scaling factor is fitted for each individual subpixel. Equation (5) can then be rewritten as

$$\left(\hat{\lambda_i}\right) = \sum\nolimits_{i=1}^{\mathrm{N}} \left[S_{e_i} - f(\boldsymbol{h}, \hat{\alpha}, \hat{n}, \lambda_i)\right]^2. \tag{7}$$

Equation (7) is subject to the constraint that the coarse grid mean of the set of scaling factors is unity ($\frac{1}{N}\sum_{i=1}^{N} \log_{10}(\hat{\lambda_i}) = 0$). This constraint is already approximated by adequately fitting of $\hat{\alpha}$ and $\hat{n}$ in the first step. Again, similar to the first step, the unsaturated HCC is scaled based on the parameters estimated for the WRC.

We recommend calculating the variance of the logarithmic $\hat{\lambda_i}$ as a parameter of sub-grid variability for further use. The sub-grid variability information var($\log_{10}\hat{\lambda_i}$) can be used in further research to perturb the soil hydraulic parameterization in ensemble runs of climate or weather prediction models.

2.4 Global application

The scaling method proposed here is applied to the parameters derived from the whole SoilGrids1km data set. In this study, every terrestrial coarse grid cell is identified with a unique ID, where the SoilGrids1km attribution to the coarse cell was performed within a GIS system. This ensures flexibility to predict parameters for any type of grid, no matter whether it is approximately isotropic, such as in the MetOffice Global Atmosphere 4.0 and Global Land 4.0 model (Walters et al., 2014), or an unstructured mesh of hexagonal/triangular grid cells in the Ocean–Land–Atmosphere Model (OLAM) (Walko and Avissar, 2008). We chose a spatial resolution of 0.25°, as used, for example, in GLDAS-NOAH (Rodell et al., 2004). One coarse grid cell contains exactly $30 \times 30 = 900$ fine-resolution pixels. The number of fine-resolution pixels used for calculating the scaling statistics is also provided with the data set, because lakes and broad rivers reduce the number of relevant pixels. In addition, for global application a land–sea mask has been established to omit irrelevant pixels. The final data set is delivered for latitudes ranging from −60 to 90°, omitting Antarctica.

2.5 Analysis procedure

In this section the procedure is explained concerning how the final data set of hydraulic parameters and scaling information was evaluated. This was done by selecting sample regions for a detailed presentation of the data set performance. Two coarse grid cells of different sub-grid heterogeneity were selected and the scaling results for the WRCs and HCCs compared discussed. The importance of considering sub-grid variability is stressed for different spatial resolutions by means of three larger regions.

2.5.1 Detailed grid cells analysis

In order to investigate the performance of the scaling approach in more detail, two coarse grid cells within Germany were selected based on an initial analysis of the sub-grid sand standard deviation (Fig. 2).

The focus on German sites is motivated by the large variation in soil texture, from a heterogeneous region in the north of Germany to a relatively homogeneous region in the German central lowlands. Moreover, the quantity of soil profile information contribution to the SoilGrids1km neural network approach is quite high in these regions. The first grid cell was selected in the south of Lower Saxony where Pleistocene morainal plains turn into Jurassic and Triassic rocks. This region exhibits small-scale differences in rocks and sediments where the soils developed from. The second region selected

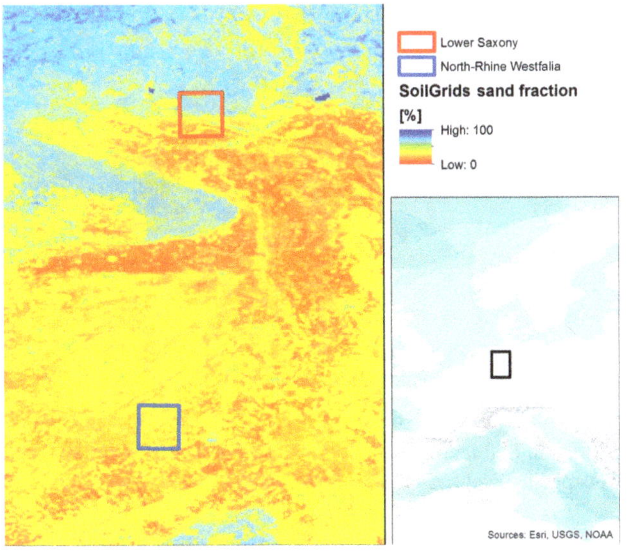

Figure 2. Location of the 0.25° test pixels in Germany, and their sand fraction based on SoilGrids1km.

is located in the southeast of North Rhine-Westphalia where the soil developed from Devonian weathered rocks as well as fluviatile sediments from the Rhine River system. This region can be regarded to be relatively homogeneous in soil texture. See also the soil texture diagrams in Fig. 3.

In order to compare our results to other commonly applied aggregation schemes, we also calculated the WRCs and HCCs (i) by averaging soil texture information and then using the ROSETTA equations; (ii) by averaging MvG parameters directly; (iii) by identifying the dominant USDA soil class for each coarse grid cell and then utilizing class-representative MvG parameters; and (iv) by identifying the dominant USDA soil class for each coarse grid cell and then utilizing the Clapp and Hornberger (1978) approach to calculate the hydraulic properties, which requires dedicated hydraulic parameters. The Clapp and Hornberger (1978) parameterization is based on the Campbell approach (Campbell, 1974) for calculation of water retention and unsaturated hydraulic conductivity. This approach has been added to illustrate the differences between the MvG and Campbell (1974) approach that is still often used in GCM. $S_{\mathrm{e_{Camp}}}$ after Campbell (1974) is given by

$$S_{\mathrm{e_{Camp}}}(h) = \frac{\theta}{\theta_{\mathrm{s}}} \begin{cases} 1 & h \geq h_{\mathrm{B}} \\ \left(\frac{1}{h_{\mathrm{B}}}|\boldsymbol{h}|\right)^{-\frac{1}{b}} & h < h_{\mathrm{B}}, \end{cases} \tag{8}$$

where h_{B} is the air entry value and b is the pore size distribution index (–). The related hydraulic conductivity function for Campbell (1974) is given by

$$K_{\mathrm{r,\,Camp}}(h) = K_{\mathrm{s}} \left(\frac{\theta(h)}{\theta_{\mathrm{s}}}\right)^{(3+2b)}. \tag{9}$$

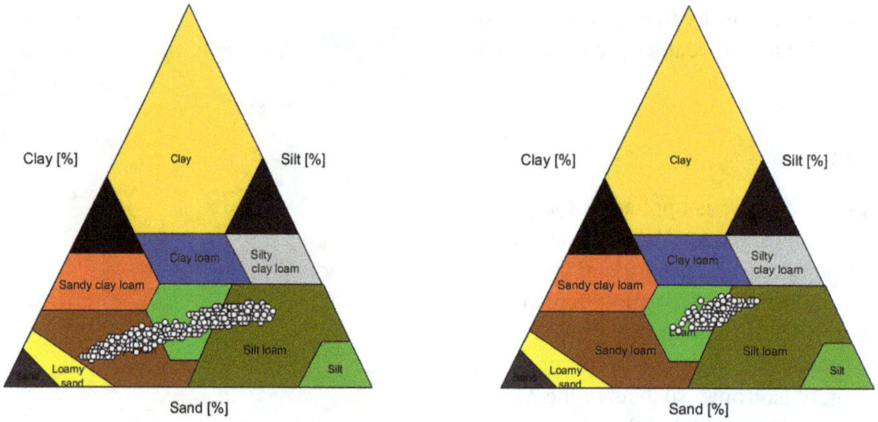

Figure 3. Soil texture triangles, illustrating the difference in soil textural variability of the Lower Saxony pixel (left) and the North Rhine-Westphalia pixel (right), according to USDA classification.

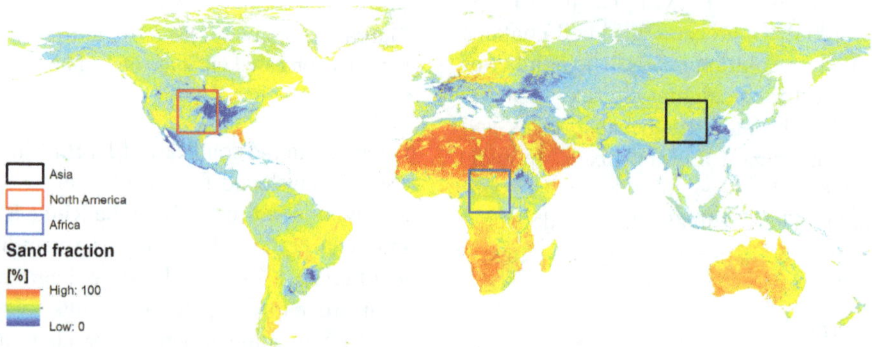

Figure 4. Example regions selected to evaluate the scaling variance loss from Warrick scaling at different spatial resolutions when neglecting the scaling variance. The background shows the sand fraction from SoilGrids1km.

Note that the Campbell approach is similar to the Brooks and Corey (1964) approach – the only difference is that the latter did not set $\theta_r = 0$ in the WRC. Also, Clapp and Hornberger (1978) did not consider θ_r in their equations.

2.5.2 Analysis of scaling variability for different spatial resolutions

In order to stress the importance of considering the subgrid scaling information provided by the proposed method, the mean var($\log_{10}\hat{\lambda}_i$) is quantified for different spatial resolutions. Three regions were analysed in more detail – namely regions in North America, central Africa and China/Mongolia consisting of 2048×2048 fine pixels from the SoilGrids1km database (see Fig. 4). For each of these larger regions a single parameter set of $\hat{\alpha}$ and $\hat{n}$ is estimated by using Eq. (5). For each fine 1 km pixel the scaling parameter $\hat{\lambda}_i$ is calculated according to Eq. (7). Different resolutions of 2, 4, 8, 16, 32, 64, 128, 256, 512 and 1024 km were applied to the resulting map of $\hat{\lambda}_i$. For each spatial resolution the cell-specific scaling parameter is averaged (mean($\log_{10}\hat{\lambda}_i$)). Finally, the variance of these averaged scaling parameters is calculated for the large regions. Herein we hypothesize that the variance of the scaling parameter is a function of spatial resolution.

This analysis uses the original SoilGrids1km spatial resolution as a reference. However, it has to be mentioned that the calculated scaling variability only refers to the information content of the SoilGrid 1 km reference, and does not necessarily provide the real soil scaling variability.

3 Results and discussion

For brevity in this paper the results for the top soil layer are discussed only. After a description of the global data set, this section discusses the results for the two example pixels in Germany and the influence of different spatial resolutions on the variability of three larger regions. This discussion is followed by an analysis of the uncertainties related to the data set.

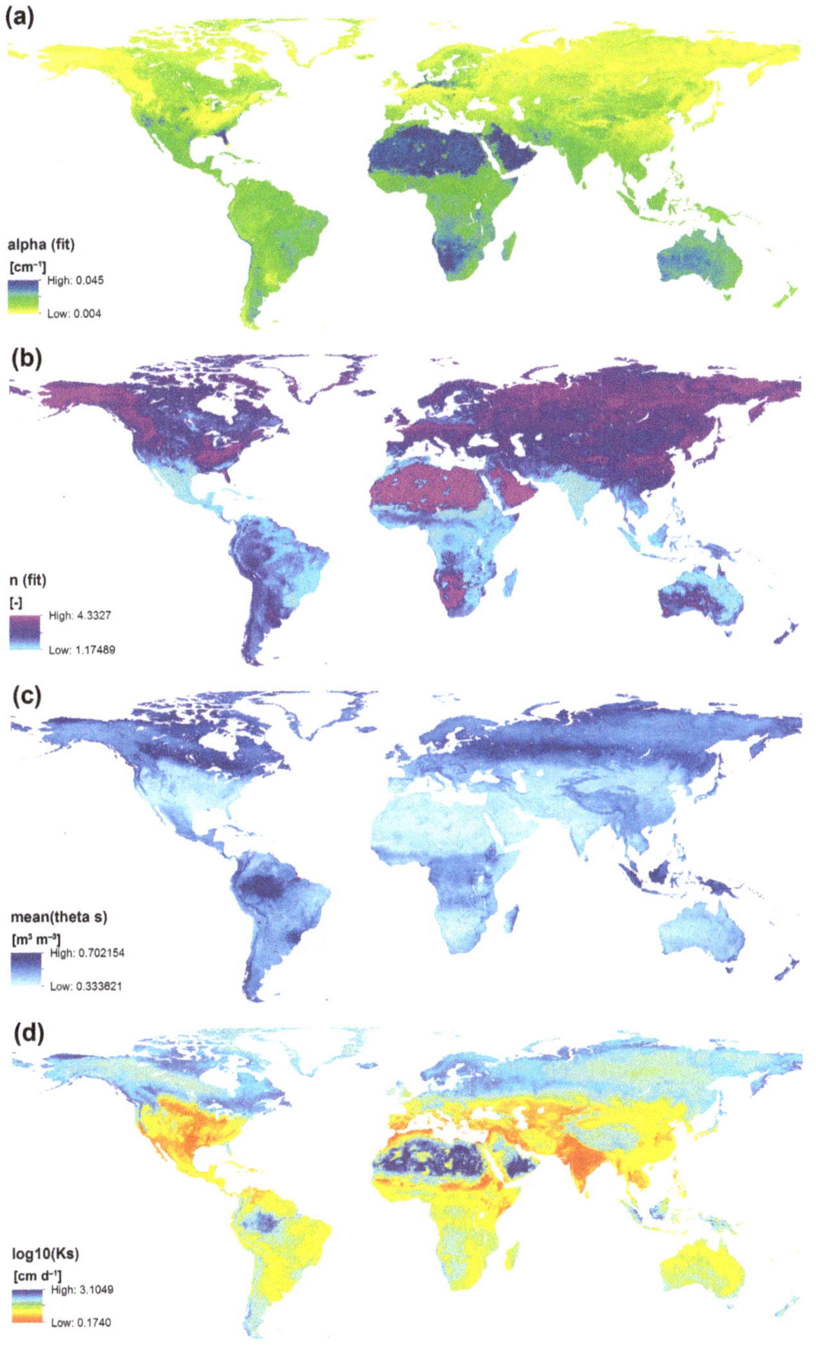

Figure 5. Global map of $\hat{\alpha}$ **(a)**, $\hat{n}$ **(b)**, $\overline{\theta_s}$ **(c)** and mean($\log_{10}(K_s)$) **(d)**, as derived when using the SoilGrids1km data set as input to the Rosetta PTF (Schaap et al., 2001), at 0.25° spatial resolution.

3.1 Global analysis

The resulting global hydraulic parameters $\hat{\alpha}$, $\hat{n}$, $\overline{\theta_s}$ as well as mean($\log_{10} K_s$) are presented in Fig. 5. Parameter $\hat{\alpha}$ ranges between 0.0036 and 0.045 cm^{-1} with a global average of 0.0143 cm^{-1} and shows a clear biogeographical or climatic related distribution. Relatively low values can be found mainly under boreal forests, but also in the North China Plain, central Europe, US Midwest and the Cordoba province in Argentina. Especially the low bulk density of boreal top soils in the SoilGrids1km database, and their widespread occurrence, reduce the global average of $\hat{\alpha}$ to that low value. Relatively high values of $\hat{\alpha}$ were found at locations with high sand fractions, such as the desert regions of the Sahara, the Namib, and the Arabian Peninsula, and to a lesser extent also

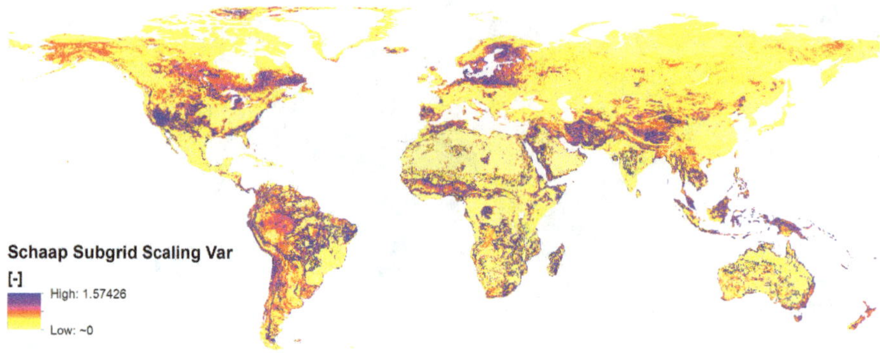

Figure 6. Global map of var($\log_{10}\hat{\lambda}_i$) calculated from SoilGrids1km data set and the Rosetta PTF (Schaap et al., 2001) for 0.25° resolution.

Table 1. Variables stored in the final data set. The variable z indicates the soil depth, i.e. $z \in [0, 5, 15, 30, 60, 100, 200]$ cm.

Variable	Units	Explanation	Variable name
Latitude	Decimal degree	Latitude in degrees north, Southern Hemisphere in negative numbers	Latitude
Longitude	Decimal degree	Longitude in degrees east, west of Greenwich in negative numbers	Longitude
$\hat{\alpha}$	cm^{-1}	Fitted α at z cm depth for MvG parameterization	alpha_fit_zcm
$\hat{n}$	–	Fitted n at z cm depth for MvG parameterization	n_fit_zcm
$\overline{\theta_s}$	$\mathrm{m}^3\,\mathrm{m}^{-3}$	Mean θ_s at z cm depth for MvG parameterization	mean_theta_s_zcm
$\overline{\theta_r}$	$\mathrm{m}^3\,\mathrm{m}^{-3}$	Mean θ_r at z cm depth for MvG parameterization	mean_theta_r_zcm
$\overline{L}$	–	Mean pore connectivity parameter at z cm depth for MvG parameterization	mean_L_zcm
(mean($\log_{10}K_s$))	$\mathrm{cm\,d}^{-1}$	Mean saturated hydraulic conductivity at z cm depth	mean_Ks_zcm
var($\log_{10}\hat{\lambda}_i$)	–	Scaling parameter variance at z cm depth	var_scaling_zcm
Valid subpixels	–	Number of valid subpixels for calculating scaling statistics for the z cm soil depth	valid_subpixels_zcm

in Australia. Smaller regions with high $\hat{\alpha}$ could be traced in Florida and the morainal plains of northern Europe and the Rocky Mountains. The global average of $\hat{n}$ is 1.547, with a range between 1.174 and 4.33 (–). The extreme high $\hat{n}$ values are found only in the non-alluvial regions of the Sahara and Rub' al Khali ("Empty Quarter", Arab Peninsula). The relatively high global average of $\hat{n}$ is caused by the same effect that caused the low $\hat{\alpha}$ average in the low bulk density of boreal top soils. Those soils typically are characterized by high organic carbon contents, behave quite differently in hydraulic sense compared to more mineral-dominated soils, and are rarely used to develop classical PTFs. Therefore, independent of the aggregation or up-scaling approaches, more research is needed to adequately parameterize boreal soils by appropriate PTFs.

The global map of mean saturated hydraulic conductivity (mean($\log_{10}K_s$)) in Fig. 5 ranges between 0.174 and 3.105 with a global average of 1.784 ($\mathrm{cm\,d}^{-1}$). Low soil-saturated hydraulic conductivities are located in India, the Sahel, the Mediterranean, central Asia, the Levant and Iran, Texas, the US prairie regions, California and south-central Canada. Highest mean($\log_{10}K_s$) are found in the Sahara and Rub' al Khali, where sandy soils dominate, but also in the upper Amazon Basin (Marthews et al., 2014) and in cold climates.

Figure 6 shows the global map of sub-grid scaling variance calculated from SoilGrids1km data set with a 0.25° grid reference. Global var($\log_{10}\hat{\lambda}_i$) ranges between ~0 and 1.574, with an average value of 0.229. This shows that large regions, e.g. Siberia, East China, southern coastal provinces of Brazil and Mexico, are relatively homogeneous within individual grid cells. The regions with particularly high sub-grid variability are typically transition zones of soils evolved from young sediments. Examples are the holocene morainal plains in northern Europe and Canada, as well as the slopes of high mountains areas of the Andes and Himalayas. Similarly, young deposits of large rivers such as the Amazon and the inner Congo Basin are characterized by high variability. These are the regions where the consideration of the sub-grid variability may have strong impacts on weather prediction and climate simulations.

The final data set is stored in netcdf format in WGS84 projection and contains the information given in Table 1. The file naming convention follows the scheme "Hydraul_Param_*BasicSoilMap*_*PTF*_slX", where *BasicSoilMap* indicates the soil map used for development of the data set (here SoilGrids1km), *PTF* indicates the pedotransfer function used (here Schaap) and *X* indicates the soil layer (sl; in the case of SoilGrids1km the layers 1–7 are available).

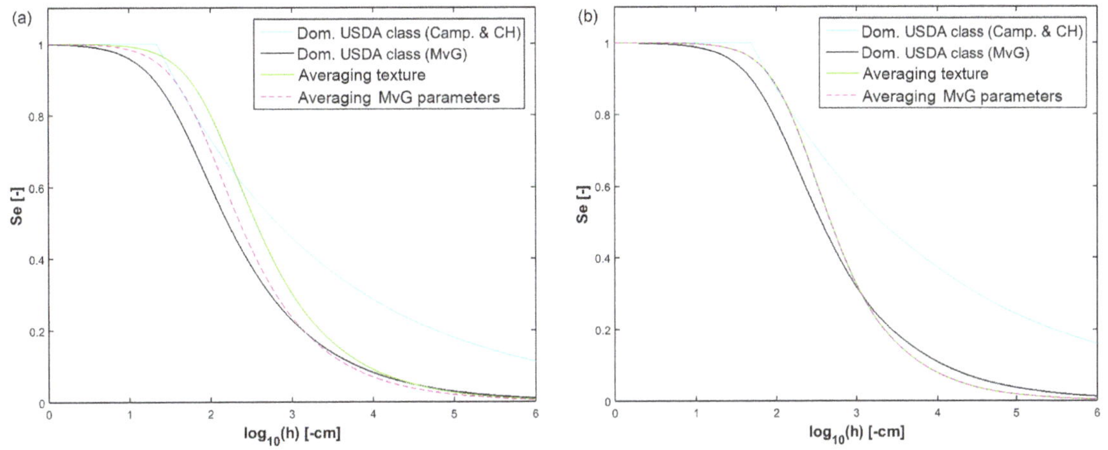

Figure 7. PF curve after applying typical scaling or aggregation methods: by dominant USDA soil class and Clapp and Hornberger parameters for the Brooks and Corey equation, by dominant USDA soil class and MvG equation, by averaging soil texture parameters and then applying ROSETTA PTFs, and by averaging MvG soil hydraulic parameters directly, for Lower Saxony (left) and North Rhine-Westphalia (right).

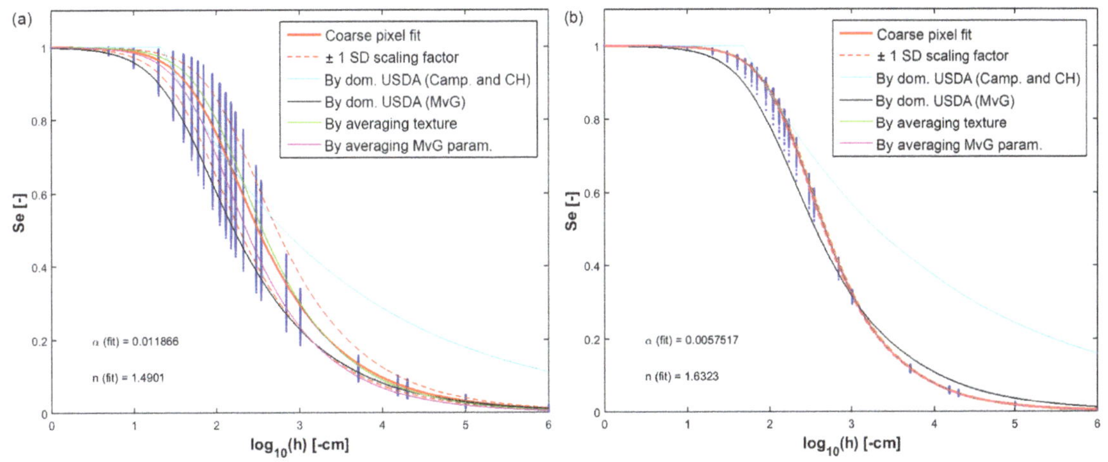

Figure 8. Retention curves for Lower Saxony (left) and North Rhine-Westphalia (right) calculated using different approaches. The coarse pixel fit is the result of the Warrick scaling approach, where $(\log_{10}\hat{\lambda}_i)$ was also estimated. Therefore, ± 1 SD$(\log_{10}\hat{\lambda}_i)$ could also be provided to identify the sub-grid uncertainty. Further retention curves were calculated by dominant USDA class (both for Brooks and Corey with Clapp and Hornberger and for MvG equation), by averaging texture and by averaging MvG parameters. Blue points indicate S_e at standard hydraulic heads for each individual subpixel.

3.2 Example pixels

The detailed analysis of the WRC of two example model grid cells after applying a range of typical aggregation methods is shown in Fig. 7. This includes aggregation by averaging soil texture information, averaging MvG soil hydraulic parameters, and by selecting the dominant USDA soil class (both with MvG and Campbell equations). Averaging soil texture and averaging MvG parameters to a coarser grid caused differences in the WRC for Lower Saxony but showed nearly the same curve for North Rhine-Westphalia. The reasons why textural and MvG parameter averaging yielded the same aggregated WRC in North Rhine-Westphalia are unclear but as Zhu and Mohanty (2002) pointed out, the reliability of MvG averaging greatly depends on the fine-scale (here 1 km) heterogeneity (textural differences), but also on the textural class. Therefore, in the heterogeneous region of Lower Saxony the difference is large which leads to 0.04 higher S_e at 2.5 $\log_{10}(h)$ when averaging texture as opposed to averaging MvG parameters. However, more drastic differences occur when using the dominant USDA soil class to predict coarse-scale WRCs, where the Clapp and Hornberger parameterization for the Campbell model results in wetter conditions for all pressure heads than when using the class-representative MvG parameters and the MvG model.

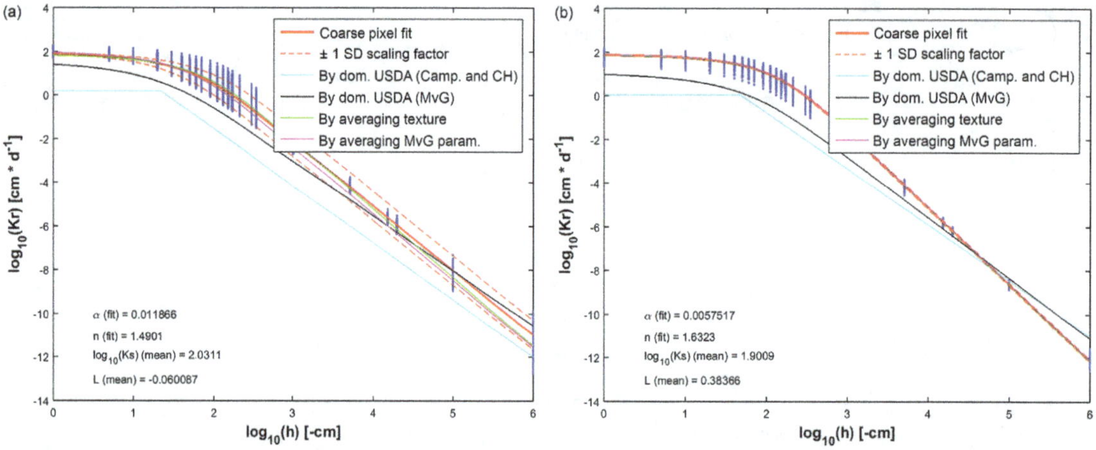

Figure 9. Hydraulic conductivity curves for Lower Saxony (left) and North Rhine-Westphalia (right). The coarse pixel fit is the result of the Warrick scaling approach, where $(\log_{10}\hat{\lambda}_i)$ was also estimated. Therefore, ± 1 SD$(\log_{10}\hat{\lambda}_i)$ could also be provided to identify the sub-grid uncertainty. Further hydraulic conductivity curves were calculated by dominant USDA class (both for Brooks and Corey with Clapp and Hornberger and for MvG equation), by averaging texture and by averaging MvG parameters. Blue points indicate $\log_{10}(K_r)$ at standard hydraulic heads for each individual subpixel.

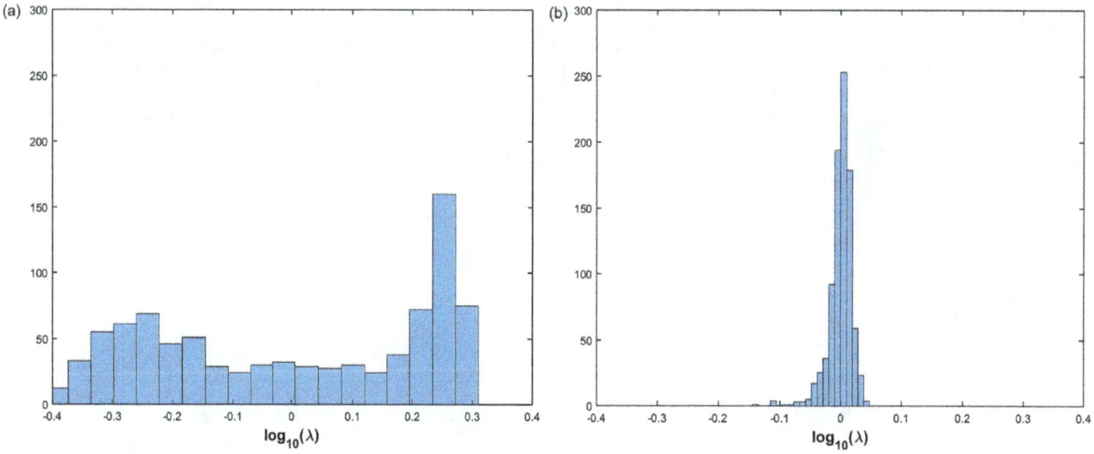

Figure 10. Histograms of the retention scaling parameter $(\log_{10}\hat{\lambda}_i)$ for Lower Saxony (left) and North Rhine-Westphalia (right).

Finally, the WRCs based on the different aggregation approaches (dominant soil class, texture averaging, Warrick scaling) are presented in Fig. 8 for the two example regions. As can be seen, the effective saturation (S_{e_i}) calculated for the 1 km sub-pixels at the standard heads of the head vector $\boldsymbol{h}$ (Eq. 3) (blue dots) reflects the natural variability in soil texture and corresponding soil hydraulic properties of the two example regions. This subscale variability is also captured by the standard deviation of the Warrick scaling approach, which is larger for Lower Saxony as compared to North Rhine-Westphalia. This indicates nicely that Warrick scaling is an appropriate approach to capture fine-scale variability and to propagate the fine-scale uncertainty into the larger scale of interest.

A similar pattern of the different scaling methods can be also found for the HCCs presented in Fig. 9. By this Warrick scaling approach, not only can large-scale modelling capture the right aggregated "mean" WRCs and HCCs but also the uncertainty can be taken into account by running the models for the retention/hydraulic characteristics spanned by the variance of the scaling factor.

Finally, we plotted the histograms of the calculated sub-grid scaling parameters (Fig. 10). Here it has to be noted that the scaling parameter λ was log-transformed with zero mean. Again, the differences in sub-grid heterogeneity of the two example regions become obvious with larger variability in λ for Lower Saxony and lower variability for North Rhine-Westphalia.

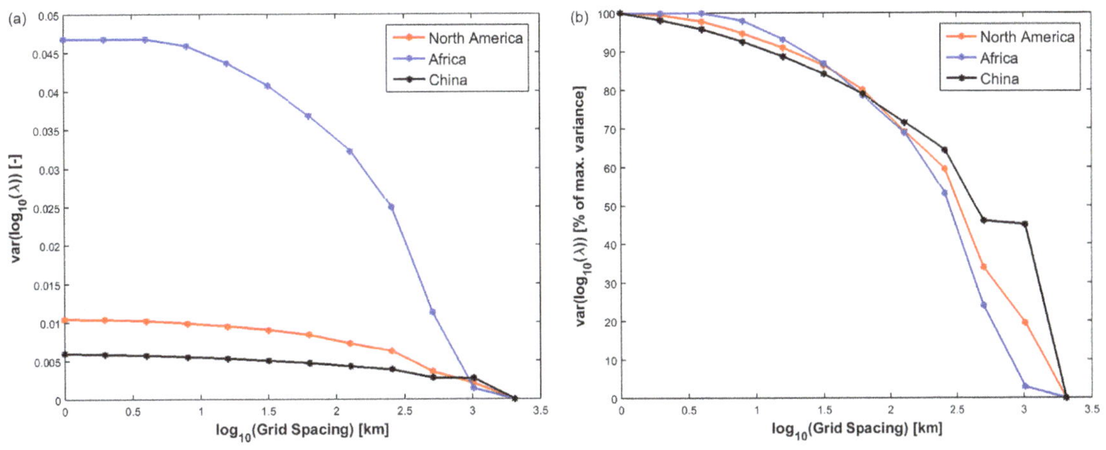

Figure 11. Retention scaling parameter variance var($\log_{10}\hat{\lambda}_i$) for different grid resolutions and regions of interest (North America, Africa and China; see also Fig. 4). (Left) absolute variance and (right) variance normalized as percentage of the maximum variance at 1 km original SoilGrids1km resolution.

3.3 Analysing scaling variability for different spatial resolutions

For the analysis of the scaling variability for different spatial resolutions the mean variance (var($\log_{10}\hat{\lambda}_i$)) was calculated for different grid resolutions and plotted in Fig. 11. For all three case sites (North America, Africa and China; see also Fig. 3) var($\log_{10}\hat{\lambda}_i$) decreased with decreasing spatial resolution; the absolute scaling parameter variability of the Africa region is largest. This can be explained by the fact that the Sahel is prone to strong seasonal dry–wet cycles, which induces large variability in soil development (Da Costa et al., 2015). On the other hand, the relative decrease of variance with coarser resolution for this region is comparable to the other ones (Fig. 11, right panel).

Interestingly, ~90 % of the variance is still maintained at 16 km grid size, ~80% at 64 km, and ~70 % at 128 km. Even at 256 km spatial resolution, more than 50 % of the variability is accounted for, but a diverging trend between the regions is detectable.

3.4 Inherent uncertainties

The use of PTFs and the practice of up-scaling of parameters from 1 km to grid scales where LSMs or related models such as GCMs are applied; both introduce uncertainty which needs to be discussed. Here it is important to mention that the scaling variance var($\log_{10}\hat{\lambda}_i$) calculated with the proposed approach denotes only the spatial sub-grid variance from 1 km resolution of SoilGrids1km towards 0.25° resolution, but this implies no information about the specific uncertainties mentioned above. Moreover, we restricted the analysis to the mean predicted values of SoilGrids1km variables such as texture and bulk density, but the full prediction uncertainty within a confidence interval could be used in an extended approach to estimate the soil hydraulic property uncertainty. Another source of uncertainty is the use of the log-transform for scaling multiplier in Warrick scaling and averaging the saturated hydraulic conductivity using log transform. The applicability of a PFT to global scale may be limited for specific conditions, because the soil database used to estimate the transfer functions is often regionally limited, so that the extrapolation to soils not included in the statistical analysis introduces large uncertainty (e.g. for boreal soils). Finally, we aimed to provide a consistent data set and therefore derived the HCC from WRC rather than calculating them independently. The new data set was developed and all calculations were made under the hypothesis that spatial variance of WRC and HCC scales with spatial resolution.

4 Data availability

The SoilGrids1km data set can be accessed at www.soilgrids.org.

5 Conclusions and outlook

Reliable soil hydraulic parameterization is important for global climate model predictions, including climate reanalyses and weather forecast models. State-of-the-art global soil maps can provide basic soil properties at sub-kilometre scale resolution. The transfer of these data towards coarser-resolution hydraulic properties has been a topic in soil/land surface up-scaling or aggregation research for several decades. In this paper we present a scaling method based on Miller–Miller similarity theory in the relaxed form of Warrick which was applied to the SoilGrids1km data set to provide parameters for the van Genuchten model of water retention curve (WRC) and the Mualem–van Genuchten model of hydraulic conductivity curve (HCC). These curves are required to solve variably saturated flow in soils using Richards' equation. In addition, the sub-grid variability of both WRC and HCC is assessed, which can be of use for model ensemble generation in climate and weather forecast models, or for down-/up-scaling approaches. The final global data set at 0.25° spatial resolution is available at https://doi.org/10.1594/PANGAEA.870605.

The new data set is presented and analysed at the global scale and in more detail by two different individual pixels differing strongly in textural composition. In comparison to aggregation by using dominant USDA soil classes, averaging soil texture and averaging soil hydraulic parameters, the curve fitting approach provides better estimates of coarse-scale water retention and conductivity curves and related parameters. Moreover, the Warrick scaling provides an indicator of sub-grid variability, which is not available from the other methods mentioned above. For three regions different spatial resolutions were analysed according to their ability to represent the soil hydraulic variability of the original SoilGrids database at 1 km resolution. For all regions a common general loss of variability was observed, with losing $\sim 10\,\%$ of the variance at 16 km grid size, $\sim 20\,\%$ at 64 km, and $\sim 30\,\%$ at 128 km. This has large implications for the application of global climate models. Process descriptions which are directly influenced by the hydraulic parameterization such as evaporation and infiltration may lose important information about extreme conditions when applying models at too coarse spatial resolution.

The presented analysis has been conducted on two-dimensional soil maps, without consideration of vertical relationships between soil layers or horizons. This approach can be easily extended towards a three-dimensional scaling that honours the vertical spatial dependency. A follow-up paper will assess the impact of this data set on water and energy fluxes at the soil surface for global simulations. Similarly, the effect of using other PTFs than Schaap et al. (2001) needs to be evaluated on the global scale as well as the uncertainties introduced during pedotransfer on the scaling parameterization. We plan to provide similar data sets for other PTFs, e.g. of Rawls and Brakensiek (1985), Wösten et al. (1999), Weynants et al. (2009) and Vereecken et al. (1989). A similar approach is planned to provide parameters for the Brooks and Corey equation.

Competing interests. The authors declare that they have no conflict of interest.

Acknowledgements. This study was supported by the Helmholtz Alliance on "Remote Sensing and Earth System Dynamics".

Edited by: Giuseppe M. R. Manzella

References

Ahuja, L. R., Naney, J. W., and Nielsen, D. R.: Scaling Soil-Water Properties and Infiltration Modeling, Soil Sci. Soc. A. J., 48, 970–973, 1984.

Ballabio, C., Panagos, P., and Monatanarella, L.: Mapping topsoil physical properties at European scale using the LUCAS database, Geoderma, 261, 110–123, https://doi.org/10.1016/j.geoderma.2015.07.006, 2016.

Balsamo, G., Albergel, C., Beljaars, A., Boussetta, S., Brun, E., Cloke, H., Dee, D., Dutra, E., Muñoz-Sabater, J., Pappenberger, F., de Rosnay, P., Stockdale, T., and Vitart, F.: ERA-Interim/Land: a global land surface reanalysis data set, Hydrol. Earth Syst. Sci., 19, 389–407, https://doi.org/10.5194/hess-19-389-2015, 2015.

Bauer, J., Weihermuller, L., Huisman, J. A., Herbst, M., Graf, A., Sequaris, J. M., and Vereecken, H.: Inverse determination of heterotrophic soil respiration response to temperature and water content under field conditions, Biogeochemistry, 108, 119–134, https://doi.org/10.1007/s10533-011-9583-1, 2012.

Best, M. J., Pryor, M., Clark, D. B., Rooney, G. G., Essery, R. L. H., Ménard, C. B., Edwards, J. M., Hendry, M. A., Porson, A., Gedney, N., Mercado, L. M., Sitch, S., Blyth, E., Boucher, O., Cox, P. M., Grimmond, C. S. B., and Harding, R. J.: The Joint UK Land Environment Simulator (JULES), model description – Part 1: Energy and water fluxes, Geosci. Model Dev., 4, 677–699, https://doi.org/10.5194/gmd-4-677-2011, 2011.

Bouma, J.: Using Soil Survey Data for Quantitative Land Evaluation, in: Advances in Soil Science, edited by: Stewart, B. A., Springer US, New York, NY, 177–213, 1989.

Brooks, R. J. and Corey, A. T.: Hydraulic properties of porous media, Colorado State University Fort Collins, CO, USA, Hydrology Papers, 3, 37 pp., 1964.

Cale, W. G., Oneill, R. V., and Gardner, R. H.: Aggregation Error in Non-Linear Ecological Models, J. Theor. Biol., 100, 539–550, 1983.

Campbell, G. S.: A Simple Method for Determining Unsaturated Conductivity From Moisture Retention Data, Soil Sci., 117, 311–314, 1974.

Clapp, R. B. and Hornberger, G. M.: Empirical Equations for Some Soil Hydraulic-Properties, Water Resour. Res., 14, 601–604, 1978.

Clausnitzer, V., Hopmans, J. W., and Nielsen, D. R.: Simultaneous Scaling of Soil-Water Retention and Hydraulic Conductivity Curves, Water Resour. Res., 28, 19–31, 1992.

Da Costa, P. Y. D., Nguetnkam, J. P., Mvoubou, C. M., Togbe, K. A., Ettien, J. B., and Yao-Kouame, A.: Old landscapes, pre-weathered materials, and pedogenesis in tropical Africa: How can the time factor of soil formation be assessed in these regions?, Quatern. Int, 376, 47–74, 2015.

Dimitrov, M., Vanderborght, J., Kostov, K. G., Jadoon, K. Z., Weihermuller, L., Jackson, T. J., Bindlish, R., Pachepsky, Y., Schwank, M., and Vereecken, H.: Soil Hydraulic Parameters and Surface Soil Moisture of a Tilled Bare Soil Plot Inversely Derived from L-Band Brightness Temperatures, Vadose Zone J., 13, 1, https://doi.org/10.2136/vzj2013.04.0075, 2014.

Ek, M., and Cuenca, R. H.: Variation in soil parameters: Implications for modeling surface fluxes and atmospheric boundary-layer development, Bound.-Lay. Meteorol., 70, 369–383, https://doi.org/10.1007/bf00713776, 1994.

Fallico, C., Tarquis, A. M., De Bartolo, S., and Veltri, M.: Scaling analysis of water retention curves for unsaturated sandy loam soils by using fractal geometry, Eur. J. Soil Sci., 61, 425–436, 2010.

Fang, Z. F., Bogena, H., Kollet, S., and Vereecken, H.: Scale dependent parameterization of soil hydraulic conductivity in 3D simulation of hydrological processes in a forested headwater catchment, J. Hydrol., 536, 365–375, 2016.

Hendrayanto, Kosugi, K., and Mizuyama, T.: Scaling hydraulic properties of forest soils, Hydrol. Process., 14, 521–538, 2000.

Hengl, T., de Jesus, J. M., MacMillan, R. A., Batjes, N. H., and Heuvelink, G. B. M.: SoilGrids1km – Global Soil Information Based on Automated Mapping, PLoS ONE, 9, e114788, https://doi.org/10.1371/journal.pone.0114788, 2014.

Hoffmann, H., Zhao, G., Asseng, S., Bindi, M., Biernath, C., Constantin, J., Coucheney, E., Dechow, R., Doro, L., Eckersten, H., Gaiser, T., Grosz, B., Heinlein, F., Kassie, B. T., Kersebaum, K. C., Klein, C., Kuhnert, M., Lewan, E., Moriondo, M., Nendel, C., Priesack, E., Raynal, H., Roggero, P. P., Rotter, R. P., Siebert, S., Specka, X., Tao, F. L., Teixeira, E., Trombi, G., Wallach, D., Weihermuller, L., Yeluripati, J., and Ewert, F.: Impact of Spatial Soil and Climate Input Data Aggregation on Regional Yield Simulations, PLoS ONE, 11, e0151782, https://doi.org/10.1371/journal.pone.0151782, 2016.

Jadoon, K. Z., Weihermuller, L., Scharnagl, B., Kowalsky, M. B., Bechtold, M., Hubbard, S. S., Vereecken, H., and Lambot, S.: Estimation of Soil Hydraulic Parameters in the Field by Integrated Hydrogeophysical Inversion of Time-Lapse Ground-Penetrating Radar Data, Vadose Zone J., 11, 4, https://doi.org/10.2136/vzj2011.0177, 2012.

Jana, R. B. and Mohanty, B. P.: Enhancing PTFs with remotely sensed data for multi-scale soil water retention estimation, J. Hydrol., 399, 201–211, 2011.

Kuhnert, M., Yeluripati, J., Smith, P., Hoffmann, H., van Oijen, M., Constantin, J., Coucheney, E., Dechow, R., Eckersten, H., Gaiser, T., Grosz, B., Haas, E., Kersebaum, K.-C., Kiese, R., Klatt, S., Lewan, E., Nendel, C., Raynal, H., Sosa, C., Specka, X., Teixeira, E., Wang, E., Weihermüller, L., Zhao, G., Zhao, Z., Ogle, S., and Ewert, F.: Impact analysis of climate data aggregation at different spatial scales on simulated net primary productivity for croplands, Eur. J. Agron., 88, 41–52, https://doi.org/10.1016/j.eja.2016.06.005, 2016.

Liang, X., Lettenmaier, D. P., Wood, E. F., and Burges, S. J.: A Simple Hydrologically Based Model of Land-Surface Water and Energy Fluxes for General-Circulation Models, J. Geophys. Res., 99, 14415–14428, 1994.

Liao, K., Xu, S., Wu, J., and Zhu, Q.: Uncertainty analysis for large-scale prediction of the van Genuchten soil-water retention parameters with pedotransfer functions, Soil Res., 52, 431–442, 2014.

Marquardt, D. W.: An Algorithm for Least-Squares Estimation of Nonlinear Parameters, J. Soc. Ind. Appl. Math., 11, 431–441, 1963.

Marthews, T. R., Quesada, C. A., Galbraith, D. R., Malhi, Y., Mullins, C. E., Hodnett, M. G., and Dharssi, I.: High-resolution hydraulic parameter maps for surface soils in tropical South America, Geosci. Model Dev., 7, 711–723, https://doi.org/10.5194/gmd-7-711-2014, 2014.

Millan, H. and Gonzalez-Posada, M.: Modelling soil water retention scaling. Comparison of a classical fractal model with a piecewise approach, Geoderma, 125, 25–38, 2005.

Miller, E. E. and Miller, R. D.: Physical Theory for Capillary Flow Phenomena, J. Appl. Phys., 27, 324–332, 1956.

Montzka, C., Moradkhani, H., Weihermüller, L., Hendricks Franssen, H.-J., Canty, M., and Vereecken, H.: Hydraulic parameter estimation by remotely-sensed top soil moisture observations with the particle filter, J. Hydrol., 399, 410–421, https://doi.org/10.1016/j.jhydrol.2011.01.020, 2011.

Ngo-Duc, T., Laval, K., Ramillien, G., Polcher, J., and Cazenave, A.: Validation of the land water storage simulated by Organising Carbon and Hydrology in Dynamic Ecosystems (ORCHIDEE) with Gravity Recovery and Climate Experiment (GRACE) data, Water Resources Resour. Res., 43, W04427, https://doi.org/10.1029/2006WR004941, 2007.

Niu, G. Y., Yang, Z. L., Mitchell, K. E., Chen, F., Ek, M. B., Barlage, M., Kumar, A., Manning, K., Niyogi, D., Rosero, E., Tewari, M., and Xia, Y. L.: The community Noah land surface model with multiparameterization options (Noah-MP): 1. Model description and evaluation with local-scale measurements, J. Geophys. Res., 116, D12109, https://doi.org/10.1029/2010JD015139, 2011

Oleson, K. W., Niu, G. Y., Yang, Z. L., Lawrence, D. M., Thornton, P. E., Lawrence, P. J., Stöckli, R., Dickinson, R. E., Bonan, G. B., Levis, S., Dai, A., and Qian, T.: Improvements to the Community Land Model and their impact on the hydrological cycle, J. Geophys. Res., 113, G01021, https://doi.org/10.1029/2007JG000563, 2008.

Pachepsky, Y. A., Shcherbakov, R. A., and Korsunskaya, L. P.: Scaling of Soil-Water Retention Using a Fractal Model, Soil Sci., 159, 99–104, 1995.

Pierce, L. L. and Running, S. W.: The Effects of Aggregating Subgrid Land-Surface Variation on Large-Scale Estimates of Net Primary Production, Landscape Ecol., 10, 239–253, 1995.

Rastetter, E. B., King, A. W., Cosby, B. J., Hornberger, G. M., Oneill, R. V., and Hobbie, J. E.: Aggregating Fine-Scale Ecological Knowledge to Model Coarser-Scale Attributes of Ecosystems, Ecol. Appl., 2, 55–70, 1992.

Rawls, W. J. and Brakensiek, D. L.: Prediction of soil water properties for hydrologic modelling, American Society of Civil Engineers, 293–299, 1985.

Reynolds, C. A., Jackson, T. J., and Rawls, W. J.: Estimating soil water-holding capacities by linking the Food and Agriculture Organization soil map of the world with global pedon databases and continuous pedotransfer functions, Water Resour. Res., 36, 3653–3662, 2000.

Rodell, M., Houser, P. R., Jambor, U., Gottschalck, J., Mitchell, K., Meng, C. J., Arsenault, K., Cosgrove, B., Radakovich, J., Bosilovich, M., Entin, J. K., Walker, J. P., Lohmann, D., and Toll, D.: The global land data assimilation system, B. Am. Meteorol. Soc., 85, 381–394, 2004.

Russo, D. and Bresler, E.: Scaling Soil Hydraulic-Properties of a Heterogeneous Field, Soil Sci. Soc. Am. J., 44, 681–684, 1980.

Schaap, M. G., Leij, F. J., and van Genuchten, M. T.: ROSETTA: a computer program for estimating soil hydraulic parameters with hierarchical pedotransfer functions, J. Hydrol., 251, 163–176, 2001.

Scharnagl, B., Vrugt, J. A., Vereecken, H., and Herbst, M.: Inverse modelling of in situ soil water dynamics: investigating the effect of different prior distributions of the soil hydraulic parameters, Hydrol. Earth Syst. Sci., 15, 3043–3059, https://doi.org/10.5194/hess-15-3043-2011, 2011.

Scheinost, A. C., Sinowski, W., and Auerswald, K.: Regionalization of soil water retention curves in a highly variable soilscape. 1. Developing a new pedotransfer function, Geoderma, 78, 129–143, 1997.

Shu, Q. S., Liu, Z. X., and Si, B. C.: Characterizing Scale- and Location-Dependent Correlation of Water Retention Parameters with Soil Physical Properties Using Wavelet Techniques, J. Environ. Qual., 37, 2284–2292, 2008.

Simmons, C. S., Nielsen, D. R., and Biggar, J. W.: Scaling of Field-Measured Soil-Water Properties, Hilgardia, 47, 77–173, 1979.

Tillotson, P. M. and Nielsen, D. R.: Scale Factors in Soil Science, Soil Sci. Soc. Am. J., 48, 953–959, 1984.

Toth, B., Weynants, M., Nemes, A., Mako, A., Bilas, G., and Toth, G.: New generation of hydraulic pedotransfer functions for Europe, Eur. J. Soil Sci., 66, 226–238, 2015.

Toth, G., Jones, A., and Montanarella, L.: LUCAS Topsoil Survey. Methodology, data and results, Publications Office of the European Union, Luxembourg, 2013.

Tuli, A., Kosugi, K., and Hopmans, J. W.: Simultaneous scaling of soil water retention and unsaturated hydraulic conductivity functions assuming lognormal pore-size distribution, Adv. Water Resour., 24, 677–688, 2001.

van Dam, J. C., Groenendijk, P., Hendriks, R. F. A., and Kroes, J. G.: Advances of modeling water flow in variably saturated soils with SWAP, Vadose Zone J., 7, 640–653, 2008.

Van Engelen, V. and Dijkshoorn, J.: Global and National Soils and Terrain Digital Databases (SOTER), Procedures Manual, version 2.0, Wageningen, The Netherlands, 192 pp., 2012.

van Genuchten, M. T.: A Closed Form Equation for Predicting the Hydraulic Conductivity of Unsaturated Soils, Soil Sci. Soc. Am. J., 44, 892–898, 1980.

Vereecken, H., Maes, J., Feyen, J., and Darius, P.: Estimating the Soil-Moisture Retention Characteristic from Texture, Bulk-Density, and Carbon Content, Soil Sci., 148, 389–403, 1989.

Vereecken, H., Weynants, M., Javaux, M., Pachepsky, Y., Schaap, M. G., and van Genuchten, M. T.: Using Pedotransfer Functions to Estimate the van Genuchten-Mualem Soil Hydraulic Properties: A Review, Vadose Zone J., 9, 795–820, 2010.

Vereecken, H., Schnepf, A., Hopmans, J. W., Javaux, M., Or, D., Roose, T., Vanderborght, J., Young, M. H., Amelung, W., Aitkenhead, M., Allison, S. D., Assouline, S., Baveye, P., Berli, M., Brüggemann, N., Finke, P., Flury, M., Gaiser, T., Govers, G., Ghezzehei, T., Hallett, P., Hendricks Franssen, H. J., Heppell, J., Horn, R., Huisman, J. A., Jacques, D., Jonard, F., Kollet, S., Lafolie, F., Lamorski, K., Leitner, D., McBratney, A., Minasny, B., Montzka, C., Nowak, W., Pachepsky, Y., Padarian, J., Romano, N., Roth, K., Rothfuss, Y., Rowe, E. C., Schwen, A., Šimůnek, J., Tiktak, A., Van Dam, J., van der Zee, S. E. A. T. M., Vogel, H. J., Vrugt, J. A., Wöhling, T., and Young, I. M.: Modeling Soil Processes: Review, Key Challenges, and New Perspectives, Vadose Zone J., 15, 5, https://doi.org/10.2136/vzj2015.09.0131, 2016.

Verhoef, A. and Egea, G.: Modeling plant transpiration under limited soil water: Comparison of different plant and soil hydraulic parameterizations and preliminary implications for their use in land surface models, Agr. Forest Meteorol., 191, 22–32, 2014.

Walko, R. L. and Avissar, R.: The Ocean-Land-Atmosphere Model (OLAM). Part I: Shallow-Water Tests, Mon. Weather Rev., 136, 4033–4044, 2008.

Walters, D. N., Williams, K. D., Boutle, I. A., Bushell, A. C., Edwards, J. M., Field, P. R., Lock, A. P., Morcrette, C. J., Stratton, R. A., Wilkinson, J. M., Willett, M. R., Bellouin, N., Bodas-Salcedo, A., Brooks, M. E., Copsey, D., Earnshaw, P. D., Hardiman, S. C., Harris, C. M., Levine, R. C., MacLachlan, C., Manners, J. C., Martin, G. M., Milton, S. F., Palmer, M. D., Roberts, M. J., Rodríguez, J. M., Tennant, W. J., and Vidale, P. L.: The Met Office Unified Model Global Atmosphere 4.0 and JULES Global Land 4.0 configurations, Geosci. Model Dev., 7, 361–386, https://doi.org/10.5194/gmd-7-361-2014, 2014.

Wang, Z. Y., Shu, Q. S., Liu, Z. X., and Si, B. C.: Scaling analysis of soil water retention parameters and physical properties of a Chinese agricultural soil, Aust. J. Soil Res., 47, 821–827, 2009.

Warrick, A. W., Mullen, G. J., and Nielsen, D. R.: Scaling Field-Measured Soil Hydraulic-Properties Using a Similar Media Concept, Water Resour. Res., 13, 355–362, 1977.

Weynants, M., Vereecken, H., and Javaux, M.: Revisiting Vereecken Pedotransfer Functions: Introducing a Closed-Form Hydraulic Model, Vadose Zone J., 8, 86–95, 2009.

Wösten, J. H. M., Lilly, A., Nemes, A., and Le Bas, C.: Development and use of a database of hydraulic properties of European soils, Geoderma, 90, 169–185, 1999.

Xue, Y. K., Zeng, F. J., and Schlosser, C. A.: SSiB and its sensitivity to soil properties – A case study using HAPEX-Mobilhy data, Global Planet Change, 13, 183–194, 1996.

Zeleke, T. B. and Si, B. C.: Wavelet-based multifractal analysis of field scale variability in soil water retention, Water Resour. Res., 43, W07446, https://doi.org/10.1029/2006WR004957, 2007.

Zhu, J. T. and Mohanty, B. P.: Spatial Averaging of van Genuchten Hydraulic Parameters for Steady-State Flow in Heterogeneous Soils: A Numerical Study, Vadose Zone J., 1, 261–272, 2002.

7

Spatially distributed water-balance and meteorological data from the rain–snow transition, southern Sierra Nevada, California

Roger Bales[1], Erin Stacy[1], Mohammad Safeeq[1], Xiande Meng[1], Matthew Meadows[1,a], Carlos Oroza[2], Martha Conklin[1], Steven Glaser[2], and Joseph Wagenbrenner[3]

[1]Sierra Nevada Research Institute, University of California, Merced, California, USA
[2]Department of Civil and Environmental Engineering, University of California, Berkeley, California, USA
[3]USDA Forest Service, Pacific Southwest Research Station, Arcata, California, USA
[a]now at: Kings River Watershed Association, Fresno, California, USA

Correspondence: Roger Bales (rbales@ucmerced.edu)

Abstract. We strategically placed spatially distributed sensors to provide representative measures of changes in snowpack and subsurface water storage, plus the fluxes affecting these stores, in a set of nested headwater catchments. The high temporal frequency and distributed coverage make the resulting data appropriate for process studies of snow accumulation and melt, infiltration, evapotranspiration, catchment water balance, (bio)geochemistry, and other critical-zone processes. We present 8 years of hourly snow-depth, soil-moisture, and soil-temperature data, as well as 14 years of quarter-hourly streamflow and meteorological data that detail water-balance processes at Providence Creek, the upper part of which is at the current 50 % rain versus snow transition of the southern Sierra Nevada, California. Providence Creek is the long-term study cooperatively run by the Southern Sierra Critical Zone Observatory (SSCZO) and the USDA Forest Service Pacific Southwest Research Station's Kings River Experimental Watersheds (KREW). The 4.6 km^2 montane Providence Creek catchment spans the current lower rain–snow transition elevation of 1500–2100 m. Two meteorological stations bracket the high and low elevations of the catchment, measuring air temperature, relative humidity, solar radiation, precipitation, wind speed and direction, and snow depth, and at the higher station, snow water equivalent. Paired flumes at three subcatchments and a V-notch weir at the integrating catchment measure quarter-hourly streamflow. Measurements of meteorological and streamflow data began in 2002. Between 2008 and 2010, 50 sensor nodes were added to measure distributed snow depth, air temperature, soil temperature, and soil moisture within the top 1 m below the surface. These sensor nodes were installed to capture the lateral differences of aspect and canopy coverage. Data are available at hourly and daily intervals by water year (1 October–30 September) in nonproprietary formats from online data repositories.

1 Introduction

Snowpack and subsurface water storage in the Sierra Nevada support ecosystem health and downstream water supply, along with recreational and aesthetic value, and other water-related services (SNEP, 1996). Two major challenges threatening these benefits are the effects of long-term forest-fire suppression and the effects of climate change. Overstocked montane coniferous forests, the result of a century of fire suppression in this region, are more prone to high-intensity wildfire and less resilient in the face of droughts (Westerling, 2016; Bales et al., 2018). Climate change will stress the balance between precipitation, subsurface water storage, and evapotranspiration, as precipitation shifts from snow to rain and atmospheric water demand increases through longer and warmer growing seasons (Bales et al., 2018). During the 2012–2015 California drought, Sierra Nevada forests experienced extensive mortality due in part to water stress and subsequent insect and fungal pathogens. This unprecedented drought , which had mean precipitation in the southern Sierra Nevada about 50 % of average and was about 1 °C warmer that during the previous decade, provides extraordinary opportunities to enumerate hydrologic mechanisms and drought response (Bales et al., 2018).

Thinning of overgrown forests can both lower the risk of high-intensity wildfire and lower forest drought stress (Stephens et al., 2012). Prior to 2002, there was little information on the hydrologic impacts of these treatments. The Kings River Experimental Watersheds (KREW) project began in part to answer these questions. Three subcatchments in Providence Creek, and a nearby one draining to Duff Creek, were assigned treatments, including mechanical thinning, prescribed burning, a combination of mechanical thinning and prescribed burning, and a control. Nearly 10 years of pretreatment data act as an additional control. At Providence, mechanical thinning was completed in 2011–2012, and prescribed burning occurred in 2015 and 2016.

Another need for the water-balance measurements of snowpack and soil-moisture storage was the lack of information on the variability of these quantities across the landscape on sub-daily timescales. For example, historical records of snowpack at a few select locations, useful as a baseline index, only capture a fraction of the variation in snow depth and snow water equivalent across the mountains (Kerkez et al., 2012; Oroza, 2017). Those historical measurement approaches prove inadequate to support sound decision making in a populous, semi-arid state under a changing climate (Cantor et al., 2018). Distributed sensor nodes that are stratified by elevation, canopy coverage, and aspect can better describe temporal and spatial patterns in the water balance needed by a new generation of forecast tools (Zhang et al., 2017; Zheng et al., 2018). The Southern Sierra Critical Zone Observatory (SSCZO) began in 2007 to quantify these measurements through distributed sensor nodes that are thus stratified. The SSCZO is also a test bed for improving the design, communication and efficacy of spatial-measurement networks (Kerkez et al., 2012; Oroza et al., 2018).

We present hydrometeorological variables in the 14-year KREW dataset for streamflow, snow depth, snow density, air temperature, relative humidity, precipitation, and wind speed and direction. These serve as a basis for additional work in the catchments on sediment, soil and stream chemistry, vegetation composition, and the impacts of treatments. We also present hydrometeorological variables in an 8-year SSCZO dataset for snow depth, soil moisture and temperature, and air temperature and humidity distributed across the landscape.

The Providence Creek catchment is one part of two larger studies. First, KREW established and maintains nested headwater catchments at Providence plus the snow-dominated Bull Creek catchments and a catchment in the adjacent Teakettle Experimental Forest, for assessing the impacts of forest-management treatments on headwater soils and catchment outputs (Hunsaker et al., 2012). Second, the SSCZO program established four focal measurement sites along an elevation transect extending over 400–2700 m elevation (Goulden et al., 2012), of which Providence is one site. Major SSCZO research questions focus on the links between climate, regolith properties, vegetation, biogeochemistry, hydrology, and the response of the mountain ecosystem and catchments to disturbance and climate change. Related studies include evaluation of the transect of eddy covariance and evapotranspiration (Goulden et al., 2012; Goulden and Bales 2014; Saksa et al., 2017; Bales et al., 2018), soil moisture (Oroza et al., 2018), hydrologic modeling (Tague and Peng, 2013; Bart et al., 2016; Son et al., 2016; Bart and Tague, 2017; Jepsen et al., 2016), biochemical studies (Liu et al., 2012; Carey et al., 2016; Aciego et al., 2017; Arvin et al., 2017; Hunsaker and Johnson, 2017), geophysical research (Hahm et al., 2014; Holbrook et al., 2014), and sediment composition (Stacy et al., 2015; McCorkle et al., 2016). Regolith water storage is further described in Klos et al. (2018). The high temporal frequency and distributed coverage make the resulting data appropriate for process studies of snow accumulation and melt, infiltration, evapotranspiration, catchment water balance, (bio)geochemistry, and other critical-zone processes.

2 Site description

The Providence Creek site is located approximately 40 miles northeast of Fresno, California, in the Sierra National Forest. The 4.6 km^2 catchment (P300) has a predominantly southwest aspect, with a moderate slope (19–22 %) and elevations of 1700–2100 m (Table 1). Instruments are installed in three subcatchments (P301, P303, P304; Fig. 1). The site has a Mediterranean climate, with cool, wet winters and dry summers that last from approximately May through October. Precipitation falls as a mix of rain and snow, and precipitation transitions from majority rainfall to majority snow typically

Table 1. Characteristics of three subcatchments and the integrating P300 catchment (Adapted from Safeeq and Hunsaker, 2016).

Site	Catchment	Drainage area, km^2	Average altitude, m	Relief, m	Average aspect, degrees	Average slope, %
Providence	P301	0.99	1979	318	208	19
Providence	P303	1.32	1905	292	233	20
Providence	P304	0.49	1899	213	249	22
Providence	P300	4.61	1883	424	223	21

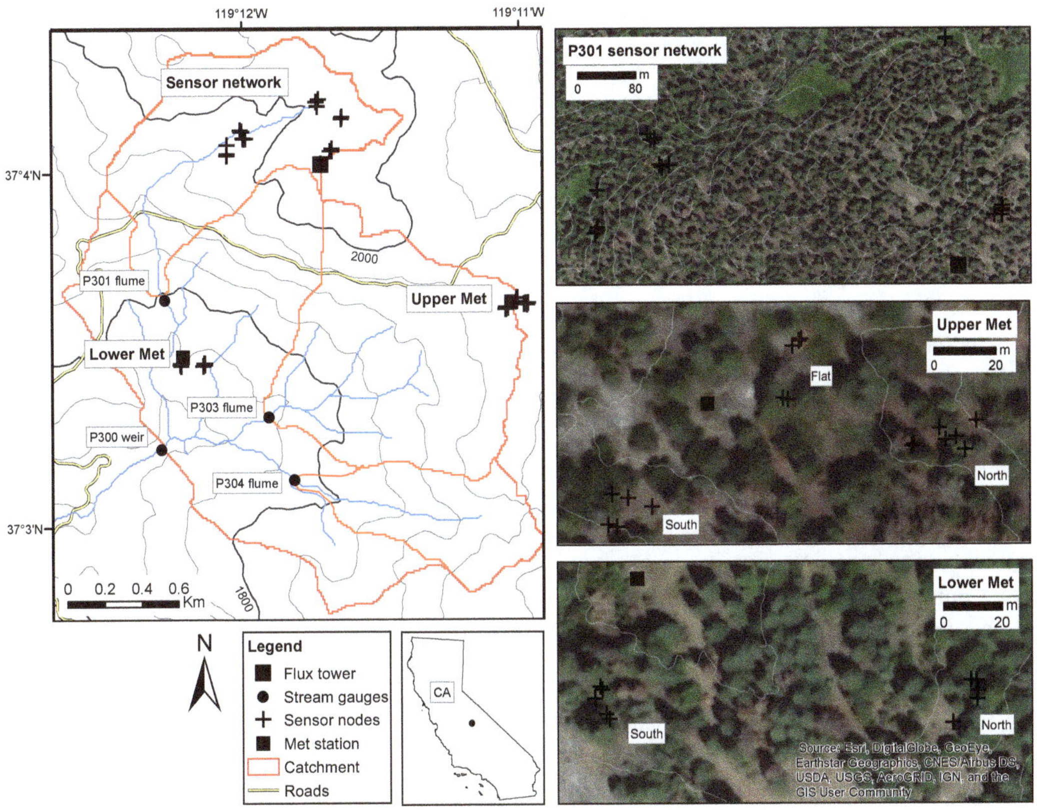

Figure 1. Map of the Providence Creek catchment, showing locations of the meteorological stations, sensor nodes, and stream-gauging stations as well as streams and catchment boundaries. Inset images show instruments and satellite imagery for the meteorological stations and instrument clusters (snow depth, soil moisture, and air temperature) at the P301 network, Upper Met, and Lower Met on 9 July 2016; slope aspect of subsites (north, south, or flat) are indicated by notes on the image. (Topographic data: EDNA filled DEM grid, U.S. Geological Survey, 2003. Satellite data: ESRI world imagery base map compiled from DigitalGlobe and other sources.).

at about 2000 m in elevation (Bales et al., 2011; Safeeq and Hunsaker, 2016). The upper part of the Providence Creek site lies at about the 50 % rain versus snow elevation.

The catchments are underlain by Dinkey Creek granodiorite and Bald Mountain leucogranite (Bateman, 1992). Soil is dominated by the Shaver, Cagwin, and Gerle series (Johnson et al., 2010). Land cover includes small areas of exposed bedrock and meadows within the dominant mature mixed-conifer forest, which primarily consists of white fir (*Abies concolor*), sugar pine (*Pinus lambertiana*), ponderosa pine (*Pinus ponderosa*), Jeffrey pine (*Pinus jeffreyi*), incense cedar (*Calocedrus decurrens*), and California black oak (*Quercus kelloggii*; Dolanc and Hunsaker, 2017).

3 Meteorological data

Meteorological stations were installed at 1975 m (Upper Met) and 1750 m (Lower Met) elevations in the Providence catchment in 2002–2003 (Table 3). Precipitation is measured with a Belfort 5-780 shielded weighing rain gauge (Belfort Instrument, Baltimore, MD, USA); the instrument is mounted 3 m above the ground. A Met One 013 wind-speed

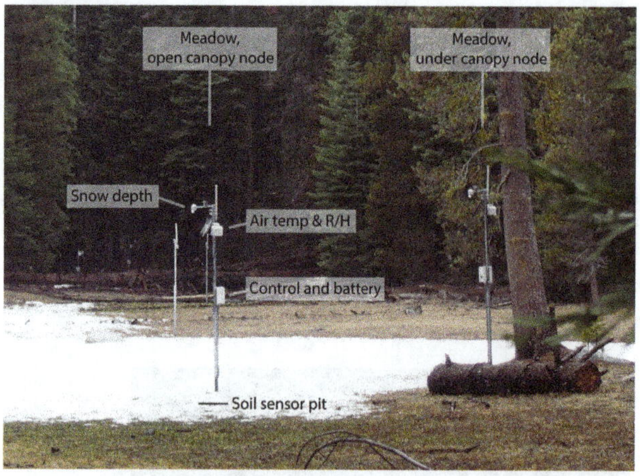

Figure 2. An example of the distributed sensor nodes, with snow-depth and air-temperature and relative-humidity sensors visible. Snow-depth sensors are located above the soil sensor pits. These nodes are in the upper P301 meadow, looking northeast.

sensor and 023 wind-direction sensor (Met One, Grants Pass, OR, USA), a Vaisala HMP45C relative-humidity and air-temperature sensor (Vaisala Corporation, Helsinki, Finland), a Kipp & Zonen CM3 pyranometer (Kipp & Zonen B.V., Delft, the Netherlands), and a Judd acoustic depth sensor (Judd Communications LLC, Salt Lake City, UT, USA) are mounted on a 6 m tower at the site. Snow water equivalent is measured at the Upper Met with a WaterSaver 3 m snow pillow (Snowsaver, Commerce City, CO, USA) with a Sensotec pressure transducer (Honeywell Inc., Columbus, OH, USA) installed approximately 4 m to the north of the weather tower. Data are recorded at each station with a Campbell CR10x (Campbell Inc., Logan, UT, USA) at 15 min intervals.

Manual measurements for instrument verification were made at twice-monthly visits unless delayed by weather. Precipitation at the weighing gauge was verified against measurement records from the snow pillow and nearby weather stations (Table 2; further information about precipitation data assurance is in Safeeq and Hunsaker, 2016).

4 Distributed-sensor clusters

4.1 Upper and Lower Met

Snow depth, soil moisture, and soil temperature are measured at 27 sensor nodes around the Upper Met and Lower Met (Bales et al., 2011; Table 3). Distance to snow or soil surface is measured in the open, at the drip edge, and under canopies with an acoustic depth sensor (Judd Communications LLC, Salt Lake City, UT, USA). Global solar radiation is measured using a LI-COR PY-200 pyranometer (LI-COR Biosciences, Lincoln, NE, USA). Soil volumetric water content and soil temperature are measured using ECHO-TM sensors (METER Group, Pullman, WA, USA) at depths of 10, 30, 60, and 90 cm below the mineral-soil surface under each snow-depth sensor. Matric potential is measured at the same depths with an MPS-1 sensor (METER Group, Pullman, WA, USA). An integrated soil volumetric water content (θ) was calculated to evaluate variation across the environment. Values from soil volumetric water content sensors were used as representative values for the soil depth above and below each sensor. The distance between sensors was evenly divided. This is a greater volume than the estimated measurement volume of the ECHO-TM sensors (approximately 715 mL), but sensor depths were chosen to represent a range of depths while remaining cost-effective. If data from a sensor were missing, depths were adjusted accordingly, with the distance between sensors evenly divided.

$$\theta_{0-100\,\mathrm{cm}}(\%) = (0.2 \cdot \theta_{10\,\mathrm{cm}} + 0.25 \cdot \theta_{30\,\mathrm{cm}} + 0.30 \cdot \theta_{60\,\mathrm{cm}} + 0.25 \cdot \theta_{90\,\mathrm{cm}}) \quad (1)$$

Instrument nodes are sited in clusters at lower Providence south-facing (LowMetS) and north-facing (LowMetN), as well as at upper Providence south-facing (UpMetS), north-facing (UpMetN) and flat aspect (UpMetF). At each cluster, 5–7 sensor nodes were installed according to tree species and canopy coverage (drip edge, under canopy, open canopy) in 2008. Data storage and sensor control are conducted at each of the five sites with a Campbell Scientific CR1000 data logger and an AM16/32B multiplexer (Campbell Scientific, Inc., Logan, UT, USA). Data are recorded at 10 min intervals, with 30 min averages reported.

4.2 P301 sensor network

In summer 2009, 23 nodes in the P301 subcatchment were instrumented with sensors to measure snow depth, air temperature, and relative humidity, as well as soil moisture, temperature, and matric potential (Fig. 2; Table 3). The same sensors are used here as in the Upper and Lower Met clusters (Sect. 4.1). Air temperature and relative humidity are measured with a SHT15DV sensor (EME Systems, Berkeley, CA, USA). Nodes are sited to capture differences in aspect (north vs. south), meadow structure (open meadow, a narrow-meadow channel, transition to forest outside of meadow), and canopy coverage. Data are collected at individual nodes with Metronome Neomote data loggers (Metronome Systems LLC., Berkeley, CA, USA) with a custom sensor wiring board at 15 min intervals. These P301 sensor-network data are available beginning in WY 2010 (1 October 2009). This installation has been the test site for two generations of wireless networking (Kerkez et al., 2012; Oroza et al., 2016, 2018).

5 Streamflow

Stream-discharge monitoring began in 2004 at subcatchments P301, P303, and P304 and in 2006 at integrating catchment P300. Subcatchment discharge is quantified with one

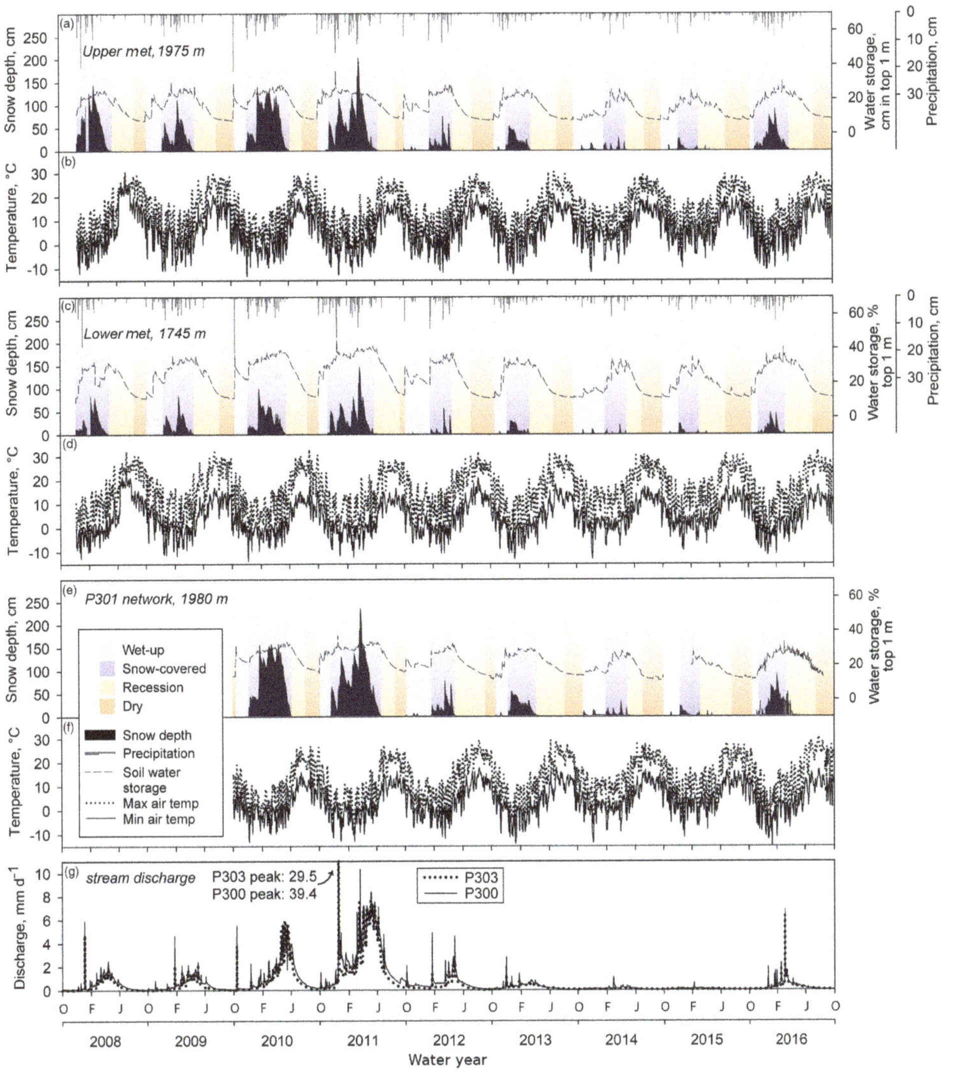

Figure 3. Partial record of **(a, c, e)** precipitation (black bars), soil water storage integrated over 1 m depth (dashed lines), and snow depth (black shaded area), **(b, d, f)** maximum and minimum air temperature (dotted and solid lines) at the Upper and Lower Met stations and P301 sensor network, and **(g)** stream discharge at subcatchment P303 and the integrating P300 catchment. Background colors in **(a, c, e)** generally indicate wet-up, snow-coverage, soil-moisture recession, and dry periods. The 10 and 15 min data have been averaged to daily values shown here.

large (61 cm for P301 and P303; 30.5 cm for P304) and one small (7.6 cm) custom-made fiberglass Parshall flume designed by the FS hydrologist (Moore Sailboats, Watsonville, CA, USA) to capture the range of flows while a 120° V-notch weir is used at P300 (Safeeq and Hunsaker, 2016). An ISCO 730 air bubbler (Teledyne Isco, Lincoln, NE, USA) is the primary stage-measurement device. Backup stage measurements were initially obtained using either an AquaRod capacitance water-level sensor (Advanced Measurements and Controls, Inc., Camano Island, WA, USA) or a Telog pressure transducer (Trimble Water, Inc., Rochester, NY, USA). Levelogger Edge M5 pressure transducers (Solinst Inc., Georgetown, ON, Canada) were installed for backup stage measurement in water year 2011. A Barologger barometer (Solinst Inc., Georgetown, ON, Canada) records barometric pressure for atmospheric corrections to stage. Stage is measured at 15 min intervals and converted to flow using the standard rating curve supplied by the flume and weir manufacturers.

6 Example data

Upper and Lower Met stations receive similar amounts of precipitation but a greater percentage falls as rain at Lower Met. The elevation difference between Upper and Lower Met (225 m) leads to a deeper and more-persistent snowpack at Upper Met (Figs. 3a–d, 4a). Wet-up at the two sites oc-

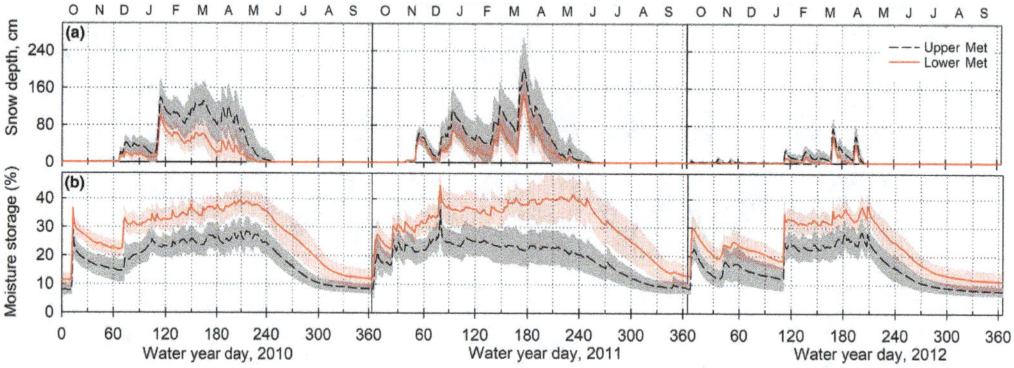

Figure 4. Measures of **(a)** snow depth and **(b)** soil water content to 1 m depth at 27 measurement nodes at Upper and Lower Met sites. Lines represent site means and shading shows ±1 standard deviation. Compared to Fig. 3, this shorter period shows the variation across the landscape for snow depth and soil moisture. Daily values are shown here, calculated from 10 min interval sampling.

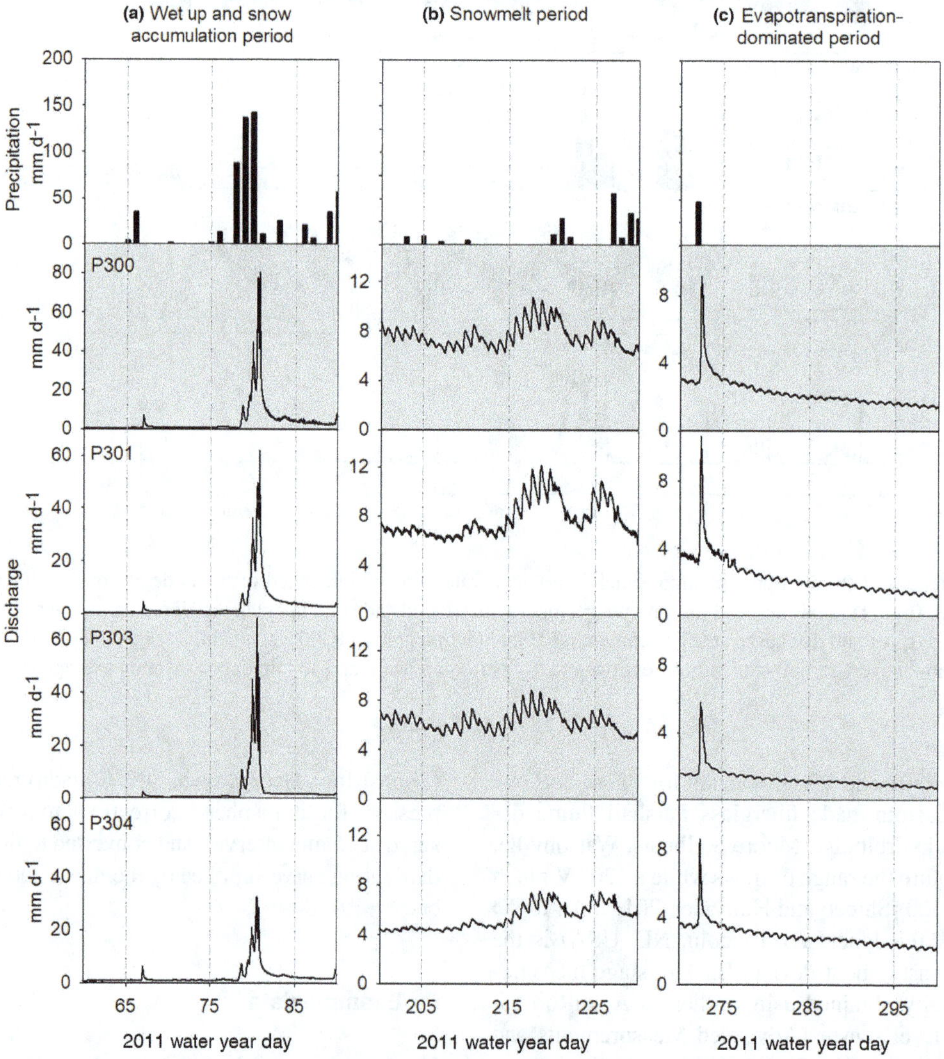

Figure 5. Selected 30-day periods in water year 2011 show hourly patterns in discharge for periods dominated by **(a)** incoming precipitation during an early season storm at the Upper Met station (29 November–29 December), **(b)** snowmelt (18 April–18 May), and **(c)** a midsummer storm during the evapotranspiration-dominant period (27 June–27 July).

Table 2. Nearby stations used for verification of the precipitation data at KREW Providence meteorological stations.

Site name (code)	Operator	Web address	Elevation, m
Tamarack Summit (TMR)	US Bureau of Reclamation	http://cdec.water.ca.gov/selectQuery.html	2301
Huntington Lake (HNT)	US Bureau of Reclamation	http://cdec.water.ca.gov/selectQuery.html	2134
Wishon Dam (WSD)	Pacific Gas and Electric	http://cdec.water.ca.gov/selectQuery.html	1996
KREW NADP (CA28)	USFS PSW Research Station	http://nadp.sws.uiuc.edu/data/sites/siteDetails.aspx?net=NTN&id=CA28	1951

Table 3. Measurement locations and explanation of site coding. Spatial files also available with data downloads.

Code	Description	UTM northing[1], m	UTM easting[1], m	Elevation, m
	Stream gauges			
P300	Integrating stream gauge for Providence Creek (P301, P303, P304)	303 993	4 103 090	1684
P301	Providence Creek stream-gauge station on P301	303 987	4 103 886	1792
P303	Providence Creek stream-gauge station on P303	304 571	4 103 274	1731
P304	Providence Creek stream-gauge station on P304	304 708	4 102 923	1768
	Meteorological stations			
Upper Prov Met	Meteorological station	305 967	4 103 683	1981
Lower Prov Met	Meteorological station	304 197	4 103 392	1753
	Lower Met north – LowMetN: snow, soil, and air sensor clusters (north aspect)			
Open	Open canopy sensor node	304 222	4 103 548	1733
CDde	*Calocedrus decurrens* (incense cedar), drip edge	304 228	4 103 562	1733
CDuc	*C. decurrens*, under canopy	304 230	4 103 562	1733
ACde	*Abies concolor* (white fir), drip edge	304 230	4 103 556	1732
ACuc	*A. concolor*, under canopy	304 230	4 103 559	1732
	Lower Met south – LowMetS: snow, soil, and air sensor clusters (south aspect)			
Open	Open canopy sensor node	304 098	4 103 556	1737
PPde	*Pinus ponderosa* (ponderosa pine), drip edge	304 100	4 103 560	1738
PPuc	*P. ponderosa*, under canopy	304 101	4 103 559	1738
ACde	*C. decurrens*, drip edge	304 102	4 103 551	1737
ACuc	*C. decurrens*, under canopy	304 103	4 103 549	1737
	Upper Met flat – UpMetF: snow, soil, and air sensor clusters, flat aspect			
Open	Open canopy sensor node	305 901	4 103 899	1983
PPde	*P. ponderosa*, drip edge	305 903	4 103 901	1983
PPuc	*P. ponderosa*, under canopy	305 904	4 103 901	1983
ACde	*A. concolor*, drip edge	305 898	4 103 883	1983
ACuc	*A. concolor*, under canopy	305 900	4 103 882	1983
	Upper Met south – UpMetS: snow, soil, and air sensor clusters (south aspect)			
Open	Open canopy sensor node	305 856	4 103 849	1977
QKde	*Quercus kelloggii* (black oak), drip edge	305 848	4 103 852	1977
QKuc	*Q. kelloggii*, under canopy	305 843	4 103 853	1977
ACde	*A. concolor*, drip edge	305 842	4 103 844	1975
ACuc	*A. concolor*, under canopy	305 845	4 103 843	1975
	Upper Met north – UpMetN: snow, soil, and air sensor clusters (north aspect)			
Open	Open canopy sensor node	305 961	4 103 876	1975
ACde	*A. concolor*, drip edge	305 940	4 103 868	1979
ACuc	*A. concolor*, under canopy	305 941	4 103 868	1979
CDde, or xxCDde	*C.decurrens*, drip edge, marked xx after tree removed during thinning	305 955	4 103 871	1977
CDuc, or xxCDuc	*C. decurrens*, under canopy, marked xx after tree removed during thinning	305 958	4 103 867	1978
PLde	*Pinus lambertiana*, drip edge	305 949	4 103 873	1977
PLuc	*P. lambertiana*, under canopy	305 951	4 103 870	1978

Table 3. Continued.

Code	Description	UTM northing[1], m	UTM easting[1], m	Elevation, m
	P301 sensor network – snow, soil, and air sensor clusters in the P301 drainage			
CZO-1	Open canopy near CZT-1	304 902	4 104 671	2014
CZO-2	South drip edge of CZT-1	304 913	4 104 671	2015
CZO-3	South under canopy of CZT-1	304 913	4 104 675	2015
CZO-4	North under canopy of CZT-1	304 913	4 104 679	2015
CZO-5	North drip edge of CZT-1	304 913	4 104 683	2015
CZO-6	West tree drip edge of upper-meadow P301 transect	304 963	4 104 840	1994
CZO-7	Open meadow of upper-meadow P301 transect	304 966	4 104 845	1993
CZO-8	East tree drip edge of upper-meadow P301 transect	304 967	4 104 850	1994
CZO-9	Meadow open canopy at P301 narrow-meadow transect	304 836	4 104 906	1991
CZO-10	South-facing open canopy at narrow-meadow transect	304 841	4 104 941	1997
CZO-26	forest–meadow interface at narrow-meadow transect	304 836	4 104 907	1991
CZO-12	South-facing *Abies concolor* under canopy at narrow-meadow transect	304 830	4 104 929	1995
CZO-29	South-facing *A. concolor* drip edge at narrow-meadow transect	304 827	4 104 929	1995
CZO-14	North-facing open canopy at P301 cedars	304 437	4 104 739	1970
CZO-30	North-facing *Calocedrus decurrens* drip edge at P301 cedars	304 441	4 104 735	1971
CZO-16	North-facing *C. decurrens* under canopy at P301 cedars	304 450	4 104 738	1973
CZO-17	South-facing open canopy at P301 cedars	304 422	4 104 780	1972
CZO-18	South-facing *A. concolor* drip edge at P301 cedars	304 426	4 104 773	1972
CZO-19	South-facing *A. concolor* under canopy at P301 cedars	304 431	4 104 774	1972
CZO-20	North-facing open canopy at P301 lower meadow	304 353	4 104 655	1961
CZO-21	North-facing *A. concolor* drip edge at P301 lower meadow	304 352	4 104 651	1961
CZO-22	North-facing *A. concolor* under canopy at P301 lower meadow	304 350	4 104 648	1961
CZO-25	North-facing open canopy forest–meadow interface at P301 lower meadow	304 352	4 104 705	1960

[1] Geographic coordinates are in Universal Transverse Mercator (UTM) projection, North American 1983 Datum, Zone 11.

curs almost simultaneously, but soil moisture at Lower Met is higher and stays wetter longer due to finer soil texture (Figs. 3a, c, 4). Measurement nodes in the P301 meadow have higher soil moisture than most other points in the network, increasing variability (Fig. 3e).

Stream discharge can peak early in the water year during large fall storms, such as in WY 2010 and 2011 (Fig. 3g). In WY 2011, peak instantaneous flows exceeded 60 mm d^{-1} in subcatchments P303 and P300 (Fig. 5a). While these storms may cause the highest instantaneous flows, the bulk of stream discharge occurs as a result of spring snowmelt (Fig. 5b). In extremely dry years such as WY 2014 or 2015, P300, P303, and P304 remained perennial, but P301 surface flow stopped. After 1 June (WY day 244), soil moisture dries to lows of 10–13 % (Figs. 3a, c, e, 4b) and stream discharge is dominated by daily evapotranspiration periods (Fig. 5c).

7 Data processing

Operating periods for each measurement site were modulated by storm cover, battery life, sensor operation and other factors (Fig. 6). Meteorological data were processed to remove noise, assure data quality, and fill gaps using nearby rain gauges (Safeeq and Hunsaker, 2016). Missing meteorological and stream-discharge data are indicated as described in the metadata files. Filled or estimated values are also flagged in the data files. For the distributed sensor nodes, all levels of data, from raw through processed, were posted on our digital library at https://eng.ucmerced.edu/snsjho/ (last access: 27 September 2018); processing steps are archived there as well as described in the metadata files.

Raw files of sensor-network data are posted as level 0 data and are made publicly available shortly after collection from the field. Further QA/QC occurs on an annual basis. After level 0 data (raw data) have been calibrated, we check and eliminate the duplicate row(s) and insert the missing row(s) based on timestamp and time interval. Outliers are then removed by running an outlier filter based on the range of anticipated values, e.g., −30 to 50 °C for air temperature, 0 to 100 % for relative humidity, and 0 to 1.0 for volumetric water content of soil. Bad data points within a reasonable range of anticipated values were found and deleted by referring to field notes and comparing with the same measurement from the nearby sites. The product of this process is level 1 data. If level 1 data have gap(s), the first step is to compare correlation with nearby measurement points to find one site that has the best correlation (an R^2 that is closest to 1.0). After identifying the most closely correlated point, a regression, typically linear regression, between these two sites is used to estimate values for the gap(s). Short gaps, or gaps in soil temperature, may be filled through linear interpolation. The time period for the correlation may vary based on the measurements and gaps. For example, it is very easy to find a good correlation (R^2 greater than 0.95) for air temperature with a nearby site over an entire year period, but for snow depth, the snow accumulation period and depletion period require sep-

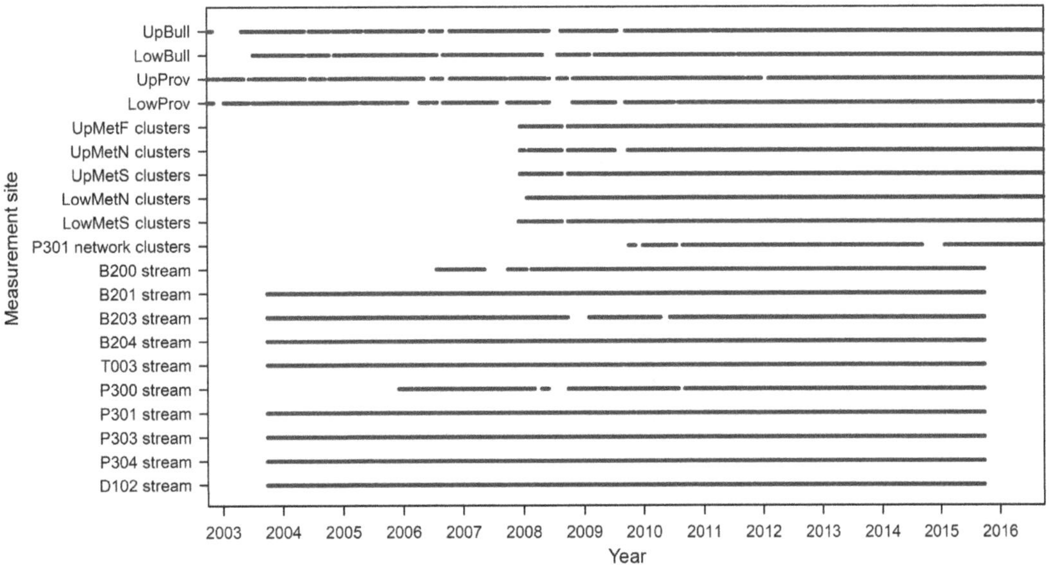

Figure 6. Operating periods for the various measurement sites. Meteorological stations and stream sites show periods when measurements were gathered (compared to periods with estimated data). Operating periods for the distributed clusters are shown where the battery voltage exceeded 11.5 V. The data archive for stream discharge currently ends at the end of WY 2015; however, measurements are continuing and it will be updated in the future.

arate correlations to get the best estimate results. Soil temperature at different depths, especially at deeper depths, will not change significantly during winter, so linear interpolation can be used to fill the gaps for this period; the results are almost the same from correlation as from a regression. Multiple neighboring nodes may be selected if needed, and different neighboring nodes may be used to fill each measurement.

Gap-filled data files have a flag column (code column) following each measurement. The flag values indicate where the measurement value is either (1) from gap filling with linear interpolation; (2) from gap filling with correlation/regression; or empty, indicating the original sensor measurement. There are also data processing notes that have the following information: how many missing points in the measurement, how many missing points were filled by linear interpolation, how many missing points were filled by correlation/regression, what nearby site was used for the regression, start time and end time for the correlation period, and parameter values for the regressions. The parameter values (a, b, R, and R^2) were used to calculate the estimate value with regression: estimate value $= a \times$ measurement from nearby site $+b$; r square is the correlation with the nearby site.

8 Data availability

Meteorological, sensor-network, and stream-discharge data are available through online data repositories. Meteorological data are available from the Forest Service Research Data Archive repository (Hunsaker and Safeeq, 2018; https://doi.org/10.2737/RDS-2018-0028, last access: 17 August 2018). Distributed snow depth, air temperature, and soil moisture and temperature are available through the California Digital Library (see https://doi.org/10.6071/Z7WC73, last access: 31 August 2018). Metadata, including process notes, data headers, and data units, are available from the data repositories. Data in the Upper and Lower Met sensor clusters are coded and sorted by site and aspect; naming codes for all measurement points are presented in Table 3. Spatial data are available in an ESRI ArcMap geodatabase available for download. Stream-discharge data are available from the Forest Service Research Data Archive repository (Hunsaker and Safeeq, 2017; https://doi.org/10.2737/RDS-2017-0037, last access: 20 August 2018). Multiple lidar flights (opentopography.org and National Ecological Observatory Network, NEON) and hyperspectral data (NEON) sets are also available for the site.

9 Summary

An 8- to 14-year meteorological and hydrologic data record is presented for a set of nested catchments in the southern Sierra Nevada. Distributed snow depth and soil temperature and moisture combined with two meteorological stations and a long-term stream-discharge record provide a means of establishing natural variability as well as testing hydrologic process models in a productive montane forest.

Author contributions. RB, MC, and SG designed the sensor networks. MM, ES, XM, and CO installed and maintained the sensor networks and processed the sensor-network data. MS and JW were responsible for the meteorological stations and stream gauges. ES and RB prepared the manuscript, with contributions from all authors.

Competing interests. Steven Glaser is a co-founder and has intellectual property associated with Metronome Systems.

Acknowledgements. We thank the Sierra National Forest, and the staff and research teams of the Southern Sierra Critical Zone Observatory and Kings River Experimental Watersheds. This research was funded in part by the USDA Forest Service Pacific Southwest Research Station; the National Fire Plan of the USDA Forest Service; NSF EAR-0619947, EAR-0725097, EAR-1239521, and EAR-1331939; and the University of California Merced.

Edited by: John Pomeroy

References

Aciego, S. M., Riebe, C. S., Hart, S. C., Blakowski, M. A., Carey, C. J., Aarons, S. M., Dove, N. C., Botthoff, J. K., Sims, K. W., and Aronson, E. L.: Dust outpaces bedrock in nutrient supply to montane forest ecosystems, Nat. Commun., 8, 14800, https://doi.org/10.1038/ncomms14800, 2017.

Arvin, L. J., Riebe, C. S., Aciego, S. M., and Blakowski, M. A.: Global patterns of dust and bedrock nutrient supply to montane ecosystems, Sci. Adv., 3, eeao1588, https://doi.org/10.1126/sciadv.aao1588, 2017.

Bales, R. C., Hopmans, J., O'Geen, A. T., Meadows, M., Hartsough, P. C., Kirchner, P., Hunsaker, C. T., and Beaudette, D.: Soil moisture response to snowmelt and rainfall in a Sierra Nevada mixed-conifer forest, Vadose Zone J., 10, 786–799, https://doi.org/10.2136/vzj2011.0001, 2011.

Bales, R. C., Goulden, M. L., Hunsaker, C. T., Hartsough, P. C., O'Geen, A. T., Hopmans, J., and Safeeq, M.: Mechanisms controlling the impact of multi-year drought on mountain hydrology, Sci. Rep. 8, 690, https://doi.org/10.1038/s41598-017-19007-0, 2018.

Bart, R. L. and Tague, C.: The impact of wildfire on baseflow recession rates in California, Hydrol. Proc., 31, 1662–1673, https://doi.org/10.1002/hyp.11141, 2017.

Bart, R. L., Tague, C. L., and Moritz, M. A.: Effect of tree-to-shrub type conversion in lower montane forests of the Sierra Nevada (USA) on streamflow, PLOS ONE, 11, e0161805, https://doi.org/10.1371/journal.pone.0161805, 2016.

Bateman, P. C.: Plutonism in the central part of the Sierra Nevada batholith, California, U.S. Geol. Survey Professional Paper 1483, 1992.

Cantor, A., Kiparksy, M., Kennedy, R., Hubbard, S., Bales, R. Pecharroman, L. C., Guivetchi, K., McCready, C., and Darling, G.: Data for water decision making: Informing the implementation of California's open and transparent water data act through research and engagement, Center for Law, Energy & the Environment Publications, 56, available at: https://scholarship.law.berkeley.edu/cleepubs/56 (last access: 18 May 2018), 2018.

Carey, C. J., Hart, S. C., Aciego, S. M., Riebe, C. S., Blakowski, M. A., and Aronson, E. L.: Microbial community structure of subalpine snow in the Sierra Nevada, California, Arct. Antarct. Alp. Res., 48, 685–701, https://doi.org/10.1657/AAAR0015-062, 2016.

Dolanc, C. R. and Hunsaker, C. T.: The transition from riparian to upland forest plant communities on headwater streams in the southern Sierra Nevada, California, United States, J. Torrey Bot. Soc., 144, 280–295, https://doi.org/10.3159/TORREY-D-15-00073.1, 2017.

Goulden, M. L. and Bales, R. C.: Vulnerability of montane runoff to increased evapotranspiration with upslope vegetation distribution, Proc. Nat. Acad. Sci. USA, 111, 14071–14075, 2014.

Goulden, M. L., Anderson, R. G., Bales, R. C., Kelly, A. E., Meadows, M., and Winston, G. C.: Evapotranspiration along an elevation gradient in the Sierra Nevada, J. Geophys. Res., 117, G03036, https://doi.org/10.1029/2012JG002027, 2012.

Hahm, W. J., Riebe, C. S., Lukens, C. E., and Araki, S.: Bedrock composition regulates mountain ecosystems and landscape evolution, Proc. Nat. Acad. Sci. USA, 111, 3338–3343, https://doi.org/10.1073/pnas.1315667111, 2014.

Holbrook, W. S., Riebe, C. S., Elwaseif, M., Hayes, J. L., Harry, D. L., Basler-Reeder, K., Malazian, A., Dosseto, A., Hartsough, P. C., and Hopmans, J. W.: Geophysical constraints on deep weathering and water storage potential in the Southern Sierra Critical Zone Observatory, Earth Surf. Proc. Land., 39, 366–380, https://doi.org/10.1002/esp.3502, 2014.

Hunsaker, C. T. and Johnson, D.W.: Concentration-discharge relationships in headwater streams of the Sierra Nevada, California, Water Resour. Res., 53, 7869–7884, https://doi.org/10.1002/2016WR019693, 2017.

Hunsaker, C. T. and Safeeq, M.: Kings River Experimental Watersheds stream discharge, Fort Collins, CO, Forest Service Research Data Archive, https://doi.org/10.2737/RDS-2017-0037, 2017.

Hunsaker, C. T. and Safeeq, M.: Kings River Experimental Watersheds meteorology data. Fort Collins, CO, Forest Service Research Data Archive, https://doi.org/10.2737/RDS-2018-0028, 2018.

Hunsaker, C. T., Whitaker, T. W., and Bales, R. C.: Snowmelt runoff and water yield along elevation and temperature gradients in California's southern Sierra Nevada, J. Am. Water Res. Assoc., 1–12, 667–678, https://doi.org/10.1111/j.1752-1688.2012.00641.x, 2012.

Jepsen, S. M., Harmon, T. C., Meadows, M. W., and Hunsaker, C. T.: Hydrogeologic influence on changes in snowmelt runoff with climate warming: Numerical experiments on a mid-elevation catchment in the Sierra Nevada, USA, Hydrology, 533, 332–342, https://doi.org/10.1016/j.jhydrol.2015.12.010, 2016.

Johnson, D. W., Hunsaker, C. F., Glass, D. W., Rau, B. M., and Roath, B. A.: Carbon and nutrient contents in soils from the Kings River Experimental Watershed, Sierra Nevada Mountains, California, Geoderma, 160, 490–502, https://doi.org/10.1016/j.geoderma.2010.10.019, 2010.

Kerkez, B., Glaser, S. D., Bales, R. C., and Meadows, M. W.: Design and performance of a wireless sensor network for catchment-scale snow and soil moisture measurements, Water Resour. Res., 48, W09515, https://doi.org/10.1029/2011WR011214, 2012.

Klos, P. Z., Goulden, M. L., Riebe, C. S., Tague, C. L., O'Geen, A. T., Flinchum, B. A., Safeeq, M., Conklin, M. H., Hart, S. C., Berhe, A. A., Hartsough, P. C., Holbrook, W. S., and Bales, R. C.: Subsurface plant-accessible water in mountain ecosystems with a Mediterranean climate, WIREs Water, 5, e1277, https://doi.org/10.1002/wat2.1277, 2018.

Liu, F., Hunsaker, C. T., and Bales, R. C.: Controls of streamflow generation in small catchments across the snow-rain transition in the Southern Sierra Nevada, California, Hydrol. Process., 27, 1959, https://doi.org/10.1002/hyp.9304, 2012.

McCorkle, E. P., Berhe, A. A., Hunsaker, C. T., Johnson, D. W., McFarlane, K. J., Fogel, M. L., and Hart, S. C.: Tracing the source of soil organic matter eroded from temperate forest catchments using carbon and nitrogen isotopes, Geology, 445, 172–184, https://doi.org/10.1016/j.chemgeo.2016.04.025, 2016.

Oroza, C. A.: Repeater placement strategy for mesh networks in complex terrain. Chapter 3 in Design and optimization of wireless-sensor networks for real-time monitoring in the Sierra Nevada and Sacramento-San Joaquin Delta, UC Berkeley PhD Thesis, 2017.

Oroza, C. A., Zheng, Z., Glaser, S. D., Tulia, D., and Bales, R. C.: Optimizing embedded sensor network design for catchment-scale snow-depth estimation using LiDAR and machine learning, Water Resour. Res., 52, 8174–8189, https://doi.org/10.1002/2016WR018896, 2016.

Oroza, C. A., Bales, R. C., Stacy, E., Zheng, Z., and Glaser, S. D.: Long-term variability of soil moisture in the southern Sierra: measurement and prediction, Vadose Zone J., 17, 170178, https://doi.org/10.2136/vzj2017.10.0178, 2018.

Safeeq, M. and Hunsaker, C.: Characterizing runoff and water yield for headwater catchments in the southern Sierra Nevada, J. Am. Water Resour. Assoc., 52, 1327–1346, https://doi.org/10.1111/1752-1688.12457, 2016.

Saksa, P., Safeeq, M., and Dymond, S.: Recent Patterns in Climate, Vegetation, and Forest Water Use in California Montane Watersheds, Forests, 8, 278, https://doi.org/10.3390/f8080278, 2017.

SNEP: Sierra Nevada Ecosystem Project, Final Report to Congress, University of California Davis, Wildland Resources Center Report 36, 1996.

Son, K., Tague, C., and Hunsaker C.: Effects of model spatial resolution on ecohydrologic predictions and their sensitivity to inter-annual climate variability, Water, 8, 321, https://doi.org/10.3390/w8080321, 2016.

Stacy, E. M., Hart, S. C., Hunsaker, C. T., Johnson, D. W., and Berhe, A. A.: Soil carbon and nitrogen erosion in forested catchments: implications for erosion-induced terrestrial carbon sequestration, Biogeosciences, 12, 4861–4874, https://doi.org/10.5194/bg-12-4861-2015, 2015.

Stephens, S. L., McIver, J. D., Boerner, R. E. J., Fettig, C. J., Fontaine, J. B., Hartsough, B. R., Kennedy, P., and Schwilk, D. W.: The Effects of Forest Fuel-Reduction Treatments in the United States, BioScience, 62, 549–560, https://doi.org/10.1525/bio.2012.62.6.6, 2012.

Tague, C. and Peng, H.: The sensitivity of forest water use to the timing of precipitation and snowmelt recharge in the California Sierra: Implications for a warming climate, J. Geophys. Res.-Biogeosci., 118, 875–887, https://doi.org/10.1002/jgrg.20073, 2013.

Westerling, A. L. R.: Increasing western US forest wildfire activity: sensitivity to changes in the timing of spring, Phil. Trans. R. Soc. B., 371, 20150178, https://doi.org/10.1098/rstb.2015.0178, 2016.

Zhang, Z., Glaser, S. G., Bales, R. C., Conklin, M., Rice, R., and Marks, D. G.: Technical report: the design and evaluation of a basin-scale wireless sensor network for mountain hydrology, Water Resour. Res., 53, 4487–4498, https://doi.org/10.1002/2016WR019619, 2017.

Zheng, Z., Molotch, N. P., Oroza, C., Conklin, M., and Bales, R. C.: Spatial snow water equivalent estimation for mountainous areas using wireless-sensor networks and remote-sensing products, Remote Sens. Environ., 215, 44–56, https://doi.org/10.1016/j.rse.2018.05.029, 2018.

8

Weekly water quality monitoring data for the River Thames (UK) and its major tributaries

Michael J. Bowes, Linda K. Armstrong, Sarah A. Harman, Heather D. Wickham, David J. E. Nicholls, Peter M. Scarlett, Colin Roberts, Helen P. Jarvie, Gareth H. Old, Emma Gozzard, Nuria Bachiller-Jareno, and Daniel S. Read

Centre for Ecology and Hydrology, Benson Lane, Crowmarsh Gifford, Wallingford, Oxfordshire, OX10 8BB, UK

Correspondence: Michael J. Bowes (mibo@ceh.ac.uk)

Abstract. The River Thames and 15 of its major tributaries have been monitored at weekly intervals since March 2009. Monitored determinands include major nutrient fractions, anions, cations, metals, pH, alkalinity, and chlorophyll *a* and are linked to mean daily river flows at each site. This catchment-wide biogeochemical monitoring platform captures changes in the water quality of the Thames basin during a period of rapid change, related to increasing pressures (due to a rapidly growing human population, increasing water demand and climate change) and improvements in sewage treatment processes and agricultural practices. The platform provides the research community with a valuable data and modelling resource for furthering our understanding of pollution sources and dynamics, as well as interactions between water quality and aquatic ecology. Combining Thames Initiative data with previous (non-continuous) monitoring data sets from many common study sites, dating back to 1997, has shown that there have been major reductions in phosphorus concentrations at most sites, occurring at low river flow, and these are principally due to reduced loadings from sewage treatment works (STWs). This ongoing monitoring programme will provide the vital underpinning environmental data required to best manage this vital drinking water resource, which is key for the sustainability of the city of London and the wider UK economy.

1 Introduction

The River Thames and its tributaries play a vital role in sustaining the ca. 13 million inhabitants of the Thames basin, including the UK capital, London. The river system supplies most of their drinking water and also provides a means of exporting human and industrial wastes, by receiving and transporting wastewater treatment effluents. The rivers also provide important recreational services, being extensively used for fishing, swimming and boating (ranging from canoeing and rowing to residential narrow boats and tourist passenger boats). The pathways alongside the length of the river are also extensively used by walkers, runners and cyclists. The river network therefore plays an important underpinning role to support both the local and UK economy. However, the River Thames basin faces growing pressures from a rapidly increasing population and water usage, which is likely to increase pollution loadings and water stress in future decades. These effects are likely to be exacerbated by future climate change, with predicted lower flows and droughts in the Thames region in the summer months and increased flooding in the winter (Bell et al., 2012; Johnson et al., 2009). To reduce these effects, major mitigation measures are being implemented. Diffuse pollution from agricultural activities is being targeted through schemes such as the Catchment Sensitive Farming initiative, and point source inputs have been reduced through large-scale investment in improved levels of sewage treatment (Kinniburgh et al., 1997), to comply with the Urban Waste Water Treatment Directive (EEC, 1991) and Water Framework Directive (CEC, 2000).

The River Thames is one of the UK's most monitored and studied rivers. Due to its importance as a source of drinking water for London, the lower River Thames has been continuously monitored for nitrate concentration stretching back to 1868, perhaps the longest continuous water quality record in the world (Howden et al., 2010). The corresponding phosphorus record for the River Thames goes back to 1936 (Haygarth et al., 2014; Powers et al., 2016). Environment Agency regulatory monitoring of phosphorus concentrations in the River Thames since the 1970s has been used to identify the major improvements in water quality due in part to the introduction of the Urban Waste Water Treatment Directive (Kinniburgh et al., 1997).

Phosphorus and nitrogen sources and dynamics in the River Thames (Neal et al., 2010a, 2000c) and its tributaries (Bowes et al., 2012c; Jarvie et al., 2006, 2002b; Neal et al., 2000a, 2004, 2006) and how nutrients interact with river ecology (House et al., 2001; Jarvie et al., 2002a; Williams et al., 2000) have been intensively studied in recent decades. These studies were based on 1- to 2-year periods of weekly monitoring by the Centre for Ecology and Hydrology (CEH), based at a small number of individual study sites, carried out between 1997 and 2008. These data are freely available from the CEH Environmental Information Data Centre portal at https://catalogue.ceh.ac.uk/documents/8e23a86b-6b54-4564-9789-23f4b4e045ea (last access: 15 November 2017) (Neal et al., 2012). Extensive biological surveys of phytoplankton (Lack, 1971; Lack and Berrie, 1976), macrophytes (Flynn et al., 2002), macroinvertebrates (Wright et al., 2002), zooplankton (May and Bass, 1998) and fish stocks (Mann et al., 1972) have also been conducted across the catchment since the 1960s.

The above studies were based on monitoring of a limited numbers of sites for 1- to 2-year durations. Therefore, step changes in water quality were often missed, and it was not possible to determine if observed changes in water quality were due to basin-wide conditions (such as weather conditions) or a specific change in catchment land use and management. Therefore, there was a strategic need to conduct continuous water quality monitoring at multiple sites across the Thames catchment, covering a large range of water quality determinands. The resulting monitoring platform was the CEH Thames Initiative. A weekly temporal resolution was adopted, as this was deemed appropriate to capture algal blooms and a selection of high-flow events (Bowes et al., 2009). This was supplemented by hourly phosphorus auto-analyser and nitrate probe data (alongside a range of other water quality data) at individual river sites (Bowes et al., 2015a; Halliday et al., 2015) (data freely available from the CEH Environmental Information Data Centre portal at https://catalogue.ceh.ac.uk/documents/db695881-eabe-416c-b128-76691b2104d8, 12 April 2017) (Bowes et al., 2015b, c). Secondly, the spatial resolution of the monitoring platform needed to cover the impacts of a range of river sizes, land use types and management options. Therefore, the Thames Initiative consists of sites along the length of the freshwater Thames, plus its major tributary subcatchments. The third feature of the Thames Initiative was that it would characterise aquatic ecology (particularly phytoplankton and bacterioplankton communities) at the same weekly frequency as the water chemistry, using flow cytometry (Read et al., 2014). The fourth feature of the Thames Initiative was that it needed to develop into a long-term (decadal) project, to identify water quality impacts in response to land use change, population pressure, sewage treatment works (STWs) improvements and variations in climate.

Since its commencement in March 2009, the Thames Initiative data have been used to quantify nutrient dynamics (Bowes et al., 2015a; Skeffington et al., 2015; Wade et al., 2012), nutrient sources (Bowes et al., 2014), phytoplankton dynamics (Bowes et al., 2012a, 2016; Read et al., 2014) and bacterioplankton biodiversity (Read et al., 2015). The resulting data sets have been extensively used as a modelling resource (Bussi et al., 2017, 2016; Hutchins et al., 2016; Whitehead et al., 2015). Other studies have used the Thames Initiative monitoring sites as a framework for investigations of organic pollutants (Nakada et al., 2017; Singer et al., 2014), heavy metals (Turner et al., 2014), emerging contaminants (Horton et al., 2017) and antimicrobial resistance (Amos et al., 2015).

The objective of this paper is to present an overview of the comprehensive data produced by the Thames Initiative research platform, detailing how samples were taken and analysed, and providing a general description of the spatial and temporal patterns in water quality. We also provide some new basic interpretation of the data, to enable future users of the data to place their studies within this basic framework.

2 Sampling and analytical methodology

2.1 Design of monitoring programme

The River Thames was monitored at weekly intervals at seven points along its length, extending from the upper Thames in the Cotswold Hills (at Hannington Wick, 46 km from the source) to Runnymede (23 km upstream of the tidal limit), covering a monitored reach of 189 km. In the lower reach, just upstream of the town of Slough, some of the water of the River Thames is channelled along the Jubilee River, which is an artificial, engineered channel that is used for flood defence for the towns of Maidenhead and Windsor. One of the seven River Thames monitoring sites is on this Jubilee River, just before it rejoins the River Thames. In addition, 15 of the major tributaries entering this monitored stretch of the River Thames were also monitored. Sites were selected to be easily accessible by road (at road bridges wherever possible,

Table 1. Sampling locations, sampling periods and land use statistics for catchments upstream of sampling sites.

Site code (Fig. 1)	River name/ monitoring location	Monitoring period	Grid reference	Catchment area (km^2)	Distance to source (km)	Base flow index	Percent land cover: Wood-land	Grass-land	Arable	Urban/ semi-urban	STW PE	STW PE density (PE km^{-2})
Cn	River Coln at Whelford	Mar 2009–Feb 2013	SU171991	135	49	0.95	15.2	38.0	42.9	2.6	5440	40
Cl	River Cole at Lynt Bridge	Mar 2009–Feb 2013	SU210980	141	33	0.54	7.2	35.9	42.0	13.1	6620	47
Le	River Leach at Lechlade	Mar 2009–Feb 2013	SU228996	77	34	0.79	9.7	22.8	64.2	2.0	1540	20
Wi	River Windrush at Newbridge	Mar 2009–Feb 2013	SP403014	362	71	0.87	13.1	34.4	45.9	4.8	46 300	128
Ev	River Evenlode at Cassington Mill	Mar 2009–Feb 2013	SP447101	427	62	0.71	14.1	31.5	48.5	4.9	40 100	94
Ch	River Cherwell at Hampton Poyle	Mar 2009–Feb 2013	SP499152	566	74	0.65	9.1	33.3	50.4	6.4	112 270	198
Ra	River Ray at Islip	Mar 2009–Feb 2013	SP527139	290	33	0.57	11.3	40.1	42.6	5.3	46 020	159
Oc	River Ock at Abingdon	Mar 2009–Feb 2013	SU495967	255	36	0.63	8.0	33.0	51.0	7.3	36 780	144
Tm	River Thame at Wheatley	Mar 2009–Feb 2013	SP612050	532	62	0.60	9.7	32.4	35.7	8.3	153 710	289
Pa	River Pang at Tidmarsh	Mar 2009–Feb 2013	SU636747	166	39	0.86	17.6	27.6	46.0	4.1	4990	30
Ke	River Kennet at Woolhampton	Nov 2009–Feb 2013	SU572667	845	79	0.87	13.2	30.0	48.7	4.4	96 380	114
En	River Enborne at Brimpton	Nov 2009–Feb 2013	SU569649	142	29	0.53	24.0	38.0	28.9	6.5	11 110	78
Lo	River Loddon at Charvil	Mar 2009–Feb 2013	SU779766	644	53	0.80	19.6	36.7	23.1	27.3	457 890	711
Wy	River Wye at Bourne End	Mar 2009–Feb 2013	SU895866	134	27	0.93	19.3	35.9	23.6	20.5	82 300	614
Cu	The Cut at Paley Street	Mar 2009–Feb 2013	SU869762	63	21	0.46	20.8	32.7	9.8	35.3	103 600	1644
TH	River Thames at Hannington	Sep 2009–Feb 2013	SU175961	566	51	0.70	10.9	40.5	38.3	8.2	237 810	420
TS	River Thames at Swinford	Mar 2009–Feb 2013	SP442085	1627	96	0.68	10.8	35.1	45.5	6.7	338 300	208
TN	River Thames at Newbridge	Mar 2009–Feb 2013	SP403013	1229	84	0.68	10.0	35.3	45.7	7.2	290 870	237
TW	River Thames at Wallingford	Mar 2009–Feb 2013	SU609902	4213	147	0.65	10.3	35.6	45.1	7.3	1 027 910	244
TSo	River Thames at Sonning	Mar 2009–Feb 2013	SU753758	5788	181	0.68	11.5	34.9	44.3	7.5	1 586 110	274
Ju	Jubilee River at Pococks	Jan 2010–Feb 2013	SU976783	7125	231	0.72	13.2	34.0	40.6	10.3	2 613 941	367
TR	River Thames at Runnymede	Mar 2009–Feb 2013	TQ006723	7170	240	0.72	13.2	34.0	40.4	10.5	2 661 370	371

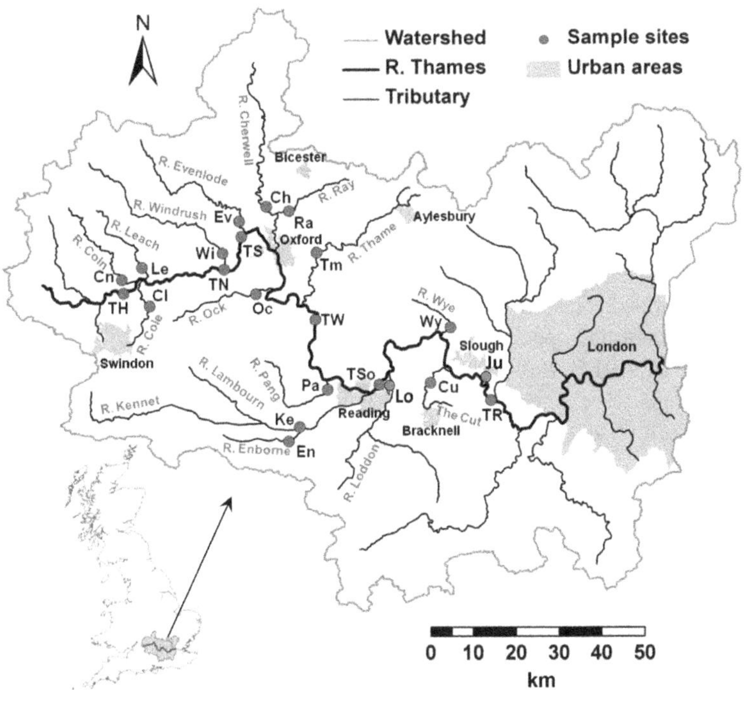

Code	Site name
Tm	Thame at Wheatley
Ra	Ray at Islip
Ch	Cherwell at Hampton Poyle
Ev	Evenlode at Cassington Mill
TS	Thames at Swinford
TN	Thames at Newbridge
Wi	Windrush at Newbridge
Le	Leach at Lechlade (Mill Lane)
Cl	Cole at Lynt Bridge
Cn	Coln at Whelford
Oc	Ock at Abingdon
Pa	Pang at Tidmarsh
Tso	Thames at Sonning
Lo	Lodden at Charvil
Cu	The Cut at Paley Street
TR	Thames at Runnymede
Wy	Wye at Bourne End
TW	Thames at Wallingford
TH	Thames at Hannington Wick
Ke	Kennet at Woolhampton
En	Emborne at Brimpton
Ju	Jubilee River at Pococks Bridge

Figure 1. Location of monitoring locations across the River Thames basin.

to allow sampling from the middle of the river) and close to their confluences with the River Thames. The locations of the sampling sites are given in Table 1 and Fig. 1. Most monitoring sites were located at, or very near to, Environment Agency flow gauging stations. The monitoring programme began at most sites at the beginning of March 2009, with monitoring of the River Thames at Hannington Wick commencing in October 2009, the rivers Kennet and Enborne in November 2009, and the Jubilee River in January 2010. Data are presented up to the end of February 2013, although the monitoring programme is ongoing. Only data from 2009 to 2013 are currently freely available and discussed in this paper, but data generated within the project after 2013 will be made available through the CEH EIDC portal in the near future.

2.2 River sampling

Bulk samples were taken from the main flow of each river, using a plastic bucket on a rope. The bucket was rinsed twice with the local river water prior to sampling, to avoid cross-contamination between sites. Sampling of the sites took place on either Monday or Tuesday of each week. The bulk samples were immediately subsampled into two 500 mL bottles for suspended solids and chlorophyll analysis, an amber-coloured glass bottle (filled to the brim and sealed to minimise degassing) for pH and alkalinity determinations, and into 60 mL bottles for total metals and total phosphorus (TP) analysis. Other subsamples were filtered immediately in the field through a 0.45 µm cellulose nitrate (Whatman WCN grade; Maidstone, UK) membrane filter into 60 mL bottles, for dissolved metals, nutrients, anions and cations analysis. All bottles were acid-washed prior to use. The water temperature of the bulk river water sample was measured in the field using an ATP multi-thermo digital thermometer (ATP Instrumentation Ltd; Ashby-de-la-Zouch, UK). Samples were stored in the dark until returned to the laboratory (within 6 h).

2.3 Analytical methods

On return to the laboratory, all samples were stored in the dark at 4 °C, prior to analysis. The pH was determined using a Radiometer Analytical PHM210 pH meter. The instrument was calibrated prior to use using pH 4, 7 and 10 buffer solutions traceable to National Institute of Standards and Technology (Gaithersburg, USA). Gran alkalinity was determined by acidimetric titration to pH 4 and 3 using 0.5 N H_2SO_4. Suspended solids concentrations were determined by filtering a known volume (approximately 500 mL) of river water through a pre-dried Whatman GF/C filter paper. The filter paper was then re-dried (16 h at 80 °C) and reweighed to determine the mass of solids in the water sample. Chlorophyll concentrations were determined by filtering a known volume of unfiltered river water (approximately 500 mL) through a Whatman GF/C filter paper. The filter paper was then extracted in 10 mL of 90 % v/v acetone/water and refrigerated overnight at 4 °C in the dark. Chlorophyll-*a* concentration was determined colorimetrically using a Beckman 750 DU spectrophotometer, using the method of Marker et al. (1980). Chlorophyll analysis was completed within 48 h of sampling, to avoid errors due to sample stability. Total phosphorus and total dissolved phosphorus (TDP) were determined by digesting an unfiltered and 0.45 µm filtered water sample (respectively) with acidified potassium persulfate in an autoclave at 121 °C for 45 min. Acidified ammonium molybdate reagent was then added to the digested samples to produce a molybdenum–phosphorus complex. This intensely blue-coloured compound was then quantified spectrophotometrically at 880 nm (Eisenreich et al., 1975). Soluble reactive phosphorus (SRP) concentrations were determined on a filtered (0.45 µm WCN-grade cellulose nitrate membrane; Whatman, Maidstone, UK) sample, using the phosphomolybdenum-blue colorimetry method of Murphy and Riley (1962), as modified by Neal et al. (2000b), using a Seal AutoAnalyzer 3 (Seal Analytical; Fareham, UK). SRP samples were analysed within 48 h, to minimise errors associated with sample instability. Prior to 16 August 2010, dissolved reactive silicon concentrations were determined by reaction with acid ammonium molybdate, to form yellow molybdosilicic acids. These were then reduced using an acidified tin (II) chloride solution to form intensely coloured silicomolybdenum blues, which were quantified spectrophotometrically using a Seal AutoAnalyzer 2 (Seal Analytical; Fareham, UK) (Mullin and Riley, 1955). From 16 August 2010 onwards, dissolved reactive silicon was determined by addition of acid ammonium molybdate, to again form yellow-coloured molybdosilicic acids, and then oxalic acid was added (to eliminate phosphate interference) followed by ascorbic acid to reduce the yellow compound to molybdenum blue, which was quantified using a Seal AutoAnalyzer 3 spectrophotometer (Skougstad et al., 1978). Ammonium concentration was determined using an indophenol-blue colorimetric method (Leeks et al., 1997) using a Seal AutoAnalyzer 3. Dissolved organic carbon (DOC) and total dissolved nitrogen were analysed by thermal oxidation using a Thermalox analyser (Analytical Sciences Ltd; Cambridge, UK) until December 2010 and with an Elementar Vario Cube (Elementar Analysensysteme GmbH; Langenselbold, Germany) from June 2011. There are large gaps in the DOC data set between May 2010 and May 2011, due to instrument breakdowns. Major dissolved anion (fluoride, chloride, bromide, nitrite, nitrate and sulfate) concentrations were determined by ion chromatography (Dionex AS50, Thermo Fisher Scientific; Waltham, USA). Total and dissolved cation concentrations were determined on unfiltered and filtered samples respectively, by acidification, followed by analysis by inductively coupled plasma optical emission spectrometry (ICP-OES)(Perkin Elmer Optima 2100; Seer Green, UK).

2.4 Quality control procedures

All chemical analyses (with the exception of suspended solids, chlorophyll and alkalinity) were carried out alongside calibration standards produced by the Wallingford Nutrient Chemistry Laboratories, using externally calibrated pipettes and balances traceable to reference standards. In addition, each batch of samples was run alongside an Aquacheck quality control standard of unknown concentration (LGC Standards, Teddington, UK). The measured concentrations of these standard Aquacheck samples were then provided to LGC Standards, who confirmed the assigned value for the sample. This "blind test" confirmed that the laboratory results were consistently close to the assigned value in terms of z score (also known as standard score) throughout the period from 2009 to 2013, and well within their range of satisfactory results set by LGC ($z =\leq 2$). The Aquacheck scheme is accredited by the United Kingdom Accreditation Service and conforms to ISO/IEC 17043.

Information of the limits of quantification for each determinand within this study, determined by Neal et al. (2012), and information on data accuracy and uncertainty are presented in Table S1 in the Supplement. Occasionally, batches failed the internal quality control procedure (> 10 % from the assigned value). These batches were rerun within a day or two, until the QC standards were met. If QC standards were unable to be met due to instrument problems, and the sample was not able to be analysed before sample degradation became an issue, then the data were omitted from the data set. After analysis, samples were stored in the dark at 4 °C until all data had been checked. Questionable data points/outliers were identified and these samples were reanalysed to determine if the unexpected result was correct.

2.5 River flow data

The water quality data sets are presented alongside the mean daily river flows for the sites on the day of sampling. The mean daily flow data were downloaded from the National River Flow Archive (NRFA, http://nrfa.ceh.ac.uk/, last access: 3 January 2017) in January 2017. Most sites were co-located at, or very close to, Environment Agency flow gauging stations. The exceptions were the River Kennet (flow gauging data taken from Theale, approximately 5 km downstream of the water quality sampling site at Woolhampton) and The Cut (flow gauging data taken from Binfield, which is approximately 6 km upstream of the water quality sampling site of Paley Street). The other exceptions were along the River Thames itself. The Thames water quality monitoring sites at Hannington Wick, Newbridge, Wallingford and Sonning had their mean daily flows estimated by interpolation of adjacent River Thames flow gauging sites data, adjusted based on the catchment area of the water quality sampling site (Bowes et al., 2014).

Table 2. Mean water quality data from March 2009 to February 2013.

Monitoring site	Daily river discharge	Water temperature ($m^3 s^{-1}$)	pH (°C)	Alkalinity ($mEq L^{-1}$)	Suspended solids ($mg L^{-1}$)	Dissolved fluoride ($mg L^{-1}$)	Dissolved chloride ($mg L^{-1}$)	Dissolved bromide ($mg L^{-1}$)	Dissolved sulfate ($mg L^{-1}$-SO_4)
River Coln at Whelford	2.13	11.76	7.99	4248	5.4	0.13	16.7	0.04	33.7
River Cole at Lynt Bridge	1.10	11.72	7.94	4342	15.0	0.19	46.5	0.08	53.4
River Leach at Lechlade	0.66	11.40	7.88	4357	3.0	0.10	16.0	0.04	35.2
River Windrush at Newbridge	9.75	12.17	7.97	4167	10.9	0.15	40.6	0.09	53.6
River Evenlode at Cassington	3.61	11.32	7.91	4027	15.5	0.12	25.7	0.05	45.7
River Cherwell at Hampton Poyle	3.70	11.47	7.91	4133	13.3	0.20	54.2	0.07	65.6
River Ray at Islip	1.94	11.27	7.66	4096	9.6	0.18	63.3	0.11	94.6
River Ock at Abingdon	1.52	11.62	7.99	4704	11.0	0.20	39.2	0.09	72.0
River Thame at Wheatley	3.33	11.54	7.83	4468	14.0	0.22	55.4	0.08	72.0
River Pang at Tidmarsh	0.53	10.82	7.90	4508	8.1	0.13	24.6	0.05	19.3
River Kennet at Woolhampton	8.78	11.27	7.99	4503	9.2	0.12	23.7	0.05	20.0
River Enborne at Brimpton	1.27	10.46	7.75	2815	9.4	0.12	34.7	0.06	26.3
River Loddon at Charvil	5.39	12.15	7.82	3207	7.2	0.12	60.6	0.09	47.8
River Wye at Bourne End	0.84	12.30	8.07	4594	13.2	0.11	42.5	0.06	20.5
The Cut at Paley Street	0.38	12.52	7.58	2450	9.2	0.17	94.7	0.11	99.7
River Thames at Hannington	4.84	11.56	7.88	3957	11.6	0.15	56.3	0.11	67.4
River Thames at Newbridge	9.75	12.17	7.97	4167	10.9	0.15	40.6	0.09	53.6
River Thames at Swinford	13.45	12.07	8.00	4070	11.5	0.14	35.4	0.08	50.2
River Thames at Wallingford	33.87	12.69	8.00	4149	15.3	0.17	46.2	0.08	67.5
River Thames at Sonning	41.33	12.31	7.96	4149	11.4	0.16	40.0	0.07	50.0
Jubilee River at Pococks	57.66	13.12	7.95	4091	8.3	0.15	44.0	0.07	47.6
River Thames at Runnymede	54.11	13.01	7.93	3999	11.7	0.15	46.5	0.08	48.9

2.6 Site characterisation

The characteristics of each sub-catchment, upstream of each sampling point, were determined using GIS (Table 1). Catchment area and distance to river source were determined using the Flood Estimation Handbook web service (https://fehweb.ceh.ac.uk/, last access: 6 February 2017). Land use percentage cover and total upstream STW population estimates (PEs) were determined in ARC GIS using the CEH Intelligent River Network (Dawson et al., 2002) and UK Land Cover Map 2000 (Fuller et al., 2002), using the RACQUEL web application. The PE is an estimated load to a STW based on the typical per capita biological oxygen demand (BOD) load from the population served by the STW, combined with estimated industrial BOD contributions to the works (Keller et al., 2006). Base flow index data were derived from NRFA flow data, obtained from the UK Hydrometric Register (Marsh and Hannaford, 2008).

The monitoring platform encompasses a wide range of river sizes, from short streams of less than 40 km in length with mean flows of less than 1 $m^3 s^{-1}$ (River Wye, Leach and Pang) to the lower reaches of the River Thames, with mean annual flows of ca. 58 $m^3 s^{-1}$ (Marsh and Hannaford, 2008) and a river length of 240 km at Runnymede. The monitoring programme also covers an extremely wide range of population pressures, with the Coln and Leach catchments in the Cotswolds region being very rural, with less than 3 % urban and semi-urban land cover (Fuller et al., 2002). In contrast, other tributaries (The Cut and River Loddon and Wye) have > 20 % urban and semi-urban land cover, as well as very high population densities connected to the wastewater treatment infrastructure (> 500 STW PE km^{-2}). The River Thames itself is impacted by sewage for the entire monitored stretch, from the large towns and cities of Swindon, Oxford, Reading and Slough, alongside hundreds of small towns and village STWs all discharging wastewater along its length.

3 Results and discussion

3.1 General water quality characteristics

Due to the chalk and limestone geology underlying most of the catchment, the River Thames and most of its tributaries have very high Gran alkalinities (in excess of 3000 $mEq L^{-1}$) and high pH (mean values ranging from of 7.66 to 8.07) (Table 2). Most of the river flows are comprised of substantial groundwater inputs, with base flow index values ranging from 0.95 (River Coln) to 0.60 (River Thame) (Table 1). Some of the catchments (such as the rivers Enborne, Ray and Cole) are overlaid with clay deposits, and these rivers have lower base flow indexes of between 0.53 and 0.57, due to increased quantities of run-off. The Cut is an artificial river that was diverted from the upper River Loddon, and it has the lowest base flow index (0.46) and also the lowest mean Gran alkalinity of 2450 $mEq L^{-1}$. Mean suspended solids concentrations across the basin were relatively low (ranging from 3.0 to 15.5 $mg L^{-1}$) (Table 2), due to the low catchment gradients and groundwater dominance of the flow regime. The lowest suspended solids concentrations were observed in the small tributaries of the Cotswolds (rivers Coln and Leach). The highest suspended solids concentrations were

Table 3. Mean nutrient and chlorophyll-*a* concentrations from March 2009 to February 2013.

Monitoring site	Soluble reactive phosphorus ($\mu g\,P\,L^{-1}$)	Total dissolved phosphorus ($\mu g\,P\,L^{-1}$)	Total phosphorus ($\mu g\,P\,L^{-1}$)	Dissolved nitrate ($mg\,NO_3\,L^{-1}$)	Dissolved nitrite ($mg\,NO_2\,L^{-1}$)	Ammonium ($mg\,NH_4\,L^{-1}$)	Total dissolved nitrogen ($mg\,N\,L^{-1}$)	Dissolved organic carbon ($mg\,L^{-1}$)	Dissolved reactive silicon ($mg\,Si\,L^{-1}$)	Chlorophyll *a* ($\mu g\,L^{-1}$)
River Coln at Whelford	61	68	84	26.2	0.06	0.04	6.3	2.0	2.6	3.0
River Cole at Lynt Bridge	228	255	306	18.5	0.06	0.05	4.5	5.6	6.4	5.8
River Leach at Lechlade	20	24	34	31.1	0.09	0.06	7.5	2.3	2.4	2.0
River Windrush at Newbridge	150	169	206	26.7	0.06	0.05	6.5	4.4	3.3	9.6
River Evenlode at Cassington	175	193	253	24.7	0.05	0.04	6.0	3.6	2.7	12.9
River Cherwell at Hampton Poyle	124	142	193	25.2	0.04	0.04	6.1	5.3	3.3	14.0
River Ray at Islip	428	472	513	33.6	0.15	0.11	8.2	8.7	3.3	7.9
River Ock at Abingdon	253	281	320	30.6	0.08	0.06	7.3	5.5	7.1	3.9
River Thame at Wheatley	580	635	712	35.1	0.13	0.24	8.6	7.5	6.5	12.4
River Pang at Tidmarsh	38	49	68	28.1	0.05	0.04	6.7	3.1	7.0	2.8
River Kennet at Woolhampton	34	45	78	24.1	0.05	0.05	5.9	3.2	6.8	8.2
River Enborne at Brimpton	115	138	182	17.1	0.06	0.08	4.2	6.4	6.9	2.5
River Loddon at Charvil	126	150	211	34.4	0.09	0.08	8.4	5.8	5.5	3.9
River Wye at Bourne End	202	234	289	27.5	0.10	0.11	6.8	3.0	6.7	3.7
The Cut at Paley Street	506	585	675	83.7	0.36	0.21	21.3	9.5	6.2	4.5
River Thames at Hannington	181	210	253	32.3	0.08	0.10	7.9	5.3	3.5	3.8
River Thames at Newbridge	150	169	206	26.7	0.06	0.05	6.5	4.4	3.3	9.6
River Thames at Swinford	116	132	171	26.0	0.05	0.05	6.3	4.0	2.9	10.8
River Thames at Wallingford	212	241	301	28.3	0.08	0.08	6.9	5.3	4.2	27.9
River Thames at Sonning	148	172	215	26.7	0.09	0.07	6.5	4.8	5.0	22.8
Jubilee River at Pococks	133	157	192	26.5	0.09	0.07	6.5	5.0	5.2	18.5
River Thames at Runnymede	154	175	222	28.1	0.10	0.09	6.9	5.1	5.1	29.5

found in some of the clay-covered catchments (rivers Thame and Cole) and the steepest catchment: the River Wye. Monitoring points along the length of the River Thames and the River Cherwell also had relatively high suspended solids concentration, but much of this would be due to high phytoplankton biomass through the spring and summer, rather than sediment entrainment/soil erosion.

3.2 Spatial data

All monitoring sites showed significant nutrient enrichment, which reflects the high population densities across much of the catchment, the large number of STWs discharging into the River Thames and its tributaries, and relatively intensive agricultural activity in the region. A total of 18 of the 22 monitoring sites had average SRP concentrations greater than $100\,\mu g\,P\,L^{-1}$, with the rivers Ray, Thame and Cut in excess of $400\,\mu g\,P\,L^{-1}$ (Table 3, Fig. 2). The monitoring sites along the River Thames had relatively consistent levels of phosphorus enrichment (with mean concentrations varying from 116 to $212\,\mu g\,SRP\,L^{-1}$ and 171 to $301\,\mu g\,TP\,L^{-1}$), due to it receiving significant STW inputs from major towns along its entire length. River Thames average phosphorus concentrations were highest in the middle reach (at Wallingford), due to its position downstream of both Oxford STW (Fig. 1) and the inputs from the phosphorus-polluted tributaries of the River Thame and Ray (Fig. 2). Only three catchments had low SRP concentrations which may be potentially limiting for primary production (Bowes et al., 2012b, 2010; McCall et al., 2017): the rivers Leach, Kennet and Pang, which had average SRP concentrations of 20, 34 and $38\,\mu g\,P\,L^{-1}$ respectively (Table 3, Fig. 2). This is probably due to the rivers Pang and Leach having the lowest STW population equivalent densities (30 and $20\,PE\,km^{-2}$ respectively), and, although the River Kennet has a higher STW-PE density ($114\,PE\,km^{-2}$), all STWs along its length have had tertiary phosphorus stripping installed (Bowes et al., 2012c), with most final effluent consents set at $< 1000\,\mu g\,P\,L^{-1}$. However, these three rivers are still highly nutrient impacted, due to their very high nitrate concentrations of between 24 and $31\,mg\,NO_3\,L^{-1}$.

All sites were heavily polluted with nitrate, with 20 of the 22 sites having nitrate concentrations in excess of $20\,mg\,NO_3\,L^{-1}$ (Table 3, Fig. 2). This is mainly due to historic agricultural contamination of the groundwaters which commonly occurs in chalk-dominated catchments (Smith et al., 2010). Due to high catchment porosity, manures and fertilisers are easily transported from the land surface and into the groundwater aquifers. Nitrate concentrations in the lower Thames have increased from ca. $8\,mg\,NO_3\,L^{-1}$ in the 1880s to $20\,mg\,NO_3\,L^{-1}$ by the 1950s (attributed to ploughing up of grassland in the 1940s, which resulted in large-scale mineralisation of organic N), and increased rapidly to ca. $34\,mg\,NO_3\,L^{-1}$, due to increased fertiliser applications and increasing arable land cover in the 1960s (Howden et al.,

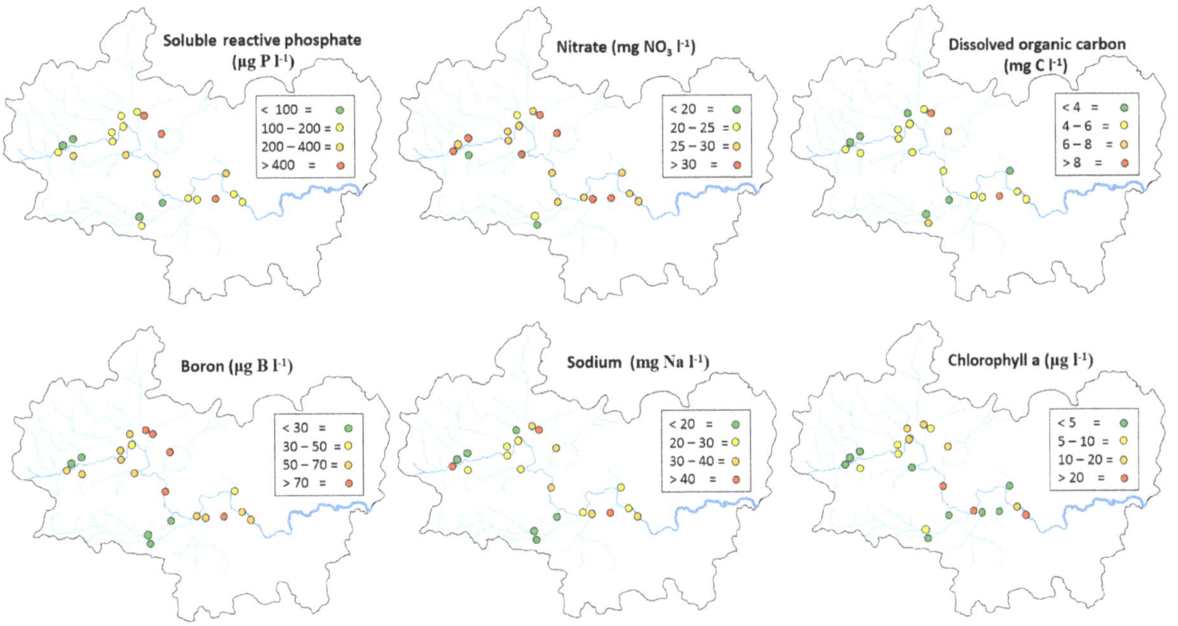

Figure 2. Spatial variation in chemical concentrations and chlorophyll *a* across the Thames basin.

2010). By far the highest average nitrate concentration was observed in The Cut, which, at 83.7 mg NO_3 L^{-1}, was over twice as high as any of the other sites. This gross nitrate pollution is most likely due to the high sewage loading at this site (with the highest STW population density of 1644 PE km^{-2}), as The Cut has the lowest base flow index and therefore the lowest groundwater inputs. The two other sites with the least groundwater inputs (base flow index values less than 0.54; the rivers Cole and Enborne) had the lowest average nitrate concentrations of 18.5 and 17.1 mg NO_3 L^{-1} respectively (Fig. 2). The average ammonium concentrations also reflect sewage inputs, with the five sites with average concentrations ≥ 1.0 mg NH_4^+ L^{-1} (River Thames at Hannington Wick, the rivers Wye, Ray, Thame, and the Cut) all having high STW-PE densities, and many were just downstream of major sewage works.

The average DOC concentrations varied from ≤ 3 mg C L^{-1} for the relatively rural rivers in the Cotswolds (rivers Leach, Coln and Evenlode) and the area west of the town of Reading (rivers Pang and Kennet) to ≥ 8 mg C L^{-1} for the sewage-impacted River Ray and Cut. The clay-dominated sub-catchments of the River Enborne and Thame also had relatively high DOC concentrations (> 6 mg C L^{-1}) and may reflect the higher rates of run-off that possibly wash greater quantities of organic material from the catchment into the river, compared to the other groundwater-dominated monitoring sites. The spatial pattern in the average dissolved boron concentrations also reflected the pattern in sewage inputs, with some of the most sewage-impacted tributaries (rivers Thame, Ray, Cherwell and The Cut) having concentrations in excess of 70 µg B L^{-1} (Table 4, Fig. 2). This is because boron is a constituent of detergents and has been used as a sewage tracer in river research in the past, although concentrations are declining rapidly due to changing detergent formulations (Neal et al., 2010b). The highest average boron concentration in the River Thames was observed in the middle reaches at Wallingford, due to its location downstream of Oxford STW and the confluences with the rivers Ray, Cherwell and Thame, all with significant boron loads. The lowest concentrations were again observed in the rural tributaries of the Cotswolds (rivers Coln and Leach) and the rivers Pang, Enborne and Kennet, west of Reading. The highest manganese, zinc and copper concentrations were observed in The Cut (Table 4), due to it having by far the highest STW population density (Table 1). Dissolved sodium concentration has a strong correlation with STW-PE density (Pearson Correlation Coefficient = 0.785), indicating the element's suitability as a sewage tracer. Its spatial pattern is similar to boron concentration (Fig. 2), with highest concentrations in The Cut, Thame, Cherwell and the upper Thames downstream of Swindon STW, and lowest in the most rural sites.

The spatial pattern in average chlorophyll concentrations was in sharp contrast to the nutrient concentrations and sewage indicators described above. Highest concentrations were observed along the middle and lower reaches on the River Thames, as well as in the larger tributaries such as the Cherwell, Evenlode and Thame. There is a strong positive relationship with the distance from the monitoring point to the river source, which is probably linked with the long transit times required to develop substantial phytoplankton biomass. The spatial and temporal patterns in these chlorophyll con-

Table 4. Mean cation concentrations from March 2009 to February 2013.

Monitoring site	Dissolved sodium ($mg L^{-1}$)	Dissolved potassium ($mg L^{-1}$)	Dissolved calcium ($mg L^{-1}$)	Dissolved magnesium ($mg L^{-1}$)	Dissolved boron ($\mu g L^{-1}$)	Dissolved iron ($\mu g L^{-1}$)	Dissolved manganese ($\mu g L^{-1}$)	Dissolved zinc ($\mu g L^{-1}$)	Dissolved copper ($\mu g L^{-1}$)
River Coln at Whelford	9	1.7	101	5.8	20	8	1.4	1.9	0.5
River Cole at Lynt Bridge	28	5.3	110	4.4	56	48	10.2	3.5	1.9
River Leach at Lechlade	8	1.5	109	5.1	25	12	2.6	1.8	0.6
River Windrush at Newbridge	27	5.5	104	5.2	53	32	6.7	3.3	1.5
River Evenlode at Cassington	16	3.6	102	4.2	51	48	6.6	2.4	1.1
River Cherwell at Hampton Poyle	36	6.2	104	7.6	73	57	6.8	3.3	1.5
River Ray at Islip	49	10.5	112	6.1	107	121	11.9	6.8	2.5
River Ock at Abingdon	25	5.9	127	4.6	62	42	7.5	3.1	1.7
River Thame at Wheatley	39	9.6	118	5.4	87	62	10.0	7.8	4.0
River Pang at Tidmarsh	12	2.9	108	3.2	21	27	2.9	3.1	1.3
River Kennet at Woolhampton	13	2.4	107	2.2	22	19	4.3	3.0	1.0
River Enborne at Brimpton	18	3.6	68	4.4	26	142	19.6	3.6	2.2
River Loddon at Charvil	39	7.5	83	5.3	57	74	17.0	5.9	2.7
River Wye at Bourne End	27	4.3	107	1.9	35	12	3.7	8.2	2.7
The Cut at Paley Street	71	13.6	85	10.1	89	91	13.3	11.0	7.5
River Thames at Hannington	41	8.1	101	5.2	65	45	8.3	5.5	1.6
River Thames at Newbridge	27	5.5	104	5.2	53	32	6.7	3.3	1.5
River Thames at Swinford	23	4.6	102	5.0	47	28	5.5	3.0	1.5
River Thames at Wallingford	30	6.5	109	5.4	77	42	7.1	4.3	4.0
River Thames at Sonning	25	5.2	105	4.5	58	37	6.9	4.2	2.9
Jubilee River at Pococks	28	5.4	102	4.4	54	29	6.1	4.1	2.8
River Thames at Runnymede	30	5.9	101	4.6	61	32	5.4	4.8	2.9

centration data across the Thames catchment are discussed at length in Bowes et al. (2012a).

3.3 Temporal data

The time series plots for the lower River Thames at Runnymede are presented as an example in Figs. 3–5, demonstrating the continuous nature of the data. The data sets from all other sites are equally complete.

3.3.1 Physical data

River water temperature follows a relatively consistent pattern of highest temperatures in the August of each year, and lowest temperatures occur between December and February (Fig. 3). The pH data vary between 7.6 and 8.2, reflecting the alkaline bedrock throughout the catchment. It is important to note that previous high-frequency monitoring of rivers across the Thames catchment have shown marked diurnal pH fluctuations of up to 0.6 pH units (Halliday et al., 2014, 2015), and therefore caution should be exercised when using data from weekly manual samples. The Gran alkalinity was high (mean = 3999 $mEq L^{-1}$), due to the chalk and limestone bedrock. Short-term reductions in Ca and alkalinity (to below 70 $mg Ca L^{-1}$ and 2600 $mEq L^{-1}$) in the Thames at Runnymede coincided with peaks in chlorophyll concentrations in 2009, 2011 and 2012 (Fig. 3). During algal blooms high rates of photosynthesis result in depletion of dissolved carbon dioxide and calcium carbonate precipitation can occur (Hartley et al., 1995; Neal et al., 2002), resulting in reductions in alkalinity and Ca concentration within the water column. Suspended solid concentrations were highest during periods of high flow, due to a combination of soil erosion inputs during the wet winter and spring periods, as well as the resuspension of bed sediment within the river channel. Suspended solid concentration peaks in the lower River Thames also coincided with peaks in chlorophyll concentration, indicating that phytoplankton biomass provided a significant proportion of the suspended solids load during spring–summer algal blooms. Peaks in chlorophyll concentrations occurred between April and June of each year, but the magnitude and duration of chlorophyll peaks varied greatly between years. The timing and magnitude of blooms has been shown to be related to SRP and dissolved silicon concentrations, sunlight duration, river flow and water temperature, and is fully described in Bowes et al. (2016). The largest and sustained chlorophyll peaks are observed in (i) the middle and lower Thames sites and (ii) the longer tributaries (River Cherwell, River Thame), especially those connected to canal systems, indicating the importance of residence time (Bowes et al., 2012a). The river flow data shows a regular pattern, typical of UK rivers, of highest flows over the winter–spring periods (December to February) and lowest flows in July to October. However, the data set also captures a winter drought in 2011–2012, followed by uncharacteristic flooding in summer 2012.

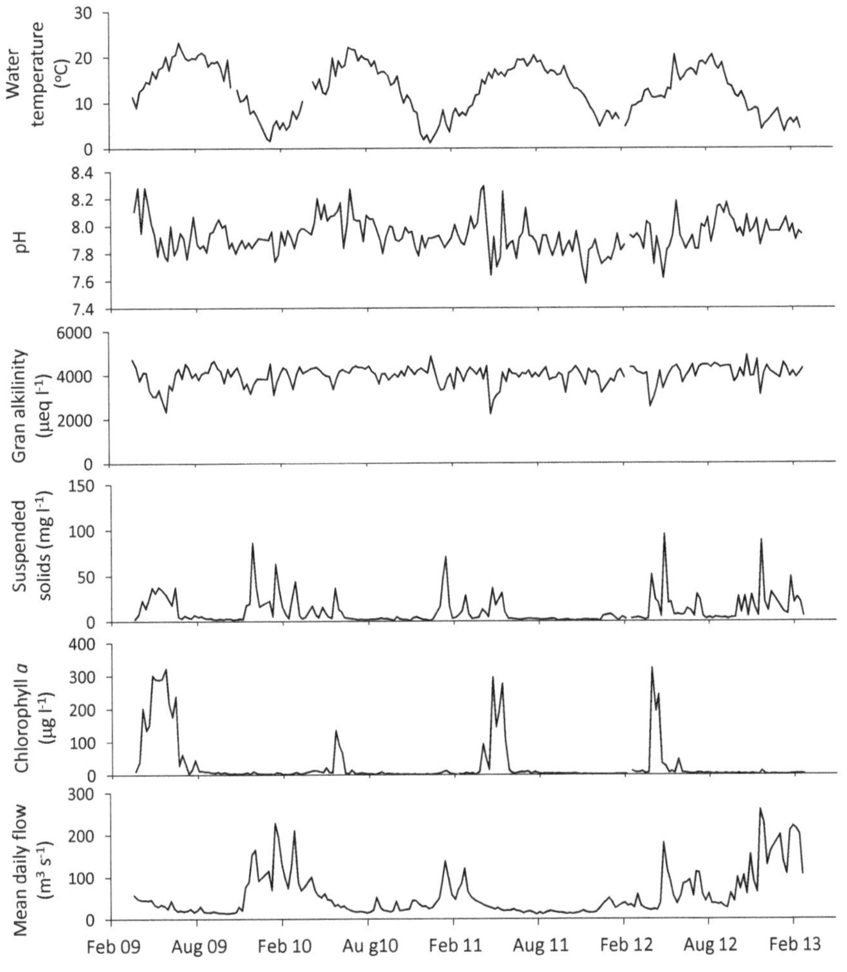

Figure 3. General water quality time series data for the downstream extent of the monitoring platform (River Thames at Runnymede).

3.3.2 Nutrient data

Phosphorus in the River Thames at Runnymede was predominantly in SRP form (Fig. 4), indicating the dominance of sewage effluent inputs to the River Thames and many of its tributaries. However, during periods of high chlorophyll concentrations in 2009, 2011 and 2013, SRP concentrations reduced to below 20 $\mu g\,P\,L^{-1}$ while TP was maintained at ca. 200 $\mu g\,L^{-1}$, indicating that dissolved phosphorus was being sequestered by the rapidly growing phytoplankton biomass (thereby becoming particulate phosphorus). The two highest peaks in TP concentration (30 April and 26 November 2012; Fig. 4) correspond with the two highest suspended solids concentrations (Fig. 3), occurring during the major storm that ended the 2011–2012 winter drought and the highest recorded river flow respectively. This indicates that there were major inputs of particulate-bound phosphorus from the catchment during these storm events, particularly following dry antecedent conditions. Nitrate concentrations remained high throughout the monitoring period, as a result of the gross pollution of groundwater aquifers typical of English chalk catchments (Smith et al., 2010). Small decreases in nitrate concentration coincided with flow peaks, indicating that these high nitrate inputs from groundwater were being diluted by rainwater/surface run-off inputs to the River Thames. Other more-sustained declines in nitrate concentrations coincided with periods of high chlorophyll concentrations and SRP depletion in spring of 2009 and 2011, due to N uptake by phytoplankton. Dissolved reactive silicon also depleted to below 1 $mg\,L^{-1}$ during these phytoplankton blooms, indicating that diatoms (with their silicon frustules) were the major component of algal biomass. Nitrite concentrations were consistently low ($< 0.05\,mg\,NO_2^-\,L^{-1}$), but were high during the winter drought period of 2011, possibly due to lack of dilution of sewage effluent inputs and low biological processing rates on the nitrite within the river channel at low water temperatures. The highest ammonium concentrations (up to 0.26 $mg\,NH_4^+\,L^{-1}$) were observed during the flooding in April 2012, probably due to in-wash of

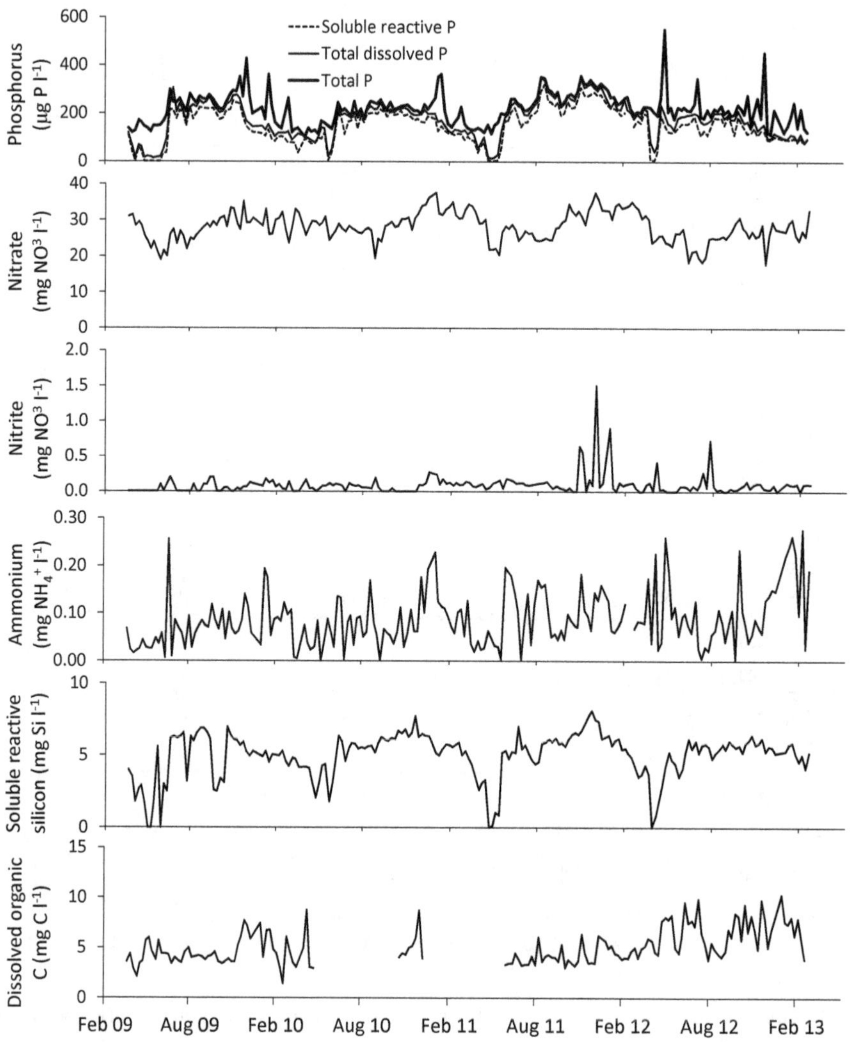

Figure 4. Nutrient concentration time series data for the downstream extent of the monitoring platform (River Thames at Runnymede).

fertilisers and manures from agricultural fields into the watercourses.

3.3.3 Cation data

Sodium, potassium, calcium, magnesium and boron loads in the River Thames at Runnymede were present almost entirely in dissolved form (Fig. 5). In contrast, iron was largely present in particulate form. Similar patterns were observed at all monitoring sites within this study. Sodium, potassium and to a lesser extent boron all showed sudden drops in concentration (Fig. 5) coinciding with periods of high flow (Fig. 3), indicating that their predominant sources (sewage effluent and groundwater) were being diluted by run-off and rainwater inputs. Particulate iron concentrations peaked during high flows, indicating that the predominant source was diffuse inputs from catchment soil erosion and resuspension of within-channel bed sediments. The calcium concentration data were relatively constant through the monitoring period, at ca. $100\,\mathrm{mg\,Ca\,L^{-1}}$, but there were sudden reductions in calcium concentrations to below $80\,\mathrm{mg\,L^{-1}}$ in the spring of each year, coinciding with reductions in alkalinity and periods of high chlorophyll concentrations, due to the precipitation of $CaCO_3$, as described above (Sect. 3.3.1).

3.4 Long-term temporal changes in nutrient concentrations

3.4.1 Phosphorus

There have been no marked changes in water quality at any of the study sites over the 2009 to 2013 monitoring period of the Thames Initiative, indicating the lack of investment in sewage treatment improvements across the catchment through this period. However, com-

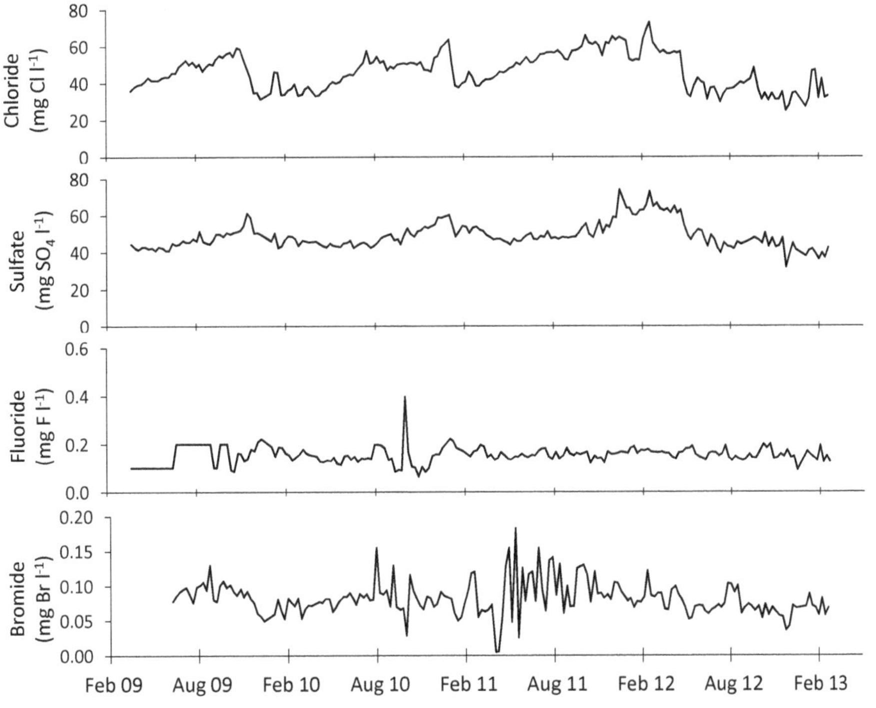

Figure 5. Cation concentration time series data for the downstream extent of the monitoring platform (River Thames at Runnymede).

parisons with past monitoring data from the same study sites (Neal et al., 2012) (https://catalogue.ceh.ac.uk/id/8e23a86b-6b54-4564-9789-23f4b4e045ea) have shown that there has been a major reduction in both total phosphorus and soluble reactive phosphorus concentrations since the late 1990s at most sites.

An example for the middle reach of the River Thames at Wallingford is presented in Fig. 6. Data prior to 2009 were collected from the River Thames at Howbery Park, which is approximately 200 m upstream of the River Thames at the Wallingford monitoring site used during the Thames Initiative research platform. There are no known inputs between these monitoring sites, and so the data from each data set were equivalent. There is a major step reduction in SRP in 1998, and another reduction between 2002 and 2006 (Fig. 6a), although the exact timing of this is unknown, as the monitoring data prior to the start of the Thames Initiative are not continuous. The Thames basin has been the focus for many mitigation measures aimed at reducing sewage effluent inputs of P (primarily through the Urban Wastewater Treatment Directive) and reducing diffuse P pollution from farming (e.g. through the Catchment Sensitive Farming initiative and adoption of agri-environment schemes). The sudden reduction in SRP concentration in winter of 1998–1999 (halving the maximum annual SRP concentration from ca. 1800 to 800 $\mu g\ SRP\ L^{-1}$) strongly suggests that there were major interventions at this time, rather than the multiple small agricultural interventions. The relationship between SRP concentration and flow (Fig. 6b) clearly shows that the reductions in SRP concentrations in both winter 1998 and prior to 2006 occurred during low-flow periods. Again, this strongly suggests that it is a constant P input (i.e. sewage effluent) that has been reduced, rather than rain-related agricultural inputs (Bowes et al., 2008). This is further confirmed by plotting the SRP concentrations against sodium concentration (a conservative sewage marker) (Neal et al., 2010b) (Fig. 6b), which shows that the 80 % reduction in SRP (from an average of 1016 $\mu g\ L^{-1}$ in 1997–1998 to an average of 212 $\mu g\ L^{-1}$ in 2009–2013) is predominantly due to reductions in sewage effluent phosphorus loadings. This is further backed up by information on the implementation of the Urban Wastewater Treatment Directive within the catchment, which introduced phosphorus stripping at all STWs greater than 10 000 population equivalent (between 1996 and March 2008).

However, these data sets also show that there has been little improvement in phosphorus concentrations in the River Thames and its tributaries since February 2009. A previous study of the Thames Initiative data using Load Apportionment Modelling (Bowes et al., 2014) has highlighted that despite past improvements (due largely to the introduction of phosphorus stripping at sewage treatment works), the Thames and most tributaries are still dominated by STW P inputs. The most effective strategy to further reduce P concentrations across the catchment would be to focus resources at further reductions in point source inputs from STWs, rather than diffuse agricultural mitigation, especially as these

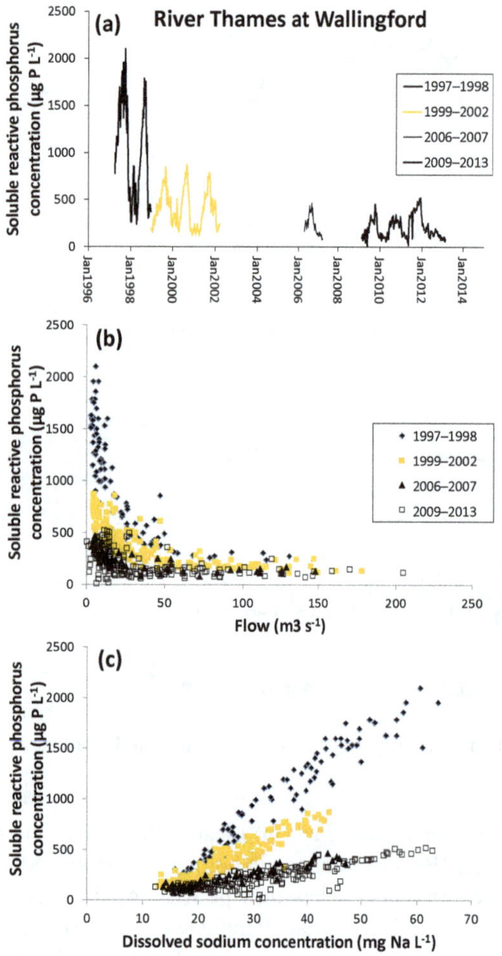

Figure 6. Changes in soluble reactive P concentration in the middle reaches of the River Thames at Wallingford from 1997 to 2013, as **(a)** a time series, **(b)** related to mean daily river flow and **(c)** related to sodium concentration.

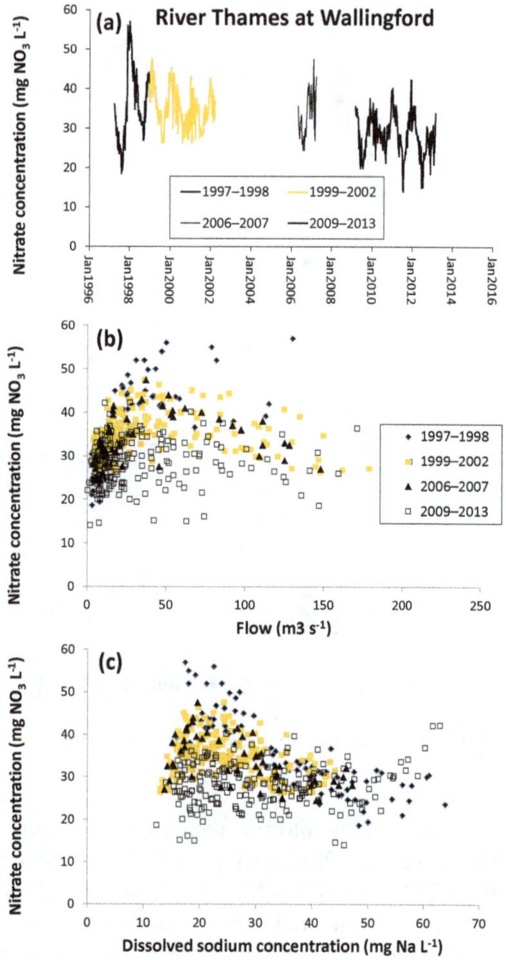

Figure 7. Changes in nitrate concentration in the middle reaches of the River Thames at Wallingford from 1997 to 2013, as **(a)** a time series, **(b)** related to mean daily river flow and **(c)** related to sodium concentration.

data sets clearly show that STW improvements have a major and immediate impact on river water quality.

3.4.2 Nitrate

The 140-year nitrate concentration data record of the lower River Thames by Howden et al. (2010) showed that nitrate concentrations have increased from $< 2\,\mathrm{mg\,NO_3{-}N\,L^{-1}}$ in the late 1860s and 1870s and reached a maximum of $> 7\,\mathrm{mg\,NO_3{-}N\,L^{-1}}$ from the late 1970s to 1990s. Since 2000, the average nitrate concentrations in the lower River Thames decreased slightly to ca. $6.8\,\mathrm{mg\,NO_3{-}N\,L^{-1}}$, suggesting that nitrate concentrations were beginning to decline. The Thames Initiative data further support this observation, showing that the average nitrate concentration in the lower River Thames at Runnymede (2009–2013) was $28.1\,\mathrm{mg\,NO_3\,L^{-1}}$ (Table 3), which is equivalent to $6.35\,\mathrm{mg\,N\,L^{-1}}$.

Combining the CEH data sets from 1997 to 2008 and the CEH Thames Initiative data (2009–2013) provides good evidence that nitrate concentrations have reached a turning point and are beginning to slowly decline. The middle River Thames at Wallingford shows nitrate concentrations peaking in 1998 and gradually reducing throughout the following years (Fig. 7a). Nitrate concentrations showed little change over time at low flows ($< 25\,\mathrm{m^3\,s^{-1}}$) but were declining at medium to high flows (25 to $150\,\mathrm{m^3\,s^{-1}}$) (Fig. 7b), suggesting that there is a reduction in diffuse, rain-related nitrate sources, but not constant inputs such as STWs. This is further supported by examining the relationship between nitrate and sodium concentration. At high sodium concentration $> 40\,\mathrm{mg\,L^{-1}}$ (indicating high sewage inputs) nitrate concentrations from the different monitoring periods are all relatively similar (Fig. 7c). When sodium concentrations are lower (particularly ca. $20\,\mathrm{mg\,L^{-1}}$), there has been a clear reduction in nitrate concentration throughout the monitoring periods. These observations suggest that the reduction in river nitrate concentration is due to reduced inputs from diffuse, rain-related sources such as agriculture run-off or groundwater nitrate concentration, and not due to reduced loadings from sewage effluent.

4 Data availability

The entire data set presented in this study is freely available through the Centre for Ecology and Hydrology's Environmental Information Data Centre data portal https://catalogue.ceh.ac.uk/eidc/documents. The data set is titled "Weekly water quality data from the River Thames and its major tributaries (2009–2013) [CEH Thames Initiative]". The digital object identifier is https://doi.org/10.5285/e4c300b1-8bc3-4df2-b23a-e72e67eef2fd (Bowes et al., 2017).

5 Conclusions

This catchment-wide biogeochemical monitoring platform provides the research community with a valuable data resource for furthering our understanding of pollution sources and dynamics, biological interactions, impacts of land use and increasing population pressures across this internationally known river catchment. The CEH Thames Initiative data can be linked with previous (non-continuous) monitoring data sets from many common study sites, dating back to 1997, and hourly physical and chemical data sets from two of its tributaries. This ongoing monitoring programme will continue to capture the impacts of increasing population densities, changes in agricultural practices and the impacts of improved sewage treatment processes, which have important implications for the sustainability of London and the UK economy.

Author contributions. Sample collection and initial preparation were undertaken by PS and CR. All laboratory analysis was carried out by HW, LA, SH, EG, and DN. Data curation was undertaken by MB and NB. MB wrote the paper and HPJ, GO, and DR provided important contributions and interpretation.

Competing interests. The authors declare that they have no conflict of interest.

Acknowledgements. We thank the Natural Environment Research Council, UK, for funding the monitoring programme, and staff at the Environmental Information Data Centre for managing the data and allocating a doi. We also thank Colin Neal, who provided advice when establishing the Thames Initiative platform.

Edited by: David Carlson

References

Amos, G. C. A., Gozzard, E., Carter, C. E., Mead, A., Bowes, M. J., Hawkey, P. M., Zhang, L., Singer, A. C., Gaze, W. H., and Wellington, E. M. H.: Validated predictive modelling of the environmental resistome, ISME J., 9, 1467–1476, https://doi.org/10.1038/ismej.2014.237, 2015.

Bell, V. A., Kay, A. L., Cole, S. J., Jones, R. G., Moore, R. J., and Reynard, N. S.: How might climate change affect river flows across the Thames Basin? An area-wide analysis using the UKCP09 Regional Climate Model ensemble, J. Hydrol., 442–443, 89–104, 2012.

Bowes, M. J., Smith, J. T., Jarvie, H. P., and Neal, C.: Modelling of phosphorus inputs to rivers from diffuse and point sources, Sci. Total Environ., 395, 125–138, 2008.

Bowes, M. J., Smith, J. T., and Neal, C.: The value of high-resolution nutrient monitoring: A case study of the River Frome, Dorset, UK, J. Hydrol., 378, 82–96, 2009.

Bowes, M. J., Lehmann, K., Jarvie, H. P., and Singer, A. C.: Investigating periphyton response to changing phosphorus concentrations in UK rivers using within-river flumes. In: BHS Third International Symposium, Managing Consequences of a Changing Global Environment, Newcastle-upon-Tyne, 2010.

Bowes, M. J., Gozzard, E., Johnson, A. C., Scarlett, P. M., Roberts, C., Read, D. S., Armstrong, L. K., Harman, S. A., and Wickham, H. D.: Spatial and temporal changes in chlorophyll *a* concentrations in the River Thames basin, UK: Are phosphorus concentrations beginning to limit phytoplankton biomass?, Sci. Total Environ., 426, 45–55, 2012a.

Bowes, M. J., Ings, N. L., McCall, S. J., Warwick, A., Barrett, C., Wickham, H. D., Harman, S. A., Armstrong, L. K., Scarlett, P. M., Roberts, C., Lehmann, K., and Singer, A. C.: Nutrient and light limitation of periphyton in the River Thames: Implications for catchment management, Sci. Total Environ., 434, 201–212, 2012b.

Bowes, M. J., Palmer-Felgate, E. J., Jarvie, H. P., Loewenthal, M., Wickham, H. D., Harman, S. A., and Carr, E.: High-frequency phosphorus monitoring of the River Kennet, UK: are ecological problems due to intermittent sewage treatment works failures?, J. Environ. Monitor., 14, 3137–3145, 2012c.

Bowes, M. J., Jarvie, H. P., Naden, P. S., Old, G. H., Scarlett, P. M., Roberts, C., Armstrong, L. K., Harman, S. A., Wickham, H. D., and Collins, A. L.: Identifying priorities for nutrient mitigation using river concentration–flow relationships: The Thames basin, UK, J. Hydrol., 517, 1–12, 2014.

Bowes, M. J., Jarvie, H. P., Halliday, S. J., Skeffington, R. A., Wade, A. J., Loewenthal, M., Gozzard, E., Newman, J. R., and Palmer-Felgate, E. J.: Characterising phosphorus and nitrate inputs to a rural river using high-frequency concentration–flow relationships, Sci. Total Environ., 511, 608–620, 2015a.

Bowes, M. J., Gozzard, E., Newman, J., Loewenthal, M., Halliday, S., Skeffington, R. A., Jarvie, H. P., Wade, A., and Palmer-Felgate, E.: Hourly physical and nutrient monitoring data for The

Cut, Berkshire (2010–2012), NERC Environmental Information Data Centre, https://doi.org/10.5285/abe4dd7c-a340-4595-a57f-8c1446ff7656, 2015b.

Bowes, M. J., Gozzard, E., Newman, J., Loewenthal, M., Halliday, S., Skeffington, R. A., Jarvie, H. P., Wade, A., and Palmer-Felgate, E.: Hourly physical and nutrient monitoring data for the River Enborne, Berkshire (2009–2012), NERC Environmental Information Data Centre, https://doi.org/10.5285/11d712e0-7456-4ea9-8af8-fe81a666e91b, 2015c.

Bowes, M. J., Loewenthal, M., Read, D. S., Hutchins, M. G., Prudhomme, C., Armstrong, L. K., Harman, S. A., Wickham, H. D., Gozzard, E., and Carvalho, L.: Identifying multiple stressor controls on phytoplankton dynamics in the River Thames (UK) using high-frequency water quality data, Sci. Total Environ., 569–570, 1489–1499, 2016.

Bowes, M. J., Armstrong, L. K., Wickham, H. D., Harman, S. A., Gozzard, E., Roberts, C., and Scarlett, P. M.: Centre for Ecology & Hydrology, Weekly water quality data from the River Thames and its major tributaries (2009–2013) [CEH Thames Initiative], NERC Environmental Information Data Centre, https://doi.org/10.5285/e4c300b1-8bc3-4df2-b23a-e72e67eef2fd, 2017.

Bussi, G., Whitehead, P. G., Bowes, M. J., Read, D. S., Prudhomme, C., and Dadson, S. J.: Impacts of climate change, land-use change and phosphorus reduction on phytoplankton in the River Thames (UK), Sci. Total Environ., 572, 1507–1519, https://doi.org/10.1016/j.scitotenv.2016.02.109, 2016.

Bussi, G., Dadson, S. J., Bowes, M. J., and Whitehead, P. G.: Seasonal and Interannual Changes in Sediment Transport Identified through Sediment Rating Curves, J. Hydrol. Eng., 22, 06016016, https://doi.org/10.1061/(ASCE)HE.1943-5584.0001466, 2017.

CEC: Directive 2000/60/EC of the European Parliament and of the Council establishing a framework for the Community action in the field of Water Policy, Official Journal of the European Communities, European Parliament, Council of the European Union, 43, 327, ISSN 0378-6978, 2000.

Dawson, F. H., Hornby, D. D., and Hilton, J.: A method for the automated extraction of environmental variables to help the classification of rivers in Britain, Aquat. Conserv.-Mar. Freshw. Ecosyst., 12, 391–403, 2002.

EEC: Urban waste water treatment, European Economic Community, Brussels, Belgium, Directive 91/271/EEC, 1991.

Eisenreich, S. J., Bannerman, R. T., and Armstrong, D. E.: A simplified phosphorus analytical technique, Environ. Lett., 9, 45–53, 1975.

Flynn, N. J., Snook, D. L., Wade, A. J., and Jarvie, H. P.: Macrophyte and periphyton dynamics in a UK Cretaceous chalk stream: the River Kennet, a tributary of the Thames, Sci. Total Environ., 282, 143–157, 2002.

Fuller, R. M., Smith, G. M., Sanderson, J. M., Hill, R. A., and Thomson, A. G.: The UK Land Cover Map 2000: Construction of a parcel-based vector map from satellite images, Cartogr. J., 39, 15–25, 2002.

Halliday, S., Skeffington, R., Bowes, M., Gozzard, E., Newman, J., Loewenthal, M., Palmer-Felgate, E., Jarvie, H., and Wade, A.: The Water Quality of the River Enborne, UK: Observations from High-Frequency Monitoring in a Rural, Lowland River System, Water, 6, 150–180, 2014.

Halliday, S. J., Skeffington, R. A., Wade, A. J., Bowes, M. J., Gozzard, E., Newman, J. R., Loewenthal, M., Palmer-Felgate, E. J., and Jarvie, H. P.: High-frequency water quality monitoring in an urban catchment: hydrochemical dynamics, primary production and implications for the Water Framework Directive, Hydrol. Process., 29, 3388–3407, 2015.

Hartley, A. M., House, W. A., Callow, M. E., and Leadbeater, B. S. C.: The Role of a Green Alga in the Precipitation of Calcite and the Coprecipitation of Phosphate in Freshwater, Internationale Revue der gesamten Hydrobiologie und Hydrographie, 80, 385–401, 1995.

Haygarth, P. M., Jarvie, H. P., Powers, S. M., Sharpley, A. N., Elser, J. J., Shen, J., Peterson, H. M., Chan, N.-I., Howden, N. J. K., Burt, T., Worrall, F., Zhang, F., and Liu, X.: Sustainable Phosphorus Management and the Need for a Long-Term Perspective: The Legacy Hypothesis, Environ. Sci. Technol., 48, 8417–8419, 2014.

Horton, A. A., Svendsen, C., Williams, R. J., Spurgeon, D. J., and Lahive, E.: Large microplastic particles in sediments of tributaries of the River Thames, UK–Abundance, sources and methods for effective quantification, Mar. Pollut. Bull., 114, 218–226, 2017.

House, W. A., Duplat, D., Denison, F. H., Henville, P., Dawson, F. H., Cooper, D. M., and May, L.: The role of macrophytes in the retention of phosphorus in the River Thame, England, Chem. Ecol., 17, 271–291, 2001.

Howden, N. J. K., Burt, T. P., Worrall, F., Whelan, M. J., and Bieroza, M.: Nitrate concentrations and fluxes in the River Thames over 140 years (1868–2008): are increases irreversible?, Hydrol. Process., 24, 2657–2662, 2010.

Hutchins, M. G., Williams, R. J., Prudhomme, C., Bowes, M. J., Brown, H. E., Waylett, A. J., and Loewenthal, M.: Projections of future deterioration in UK river quality are hampered by climatic uncertainty under extreme conditions, Hydrol. Sci. J., 61, 2818–2833, 2016.

Jarvie, H. P., Neal, C., Warwick, A., White, J., Neal, M., Wickham, H. D., Hill, L. K., and Andrews, M. C.: Phosphorus uptake into algal biofilms in a lowland chalk river, Sci. Total Environ., 282, 353–373, 2002a.

Jarvie, H. P., Neal, C., Williams, R. J., Neal, M., Wickham, H. D., Hill, L. K., Wade, A. J., Warwick, A., and White, J.: Phosphorus sources, speciation and dynamics in the lowland eutrophic River Kennet, UK, Sci. Total Environ., 282, 175–203, 2002b.

Jarvie, H. P., Neal, C., Juergens, M. D., Sutton, E. J., Neal, M., Wickham, H. D., Hill, L. K., Harman, S. A., Davies, J. J. L., Warwick, A., Barrett, C., Griffiths, J., Binley, A., Swannack, N., and McIntyre, N.: Within-river nutrient processing in Chalk streams: The Pang and Lambourn, UK, J. Hydrol., 330, 101–125, 2006.

Johnson, A. C., Acreman, M. C., Dunbar, M. J., Feist, S. W., Giacomello, A. M., Gozlan, R. E., Hinsley, S. A., Ibbotson, A. T., Jarvie, H. P., Jones, J. I., Longshaw, M., Maberly, S. C., Marsh, T. J., Neal, C., Newman, J. R., Nunn, M. A., Pickup, R. W., Reynard, N. S., Sullivan, C. A., Sumpter, J. P., and Williams, R. J.: The British river of the future: How climate change and human activity might affect two contrasting river ecosystems in England, Sci. Total Environ., 407, 4787–4798, 2009.

Keller, V., Fox, K., Rees, H. G., and Young, A. R.: Estimating population served by sewage treatment works from readily available GIS data, Sci. Total Environ., 360, 319–327, 2006.

Kinniburgh, J. H., Tinsley, M. R., and Bennett, J.: Orthophosphate Concentrations in the River Thames, Water Environ. J., 11, 178–185, 1997.

Lack, T. J.: Quantitative studies on the phytoplankton of the Rivers Thames and Kennet at Reading, Freshw. Biol., 1, 213–224, 1971.

Lack, T. J. and Berrie, A. D.: Phytoplankton production in the Rivers Thames and Kennet at Reading, England during 1970, Blackwell Scientific Publications, Oxford, UK, 1976.

Leeks, G. J. L., Neal, C., Jarvie, H. P., Casey, H., and Leach, D. V.: The LOIS river monitoring network: strategy and implementation, Sci. Total Environ., 194–195, 101–109, 1997.

Mann, K. H., Britton, R. H., Kowalczewski, A., Lack, T. J., Mathews, C. P., and McDonald, I.: Productivity and energy flow at all trophic levels in the River Thames England, IBP-UNESCO, Krakow, Poland, 1972.

Marker, A. F. H., Nusch, E. A., Rai, H., and Riemann, B.: The measurement of photosynthetic pigments in freshwaters and standardisation of methods: Conclusions and recommendations, Arch. Hydrobiol. Beih., 14, 91–106, 1980.

Marsh, T. J. and Hannaford, J.: UK Hydrometric Register. Hydrological data UK series, Centre for Ecology and Hydrology, 210 pp., 2008.

May, L. and Bass, J. A. B.: A study of rotifers in the River Thames, England, April–October, 1996, Hydrobiologia, 387, 251–257, 1998.

McCall, S. J., Hale, M. S., Smith, J. T., Read, D. S., and Bowes, M. J.: Impacts of phosphorus concentration and light intensity on river periphyton biomass and community structure, Hydrobiologia, 792, 315–330, https://doi.org/10.1007/s10750-016-3067-1, 2017.

Mullin, J. B. and Riley, J. P.: The colourometric determination of silicate with special reference to sea and natural waters, Anal. Chim. Acta, 12, 31–36, 1955.

Murphy, J. and Riley, J. P.: A modified single solution method for the determination of phosphorus in natural waters, Anal. Chem. Acta, 12, 31–36, 1962.

Nakada, N., Hanamoto, S., Jurgens, M. D., Johnson, A. C., Bowes, M. J., and Tanaka, H.: Assessing the population equivalent and performance of wastewater treatment through the ratios of pharmaceuticals and personal care products present in a river basin: Application to the River Thames basin, UK, Sci. Total Environ., 575, 1100–1108, 2017.

Neal, C., Jarvie, H. P., Howarth, S. M., Whitehead, P. G., Williams, R. J., Neal, M., Harrow, M., and Wickham, H.: The water quality of the River Kennet: initial observations on a lowland chalk stream impacted by sewage inputs and phosphorus remediation, Sci. Total Environ., 251, 477–495, 2000a.

Neal, C., Neal, M., and Wickham, H.: Phosphate measurement in natural waters: two examples of analytical problems associated with silica interference using phosphomolybdic acid methodologies, Sci. Total Environ., 251, 511–522, 2000b.

Neal, C., Williams, R. J., Neal, M., Bhardwaj, L. C., Wickham, H., Harrow, M., and Hill, L. K.: The water quality of the River Thames at a rural site downstream of Oxford, Sci. Total Environ., 251, 441–457, 2000c.

Neal, C., Jarvie, H. P., Williams, R. J., Neal, M., Wickham, H., and Hill, L.: Phosphorus–calcium carbonate saturation relationships in a lowland chalk river impacted by sewage inputs and phosphorus remediation: an assessment of phosphorus self-cleansing mechanisms in natural waters, Sci. Total Environ., 282–283, 295–310, 2002.

Neal, C., Jarvie, H. P., Wade, A. J., Neal, M., Wyatt, R., Wickham, H., Hill, L., and Hewitt, N.: The water quality of the LOCAR Pang and Lambourn catchments, Hydrol. Earth Syst. Sci., 8, 614–635, https://doi.org/10.5194/hess-8-614-2004, 2004.

Neal, C., Neal, M., Hill, L., and Wickham, H.: River water quality of the River Cherwell: An agricultural clay-dominated catchment in the upper Thames Basin, southeastern England, Sci. Total Environ., 360, 272–289, 2006.

Neal, C., Jarvie, H. P., Williams, R., Love, A., Neal, M., Wickham, H., Harman, S., and Armstrong, L.: Declines in phosphorus concentration in the upper River Thames (UK): Links to sewage effluent cleanup and extended end-member mixing analysis, Sci. Total Environ., 408, 1315–1330, 2010a.

Neal, C., Williams, R. J., Bowes, M. J., Harrass, M. C., Neal, M., Rowland, P., Wickham, H., Thacker, S., Harman, S., Vincent, C., and Jarvie, H. P.: Decreasing boron concentrations in UK rivers: insights into reductions in detergent formulations since the 1990s and within-catchment storage issues, Sci. Total Environ., 408, 1374–1385, 2010b.

Neal, C., Bowes, M., Jarvie, H. P., Scholefield, P., Leeks, G., Neal, M., Rowland, P., Wickham, H., Harman, S., Armstrong, L., Sleep, D., Lawlor, A., and Davies, C. E.: Lowland river water quality: a new UK data resource for process and environmental management analysis, Hydrol. Process., 26, 949–960, 2012.

Powers, S. M., Bruulsema, T. W., Burt, T. P., Chan, N. I., Elser, J. J., Haygarth, P. M., Howden, N. J. K., Jarvie, H. P., Lyu, Y., Peterson, H. M., Sharpley, Andrew N., Shen, J., Worrall, F., and Zhang, F.: Long-term accumulation and transport of anthropogenic phosphorus in three river basins, Nat. Geosci., 9, 353–356, 2016.

Read, D. S., Bowes, M. J., Newbold, L. K., and Whiteley, A. S.: Weekly flow cytometric analysis of riverine phytoplankton to determine seasonal bloom dynamics, Environ. Sci.-Proc. Imp., 16, 594–603, 2014.

Read, D. S., Gweon, H. S., Bowes, M. J., Newbold, L. K., Field, D., Bailey, M. J., and Griffiths, R. I.: Catchment-scale biogeography of riverine bacterioplankton, ISME J., 9, 516–526, 2015.

Singer, A. C., Järhult, J. D., Grabic, R., Khan, G. A., Lindberg, R. H., Fedorova, G., Fick, J., Bowes, M. J., Olsen, B., and Söderström, H.: Intra- and Inter-Pandemic Variations of Antiviral, Antibiotics and Decongestants in Wastewater Treatment Plants and Receiving Rivers, PLoS One, 9, e108621, https://doi.org/10.1371/journal.pone.0108621, 2014.

Skeffington, R. A., Halliday, S. J., Wade, A. J., Bowes, M. J., and Loewenthal, M.: Using high-frequency water quality data to assess sampling strategies for the EU Water Framework Directive, Hydrol. Earth Syst. Sci., 19, 2491–2504, https://doi.org/10.5194/hess-19-2491-2015, 2015.

Skougstad, M. W., Fishman, M. J., Friedman, L. C., Erdmann, D. E., and Duncan, S. S.: Methods for analysis of inorganic substances in water and fluvial sediments, U.S. Geological Survey, Report no. 78-679, 545 pp., 1978.

Smith, J. T., Clarke, R. T., and Bowes, M. J.: Are groundwater nitrate concentrations reaching a turning point in some chalk aquifers?, Sci. Total Environ., 408, 4722–4732, 2010.

Turner, G. S. C., Mills, G. A., Bowes, M. J., Burnett, J. L., Amos, S., and Fones, G. R.: Evaluation of DGT as a long-term water quality monitoring tool in natural waters; uranium as a case study, Environ. Sci.-Proc. Imp., 16, 393–403, https://doi.org/10.1039/C3EM00574G, 2014.

Wade, A. J., Palmer-Felgate, E. J., Halliday, S. J., Skeffington, R. A., Loewenthal, M., Jarvie, H. P., Bowes, M. J., Greenway, G. M., Haswell, S. J., Bell, I. M., Joly, E., Fallatah, A., Neal, C., Williams, R. J., Gozzard, E., and Newman, J. R.: Hydrochemical processes in lowland rivers: insights from in situ, high-resolution monitoring, Hydrol. Earth Syst. Sci., 16, 4323–4342, https://doi.org/10.5194/hess-16-4323-2012, 2012.

Whitehead, P. G., Bussi, G., Bowes, M. J., Read, D. S., Hutchins, M. G., Elliott, J. A., and Dadson, S. J.: Dynamic modelling of multiple phytoplankton groups in rivers with an application to the Thames river system in the UK, Environ. Model. Softw., 74, 75–91, 2015.

Williams, R. J., White, C., Harrow, M. L., and Neal, C.: Temporal and small-scale spatial variations of dissolved oxygen in the Rivers Thames, Pang and Kennet, UK, Sci. Total Environ., 251, 497–510, 2000.

Wright, J. F., Gunn, R. J. M., Winder, J. M., Wiggers, R., Vowles, K., Clarke, R. T., and Harris, I.: A comparison of the macrophyte cover and macroinvertebrate fauna at three sites on the River Kennet in the mid 1970s and late 1990s, Sci. Total Environ., 282, 121–142, 2002.

9

The Global Streamflow Indices and Metadata Archive (GSIM) – Part 1: The production of a daily streamflow archive and metadata

Hong Xuan Do[1], Lukas Gudmundsson[2], Michael Leonard[1], and Seth Westra[1]

[1]School of Civil, Environmental and Mining Engineering, University of Adelaide, Adelaide, Australia
[2]ETH Zürich, Institute for Atmospheric and Climate Science, Zürich, Switzerland

Correspondence: Hong Xuan Do (hong.do@adelaide.edu.au)

Abstract. This is the first part of a two-paper series presenting the Global Streamflow Indices and Metadata archive (GSIM), a worldwide collection of metadata and indices derived from more than 35 000 daily streamflow time series. This paper focuses on the compilation of the daily streamflow time series based on 12 free-to-access streamflow databases (seven national databases and five international collections). It also describes the development of three metadata products (freely available at https://doi.pangaea.de/10.1594/PANGAEA.887477): (1) a GSIM catalogue collating basic metadata associated with each time series, (2) catchment boundaries for the contributing area of each gauge, and (3) catchment metadata extracted from 12 gridded global data products representing essential properties such as land cover type, soil type, and climate and topographic characteristics. The quality of the delineated catchment boundary is also made available and should be consulted in GSIM application. The second paper in the series then explores production and analysis of streamflow indices. Having collated an unprecedented number of stations and associated metadata, GSIM can be used to advance large-scale hydrological research and improve understanding of the global water cycle.

1 Introduction

Streamflow observations with global coverage are essential to make progress in the science of large-scale hydrology. For example, global datasets provide particular value when evaluating global hydrological models (Gudmundsson et al., 2012; Huang et al., 2016; Ward et al., 2013), producing runoff estimation data products (Fekete et al., 2002a, b; Gudmundsson and Seneviratne, 2015; Vörösmarty et al., 1989), investigating large-scale weather patterns and their relation to hydrological extremes (Wanders and Wada, 2015; Ward et al., 2014), and detecting changes in the global hydrological extremes over space and time (Do et al., 2017; Gudmundsson et al., 2017; Kundzewicz et al., 2012; Milly et al., 2002), amongst numerous other applications.

Despite the fundamental, widespread, and varied applications that streamflow observations support, there are many obstacles to the existence and utility of a large-scale streamflow archive. Firstly, there are threats to the quantity of data, such as political sensitivities (Nelson, 2009), cost recovery and strict access policies (Hannah et al., 2011), unavailability in an electronic format, consistency of data formats, limited documentation, missing metadata, and a lack of resources for database maintenance and updating. Secondly, there are difficulties associated with the quality of the data in many regions, such as poor spatial coverage, poor quality control, variable quality control between regions, inconsistent metadata, imprecise geographic coordinates of the site, changes in the density of stream gauges, and variable record lengths. Lastly, even in locations where there are abundant and high-quality streamflow observations, there can be questions over its utility in specific research such as climate sensitivity analysis due to the manifestation of human impacts – for example, urbanization, land-use changes, channelization, and upstream dams (Hannah et al., 2011).

To date, the Global Runoff Data Base (GRDB) maintained by the Global Runoff Data Centre has been the primary dataset used in large-scale hydrological studies, with more than 9000 stations available to the research community (GRDC, 2015). The Global Runoff Data Centre (GRDC) database operates under the auspices of the UN – World Meteorological Organization (WMO), and its database is supported on a voluntary basis so that the number of data submissions depends on contributions by national authorities. However, although numerous countries have databases of acceptable quality, data supply remains resource intensive and the GRDB remains sparse in some regions. For example, the latest catalogue of the GRDB database (version 5 December 2017) shows that out of 7238 daily time series, there are only 637 stations over South America and only 642 stations over Asia. Moreover, many stations in regions such as Asia and Russia have not been updated for many years and are missing otherwise available data at the end of their records.

The Global Streamflow Indices and Metadata (GSIM) project has been initiated in order to address the demand for a global streamflow database (Bierkens, 2015; Fekete et al., 2015; Hannah et al., 2011; Kundzewicz et al., 2013; Merz et al., 2012; Milly et al., 2015). The approach of this project is not to collect high-quality data from referenced hydrological networks, which have been conducted in other studies (Addor et al., 2017; Burn et al., 2012; Hannaford and Marsh, 2006; Hodgkins et al., 2017; Whitfield et al., 2012) to support research that requires assumptions regarding the minimum impact of human interference on streamflow, such as the investigation of climate change implication for changes in extreme events. Instead, the activities of the GSIM project have been to collate publicly available data, apply basic consistency to the formatting, and establish a standardized set of metadata. In so doing, GSIM intends to promote more widespread use of streamflow data, facilitate improved research outcomes through increased spatial coverage and gauge density, and tackle ongoing challenges for the hydrological community, for example, addressing fundamental issues of data quality, identifying additional data sources, lobbying for continuity of data networks, and developing a method for improved governance and maintenance of streamflow data at the global scale.

To maximize the value of the streamflow dataset for a wide range of applications, the GSIM project also seeks to provide information on catchment characteristics upstream of the streamflow gauging station. This necessitates a consistent approach to delineating the upstream catchment boundary for every gauge station, and this is achieved using data from a global digital elevation model (DEM). This is because, with the exception of the GRDB databases, catchment boundaries representing the direct drainage area of stations were unavailable. Filling in this missing element of metadata is important to facilitate further analysis of the streamflow observations with respect to a wide and ever-increasing variety of spatial datasets. Although there have been previous efforts in producing catchment boundaries for a smaller number of stations (Addor et al., 2017; Arsenault et al., 2016; Lehner, 2012; Schaake et al., 2006), similar work at this magnitude has not been undertaken. This task is complicated by a lack of precision in the supplied geographic coordinates of a given site; for example, when a catchment boundary is extracted, the corresponding calculated area may not match the reported area of the catchment and a procedure for checking minor shifts in the coordinates is needed to improve identification of the likely catchment boundary. The quality of the delineated catchment boundary is also made available to GSIM users and should be considered prior to using this data product and any accompanying information.

The availability of catchment boundaries for each gauge enables the association of environmental variables with each gauge by extracting them from corresponding global-scale gridded products. As part of the GSIM project, a number of global data products are provided as an additional dataset so that a user can readily filter the GSIM dataset according to specific interests, for example, by climate type, soil type, land-use type, irrigation area, and population density. Other potential applications of this auxiliary information might include a comparison to a database of dams for identifying upstream impacts; to remotely sensed estimates of forest cover or urban extent for determining land-use change; to population demographics for improving estimates of flood exposure; and to hydrological model outputs for evaluating model performance.

Finally, to facilitate benefits of this project to the broader community, indices characterizing water-balance aspects, hydrological extremes, and features of the seasonal cycle have been derived from the GSIM time series and will be made publicly available. To ensure standardized quality for the derived indices, a quality control procedure coupling the information provided by data providers and a data-driven approach was also applied.

This is the first paper of a two-part series detailing the production of GSIM and corresponding data products. This paper outlines the provenance of daily streamflow time series (Sect. 2), procedures for reformatting and combining the time series (Sect. 3), the development of metadata associated with each gauge (Sect. 4), an overall summary of the GSIM time series and metadata (Sect. 5), and data availability (Sect. 6). As the time-series database cannot be made available online due to varieties of terms and conditions from data providers, the second paper in this series (Gudmundsson et al., 2018) is dedicated to the production of streamflow time-series indices, including (1) checks for data quality, (2) the production of streamflow time-series indices, and (3) homogeneity assessment of the derived indices.

2 Daily streamflow data and where to find them

GSIM is a compilation of 12 databases that have either open-access or restricted-access policies, and that collectively represent a total of 35 002 stations. The spatial distribution and the number of stations available in each database are illustrated in Fig. 1 (continental-scale figures are also provided as a Supplement). A summary of the data sources is also provided in Table 1 and detailed information on each database is elaborated upon in the following sections. The list of databases identified as part of GSIM is not exhaustive of all possible data sources, only of those that were known to the authors and readily accessible within the project time frame. Where additional data are available in a convenient format, it may be possible to further augment GSIM in the future.

The various data sources were classified as either a "research database" or a "national database". The reasons for this classification are further outlined in Sect. 3, but relate to issues when merging databases and removing duplicate gauges. The data sources include the following.

1. Research databases: databases with daily streamflow data that have been compiled on an ad hoc basis from a variety of original sources by research organizations. This category includes five different databases: the Global Runoff Data Base (GRDB); the European Water Archive (EWA); the China Hydrological Data Project (CHDP) data archive; the GEWEX Asian Monsoon Experiment – Tropics (GAME) data archive; and the Regional Hydrographic Data Network for the Arctic Region (ARCTICNET) data archive.

2. National databases: databases with daily streamflow data made publicly available by national water authorities as part of water-related regulations. This category includes seven databases: the USGS Water Data for USA database (USGS); Canada's National water data archive (HYDAT); Japan's Ministry of Land and Infrastructure database Water Information System (MLIT); Spain's digital hydrological yearbook database (Anuario de aforos digital 2010–2011, AFD); Australia's Bureau of Meteorological Water Data Online database (BOM); India's Water Resources Information System database (WRIS); and Brazil's National Water Agency database (ANA).

2.1 The Global Runoff Data Base (GRDB)

The daily streamflow dataset received from the GRDC (6313 stations with more than 10 years on record; see also Gudmundsson and Seneviratne, 2016) is referred to as the GRDB in this project. To date, the GRDB has been the largest and most extensively used dataset for streamflow analysis at regional and global scales. It was thus considered as the starting point and "base" for the GSIM project. Indeed, it was awareness of data not available from the GRDB that prompted the initial search for additional sources of data to complement the database.

The GRDC was initiated in 1988 by the WMO and is now maintained at the German Federal Institute of Hydrology in Koblenz. The GRDC provides free and unrestricted access to all hydrological data and products, although the data policy indicates that requests for data must reach the GRDC in written form to ensure data users do not redistribute the time series. More detail about the GRDC data policy, and the procedure for obtaining its time series, are outlined at http://www.bafg.de/GRDC/EN/01_GRDC/12_plcy/data_policy_node.html (last access: 23 June 2017).

2.2 The European Water Archive

The European Water Archive (referred to as the EWA in this paper) is one of the most comprehensive streamflow time-series archives in Europe, with more than 3000 river gauging stations distributed across 29 countries. This archive is also currently held by the GRDC and available under the GRDC data policy (http://www.bafg.de/GRDC/EN/04_spcldtbss/42_EWA/ewa_node.html, last access: 3 January 2018). The EWA stations used in this paper were selected using the same criteria as Gudmundsson and Seneviratne (2016), with a total of 3731 daily records.

2.3 The China Hydrology Data Project

The China Hydrology Data Project (CHDP) aims to digitize an arrangement of hydrological measurements taken at Chinese stations. These measurements (including daily discharge) were originally only available in book form (Henck et al., 2010). The original data were collected by the Chinese Hydrology Bureau and published in annual yearbooks. At the time GSIM began, discharge data were only available for the Yunnan-Tibet International Rivers, which corresponded to 163 stations until 1987. This project has been terminated since the 2000s and thus no further update is available. The data and metadata were obtained directly from the author of the project. Detailed information can be viewed at http://www.oberlin.edu/faculty/aschmidt/chdp/index.html (last access: 23 June 2017).

2.4 The GEWEX Asian Monsoon Experiment – Tropics project

The GEWEX Asian Monsoon Experiment – Tropics project (GAME) was initiated in 1996 to monitor several hydroclimatological variables over the humid temperate area in South-East Asia. As one of several important activities in this project, many hydrological observation datasets were collected, including streamflow data. Available streamflow data were provided by the Royal Irrigation Department of Thailand, and comprised 129 time series spanning a period from 1980 to 2000. Daily discharge data and associ-

Table 1. Basic information of daily streamflow databases included in the GSIM project.

Database (referred name)	Database category	Spatial coverage	Data access information
Global Runoff Data Base (GRDB)	Research database	Global	www.bafg.de/GRDC/ (last access: 23 June 2017) Archived database can be obtained via written request to the Global Runoff Data Centre. This database is updated when new data are submitted by national suppliers.
European Flow Regimes from International Experimental and Network Data (EWA)	Research database	European	http://ne-friend.bafg.de/servlet/is/7413/ (last access: 23 June 2017) Data can be obtained via written request to the Global Runoff Data Centre. This database has been frozen since October 2014 and is being integrated into the GRDB database.
A Regional, Electronic, Hydrographic Data Network for Russia (ARCTICNET)	Research database	Russia	http://www.russia-arcticnet.sr.unh.edu/ (last access: 23 June 2017) Archived and closed historic database. Part of this data archive has been included in the databases of the Global Runoff Data Centre and updated based on data deliveries.
China Hydrology Database Project (CHDP)	Research database	China	http://www.oberlin.edu/faculty/aschmidt (last access: 23 June 2017) Archived and closed historic database can be obtained via written request to the author of the database.
GEOSS ana MAHASRI Experiment in Tropics (GAME)	Research database	Thailand	http://hydro.iis.u-tokyo.ac.jp/GAME-T/GAIN-T/routine/rid-river/disc_d.html (last access: 23 June 2017) Archived and closed historic database
US National Water Information System (USGS)	National database	USA	http://waterdata.usgs.gov/nwis (last access: 23 June 2017) Individual time series can be downloaded from the data portal (updated regularly).
Canada National Water Data Archive (HYDAT)	National database	Canada	https://ec.gc.ca/rhc-wsc/ (last access: 23 June 2017) Archived database. The archive is updated quarterly by the data authority.
Brazil National Water Agency (ANA)	National database	Brazil	http://hidroweb.ana.gov.br/ (last access: 23 June 2017) Individual time series can be downloaded from the data portal (updated regularly).
Japan Water Information System (MLIT)	National database	Japan	http://www1.river.go.jp/ (last access: 23 June 2017) Individual time series can be downloaded from the data portal (updated regularly).
Anuario de aforos digital 2010–2011 (AFD)	National database	Spain	http://ceh-flumen64.cedex.es/anuarioaforos (last access: 23 June 2017) Archived database, DVD available from Spanish authorities (updated annually)
Australia Water Data Online (BOM)	National database	Australia	http://www.bom.gov.au/waterdata/ (last access: 23 June 2017) Individual time series can be downloaded from the data portal (updated regularly).
Water Resources Information System of India (I-WRIS)	National database	India	http://www.india-wris.nrsc.gov.in/wris.html (last access: 23 June 2017) Individual time series can be downloaded from the data portal (updated regularly).

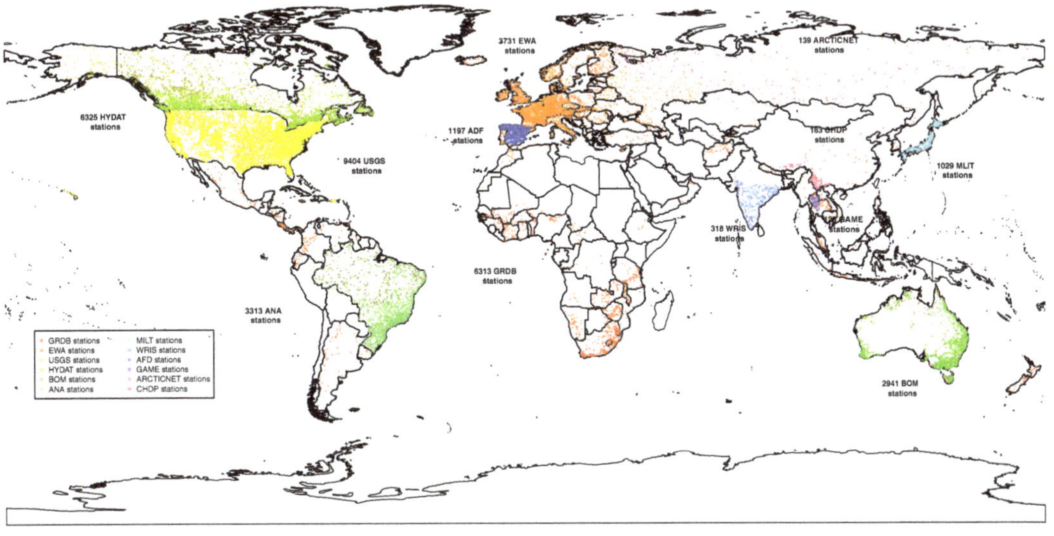

Figure 1. The distribution of stations from original data sources.

ated metadata were archived and can be accessed online at http://hydro.iis.u-tokyo.ac.jp/GAME-T/GAIN-T/routine/rid-river/index.html (last access: 23 June 2017).

2.5 The ARCTICNET project

A regional hydrometeorological data network for the pan-Arctic Region project is a regional database that can be accessed via the Internet and is referred to as ARCTICNET in this paper. The database is designed to support hydrological sciences and water resource assessments over this region with the goal of estimating the contemporary water and constituent balances for the pan-Arctic drainage system. ARCTICNET is a static dataset and some time series have been included in the databases of the GRDC and updated based on data deliveries. Although most data provided in the data portal are at monthly resolution, there are 139 high-quality daily streamflow time series across Russia that are also available which have not been fully integrated into GRDB. Although ARCTICNET's future status is likely to be a part of the GRDB, these stations have still been considered in GSIM production and are referred to as the ARCTICNET database in this paper. These time series, along with their metadata, were archived and can be downloaded at http://www.r-arcticnet.sr.unh.edu/v4.0/index.html (last access: 23 June 2017).

2.6 The USGS database

The USGS National Data Services for the US provide access to water resources data collected at approximately 1.5 million sites in all 50 states of the USA, also including the District of Columbia, Puerto Rico, the Virgin Islands, Guam, American Samoa, and the Commonwealth of the Northern Mariana Islands. All time series and associated metadata can be queried from the data portal http://waterdata.usgs.gov/nwis (last access: 23 June 2017). To ensure the queried data have sufficient geographic metadata (critical for the present project), the stations listed in the Geospatial Attributes of Gages for Evaluating Streamflow, version II (GAGES II) database were used (Falcone, 2011). The time series from 9404 stream gauges obtained from the USGS data portal are referred to as the USGS database in this paper.

2.7 The HYDAT database

Canada's National Water Data Archive (HYDAT) is a database containing daily observed hydrometric data from publicly funded gauges in Canada. Also available in the HYDAT database are metadata about the hydrometric stations, such as latitude and longitude, catchment area, record length, as well as information regarding flow conditions (current status, regulated or natural regime). The database is updated four times per year and currently contains data for 6325 streamflow stations across Canada. The raw data, as well as an extractor executable, are publicly available from Environment Canada's website at https://ec.gc.ca/rhc-wsc/default.asp?lang=En&n=9018B5EC-1 (last access: 23 June 2017).

2.8 The ANA database

The HIDROWEB data portal was organized by the Brazilian National Water Agency (ANA). It provides a database with all the information collected by Brazil's hydrometeorological network. Streamflow data and associated metadata were made publicly available by Brazil's national water regulations, and have been used extensively to monitor critical events, such as floods and droughts. Individual time series and their associated metadata can be viewed or downloaded at http://hidroweb.ana.gov.br (last access: 23 June 2017). The 3313 stations downloaded from this website are referred to as the ANA in this paper.

2.9 The AFD database

Spanish streamflow data were retrieved from the digital hydrological yearbook (Anuario de aforos digital 2010–2011, AFD), which provides observations until 2013–2014 and is freely accessible online at http://ceh-flumen64.cedex.es/anuarioaforos/default.asp (last access: 23 June 2017). For the GSIM, we used the time series that was used to develop the E-RUN dataset (Gudmundsson and Seneviratne, 2016). The original DVD containing the full database was obtained directly from the Spanish authorities via a written form request. This collection contains streamflow data from 1197 gauging stations, and is referred to as ADF in this paper.

2.10 The MLIT database

In Japan, the Ministry of Land, Infrastructure, Transport and Tourism is responsible for organizing hydrological data. All records are disseminated at http://www1.river.go.jp/ (last access: 23 June 2017). As of 2010, the database kept records of all river stations (at both discharge and gauge level). The composition of the 15-digit station IDs is outlined in the file http://www1.river.go.jp/kitei_sosoku.pdf (PDF), and can be used to query and download time series, along with its metadata. As the whole database is recorded in Japanese, the translateR package (Lucas and Tingley, 2016) was used to translate the metadata into English. The time series downloaded from the Japanese water data portal (1029 stations in total) is referred to as MLIT in this paper.

2.11 The BOM database

As part of the water reform programme established in Australia, Water Data Online was created to provide free access to nationally consistent, current and historical water information. It can be accessed at http://www.bom.gov.au/waterdata (last access: 23 June 2017). Water Data Online also contains historical data from some stations that are no longer operational. Users can view or download individual streamflow time series from the data portal, along with standardized data and reports. The time series measured at 2941 stations obtained from Water Data Online is referred to as the BOM database in this project.

2.12 The WRIS database

The Generation of Database and Implementation of Web Enabled Water Resources Information System in the Country project (India-WRIS WebGIS) was initiated as a joint venture of the Indian Central Water Commission (CWC) and the Indian Space Research Organization (ISRO). Unclassified data can be accessed online and free of charge at http://www.india-wris.nrsc.gov.in/wris.html (last access: 23 June 2017), while the metadata are documented at http://www.cwc.nic.in/main/downloads/HydrologicalnetworkdetailsofCWC.pdf (last access: 23 June 2017). All 318 stations were downloaded from the website. They are referred to as the WRIS database in this paper.

The production of time series and metadata for GSIM comprises several stages due to the range of data formats and significant variation in the quality of metadata across data sources. To ensure GSIM is presented in a transparent manner, the following sections outline procedures that are used to collate the time series across (Sect. 3), and to produce the metadata (Sect. 4).

3 Procedure for combining databases

Several of the identified data sources share common spatial domains, where typically the research databases may contain a subset of gauges from the national databases. It is therefore important to correctly identify duplicate time series when merging the databases. To maximize the quality of combined time series and minimize the requirement to combine time series, this task is conducted following three sequential steps: Step 1 – pre-processing the data to a common structure; Step 2 – replacing all GRDB stations in countries that have a national database; and Step 3 – identifying remaining duplicates. From the 35 002 gauges, 3197 (2958 and 239 gauges from the GRDB and EWA databases, respectively) were replaced by national databases in Step 2, and 846 cases of "very likely identical" stations were identified and removed in Step 3, leaving 30 959 "duplication-free" time series available in the GSIM.

3.1 Pre-processing the time series into a singular data structure

One of the major challenges in producing consistent streamflow indices is that data from different sources have different structures and storage formats. For example, the MLIT database divides streamflow records at one location into separate text files, and each file contains streamflow measurements for 1 year. In comparison, the HYDAT archive includes streamflow measurements from all available stations in a single matrix.

To address the varying standards of data management, the first step in combining the databases was to reformat all the streamflow records to ensure that each time series is kept in a consistent format. Using the GRDB as a guide, it was decided to store all data for a given site in a single text file with three columns: (a) date of measurement, (b) value of measurement, and (c) original quality flags (if available), and with basic metadata (station name, ID, etc.) stored in the header of the file. All additionally derived metadata (i.e. from global gridded products) are stored in the station catalogue. The streamflow measurements were also converted into consistent units (cubic metres per second).

Metadata that have special characters in foreign language sources were also pre-processed into the ASCII encoding system. For river names and station names that are recorded in Spanish (ADF) or Portuguese (ANA), the special characters were replaced by plain alphabetic characters using the core function iconv() of the R programming language. For river names and station names that are recorded in Japanese characters (MLIT), R package "translateR" (Lucas and Tingley, 2016) was used with the Google Translate API for this task. Although there are some limitations related to this toolset (e.g. some Japanese characters remaining untranslated and requiring manual translation; inconsistency in the translated results using the same original Japanese characters), this option was chosen to enable an automated and expedient translation. As a result, any text-related metadata associated with Japanese stations should be treated with care.

3.2 Replace the GRDB stations with national databases, if applicable

The streamflow records hosted by the GRDC (the GRDB and EWA databases) are themselves originally provided by national water agencies, and have been undergone quality control procedures by the GRDC. In cases that the supplied data contain errors, the GRDC informs data suppliers to improve the quality of their database. In term of data availability, time series downloaded directly from the national data portal usually represents the latest version of streamflow observation, and thus it seemed appropriate to replace stations hosted by the GRDC for countries where an equivalent national database was available. While this approach is efficient, there is a potential downside of removing GRDB stations that were not otherwise present in the national data depositories, perhaps due to differences in maintenance of the databases. Nonetheless, the number of stations available in the GRDB and EWA databases is much lower than that available in national databases for all countries (see Table 2). As a result of this step, 2958 stations located in seven countries (Australia, Brazil, Canada, India, Japan, Spain, and the United States) were removed from the GRDB collection. In addition, 239 stations located in Spain were removed from the EWA archive.

3.3 Identify and remove duplicates in research databases

The method of de-duplicating time series involves identification of duplicates where two data sources have overlapping coverage and potential merging of two records at a duplicated site to create a unified record. The de-duplication step was generally undertaken between the GRDB and a "paired" dataset (e.g. GRDB and GAME). The only exceptions to this step are for GRDB, EWA, and ARCTICNET, as these three datasets share Russia as a common spatial domain.

Table 2. Number of stations in countries where national databases are available.

Country	Database		
	EWA	GRDB	National
Australia	–	358	2941 (BOM)
Brazil	–	439	3313 (ANA)
Canada	–	1029	6325 (HYDAT)
India	–	0	318 (WRIS)
Japan	–	151	1029 (MLIT)
Spain	239	0	1197 (ADF)
United States	–	981	9404 (USGS)

The techniques adopted for combining research databases were based on the de-duplication procedures developed in Gudmundsson and Seneviratne (2016), which consists of three sequential steps.

1. Identification of "duplication candidates" using metadata similarity. This step aims to identify time series with a high level of similarity in metadata (either within one database or across different databases). We used three similarity metrics to identify potential time series: (1) Jaro–Winkler distances, a metric representing the alphanumeric similarity of strings (Christen, 2012), applied to river names of two records; (2) Jaro–Winkler distances between station names of two records; and (3) geographical proximity estimated from geographical coordinates between two records. These metrics were normalized to have the same range between 0 and 1, where a value of 0 indicates identical metadata (e.g. the same geographic coordinates). This similarity analysis was run for each pair in the pool of stations, and any pair with an average value below 0.25 was identified as candidate duplicate records.

2. Classifications of duplication candidates using time-series similarity. This step aims to decide whether a specific pair of duplication candidates is likely to be identical. The overlapping period and correlation coefficient were used as criteria for making a decision. Firstly, all duplication candidates that do not share any overlap in their period of record are kept in the final GSIM collection, as they can represent separate time series even if they measured discharge at the same geographical location (e.g. due to reconstruction of the gauging station). Secondly, any time series with a correlation coefficient (R^2) lower than 0.90 was automatically identified as "very likely different" (26 pairs), whereas $R^2 > 0.99$ indicates "very likely identical" time series (786 pairs). Finally, candidates with $0.90 \leq R^2 \leq 0.99$

Table 3. Basic metadata available from data sources.

Database	Station ID	Station name	River name	Geographical coordinates	Station elevation	Drainage area	Catchment boundary
GRDB	X	X	X	X	X	X	X
EWA	X	X	X	X	X	X	X
CHDP	X	X	X	X	–	X	–
GAME	X	X	X	X	X	X	–
ARCTICNET	X	X	X	X	X	X	–
USGS	X	X	–	X	X	X	–
HYDAT	X	X	–	X	–	X	–
ANA	X	E	E	X	X	X	–
ADF	X	E	E	X	X	X	–
MLIT	X	E	E	X	–	–	–
BOM	X	X	–	X	–	–	–
WRIS	X	X	X	X	X	X	–

X: metadata available; –: metadata are unavailable; E: metadata are not available in English.

(65 pairs) were visually inspected and manually classified as "very likely identical" (60 pairs) or "very likely different" (five pairs). All time series in the "very likely different" category were retained while stations of the "very likely identical" category were processed using the de-duplication procedure (see below).

3. De-duplication of identical time series: regardless of whether identical time series come from either the same database or from different databases, records with the greater number of data points in the streamflow time series were kept while the other(s) were discarded. Although this approach has the downside of truncating the length of useful records, the number of time series that could be influenced by this approach is relatively low (846 time series, corresponding to 2.8 % of the total number of available streamflow records).

A visual example of the de-duplication procedure is provided in Fig. 2. The left panel demonstrates a case of "very likely identical" stations, when station number 2964035 in the GRDB database was identified as an identical gauge to W.16 in the GAME archive, based on the similarities between the provided metadata and correlation coefficient. The time series representing station "GAME_W.16" was kept in the final collection, while time series "GRDB_2964035" was removed. The right panel in Fig. 2 demonstrates a case of a "duplication candidate" with correlation coefficient of 0.92 (time series "GRDB_6123645" and "EWA_9110028"). These time series were visually inspected, assigned a "very likely different" label, and both time series were kept in the final collection.

4 Production of the GSIM metadata

Providing a consistent set of metadata for each site has been a significant undertaking for GSIM. This section outlines three main stages to developing the GSIM metadata: (1) consolidating all available basic metadata; (2) consistently delineating catchment boundaries for each site; and (3) developing a supplementary set of catchment-scale metadata based on the delineated boundaries.

4.1 Consolidating basic metadata from available sources

Following the GRDB format, each time series was accompanied by basic metadata, including

1. station ID,
2. station name,
3. river name of gauging location,
4. geographical coordinates of station,
5. elevation of station,
6. drainage area, and
7. catchment boundary from original data sources.

These data are useful for filtering stations according to specific criteria and analysis objectives. Moreover, the availability of a catchment boundary for the gauge enables additional catchment-scale metadata to be derived as necessary. However, not all of these basic metadata were available for all data sources. For example, the catchment boundary was only available for parts of the GRDB and EWA stations, the drainage area was unavailable in the BOM and MLIT databases, and though several data sources included river names in station names (BOM, HYDAT, USGS), these metadata were unavailable in English for other sources (MLIT, ANA, ADF). Table 3 further outlines the availability of basic metadata for each source.

The method for consolidating basic metadata for each station follows three steps.

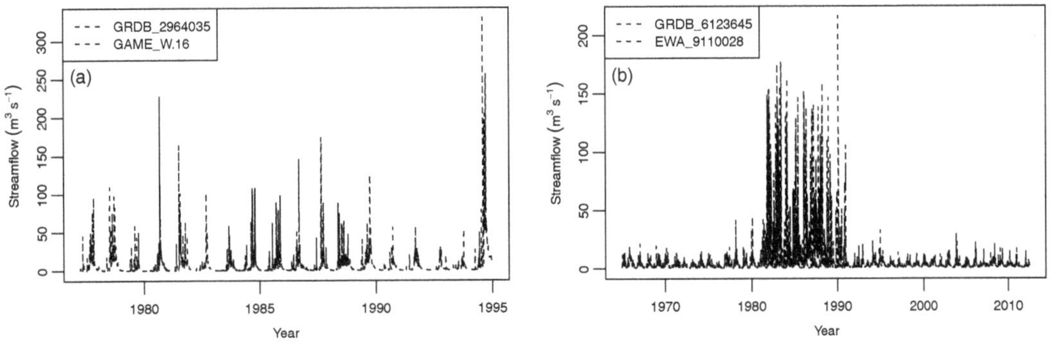

Figure 2. Examples of visually inspected duplication-candidate time series. **(a)** Two stations that were labelled "very likely identical" stations. **(b)** Two stations that were labelled "very likely different" stations.

Step 1. Transfer and review metadata available from original sources

The transfer of all existing metadata required a range of simple consistency checks and conforming rules, including the following.

1. Reviewing the geographical coordinates of all stations. Stations with unreasonable locations (e.g. located in the middle of the North Atlantic Ocean without any land mass, identified from Google Earth) were marked to be excluded from the subsequent delineation procedure (24 stations).

2. Separating the river name from the station name. Several sources use a consistent format for the station name consisting of two parts: the name of the station followed by the name of the water body. This pattern used a formula with "linking words" such as "at", "upstream" and "downstream". Taking station "BOM_406219" with original station name "Campaspe River at Lake Eppalock (Head Gauge)" as an example, the position of linking word "at" was identified and used to extract "river" metadata (Campaspe River) from the full station name.

3. Retaining the metadata of duplicated time series with the most data points in contrast to the other time series being removed. While this step may mistakenly remove some information, it is expedient and reflects the typical result of de-duplicated records that longer time series were kept while the shorter time series were removed.

Step 2. Generate "database-merging" information

This step documents a summary of efforts taken in creating a consistent set of GSIM metadata, and allows a user to check steps that were taken or to identify better procedures using alternative time series or metadata obtained from original sources. There are 12 fields documented for this purpose:

1. an indication of whether the time-series de-duplication procedure was used (one field),

2. which database and station were kept to construct the GSIM time series (two fields),

3. which station was removed and the corresponding database (three fields),

4. the value of metrics that represent similarities in the time-series metadata (five fields), and

5. the number of overlapping days, if applicable (one field).

Step 3. Generate information about data availability

The last step in compiling basic metadata for GSIM was to generate metrics that represent data availability for each GSIM time series, including the temporal coverage (i.e. the first and final years), the number of available daily observations, the number of missing data points, and the proportion of missing data points.

4.2 Catchment delineation procedure

With the ever-increasing availability of remote-sensing and modelled data products at global and continental scales, the provision of catchment boundaries is an important mechanism for extending the utility of GSIM. Although catchment boundaries can be generated easily using standard delineation algorithms in GIS packages, it requires a global coverage DEM dataset and reliable location to represent the outlet of each drainage area, which were unfortunately not readily available for GSIM project. This section describes the DEM products, and the algorithm to identify the "best outlet location" associated with each station that has been used in GSIM project.

The main DEM product used for GSIM was HydroSHEDS (http://hydrosheds.org, last access: 23 June 2017), which is

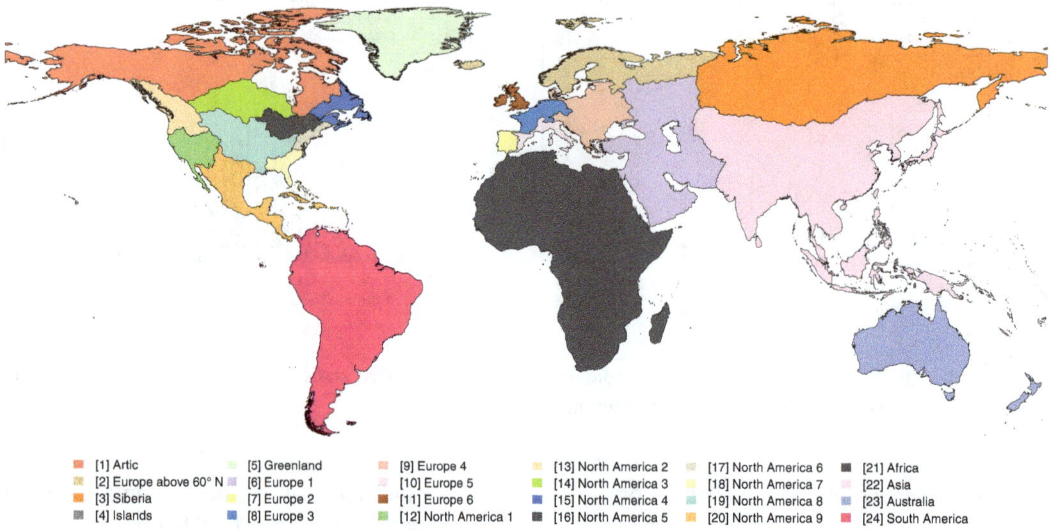

Figure 3. GSIM regions for catchment delineation and metadata extraction procedures.

available at 15 arcsec resolutions (Lehner et al., 2006), and has been used extensively in large-scale hydrological studies (Do et al., 2017; Lehner and Grill, 2013; Lehner et al., 2008; Wood et al., 2011). To address a limitation in the coverage of HydroSHEDS (no information in regions above 60° N, and some islands), the Viewfinder Panoramas elevation product at 15 arcsec resolutions was used (http://viewfinderpanoramas.org, last access: 25 June 2017) for those locations. This dataset has been used in several studies as an alternative DEM product to overcome similar data coverage issues (Barr and Clark, 2012; Fredin et al., 2012; Sil and Sitharam, 2016; Yamazaki et al., 2015). As there were more than 30 000 stations needing to be delineated, the HydroBASINS dataset was used, dividing the world into 24 regions, so that the task of delineation could be performed in parallel. The regions are shown in Fig. 3 and are generally independent in terms of drainage areas (Lehner and Grill, 2013). North America and Europe were specifically broken into more regions to address their relatively higher density of gauges. To maintain consistency when delineating boundaries, only one DEM product was used per GSIM region. As the quality of the Viewfinder Panoramas is not as clearly documented as for HydroSHEDS, its use was kept to a minimum. This resulted in five regions using Viewfinder DEM and 19 regions using HydroSHEDS (see Table 4).

Other challenges in the catchment delineation procedure are possible errors in the geographical coordinates representing the catchment outlet, such as typos in reported coordinates (e.g. 13.47° N instead of 14.47° N) or swapped order of the coordinate digits (e.g. 103.45° E instead of 103.54° E). These errors can lead to unreliable results of the delineation procedure, and so an algorithm to identify a location that represents catchment outlets well was also applied. This is described below.

Case 1. Reported station coordinates adopted as the outlet

If there was no information about a drainage area in the station metadata, the geographical coordinates of the station available from the data source were used as the outlet of the delineation process. There are automated techniques for repositioning outlets, such as choosing cells with the greatest flow accumulation within a search distance (Snap Pour Point ArcGIS tool), or finding the nearest cell possessing a flow-accumulation value above a specified threshold (Lindsay et al., 2008). Nonetheless, without information on the catchment area, it is impossible to assess the quality of the delineated catchment. Even if a repositioning technique were adopted, delineated catchment boundaries should be used with caution in this case, and therefore the original geographical coordinates was used to represent "best outlet location".

Case 2. Application of an automated repositioning algorithm

For stations with available information on catchment area, the automated repositioning procedure documented in GRDC report number 41 (Lehner, 2012) was used with some minor adjustments, and is summarized below.

1. The catchment area was estimated using the flow-accumulation dataset derived from the DEM products. This calculation was repeated for all pixels of the HydroSHEDS/Viewfinder gridded river network within a search radius of 5 km from the geographical coordinates of a specific station.

2. The estimated area values were compared with the reported area in the original metadata. All pixels were coded with the absolute value of their area differences

Table 4. DEM products used for each GSIM region.

Region	Description	DEM product
Arctic (region 1)	Represents the distant part of North America (including Alaska, most parts of Canada, and the eastern part of Autonomous Province, Russia)	Viewfinder DEM 15s
Europe above 60° N (region 2)	Represents countries located above 60° N (e.g. Sweden, Denmark, Norway, part of Germany, part of Russia)	Viewfinder DEM 15s
Siberia (region 3)	Represents areas above the 60° N part of Asia	Viewfinder DEM 15s
Islands (region 4)	Represents some islands across the Pacific Ocean (e.g. Honolulu, US) and Atlantic Ocean	Viewfinder DEM 15s
Greenland (region 5)	Represents land mass of Greenland	Viewfinder DEM 15s
Europe 1 to Europe 6 (six regions, from region 6 to region 11)	Represent most European countries (below 60° N)	HydroSHEDS DEM 15s
North America 1 to North America 9 (nine regions, from region 12 to region 20)	Represent US (except Alaska) and the southern part of Canada (below 60° N). It also includes central America for simplicity in processing catchment boundaries.	HydroSHEDS DEM 15s
Africa (region 21)	Represents Africa region	HydroSHEDS DEM 15s
Asia (region 22)	Represents Asia region (part of Kazakhstan, China, Mongolia, and Russia)	HydroSHEDS DEM 15s
Australia (region 23	Represents Australia, New Zealand, and some Pacific islands	HydroSHEDS DEM 15s
South America (region 24)	Represents South America	HydroSHEDS DEM 15s

(in %, with reported area in the metadata used as a reference). Pixels with area differences of more than 50 % were excluded. This procedure provided an area-based ranking scheme (RA) ranging from 0 to 50, where 0 indicates perfect agreement in catchment areas.

3. The distance to the original location of the station (geographical coordinates reported in the original metadata) was calculated for each pixel and normalized to reach 50 at the maximum distance of 5 km. This procedure provided a distance-based ranking scheme (RD) ranging from 0 to 50, where 0 indicates perfect agreement in station locations.

4. The final ranking scheme (R) was calculated as a combination of RA and RD, where distance rank was weighted twice as high ($R = \mathrm{RA} + 2\mathrm{RD}$) to penalize pixels that were further away from the original location.

5. The outlet was automatically relocated to the position of the pixel showing the lowest ranking value, and geographical coordinates of the pixel centroid were defined as the "best" outlet for this specific catchment.

6. In the original technical document (Lehner, 2012), a manual procedure was adopted for stations with differences in area above 50 (i.e. the search algorithm cannot find any pixel with an area difference less than 50 % within the 5 km search radius), or for stations that had no reported area in the data catalogue. This manual inspection process was infeasible given the scope of the GSIM project, having over 30 000 catchments being delineated and where river names were not available (or potentially inaccurately translated) for many stations.

A Python script was developed to automatically call the "best outlet location" algorithm and the catchment delineation toolset available in ArcGIS software (Jenson and Domingue, 1988) for each gauge using the chosen DEM data product. The delineated catchment boundary for each station was assigned a quality flag according to the discrepancy between reported drainage area and delineated catchment boundary area. There are four quality categories associated with the catchment boundary:

1. "High" quality: Area difference less than 5 %

2. "Medium" quality: Area difference from 5 % to less than 10 %

3. "Low" quality: Area difference from 10 % to less than 50 %

4. "Caution" quality: Area difference greater than or equal to 50 %, or the reported catchment area was not available in the GSIM catalogue.

Figure 4 demonstrates an example where the repositioning algorithm was used. Here the "best outlet location" was determined to be 4.8611 km away from the original location, which is defined by the reported geographical coordinates in the metadata (for station AR_0000007). The reported area in the metadata is 340 km^2, while the area of the delineated catchment boundary using the original coordinates was only 0.8 km^2, which is significantly lower than the correct number. On the other hand, the delineated catchment boundary using the "best outlet location" has an area of 363 km^2, indicating a better estimation of the upstream catchment boundary for this particular station.

4.3 Extraction of catchment-scale metadata

An important aspect of large-scale hydrology is the ability to exploit gridded datasets at the global scale (Bierkens, 2015; Bierkens et al., 2015; Gudmundsson and Seneviratne, 2015; Seneviratne et al., 2012; Ward et al., 2015). Having developed catchment boundaries for each GSIM station enabled a supplementary set of catchment-scale metadata to be derived with relative ease. A key feature is that the catchment boundaries and the subsequent metadata relates to the upstream contributing area that influences a gauge, rather than to the catchment (or arbitrarily defined sub-catchment) that contains the gauge and therefore includes a non-influencing downstream region.

In developing the catchment-scale metadata, a standard set of variables have been identified with a view to supporting a range of applications such as filtering stations according to characteristic features, performing analyses of streamflow according to explanatory features of a catchment, or classifying stations according to the (in)significance of human impact. As summarized in Table 5, a total of 12 global data products were used to derive 19 elements of catchment-scale metadata. These products were chosen to represent five main categories of catchment characteristics: (1) topography, (2) human impact, (3) climate type, (4) vegetation type, and (5) soil profile. Because the global data products have varying resolution and structure, the following method was used to derive the catchment-scale metadata.

1. Delineated catchment boundaries associated with each stream gauge were used to mask the subset of pixels from the resampled dataset.

2. If more than 30 % of the catchment area was not covered by a specific global data product, a "No data" code was given.

3. Metadata representing the characteristics of the upstream catchment for each streamflow gauge were calculated from the gridded data masked in step (1). There were three types of metrics calculated during this step.

   a. A single value. Used only for the elevation at the geographical coordinates of the gauge (i.e. the catchment outlet), number of large dams located within the catchment boundary, and total volume of corresponding reservoir.

   b. Average, min, max, and quartile values. Used for continuously varying data such as a slope or topography index. These metrics allow an idea of central tendency as well as spread of extracted data within each catchment boundary.

   c. Percentages of different classes of catchment characteristics. Used for categorical data. For example, there are 16 classes in the global lithology dataset, and the co-presence of more than one type of lithology occurs very often across all catchments. The percentages of each lithology class were therefore calculated and recorded for all available catchments. To make the results presentable in a final catchment-scale metadata matrix, an aggregated metric was calculated to indicate that there is a dominant class within the catchment boundary (i.e. more than 50 % of all available pixels). If there is no dominant class within the catchment boundary, a "No dominant class" string is provided.

5 Overview of the GSIM archive

This section summarizes the GSIM archive, including the availability of time series combined from 12 original data sources, the associated data products, and documentation outlining data quality (Sect. 5.1). The whole time-series database cannot be made available online due to data policies from a number of original data sources, some of which apply very strict terms and conditions regarding the redistribution of streamflow time series. To address this limitation and maintain the usefulness of GSIM to the research community, three metadata products have been developed and the availability of these data products is further discussed in Sect. 5.2.

5.1 Time-series availability

From the 35 002 time-series records obtained from 12 different sources, the final GSIM time-series archive holds a total of 30 959 unique stations, of which 30 935 stations have associated catchment shapefiles and catchment-scale metadata (24 stations were removed from this process due to suspect geographical locations). Most data sources are still active and being updated by the data authorities. GSIM, however, also included 425 "static" time series (from the ARCTICNET, GAME, and CHDP databases) that have been frozen since the early 2000s as these stations have improved the gauge

Table 5. Global data products used in GSIM and derived catchment-scale metadata.

Variables	Data sources	Spatial resolution	Reference period	Extracted metadata
Elevation	HydroSHEDS http://hydrosheds.org/ (last access: 23 June 2017) ViewFinder http://viewfinderpanoramas.org/ (last access: 23 June 2017)	15 arcsec × 15 arcsec	–	(1) Gauge elevation (2a–f) Average, minimum, maximum, first quartile, second quartile, and third quartile values of catchment elevation
Slope	Derived from HydroSHEDS and ViewFinder DEM by authors	15 arcsec × 15 arcsec	–	(3a–f) Average, minimum, maximum, first quartile, second quartile, and third quartile values of catchment slope
Topographic index	High-resolution global topographic index values (Marthews et al., 2015) https://catalogue.ceh.ac.uk/documents/ce391488-1b3c-4f82-9289-4beb8b8aa7da (last access: 23 June 2017)	15 arcsec × 15 arcsec	–	(4a–f) Average, minimum, maximum, first quartile, second quartile, and third quartile values of catchment topographic index
Drainage density	GRIN – Global River Network (Schneider et al., 2017) https://www.metis.upmc.fr/fr/node/375 (last access: 23 June 2017)	7.5 arcmin × 7.5 arcmin	–	(5a–f) Average, minimum, maximum, first quartile, second quartile, and third quartile values of catchment drainage density (km^{-1})
Dams	Global Reservoir and Dam (GRanD), version 1 (Lehner et al., 2011) http://sedac.ciesin.columbia.edu/data/set/grand-v1-dams-rev01 (last access: 23 June 2017)	6862 datapoints storage capacity of more than 0.1 km3	–	(6) Number of dams upstream (7) Total upstream storage volume
Population	Gridded Population of the World (GPW) version 4 (CIESIN, 2016) http://sedac.ciesin.columbia.edu/data/set/gpw-v4-population-count (last access: 23 June 2017)	30 arcsec × 30 arcsec	2005–2014	(8a–f) Average, minimum, maximum, first quartile, second quartile, and third quartile values of catchment population (2010) (9) 2010 Population count
Urbanization	Night Light Development Index (NLDI) dataset (Elvidge et al., 2012) http://www.soc-geogr.net/7/23/2012/sg-7-23-2012.html (last access: 23 June 2017)	0.25 arcdeg × 0.25 arcdeg	2006	(10a–f) Average, minimum, maximum, first quartile, second quartile, and third quartile values of NLDI over catchment
Irrigation	Historical Irrigation Dataset (Siebert et al., 2015) https://mygeohub.org/publications/8/2 (last access: 23 June 2017)	5 arcmin × 5 arcmin	2005	(11a–f) Average, minimum, maximum, first quartile, second quartile, and third quartile values of catchment Irrigated area (2005)
Climate type	World map of Koppen–Weiger climate classification system (Rubel and Kottek, 2010) http://koeppen-geiger.vu-wien.ac.at (last access: 23 June 2017)	5 arcmin × 5 arcmin	1951–2000	(12) Type of catchment climate (Koppen–Weiger) if one type present over more than 50 % catchment area, or "No dominant type"
Land cover	The Climate Change Initiative Land Cover (CCI-LC) dataset http://maps.elie.ucl.ac.be/CCI/viewer/download.php (last access: 23 June 2017)	7.5 arcsec × 7.5 arcsec	2015	(13) Type of catchment land cover (UN Land Cover Classification System) for 2015 if one type present over more than 50 % catchment area, or "No dominant type"
Lithological	The Global Lithological Map v1.0 (GLiM) dataset (Hartmann and Moosdorf, 2012) https://www.clisap.de/research/b:-climate-manifestations-and-impacts/crg-chemistry-of-natural-aqueous-solutions/global-lithological-map/ (last access: 23 June 2017)	0.5 arcdeg × 0.5 arcdeg	–	(14) Type of catchment lithology if one type present over more than 50 % catchment area or "No dominant type"
Soil profile	Soil grid 250 m (Hengl et al., 2017) https://soilgrids.org (last access: 23 June 2017)	7.5 arcsec × 7.5 arcsec	–	(15) Type of catchment soil class (World Reference Base) if one type present over more than 50 % catchment area or multiple types "No dominant type". (16a–f) Average, minimum, maximum, first quartile, second quartile, and third quartile values of weight percentage of sand over the catchment (17a–f) Average, minimum, maximum, first quartile, second quartile, and third quartile values of weight percentage of silt over the catchment (18a–f) Average, minimum, maximum, first quartile, second quartile, and third quartile values of weight percentage of clay over the catchment (19a–f) Average, minimum, maximum, first quartile, second quartile, and third quartile values of bulk content of soil over the catchment ($kg\,m^{-3}$)

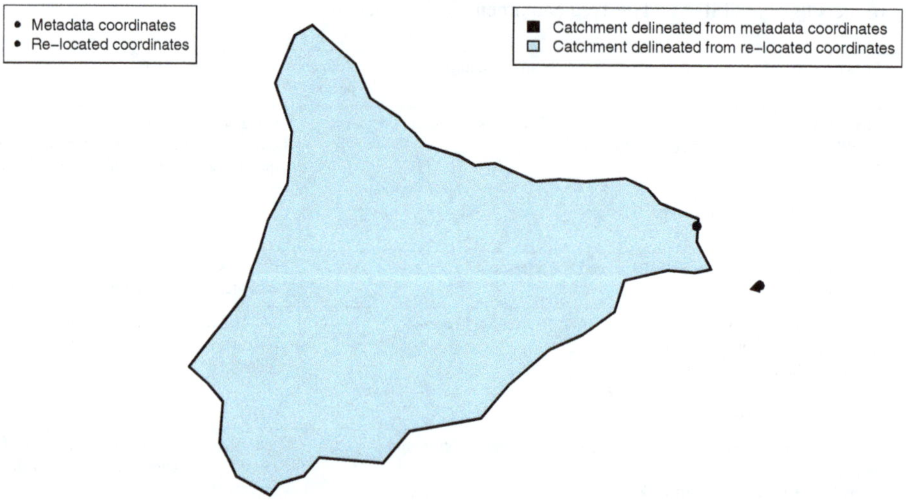

Figure 4. Example of improvement in quality of a catchment boundary using re-located geographical coordinates (for station AR_0000007).

density in regions with sparse streamflow observation systems (Russia, China, and Thailand, respectively). In addition, 2735 EWA stations (frozen since October 2014) were also included into GSIM as these time series have not been completely mirrored into GRDB database at the time GSIM was initiated. As these "static" time series have been frozen and no further update were provided, GSIM users are advised to use them with caution as the data may contain errors and/or have been replaced or updated.

As shown in Table 6, it is apparent that spatial coverage of the stations in the GSIM database varies significantly across continents, with North America and Europe having the greatest number of stations. Including the national databases such as MLIT (Japan), ANA (Brazil), BOM (Australia), and IWRIS (India) has significantly improved the observational network over the regions of Asia, South America, and Oceania (top panel of Fig. 5), some of which have recorded streamflow since the mid-20th century and were still operating at the time the GSIM database was initiated. This suggests that the national databases that are currently available should be given more attention in order to improve the quality and quantity of international archives.

Regarding temporal coverage, streamflow records across the globe are generally available for the second half of the 20th century (as shown in the bottom panel of Fig. 5). Regardless of missing data criteria, the number of available data gradually rises to its peak in the late 1970s to early 1980s, followed by a mild decrease in the late 1980s as also discussed by Hannah et al. (2011) and a secondary peak in the late 2000s. While the overall database has over 30 000 gauges, it is clear from Fig. 5 that from the 1960s onwards there are approximately from 10 000 to 15 000 gauges simultaneously active. This represents a significant increase in availability compared to the GRDB dataset, which had a total of approximately 9000 gauges and with a similar drop-off in available gauges depending on the filtering criteria applied.

5.2 Data products of GSIM

5.2.1 GSIM catalogue

The GSIM catalogue is designed for users to easily filter stations according to their purpose of application, and where necessary to transparently identify steps taken in the development of GSIM. The total number of 27 fields included in this document can be divided into three groups, namely the following.

1. Basic metadata. This group provides station identification, including a unique GSIM number, the name of the river, the name of the station, the elevation of the gauge, the provided geographical coordinates, and the catchment area.

2. Database merging metadata. This group of fields provides the identity of the numbers of original source(s), and if applicable the similarity metrics between duplicates.

3. Data availability metadata. This group of fields provides an overview of the data availability of each time series. These statistics were generated from the time-series data and can be used to filter station information, such as temporal coverage, data length, and the fraction of missing data.

As illustrated in Table 7, source datasets had significant gaps in the metadata, especially in cases of gauge elevation (not available in CHDP, GAME, HYDAT, BOM, and MLIT) and catchment area (not available in BOM and MLIT). In addition, the geographical coordinates of all stations were not

Table 6. Summary statistics of GSIM time series.

Continent	Number of stations	Average temporal coverage (years)	Shortest record (years)	Longest record (years)	Year of earliest entry	Year of latest entry
Africa	949	33.8	1	110	1903	2015
Europe	5778	40.3	1	208	1806	2016
Asia	1915	22.2	1	79	1921	2015
North America	15 884	42.9	1	156	1860	2016
South America	3449	29.3	1	116	1901	2016
Australia and Oceania	2984	31.4	1	131	1886	2016
Global	30 959	38.2	1	208	1806	2016

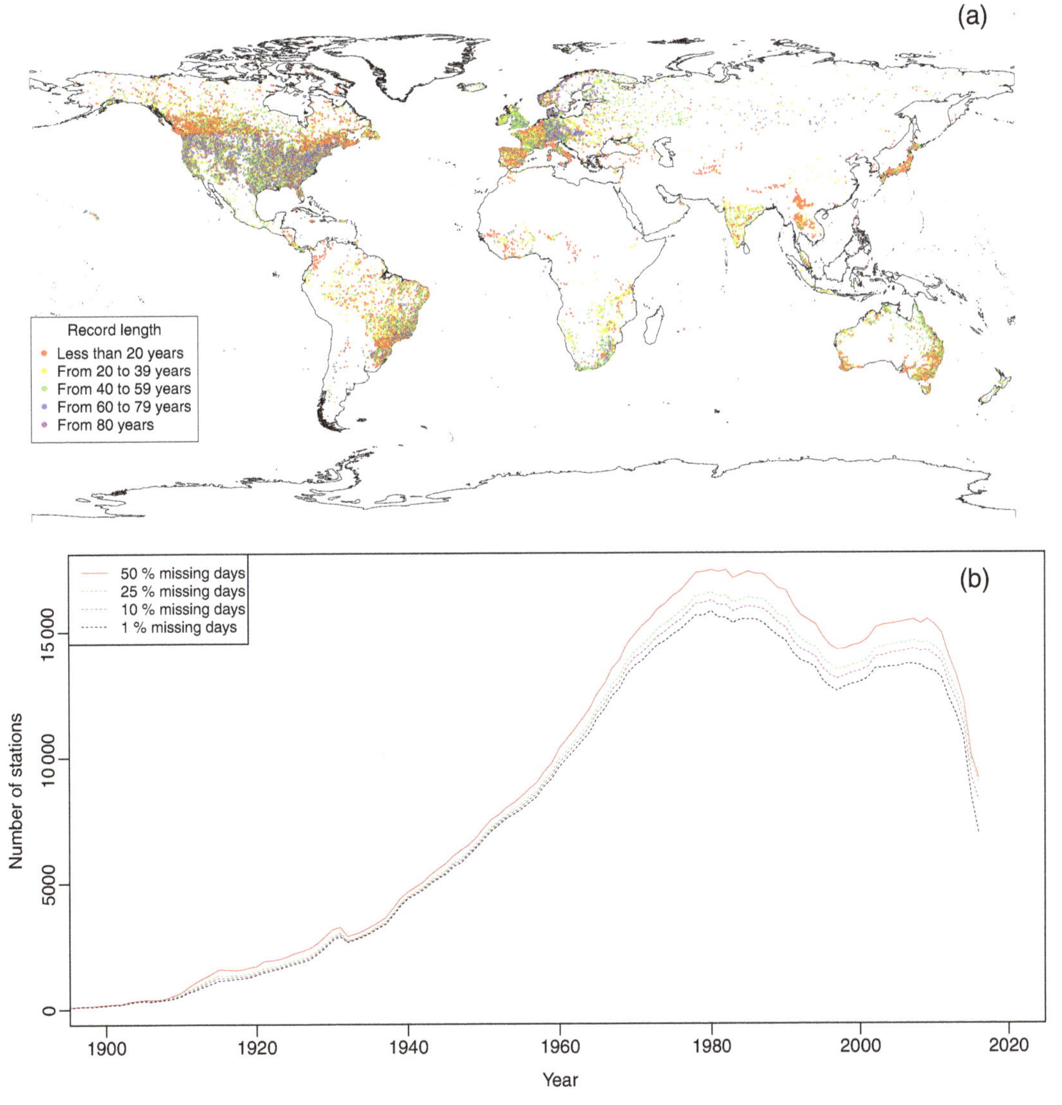

Figure 5. Availability of GSIM time series. **(a)** illustrates the length of record at each station, and **(b)** illustrates the number of available time series over time for four different missing data criteria.

Table 7. The percentage of stations accompanied by all basic metadata.

Dataset	Station ID	River name	Station name	Latitude	Longitude	Altitude	Catchment area
ADF	100	100	100	100	100	96.2	99.3
ANA	100	99.9	100	100	100	69	99
ARCTICNET	100	100	100	99.3	99.3	99.3	100
BOM	100	100	100	100	100	0	0
CHDP	100	99.4	100	100	100	0	84
EWA	100	100	100	100	100	98.5	94.5
GAME	100	100	100	100	100	0	100
GRDB	100	100	100	100	100	67	100
HYDAT	100	100	100	100	100	0	85.8
MLIT	100	100	100	100	100	0	0
USGS	100	100	100	100	100	93.7	25.5
WRIS	100	100	100	100	100	81.6	97.4
GSIM	100	99.9	100	99.9	99.9	50.4	73.8

Table 8. Percentages of available catchment-scale characteristics.

Catchment characteristics	Number of stations	Availability percentage
Climate classification	30 773	99.5
Drainage density	29 574	95.6
Elevation	30 932	99.9
Irrigation area	30 857	99.7
Land cover classification	30 888	99.8
Lithology type	30 154	97.5
Nightlight Development Index	23 096	74.7
Population count	30 894	99.9
Population density	30 800	99.6
Slope	30 862	99.8
Soil bulk density	30 812	99.6
Soil classification	30 764	99.4
Clay content	30 768	99.5
Clay content	30 695	99.2
Silt content	30 828	99.7
Topographic index	30 725	99.3

correctly recorded for all stations, with 24 removed as having suspect locations and 4871 shifted coordinates as part of the procedure for aligning catchment outlets with reported catchment areas.

5.2.2 Quality of catchment boundary

The catchment boundary is the second metadata product that is available through GSIM. Of all GSIM stations, 12 150 (39 %) were not associated with any information about drainage areas (including all MLIT and BOM stations); thus, a "Caution" flag is attached to upstream catchments of these stations. Another 24 stations with suspected geographical coordinates of stations were removed, and the final 18 785 stations were processed to identify the "best outlet" location to represent the outlet for delineating upstream catchments. The distribution and quality of the delineated catchments of these stations are provided in Fig. 6 (figures at continental scale are also provided as a Supplement).

As illustrated in the top panel, "Caution" catchments using "best" outlets (identified using the method outlined in Sect. 4.2) are generally located across all GSIM regions. However, the "Caution" flag appears more frequently over regions above 60° N. Further checks would be required to improve the association of catchment boundaries with stations. Unfortunately, the biggest caveat that applies to the GSIM database, as with any global database, is that the metadata were collated from a number of sources with varying standards of documentation and quality assurance and with limited capacity for additional checking other than automated procedures. Therefore, there is likely to be a non-trivial degree of error in the metadata for both geographical location and drainage area. Another issue that may lead to unreliable results of the delineation process is error in the DEM products. This potential error has been documented (Lehner, 2012; Lehner et al., 2006), and lower-quality DEM products generally exist for regions above 60° N due to the lower quality of the original elevation products used to derive the DEM datasets. Another note for the use of delineated catchments is that very small catchments (area less than 50 km^2) should be handled with care, as the "best" outlets could be located incorrectly while still delivering "acceptable" discrepancies as part of the automated procedure.

Nonetheless, the quality of delineated catchments is quite positive (as illustrated in the lower panels of Fig. 6). Of all 18 785 catchments that had reported drainage area in the GSIM catalogue, 68.25, 11.8, and 15.92 % of catchments have "High" quality (area discrepancy of less than 5 %), "Medium" quality (area discrepancy from 5 % to less than 10 %), and "Low" quality (area discrepancy from 10 to less than 50 %), respectively, while there are only 4.03 % catch-

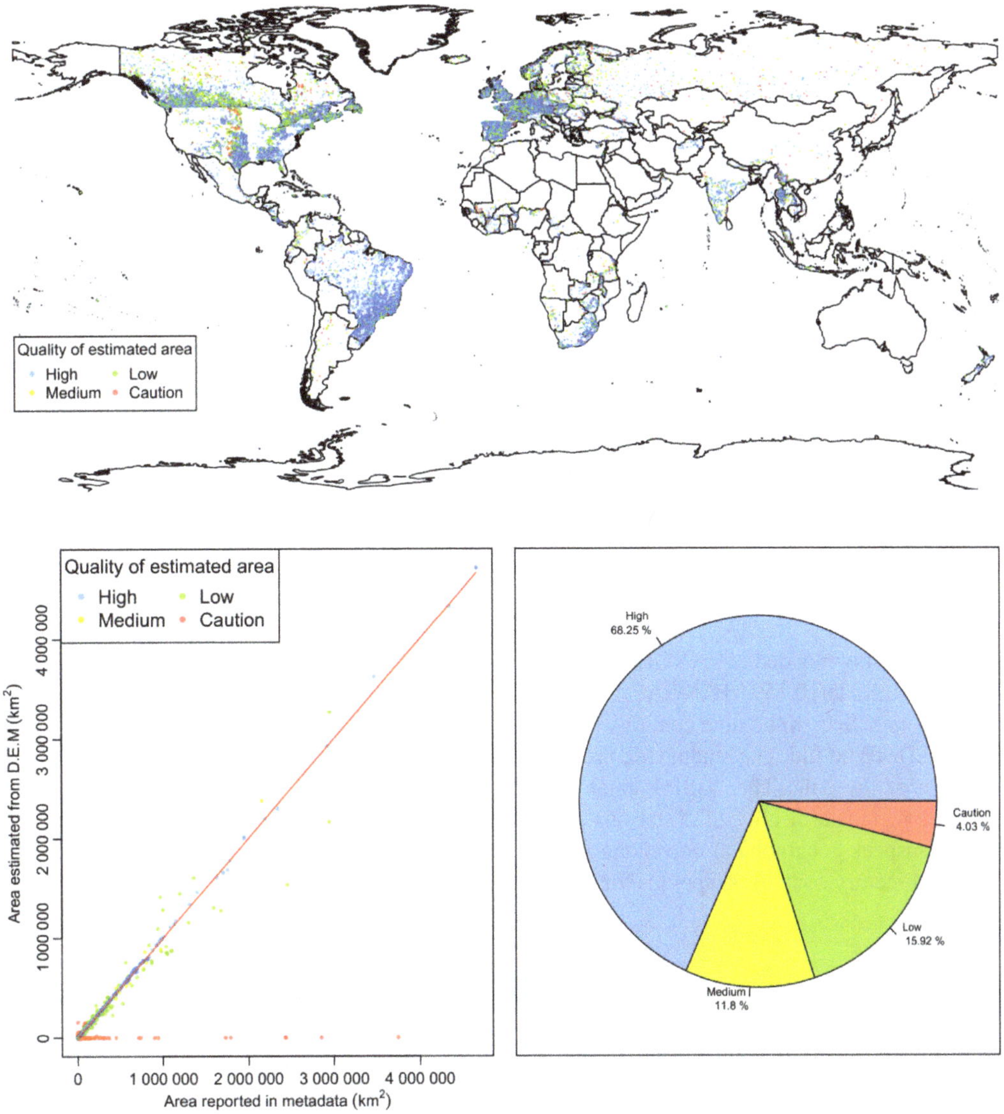

Figure 6. Quality of the delineated catchment boundary according to the categories of high, medium, low, and caution identified in Sect. 4.2 (for 18 785 stations that have reported drainage area and reasonable geographical coordinates).

ments with "Caution" quality (area discrepancy of more than or equal to 50 %).

5.2.3 Catchment-scale characteristics

The final data product that has been made available is the auxiliary information extracted from 12 global coverage datasets representing many characteristics associated with GSIM stations. Overall, the spatial coverage of original data products (mostly satellite-based is quite good (see Table 8), with just a small fraction of catchments (less than 10 %) that have more than 30 % of their areas not covered by these datasets. The exception is the Nightlight Development Index (NLDI – computed from the 2006 Nightlights dataset, Ziskin et al., 2010, and the 2006 Landscan gridded population, Bhaduri et al., 2002). This dataset does not have approximately 25.3 % of catchments covered, for more than 70 % of their areas.

It is important to note that while these catchment-scale characteristics are consistent products available for all stations, documentation for the original source data should be consulted during application to appreciate the limitations and appropriateness of each variable. For example, the GRanD database is not exhaustive of all dams worldwide and there can be ambiguities over the affiliated dates (e.g. whether they represent conception, construction, or commissioning). Furthermore, the extent of the overlapping period between temporal coverage of streamflow time series and remote sensing based datasets needs to be carefully assessed in cause–

effect studies. Similarly, it is likely that there will be updated or new data gridded datasets available over time so that applications should consider the appropriateness of the information used. The availability of metadata products emerging from the GSIM project demonstrates the possibility of using reported global data products to extract catchment-scale characteristics associated with each station with reasonable quality, enabling many potential applications from this rich information.

6 Data availability

The data described in this paper are available as a compressed zip archive containing (i) a readme file, (ii) metadata of all GSIM stations obtained from original data sources and time series, (iii) quality of catchment boundary and catchment characteristics extracted from 12 global data products, (iv) a list of stations with suspect geographical coordinates, and (v) catchment boundaries for 30 935 stations that have a reasonable geographical location.

The data can be freely downloaded at PANGEA data depository https://doi.pangaea.de/10.1594/PANGAEA.887477 (Do et al., 2018). The uploaded zip archive contains two directories and one README.txt file. The readme file provides a detailed description of the data. The "GSIM_catalogue" directory contains the metadata of all GSIM stations and a list of stations with suspect geographical coordinates. The "GSIM_catchments" directory contains shapefiles for 30 935 stations.

7 Conclusions

In situ observations of daily streamflow with global coverage are crucial to understanding large-scale freshwater resources that are fundamental for societal development. The GSIM archive, designed as an expansion of the GRDB database, has demonstrated the possibility of significantly improving the coverage and density of the global streamflow observational datasets using free-to-access databases. The development of the GSIM database was not possible without the tremendous investment in the production and ongoing maintenance of original data sources of GSIM. This fact emphasizes the key role of data authorities and international initiatives in enabling advances in large-scale hydrology by making data publicly available to the community.

While the activities of GSIM have been extensive in searching out and collating databases, they are by no means exhaustive (e.g. since submission we have been notified of additional potential candidates for inclusion such as the Mekong River Commission database, Chile national water database, and Argentina national water database). It is the authors' intention that this project will stimulate further efforts toward the development of coordinated and consistent representation of global streamflow observations. For this reason, the process of developing the archive was designed with automation in mind. With the exception of needing to visually inspect some cases of duplicated time series, the archive was automated using scripts in the R and Python programming languages.

Although the GSIM database was compiled from data sources that can be obtained free of charge via a data portal or by submitting written requests to data authorities, there are some strict conditions related to the redistribution of unprocessed data. Therefore, it is impossible to make the whole GSIM collection publicly available. In addition, with the main aim of harvesting as much data as possible, the GSIM database is not focused on collecting high-quality datasets such as referenced hydrological networks that are available in many countries (Whitfield et al., 2012), and thus the data quality may vary significantly across the available time series. To address these limitations and increase the usefulness of the GSIM database, we conducted a set of quality checking procedures for all GSIM time series. These quality-assured records were then used to produce a dedicated set of indices capturing important aspects of the daily dynamics from GSIM time series, and to explore potential applications of GSIM in large-scale hydrology. Detailed information about this work and associated distributed data is described in the second part of our series on GSIM (Gudmundsson et al., 2018a, b).

With the GSIM archive and production information made publicly available in a transparent manner, this project serves the broader hydrology community with improved coverage and quality of streamflow information. This project has yielded a significant increase in the availability of streamflow observations through the process of collating readily accessed online data, and with ongoing efforts there will be opportunities for further extension. Streamflow observations represent an underutilized resource, in part due to access limitations, but also due to challenges in accounting for human impacts in the observed record. These challenges notwithstanding, ongoing advances in global-scale hydrological models and ever-increasing access to remote-sensed products indicate that wider access to streamflow data has the potential to significantly enhance our knowledge of global water resources.

Competing interests. The authors declare that they have no conflict of interest.

Acknowledgements. The authors would like to express their appreciation to all the national agencies and institutions that made the streamflow data available for this study. We would like to thank

Sonia I. Seneviratne for her discussions and support on the collation of the GSIM archive. Hong Xuan Do receives financial support from the Australia Award Scholarship (AAS). Seth Westra's time was supported by Australian Research Council Discovery project DP150100411. The authors also wish to thank two reviewers for their constructive comments and suggestions. The authors would like to express their sincere thanks to Danlu Guo for her support in collecting the MLIT database. This work was supported with supercomputing resources provided by the Phoenix HPC service at the University of Adelaide.

Edited by: David Carlson

References

Addor, N., Newman, A. J., Mizukami, N., and Clark, M. P.: The CAMELS data set: catchment attributes and meteorology for large-sample studies, Hydrol. Earth Syst. Sci., 21, 5293–5313, https://doi.org/10.5194/hess-21-5293-2017, 2017.

Arsenault, R., Bazile, R., Ouellet Dallaire, C., and Brissette, F.: CANOPEX: A Canadian hydrometeorological watershed database, Hydrol. Process., 30, 2734–2736, 2016.

Barr, I. D. and Clark, C. D.: An updated moraine map of Far NE Russia, J. Maps, 8, 431–436, 2012.

Bhaduri, B., Bright, E., Coleman, P., and Dobson, J.: LandScan, Geoinformatics, 5, 34–37, 2002.

Bierkens, M. F. P.: Global hydrology 2015: State, trends, and directions, Water Resour. Res., 51, 4923–4947, 2015.

Bierkens, M. F. P., Bell, V. A., Burek, P., Chaney, N., Condon, L. E., David, C. H., de Roo, A., Döll, P., Drost, N., Famiglietti, J. S., Flörke, M., Gochis, D. J., Houser, P., Hut, R., Keune, J., Kollet, S., Maxwell, R. M., Reager, J. T., Samaniego, L., Sudicky, E., Sutanudjaja, E. H., van de Giesen, N., Winsemius, H., and Wood, E. F.: Hyper-resolution global hydrological modelling: what is next?, Hydrol. Process., 29, 310–320, 2015.

Burn, D. H., Hannaford, J., Hodgkins, G. A., Whitfield, P. H., Thorne, R., and Marsh, T.: Reference hydrologic networks II. Using reference hydrologic networks to assess climate-driven changes in streamflow, Hydrolog. Sci. J., 57, 1580–1593, 2012.

Christen, P.: Data matching: concepts and techniques for record linkage, entity resolution, and duplicate detection, Springer Science & Business Media, 2012.

CIESIN: Gridded Population of the World, Version 4 (GPWv4): Population Density Adjusted to Match 2015 Revision UN WPP Country Totals, NASA Socioeconomic Data and Applications Center (SEDAC), Palisades, NY, 2016.

Do, H. X., Westra, S., and Michael, L.: A global-scale investigation of trends in annual maximum streamflow, J. Hydrol., 552, 28–43, https://doi.org/10.1016/j.jhydrol.2017.06.015, 2017.

Do, H. X., Gudmundsson, L., Leonard, M., and Westra, S.: The Global Streamflow Indices and Metadata Archive – Part 1: Station catalog and Catchment boundary, PANGAEA, https://doi.org/10.1594/PANGAEA.887477, 2018.

Elvidge, C. D., Baugh, K. E., Anderson, S. J., Sutton, P. C., and Ghosh, T.: The Night Light Development Index (NLDI): a spatially explicit measure of human development from satellite data, Soc. Geogr., 7, 23–35, 2012.

Falcone, J. A.: GAGES-II: Geospatial attributes of gages for evaluating streamflow, US Geological Survey, 2011.

Fekete, B. M., Vörösmarty, C., and Grabs, W.: Global Composite Runoff Fields on Observed River Discharge and Simulated Water Balances/Water System Analysis Group. University of New Hampshire, and Global Runoff Data Centre, Koblenz, Federal Institute of Hydrology (BfG), Koblenz, Germany, Federal Institute of Hydrology (BfG), 2002a.

Fekete, B. M., Vörösmarty, C. J., and Grabs, W.: High-resolution fields of global runoff combining observed river discharge and simulated water balances, Global Biogeochem. Cy., 16, 15-11–15-10, 2002b.

Fekete, B. M., Robarts, R. D., Kumagai, M., Nachtnebel, H.-P., Odada, E., and Zhulidov, A. V.: Time for in situ renaissance, Science, 349, 685–686, 2015.

Fredin, O., Rubensdotter, L., van Welden, A., Larsen, E., and Lyså, A.: Distribution of ice marginal moraines in NW Russia, J. Maps, 8, 236–241, 2012.

GRDC: Report of the Twelfth Meeting of the GRDC Steering Committee, Koblenz, Germany, 18–19 June 2014, Global Runoff Data Centre (GRDC), Koblenz, Germany, 23 pp., 2015.

Gudmundsson, L. and Seneviratne, S. I.: Towards observation-based gridded runoff estimates for Europe, Hydrol. Earth Syst. Sci., 19, 2859–2879, https://doi.org/10.5194/hess-19-2859-2015, 2015.

Gudmundsson, L. and Seneviratne, S. I.: Observation-based gridded runoff estimates for Europe (E-RUN version 1.1), Earth Syst. Sci. Data, 8, 279–295, https://doi.org/10.5194/essd-8-279-2016, 2016.

Gudmundsson, L., Tallaksen, L. M., Stahl, K., Clark, D. B., Dumont, E., Hagemann, S., Bertrand, N., Gerten, D., Heinke, J., Hanasaki, N., Voss, F., and Koirala, S.: Comparing Large-Scale Hydrological Model Simulations to Observed Runoff Percentiles in Europe, J. Hydrometeorol., 13, 604–620, 2012.

Gudmundsson, L., Seneviratne, S. I., and Zhang, X.: Anthropogenic climate change detected in European renewable freshwater resources, Nat. Clim. Change, 7, 813–816, https://doi.org/10.1038/nclimate3416, 2017.

Gudmundsson, L., Do, H. X., Leonard, M., and Westra, S.: The Global Streamflow Indices and Metadata Archive (GSIM) – Part 2: Quality control, time-series indices and homogeneity assessment, Earth Syst. Sci. Data, 10, 787–804, https://doi.org/10.5194/essd-10-787-2018, 2018a.

Gudmundsson, L., Do, H. X., Leonard, M., and Westra, S.: The Global Streamflow Indices and Metadata Archive (GSIM) – Part 2: Time Series Indices and Homogeneity Assessment, PANGAEA, https://doi.org/10.1594/PANGAEA.887470, 2018b.

Hannaford, J. and Marsh, T.: An assessment of trends in UK runoff and low flows using a network of undisturbed catchments, Int. J. Climatol., 26, 1237–1253, 2006.

Hannah, D. M., Demuth, S., van Lanen, H. A. J., Looser, U., Prudhomme, C., Rees, G., Stahl, K., and Tallaksen, L. M.: Large-scale river flow archives: importance, current status and future needs, Hydrol. Process., 25, 1191–1200, 2011.

Hartmann, J. and Moosdorf, N.: The new global lithological map database GLiM: A representation of rock properties at the Earth surface, Geochem. Geophy. Geosy., 13, Q12004, https://doi.org/10.1029/2012GC004370, 2012.

Henck, A. C., Montgomery, D. R., Huntington, K. W., and Liang, C.: Monsoon control of effective discharge, Yunnan and Tibet, Geology, 38, 975–978, 2010.

Hengl, T., de Jesus, J. M., Heuvelink, G. B., Gonzalez, M. R., Kilibarda, M., Blagotić, A., Shangguan, W., Wright, M. N., Geng, X., and Bauer-Marschallinger, B.: SoilGrids250m: Global gridded soil information based on machine learning, PloS one, 12, e0169748, https://doi.org/10.1371/journal.pone.0169748, 2017.

Hodgkins, G. A., Whitfield, P. H., Burn, D. H., Hannaford, J., Renard, B., Stahl, K., Fleig, A. K., Madsen, H., Mediero, L., Korhonen, J., Murphy, C., and Wilson, D.: Climate-driven variability in the occurrence of major floods across North America and Europe, J. Hydrol., 552, 704–717, 2017.

Huang, S., Kumar, R., Flörke, M., Yang, T., Hundecha, Y., Kraft, P., Gao, C., Gelfan, A., Liersch, S., Lobanova, A., Strauch, M., van Ogtrop, F., Reinhardt, J., Haberlandt, U., and Krysanova, V.: Evaluation of an ensemble of regional hydrological models in 12 large-scale river basins worldwide, Climatic Change, 141, 381–397, https://doi.org/10.1007/s10584-016-1841-8, 2016.

Jenson, S. K. and Domingue, J. O.: Extracting topographic structure from digital elevation data for geographic information-system analysis, Photogramm. Eng. Rem. S., 54, 1593–1600, 1988.

Kundzewicz, Z. W., Plate, E. J., Rodda, H. J., Rodda, J. C., Schellnhuber, H. J., and Strupczewski, W. G.: Changes in flood risk–setting the stage. In: Changes in flood risk in Europe, CRC Press, 2012.

Kundzewicz, Z. W., Kanae, S., Seneviratne, S. I., Handmer, J., Nicholls, N., Peduzzi, P., Mechler, R., Bouwer, L. M., Arnell, N., Mach, K., Muir-Wood, R., Brakenridge, G. R., Kron, W., Benito, G., Honda, Y., Takahashi, K., and Sherstyukov, B.: Flood risk and climate change: global and regional perspectives, Hydrolog. Sci. J., 59, 1–28, 2013.

Lehner, B.: Derivation of watershed boundaries for GRDC gauging stations based on the HydroSHEDS drainage network, Report 41 in the GRDC Report Series, available at: http://www.bafg.de/GRDC/EN/02_srvcs/24_rprtsrs/report_41.html?nn=201764 (last access: 23 June 2017), 2012.

Lehner, B. and Grill, G.: Global river hydrography and network routing: baseline data and new approaches to study the world's large river systems, Hydrol. Process., 27, 2171–2186, 2013.

Lehner, B., Liermann, C. R., Revenga, C., Vörösmarty, C., Fekete, B., Crouzet, P., Döll, P., Endejan, M., Frenken, K., and Magome, J.: Global reservoir and dam (grand) database, Technical Documentation, Version, 1, 2011.

Lehner, B., Verdin, K., and Jarvis, A.: HydroSHEDS technical documentation, version 1.0, World Wildlife Fund US, Washington, DC, 1–27, 2006.

Lehner, B., Verdin, K., and Jarvis, A.: New Global Hydrography Derived From Spaceborne Elevation Data, Eos, Transactions American Geophysical Union, 89, 93–94, 2008.

Lindsay, J. B., Rothwell, J. J., and Davies, H.: Mapping outlet points used for watershed delineation onto DEM-derived stream networks, Water Resour. Res., 44, W08442, https://doi.org/10.1029/2007WR006507, 2008.

Lucas, C. and Tingley, D.: translateR, available at: https://cran.r-project.org/web/packages/translateR/index.html (last access: 23 June 2017), 2016.

Marthews, T. R., Dadson, S. J., Lehner, B., Abele, S., and Gedney, N.: High-resolution global topographic index values for use in large-scale hydrological modelling, Hydrol. Earth Syst. Sci., 19, 91–104, https://doi.org/10.5194/hess-19-91-2015, 2015.

Merz, B., Kundzewicz, Z., Delgado, J., Hundecha, Y., and Kreibich, H.: Detection and attribution of changes in flood hazard and risk, Changes in flood risk in Europe, IAHS Special Publication, 10, 435–458, 2012.

Milly, P. C. D., Wetherald, R. T., Dunne, K., and Delworth, T. L.: Increasing risk of great floods in a changing climate, Nature, 415, 514–517, 2002.

Milly, P. C. D., Betancourt, J., Falkenmark, M., Hirsch, R. M., Kundzewicz, Z. W., Lettenmaier, D. P., Stouffer, R. J., Dettinger, M. D., and Krysanova, V.: On Critiques of "Stationarity is Dead: Whither Water Management?", Water Resour. Res., 51, 7785–7789, 2015.

Nelson, B.: Data sharing: Empty archives, Nature, 461, 160–163, https://doi.org/10.1038/461160a, 2009.

Rubel, F. and Kottek, M.: Observed and projected climate shifts 1901–2100 depicted by world maps of the Köppen-Geiger climate classification, Meteorol. Z., 19, 135–141, 2010.

Schaake, J., Cong, S., and Duan, Q.: The US MOPEX data set, IAHS publication, 307, 9–28, 2006.

Schneider, A., Jost, A., Coulon, C., Silvestre, M., Théry, S., and Ducharne, A.: Global-scale river network extraction based on high-resolution topography and constrained by lithology, climate, slope, and observed drainage density, Geophys. Res. Lett., 44, 2773–2781, 2017.

Seneviratne, S. I., Nicholls, N., Easterling, D., Goodess, C. M., Kanae, S., Kossin, J., Luo, Y., Marengo, J., McInnes, K., and Rahimi, M.: Changes in climate extremes and their impacts on the natural physical environment, Managing the risks of extreme events and disasters to advance climate change adaptation, 109–230, 2012.

Siebert, S., Kummu, M., Porkka, M., Döll, P., Ramankutty, N., and Scanlon, B. R.: A global data set of the extent of irrigated land from 1900 to 2005, Hydrol. Earth Syst. Sci., 19, 1521–1545, https://doi.org/10.5194/hess-19-1521-2015, 2015.

Sil, A. and Sitharam, T.: Detection of Local Site Conditions in Tripura and Mizoram Using the Topographic Gradient Extracted from Remote Sensing Data and GIS Techniques, Nat. Hazards Rev., 18, 04016009, https://doi.org/10.1061/(ASCE)NH.1527-6996.0000228, 2016.

Vörösmarty, C. J., Moore, B., Grace, A. L., Gildea, M. P., Melillo, J. M., Peterson, B. J., Rastetter, E. B., and Steudler, P. A.: Continental scale models of water balance and fluvial transport: an application to South America, Global Biogeochem. Cy., 3, 241–265, 1989.

Wanders, N. and Wada, Y.: Decadal predictability of river discharge with climate oscillations over the 20th and early 21st century, Geophys. Res. Lett., 42, 10689–610695, 2015.

Ward, P. J., Jongman, B., Weiland, F. S., Bouwman, A., van Beek, R., Bierkens, M. F., Ligtvoet, W., and Winsemius, H. C.: Assessing flood risk at the global scale: model setup, results, and sensitivity, Environ. Res. Lett., 8, 044019, https://doi.org/P10.1088/1748-9326/8/4/044019, 2013.

Ward, P. J., Eisner, S., Flörke, M., Dettinger, M. D., and Kummu, M.: Annual flood sensitivities to El Niño-Southern Oscillation at the global scale, Hydrol. Earth Syst. Sci., 18, 47–66, https://doi.org/10.5194/hess-18-47-2014, 2014.

Ward, P. J., Jongman, B., Salamon, P., Simpson, A., Bates, P., De Groeve, T., Muis, S., de Perez, E. C., Rudari, R., Trigg, M. A., and Winsemius, H. C.: Usefulness and limitations of global flood risk models, Nat. Clim. Change, 5, 712–715, 2015.

Whitfield, P. H., Burn, D. H., Hannaford, J., Higgins, H., Hodgkins, G. A., Marsh, T., and Looser, U.: Reference hydrologic networks I. The status and potential future directions of national reference hydrologic networks for detecting trends, Hydrolog. Sci. J., 57, 1562–1579, 2012.

Wood, E. F., Roundy, J. K., Troy, T. J., van Beek, L. P. H., Bierkens, M. F. P., Blyth, E., de Roo, A., Döll, P., Ek, M., Famiglietti, J., Gochis, D., van de Giesen, N., Houser, P., Jaffé, P. R., Kollet, S., Lehner, B., Lettenmaier, D. P., Peters-Lidard, C., Sivapalan, M., Sheffield, J., Wade, A., and Whitehead, P.: Hyperresolution global land surface modeling: Meeting a grand challenge for monitoring Earth's terrestrial water, Water Resour. Res., 47, W05301, https://doi.org/10.1029/2010WR010090, 2011.

Yamazaki, D., Trigg, M. A., and Ikeshima, D.: Development of a global ~ 90 m water body map using multi-temporal Landsat images, Remote Sens. Environ., 171, 337–351, 2015.

Ziskin, D., Baugh, K., Hsu, F.-C., and Elvidge, C. D.: Methods used for the 2006 radiance lights, Proceedings of the Asia-Pacific Advanced Network, 30, 131–142, 2010.

10

The Global Streamflow Indices and Metadata Archive (GSIM) – Part 2: Quality control, time-series indices and homogeneity assessment

Lukas Gudmundsson[1], Hong Xuan Do[2], Michael Leonard[2], and Seth Westra[2]

[1]Institute for Atmospheric and Climate Science, Department of Environmental Systems Science, ETH Zurich, Universitaetstrasse 16, Zurich 8092, Switzerland

[2]School of Civil, Environmental and Mining Engineering, University of Adelaide, Adelaide, Australia

Correspondence: Lukas Gudmundsson (lukas.gudmundsson@env.ethz.ch)

Abstract. This is Part 2 of a two-paper series presenting the Global Streamflow Indices and Metadata Archive (GSIM), which is a collection of daily streamflow observations at more than 30 000 stations around the world. While Part 1 (Do et al., 2018a) describes the data collection process as well as the generation of auxiliary catchment data (e.g. catchment boundary, land cover, mean climate), Part 2 introduces a set of quality controlled time-series indices representing (i) the water balance, (ii) the seasonal cycle, (iii) low flows and (iv) floods. To this end we first consider the quality of individual daily records using a combination of quality flags from data providers and automated screening methods. Subsequently, streamflow time-series indices are computed for yearly, seasonal and monthly resolution. The paper provides a generalized assessment of the homogeneity of all generated streamflow time-series indices, which can be used to select time series that are suitable for a specific task. The newly generated global set of streamflow time-series indices is made freely available with an digital object identifier at https://doi.pangaea.de/10.1594/PANGAEA.887470 and is expected to foster global freshwater research, by acting as a ground truth for model validation or as a basis for assessing the role of human impacts on the terrestrial water cycle. It is hoped that a renewed interest in streamflow data at the global scale will foster efforts in the systematic assessment of data quality and provide momentum to overcome administrative barriers that lead to inconsistencies in global collections of relevant hydrological observations.

1 Introduction

Although terrestrial freshwater is an essential component of the Earth system and a prerequisite for societal development, the availability of relevant in situ observations at the global scale has been limited. Until now, most relevant in situ observations have been held by national and regional authorities, and despite their best efforts, international data centres only have access to a small subset of the full observed record (Do et al., 2018a). This situation stands in contrast to the fact that monitoring data are increasingly being made publicly available through regional and national authorities (Do et al., 2018a). In this paper series, we present an international collection of river and streamflow observations that covers more than 30 000 stations around the globe, highlighting the fact that these are among the best monitored variables of the terrestrial water cycle (Fekete et al., 2012, 2015; Gudmundsson and Seneviratne, 2015; Hannah et al., 2011). Part 1 of the paper series (Do et al., 2018a) documents the data-collection process together with a meta-database that allows users to recreate the collection from the original data sources. In addition, Part 1 of this paper series also presents auxiliary data including catchment boundaries delineated from global digital elevation models as well as selected properties (e.g. land cover, climate) of these catchments.

While the data collection outlined in Part 1 (Do et al., 2018a) increases the spatial and temporal availability of streamflow records at the global scale, it is important to also consider the quality of the data. This is especially relevant for this merged data product combining information from several databases, which might have been set up with different objectives. Furthermore, data contained in individual databases may stem from different sources, often with unknown quality control procedures. In addition, changes in instrumentation as well as human impacts such as stream straightening or flow regulations can have pronounced effects on the observed record. Establishing a database of quality controlled streamflow observations is therefore essential for many applications, including e.g. the need to evaluate the increasing number of continental- and global-scale hydrological and land-surface models that have emerged in recent decades (Beck et al., 2017; Gudmundsson et al., 2012a, b; Haddeland et al., 2011; Zaitchik et al., 2010) and the assessment of human impacts on the terrestrial water cycle (Alkama et al., 2013; Barnett et al., 2008; Destouni et al., 2013; Gudmundsson et al., 2017; Hegerl et al., 2015; Hidalgo et al., 2009; Jaramillo and Destouni, 2015; Oliveira et al., 2011). While there have been significant efforts in the climatological community to share and standardize transnational weather observations as well as derivative data products (Alexander et al., 2006; Becker et al., 2013; Dee et al., 2011; Harris et al., 2014; Haylock et al., 2008; Poli et al., 2016), the hydrological community has traditionally been reticent to adopt regional or global approaches, instead focussing predominantly on the catchment scale. A more concerted and coordinated effort to understand the quality of streamflow observations across the globe provides significant opportunities for fostering hydrological research in support of understanding of global water budgets. This paper initiates the process of evaluating, analysing and documenting the quality of observed streamflow time series, providing a method for increasing the reliability and ongoing value of the database. To do so, this paper expands on previous research (Gudmundsson and Seneviratne, 2016) and applies a set of transparent and reproducible methods to evaluate the quality of the considered records.

One limitation of the newly assembled collection of daily river flow and streamflow time series is that publication of unprocessed daily values is restricted for some of the original data sources. To nevertheless be able to publish relevant information on observational streamflow, we therefore present here processed data in the form of time-series indices that capture essential aspects of (i) the water balance, (ii) seasonality, (iii) low flows and (iv) floods. The approach of publishing time-series indices instead of raw daily values is adapted from the CCl/WCRP/JCOMM Expert Team on Climate Change Detection and Indices (ETCCDI) (https://www.wcrp-climate.org/data-etccdi), which has developed this approach to make relevant climate information publicly available in cases where access to raw daily values is restricted. The ETCCDI has focussed on indices characterizing changes in extreme precipitation and temperature, based on a core collection of indices proposed by Frich et al. (2002). Both Klein Tank et al. (2009) and Zhang et al. (2011) provide additional background on the usage and computation of the ETCCDI indices. Klein Tank et al. (2009) also provide guidelines for quality control of the raw daily input data, index computation and assessment of time-series homogeneity.

The use of time-series indices for characterizing the temporal evolution of selected river flow characteristics is also common practice in the hydrological literature. Typically used time-series indices include mean annual flows (e.g. Kumar et al., 2009; Milly et al., 2005; Small et al., 2006; Stahl et al., 2010, 2012), indices that can be used to characterize changes in the seasonal cycle (e.g. Blöschl et al., 2017; Cunderlik and Ouarda, 2009; Ehsanzadeh and Adamowski, 2010; Hidalgo et al., 2009; Moore et al., 2007; Rauscher et al., 2008; Regonda et al., 2005; Stewart et al., 2005), time series of annual percentiles (e.g. Gudmundsson et al., 2011; Lins and Slack, 1999; Zhang et al., 2001), flood indices (e.g. Blöschl et al., 2017; Hodgkins et al., 2017; Kumar et al., 2009; Kundzewicz et al., 2005; Lins and Slack, 1999; McCabe and Wolock, 2002; Small et al., 2006; Svensson et al., 2005; Zhang et al., 2001) and low-flow indicators (e.g. Hisdal et al., 2001; Lins and Slack, 1999; McCabe and Wolock, 2002; Small et al., 2006; Stahl et al., 2010, 2012; Svensson et al., 2005; Tallaksen et al., 1997; Zhang et al., 2001).

In addition, several studies have focussed on collections of hydrological signatures (or flow characteristics) that are designed to summarize long-term properties of observed river flow and streamflow (e.g. 2013; Beck et al., 2015; Olden and Poff, 2003; Sawicz et al., 2011, 2014; Westerberg et al., 2016). These hydrological signatures include e.g. mean annual flow, flow percentiles, characteristics of the flow duration curves, indications of seasonality and the base flow index. These signatures are typically derived from all daily values in a long time window (e.g. the base flow index computed from all daily values from 1985 to 2010). This is an important structural difference if compared to time-series indices, which are typically computed every year, every season or every month (e.g. time series of annual maxima) and thus also allow for an assessment of changing hydrological conditions over time.

The following sections build upon these efforts and present a collection of quality controlled river and streamflow time-series indices. To do so, we first introduce an approach to check the quality of individual daily observations using a combination of information provided with the original data and data-driven procedures. Subsequently we present a collection of time-series indices that can be computed for yearly, seasonal and monthly resolution. An assessment of the statistical homogeneity of the newly derived indices is provided to allow users to filter the published data according to their own eligibility criteria. Given that each application may warrant a different assessment of the trade-off between the quantity and quality of available data, the presented collection of streamflow time-series indices has sought to avoid predefined eligibility criteria (such as predefining a base period

Table 1. Quality flags of daily values of all databases that enter the GSIM collection (see Do et al., 2018a).

Database	Quality code
GRDB	Not recommended by data provider. There are four flags: −999 – missing data, no correction 1 – corrected data, no method specified 99 – usage not recommended by the provider 900 – calculated from daily water level Note: in recent updates GRDC does not provide quality flags.
EWA	Not recommended by data provider (similar to GRDB)
ARCTICNET	Quality flag not provided
GAME	Quality flag not provided
CHDP	Quality flag not provided
USGS	Flags were provided for each data point. There are four categories: A: value has been validated to be published A:e: value was estimated and validated to be published P and P:e: Provisional data
BOM	Flags were provided for each data point. There are five categories documented: A (flag 10): best available data B (flag 90): compromised to represent the parameter C (flag 110): estimated value E (flag 140): quality is not known F (flag 210): poor quality or missing Flag "−1" also presents to indicate missing value
HYDAT	Quality flags were only provided for some data points. There are five categories documented: A: Partial Day (numeric value 1) B: Ice Conditions (numeric value 2) D: Dry (numeric value 3) E: Estimated (numeric value 4) S: Sample(s) collected this day (numeric value 5)
WRIS	Qc flag not provided
ANA	Quality flags were only provided for some data points. Flags were described in Portuguese. Below are the authors' interpretation. 0: No description 1: Real measurement 2: Estimated 3: Doubtful 4: Dry
MLIT	Quality flag not provided
AFD	Quality flag not provided

or presupposing only high-quality sites). The paper closes with an open invitation to the hydrological and Earth science communities on how to best facilitate activities that might lead to sustained collation, curation and improvement of global streamflow data.

2 Quality control (QC) of daily values

2.1 Strategy for QC of daily values

As the considered data stem from several sources, some of which have a complex history, it is difficult to a priori judge the quality of individual records. Ideally, each of the considered series would be accompanied by detailed information on the station properties (e.g. information on sensors or the design of the gauging weir) and on the credibility of indi-

Table 2. Translation of daily quality control (QC) flags of the original databases (Table 1) to standardized values prior to the calculation of indices. Note that the Global Runoff Data Centre advises not to consider the QC flags in the GRDB and EWA files. Note also that some databases (HYDAT, ANA) do not provide QC flags for all daily data.

Dataset	QC flag not provided	QC flags are not recommended by data provider	Reliable	Suspect
GRDB	–	All data points	–	–
EWA	–	All data points	–	–
ARCTICNET	All data points	–	–	–
GAME	All data points	–	–	–
CHDP	All data points	–	–	–
USGS	–	–	"A" and "A:e" (approved data)	"P" and "P:e" data (provisional data)
BOM	–	–	A (table below)	B, C, D, F (table below)
HYDAT	Other data points	–	B, D, S	A, E
WRIS	All data points	–	–	–
ANA	0, no value	–	1, 4	2, 3
MLIT	All data points	–	–	–
AFD	All data points	–	–	–

vidual daily values. However, this information is often not available or difficult to access and only some of the original data sources provide daily quality flags (Table 1). In addition, the large number of languages involved and the sheer quantity of gauging stations render a detailed manual assessment unfeasible. Nevertheless, it is essential to apprise the quality of individual observations prior to any assessment. As some of the considered time series come with daily quality flags (usually based on simple plausibility checks), while others do not, the two cases are treated separately.

2.2 Quality control of daily values if reliable flags are provided

As noted in Do et al. (2018a), some of the considered databases provide quality control (QC) flags for daily values that distinguish between reliable and suspect observations (Table 1). To allow for a combined assessment, the original QC flags were translated into a common set that distinguishes *suspect* from *reliable* values (Table 2). This step is necessary for consistency, since some databases provide a variety of QC flags to indicate suspect cases, but neither the same flags nor the level of fidelity are available across all databases. Regarding the Global Runoff Data Centre (GRDC), while QC flags are available in the EWA and GRDB files entering the presented collection, the GRDC advised not to use them. In these cases, the time series are treated as if no QC flags were provided. Note also that the GRDC has discontinued QC flagging in the latest version of the data. Some databases do not provide QC flags for every time step (Table 2); in these cases time steps without original QC flags were assumed to be *reliable* as long as at least one time step was flagged in the respective time series.

2.3 Quality control of daily values if no reliable flags are available

For original time-series files for which no QC flags are available or for which there is advice against using available QC flags by the data providers (GRDB and EWA), automated techniques can be used to classify the reliability of individual daily data points using simple and reproducible tests focussing on the plausibility of individual values. The following three criteria are based on a previously used procedure (Gudmundsson and Seneviratne, 2016), were developed on the basis of techniques described in Reek et al. (1992) and the ECA & D Project Team and Royal Netherlands Meteorological Institute (2013; later referred to as EAC&D13), and were further refined using suggestions on outlier detection for index calculation by Klein Tank et al. (2009):

1. Days for which $Q < 0$ are flagged as *suspect*, where Q denotes a daily streamflow value. The rationale underlying this rule is that streamflow values smaller than zero are non-physical (Gudmundsson and Seneviratne, 2016).

2. Daily values with more than 10 consecutive equal values larger than zero are flagged as *suspect*. This rule is motivated by the fact that many days with consecutive streamflow values often occur due to instrument failure (e.g. damaged sensors, ice jams) or flow regulations. The threshold of 10 days is a compromise chosen to account for the possibility that consecutive equal observations may reflect the truth e.g. if day-to-day fluctuations are below the sensitivity of the employed sensor (Gudmundsson and Seneviratne, 2016).

3. Based on a previously suggested approach for evaluating temperature series (Klein Tank et al., 2009), daily streamflow values are declared as outliers if values of $\log(Q + 0.01)$ are larger or smaller than the mean value

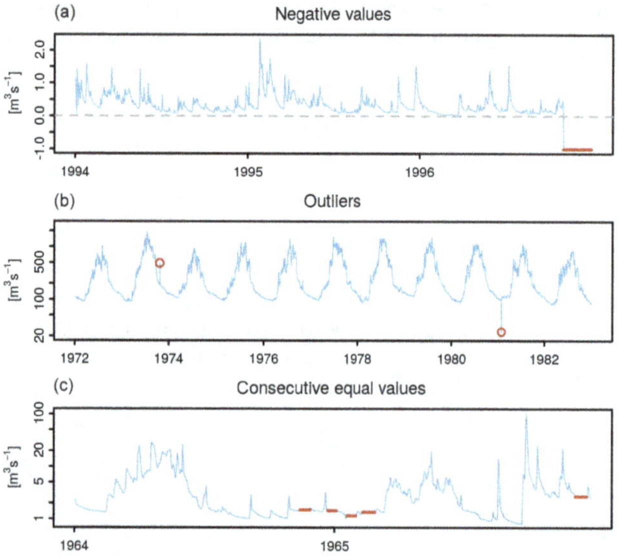

Figure 1. Three example time series illustrating issues detected by the three daily quality control criteria (highlighted in red). The first panel shows negative values at the end of the time series of Rohr at Rohrhardsberg, Germany. The second panel shows two outliers detected in the time series of Vakhsh at Gram, Tajikistan. The third panel shows instances of more than 10 consecutive equal values found in the time series of Tanara at Ponte di Nava, Italy. Note that all time series were trimmed for visualization purposes. Note also the logarithmic axis in panels two and three.

of $\log(Q+0.01)$ plus or minus 6 times the standard deviation of $\log(Q+0.01)$ computed for that calendar day for the entire length of the series. The mean and standard deviation are computed for a 5-day window centred on the calendar day to ensure that a sufficient amount of data is considered. The log-transformation is used to account for the skewness of the distribution of daily streamflow values and 0.01 was added because the logarithm of zero is undefined. Outliers are flagged as *suspect*. The rationale underlying this rule is that unusually large or small values are often associated with observational issues. The 6 standard-deviation threshold is a compromise, aiming at screening out outliers that could come from instrument malfunction, while not flagging extreme floods or low flows.

An example of the outcome of this automated quality control of daily observations is shown in Fig. 1, which displays daily streamflow observations at three locations and highlights time steps that did not pass the three above-mentioned criteria. Note that the outlier detection (middle panel) did not screen out extreme floods or low flows, but only values that were unusually large or small for the respective time of the year, where one case involves a spurious large flow and the other a spurious small flow.

3 Streamflow indices

3.1 General considerations, design rules and reliability

3.1.1 General considerations

Table 3 describes a set of streamflow time-series indices that are designed to facilitate the analysis of (i) changes in the regional water balance, (ii) changes in the seasonal cycle, (iii) floods, and (iv) low flows. Many of the considered indices have been previously used in the scientific literature and Table 4 presents, wherever possible, a selection of relevant references and additional information. Note also that index selection was limited to those that can be computed without a base period, which excludes many; examples include "the number of days in a year, or season, for which daily values exceed a time-of-year-dependent threshold" (Zhang et al., 2005), drought deficit volumes (Loon and Anne, 2015; Tallaksen et al., 1997) and anomalies with respect to a climatological normal (McKee et al., 1993; Shukla and Wood, 2008). There are two reasons for excluding these indices: first, regional differences in temporal coverage hinder an unambiguous identification of a common base period that can be used around the globe. Second, it is now well established that indices that depend on a base period are prone to inhomogeneities if the base period is shorter than the considered series (Sippel et al., 2015; Zhang et al., 2005). Although both analytical (Sippel et al., 2015) and non-parametric (Zhang et al., 2005) solutions exist to mitigate this problem, we chose not to include indices that require a base period. This is because the available solutions either depend on strong normality assumptions (Sippel et al., 2015) or are computationally intensive (Zhang et al., 2005), which implies that the time-series indices cannot be easily extended when new data become available. Finally, it is noteworthy to mention that indices are easier to update when they do not have a base period, as they can be computed without knowledge of previous values.

3.1.2 Design rules for index calculation

The design rules for calculating time-series indices closely follow the recommendations of ECA&D13. Before index calculation, all daily values that are flagged as *suspect* by the daily QC procedure are set to missing, and indices are computed using the remaining data points. All indices are computed on yearly time steps, while some indices are also computed with seasonal and monthly resolution. Seasons are defined as December–January–February (DJF), March–April–May (MAM), June–July–August (JJA) and September–October–November (SON). The reason for not computing all indices for seasonal and monthly resolutions is related either to the fact that some indices are only defined on annual timescales, or to the amount of data required for reliable computation. All considered indices are described in Tables 3 and 4.

Table 3. Definition of time-series indices contributing to the GSIM archive. Abbrev. Indicates the abbreviation of the index name used throughout this paper as well as in the database. Resol. indicates the time resolution for which the index is computed, which can take values of Y (yearly), seasonal (S) and monthly (M).

Title	Abbrev.	Units	Resol.	Definition
Mean daily streamflow	MEAN	($m^3\,s^{-1}$)	Y, S, M	Arithmetic mean of daily streamflow.
Standard deviation of daily streamflow	SD	($m^3\,s^{-1}$)	Y, S, M	Standard deviation of daily streamflow.
Coefficient of variation of daily streamflow	CV	(–)	Y, S, M	Standard deviation of daily streamflow divided by the mean daily streamflow (SD/MEAN).
Interquartile range of daily streamflow	IQR	($m^3\,s^{-1}$)	Y, S, M	75th–25th percentile of daily streamflow.
Minimum daily streamflow	MIN	($m^3\,s^{-1}$)	Y, S, M	Minimum value of daily streamflow.
Maximum daily streamflow	MAX	($m^3\,s^{-1}$)	Y, S, M	Maximum value of daily streamflow.
Minimum 7-day mean streamflow	MIN7	($m^3\,s^{-1}$)	Y, S, M	Minimum 7-day arithmetic mean streamflow. For computation, the complete daily time series are first smoothed with a backward looking moving average with a 7-day window. Subsequently, the minimum value for each yearly, seasonal or monthly period is determined.
Maximum 7-day mean streamflow	MAX7	($m^3\,s^{-1}$)	Y, S, M	Maximum 7-day arithmetic mean streamflow. For computation, the complete daily time series are first smoothed with a backward looking moving average with a 7-day window. Subsequently, the maximum value for each yearly, seasonal or monthly period is determined.
10th, 20th, 30th, 40th, 50th, 60th, 70th, 80th and 90th percentiles of daily streamflow	P10, P20, P30, P40, P50, P60, P70, P80, P90	($m^3\,s^{-1}$)	Y, S	Percentile values of daily streamflow computed for each yearly and seasonal period, where low percentiles (e.g. 10th percentile) correspond to low flows.
Centre timing	CT	(doy)	Y	The day of the year (doy) at which 50 % of the annual flow is reached. The index is computed for calendar years, where 1 denotes 1 January.
Day of minimum streamflow	DOYMIN	(doy)	Y	The day of the year (doy) at which the minimum flow occurred, where 1 denotes 1 January. The maximum value is 365 for normal years and 366 for leap years.
Day of maximum streamflow	DOYMAX	(doy)	Y	The day of the year (doy) at which the maximum flow occurred, where 1 denotes 1 January. The maximum value is 365 for normal years and 366 for leap years.
Day of minimum 7-day mean streamflow	DOYMIN7	(doy)	Y	Day of the year (doy) at which the minimum 7-day arithmetic mean streamflow occurred, where 1 denotes 1 January. The maximum value is 365 for normal years and 366 for leap years. For computation, the daily time series is first smoothed using a backward looking moving average with a 7-day window length. Subsequently, the day of the minimum of each year is determined.
Day of maximum 7-day mean streamflow	DOYMAX7	(doy)	Y	Day of the year (doy) at which the maximum 7-day arithmetic mean streamflow occurred, where 1 denotes 1 January. The maximum value is 365 for normal years and 366 for leap years. For computation, the daily time series is first smoothed using a backward looking moving average with a 7-day window length. Subsequently, the Julian day of the maximum of each year is determined.
Gini coefficient	GINI	(–)	Y	For daily runoff values q of each year, that are sorted with index i in increasing order such that $q_i \leq q_{i+1}$ GINI is defined as $\frac{1}{n}\left(n+1-2\left(\frac{\sum_{i=1}^{n}(n+1-i)q_i}{\sum_{i=1}^{n}q_i}\right)\right)$, where n is the number data points available for that year. The Gini coefficient ranges from 0 to 1. Values of 0 indicate uniform distribution of flows throughout the time period (i.e. year), whereas values close to 1 indicate that all the flows occur on a single day.

Table 4. Commentary and literature supporting the GSIM indices.

Abbrev.	Commentary
MEAN	Mean daily streamflow is a commonly used water-balance measure and often used as a proxy for renewable freshwater resources (Oki and Kanae, 2006; Shiklomanov et al., 2004; Vörösmarty et al., 2000). Observed time series of mean yearly or monthly streamflow has e.g. been subject to trend analysis at regional to continental scales (e.g. Kumar et al., 2009; Lettenmaier et al., 1994; Lins and Slack, 1999; Milly et al., 2005; Small et al., 2006; Stahl et al., 2010, 2012).
SD	The standard deviation of daily streamflow provides information on the total variability for each yearly, seasonal and monthly time step. This index therefore includes information related to floods and low flows as well as the amplitude of the annual cycle (yearly only). We are not aware of any study analysing time series of the standard deviation of daily streamflow.
CV	The coefficient of variation of daily streamflow is a relative measure of daily variability. In contrast to SD, CV is independent of the mean flow and does hence allow for an isolated assessment of day-to-day streamflow variability. We are not aware of any study analysing time series of the coefficient of variation of daily streamflow.
IQR	The inter quartile range is a measure of day-to-day streamflow variability. Through its definition as the difference between the 75th and 25th percentiles, the IQR provides information on the width of the centre of the distribution and is less sensitive to extreme outliers than SD or CV. We are not aware of any study analysing time series of the standard deviation of daily streamflow.
MIN	Minimum daily streamflow is a regularly used low-flow indicator. Especially the yearly minimum has been used widely as it is an easy to interpret measure and lends itself to analysis in the framework of the generalized extreme value distribution (Tallaksen and van Lanen, 2004). Annual minimum streamflow series are also commonly subject to large-scale trend analysis (Kumar et al., 2009; Lins and Slack, 1999; McCabe and Wolock, 2002; Zhang et al., 2001).
MAX	Maximum daily streamflow is a widely used indicator for high flows and floods. Especially annual maximum time series are regularly considered as they allow for a straightforward interpretation and can easily be analysed through the generalized extreme value distribution (Katz et al., 2002). Time series of annual maximum streamflow have been subject to regional and global trend assessments (e.g. Do et al., 2017; Hall et al., 2015; Kumar et al., 2009; Kundzewicz et al., 2005; Lins and Slack, 1999; McCabe and Wolock, 2002; Small et al., 2006; Zhang et al., 2001).
MIN7	Time series of minimum 7-day mean streamflow have been repeatedly used as a low-flow and drought metric. Through the smoothing operation, MIN7 is less sensitive to small day-to-day fluctuations, but focusses on sustained periods with limited water availability. MIN7 time series have e.g. been subject to large scale trend assessments (Kumar et al., 2009; Small et al., 2006; Stahl et al., 2010; Svensson et al., 2005).
MAX7	Time series of 7-day mean maximum streamflow do not focus on the highest water levels ever recorded, but rather on sustained periods of very high flow. Time series of MAX7 have e.g. been used to assess streamflow trends in India (Kumar et al., 2009; Stahl et al., 2012).
P10, P20, P30, P40, P50, P60, P70, P80, P90	Percentiles of daily streamflow provide together with MIN and MAX an approximation of the empirical cumulative distribution function (ECDF) of daily streamflow for each considered seasonal or yearly time period. These indices are not provided on monthly resolution, as it appears to be excessive to compute percentiles in 10 % steps based on 28 to 31 daily values. Note also that an alternative definition of the ECDF is also referred to as the flow-duration curve (FDC) in the hydrological literature. The difference between the ECDF and the FDC is that the FDC uses an inverse definition of percentiles (exceedance frequencies), such that high values correspond to low flows (Tallaksen and van Lanen, 2004; Vogel and Fennessey, 1994). Besides approximations of the ECDF, the percentile series can be used to characterize "moderate extremes" (Zhang et al., 2011), i.e. very high or very low values that can occur several times each year and are hence more robust to quantify. Sets of annual percentile series have for example been used to investigate regional low- and high-flow dynamics in Europe (Gudmundsson et al., 2011) and have been subject to regional-scale trend assessments (Lins and Slack, 1999; McCabe and Wolock, 2002; Zhang et al., 2001).
CT	The centre timing is an index that is sensitive to changes in the seasonal cycle. Lower values indicate that more than half of the annual discharge has occurred earlier in the year. That means, that values smaller or equal than 182 would correspond to a year for with at least half of the streamflow volume has occurred in the first half of the year. Note that CT is usually defined for hydrological years in the literature and that the precise definition of CT can vary between studies (Hidalgo et al., 2009; Moore et al., 2007; Rauscher et al., 2008; Regonda et al., 2005; Stewart et al., 2005). Here we compute CT for calendar years to ensure consistency with the remaining indices and because the definition of the hydrological year depends on local climate conditions. Time series of CT have been used to assess changes in the timing of the seasonal cycle of streamflow in several regional studies (Hidalgo et al., 2009; Moore et al., 2007; Rauscher et al., 2008; Regonda et al., 2005; Stewart et al., 2005).
DOYMIN	The timing of annual minimum flow can provide valuable information on the processes underlying low flows. For example, in snowy regions, the minimum flow often occurs in the winter months, whereas in other regions minimum flows occur in the season with low precipitation and large atmospheric water demand. We are not aware of any study that is explicitly analysing time series of DOYMIN.

Table 4. Continued.

Abbrev.	Commentary
DOYMAX	The timing of annual maximum streamflow can be a valuable indicator for the flood generating processes. In cold regions annual, maximum flow is often associated with snowmelt, while in other regions it may be associated with intense convective precipitation during the warm season or soil moisture. Time series of DOYMAX have for example been used to assess trends in the timing of floods in Europe (Blöschl et al., 2017) and Canada (Cunderlik and Ouarda, 2009).
DOYMIN7	Overall the interpretation of DOYMIN7 is analogous to the interpretation of DOYMIN. Note, however, that DOXMIN7 is representative of a 7-day period of sustained low flows and is less sensitive to outliers. We are not aware of any study that is explicitly analysing time series of DOYMIN.
DOYMAX7	Generally, the interpretation of DOYMAX7 is analoguous to the interpretation of DOYMAX, although DOYMAX7 represents a 1-week period of sustained high flows and is less sensitive to outliers. We are not aware of any study that is explicitly analysing time series of DOYMAX7.
GINI	The Gini coefficient is a metric that was originally established in economic sciences as a measure of economic inequality (Ceriani and Verme, 2012). It is a measure of dispersion that is not dependent on the absolute value of the variable under consideration and can be interpreted as a measure of the variability implied by the flow duration curve. It is therefore, like the CV, a relative variability measure that can easily be compared among different regions. Although we are not aware of any study investigating annual GINI time series derived from streamflow, relevant applications to observed precipitation (Rajah et al., 2014) and global hydrological model output (Masaki et al., 2014) are emerging.

3.1.3 Reliability of index values

Not all daily time steps have observations, and some daily observations have been flagged as *suspect* and were therefore removed. Consequently yearly, seasonal and monthly index values are not equally reliable. To allow users to judge the reliability of index values at individual time steps, the number of daily values used for index calculation at each time step is provided. Based on the recommendations of ECA&D13, the following rules for daily data availability can be applied to identify *reliable* index values.

1. Index values at a *yearly* time step are reliable if at least 350 daily observations are declared *reliable*.
2. Index values at a *seasonal* time step are reliable if at least 85 daily observations are declared *reliable*.
3. Index values at a *monthly* time step are reliable if at least 25 daily observations are declared *reliable*.

Note, however, that these are very conservative rules which may be relaxed depending on the needs of specific applications.

3.2 Example time series

To provide a first impression of the considered indices, Fig. 2 shows all indices at annual resolution for Wiese at Zell, located in south-western Germany. In addition, Fig. 3 shows the MEAN at monthly, seasonal and yearly resolutions of the same river.

3.3 Temporal coverage of yearly, seasonal and annual indices

Figure 4a displays the number of years covered by all considered time series, highlighting both large variations in station density and time-series length, which is consistent with the availability of the original daily time series (Do et al., 2018a). To better appraise regional differences in temporal coverage, Fig. 4b shows the distribution of the number of years that are typically available for each station for major continental regions. The median time-series length is longest for North America and Europe and shortest for Oceania and Asia. The above-mentioned daily quality control (Sect. 2) as well as ECA&D13 criteria for judging the reliability of yearly, seasonal or monthly index values (Sect. 3.1.3) imply that the space–time coverage of the index data is not equal to the coverage of the original daily time series. Figure 4c shows the distribution of the fraction of time steps that were classified as reliable for the considered continental regions and for yearly, seasonal and monthly resolutions. Overall the figure highlights that the fraction of reliable time steps is largest for the Americas, Europe and Asia, while it is lowest for Oceania and Africa. Furthermore, it should be noted that the fraction of reliable time steps is lowest for yearly indices. This is related to the fact that full years are deemed unreliable when fewer than 350 valid observations are used for computation (following the ECA&D13 rules). Note however that the relatively strict ECA&D13 rules can be relaxed and should be adapted depending on user needs.

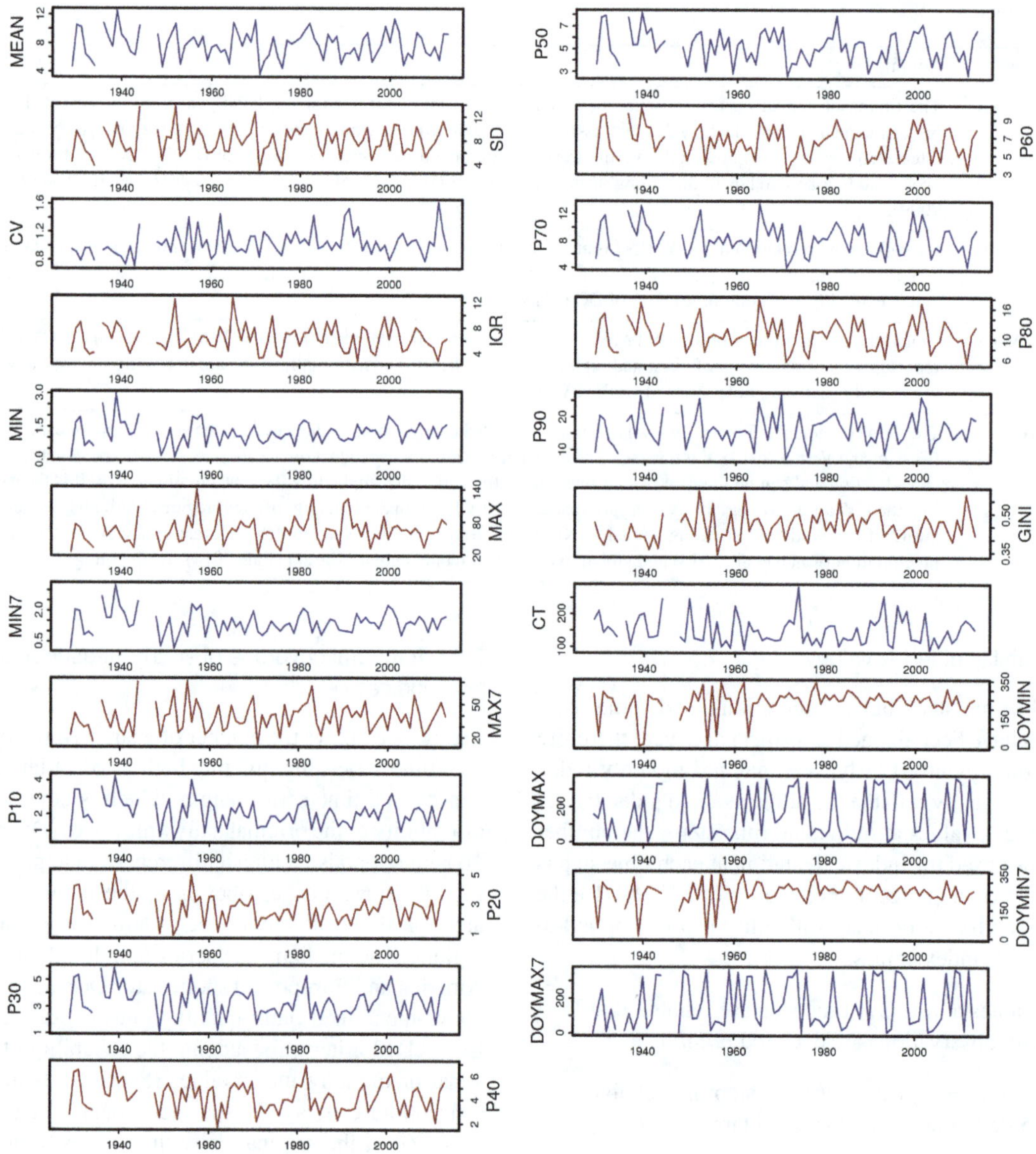

Figure 2. All considered indices at yearly resolution, shown for the River Wiese at Zell, south-western Germany. Yearly values are only displayed if they contain at least 350 reliable daily observations. See the text for details on units, interpretation and reliability classification.

4 Homogeneity assessment

4.1 Methods for homogeneity assessment

4.1.1 Homogeneity tests

Any environmental time series can be subject to inhomogeneities, i.e. unnatural sudden shifts in their statistical moments. In the simplest case, such inhomogeneities could be a jump in the mean between two time periods (see Fig. 5, top), but also changes in variability (e.g. reduced peak flows) or shifts in higher-order moments. The reasons for such inhomogeneities in streamflow time series are manifold, but they can "be related to changes in instrumentation, gauge restoration, recalibration of rating curves, flow regulation or channel engineering" (Gudmundsson and Seneviratne, 2016). As all the above-mentioned factors can be detrimental to a scientific investigation, it is essential to check time series against inhomogeneities. Here we apply a previously utilized collection of tests (Gudmundsson and Seneviratne, 2016), which is recommended by ECA&D13 and has been thoroughly tested for temperature and precipitation indices (Wijngaard et al., 2003). This collection of tests contains (i) the standard normal homogeneity test (Alexandersson, 1986), (ii) the Buishand range test (Buishand, 1982), (iii) the Pettitt test (Pettitt, 1979), and (iv) the von Neumann ratio test (von Neumann, 1941). For the application of the above-mentioned collection

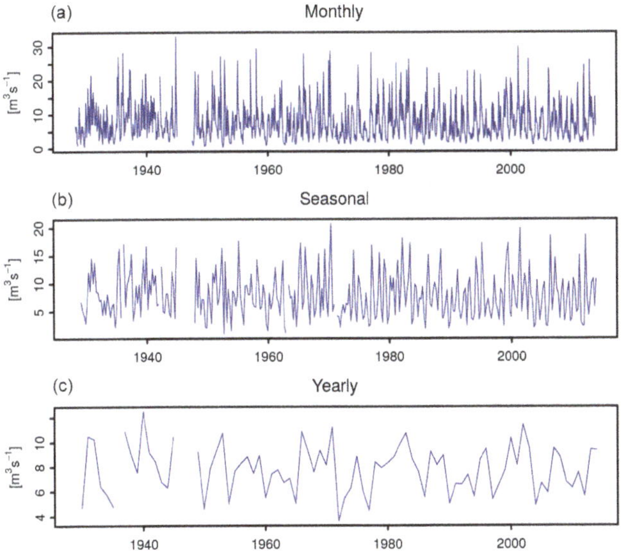

Figure 3. Monthly, seasonal, and yearly MEAN for the River Wiese at Zell, south-western Germany. Index values are only displayed if they fulfil the ECA&D13 data availability criteria. See the text for details.

of tests, we rely on tables that provide critical values of the test statistics for a given sample size that have been determined using Monte Carlo methods (ECA&D13). These tables only report critical values for a sample size of 20 and larger. Therefore, the tests can only be applied if at least 20 yearly, monthly or seasonal time steps are available. Prior to homogeneity testing, yearly, seasonal and monthly index values that are classified as unreliable according to ECA&D13 (see Sect. 3.1.3) are set to missing. Missing values were removed after pre-whitening of yearly, seasonal and monthly index time series (see Sect. 4.1.2).

4.1.2 Pre-whitening

As the considered homogeneity tests rely at least on the assumption that the data are stationary, independent and identically distributed, all indices are pre-processed (pre-whitened), aiming to reduce effects of (i) trends, (ii) seasonality, and (iii) serial correlation. For the pre-whitening procedure, linear trends and mean seasonal cycles were removed using a linear least-squares regression model which captures both the trend and the mean values as $x = b + at$, where b is the intercept, a is the trend and t is time.

1. For *yearly* indices, the linear model is fitted to and subtracted from the complete time series. This results in a time series with zero mean and no linear trend.

2. For *seasonal* indices, the linear model is fitted to and subtracted from the time series for each season (DJF, MAM, JJA, SON) individually. This results in a time series with seasonal resolution in which each season has a zero mean and no linear trend.

3. For *monthly* indices, the linear model is fitted to and extracted from the time series for each month (January, February, etc.) individually. This results in time series with monthly resolution in which each month has a zero mean and no linear trend.

As the detrended and de-seasonalized time series may still exhibit serial correlation, they were further pre-whitened by fitting a lag-1 autoregressive model and then obtaining the residuals, which are then subjected to the homogeneity analysis (Burn and Elnur, 2002; Chu et al., 2013; Gudmundsson and Seneviratne, 2016). The lag-1 autoregressive model is fitted using maximum likelihood estimation.

4.1.3 Classification of station homogeneity

To effectively combine the information of the four considered homogeneity tests, we classify the homogeneity of yearly, monthly and seasonal time-series indices following recommendations of ECA&D13:

1. *useful:* one or no tests reject the null hypothesis at the 1 % level;

2. *doubtful*: two tests reject the null hypothesis at the 1 % level;

3. *suspect*: three or four tests reject the null hypothesis at the 1 % level.

Note, however, that depending on the application, these rules may be either too relaxed or too conservative. In addition, we also introduce the following categories to account for special circumstances that can occur in this large-scale application:

4. *not sufficient data:* less than 20 yearly, seasonal or monthly reliable index values are available;

5. *constant:* all yearly, seasonal or monthly time steps have the same value;

6. *error:* an error (e.g. numerical convergence issue) occurred at any processing step.

4.2 Homogeneity testing of all yearly, seasonal and monthly time-series indices

The homogeneity analysis is applied to all indices at yearly, seasonal and monthly resolution. The rationale for applying the four tests to all indices individually is that inhomogeneities at a particular location might be relevant only for a subset of indices, while other indices are not affected. For example, it is possible that a change in instrumentation will affect peak flows, while low flows are not affected. For this homogeneity assessment, all yearly, seasonal and monthly

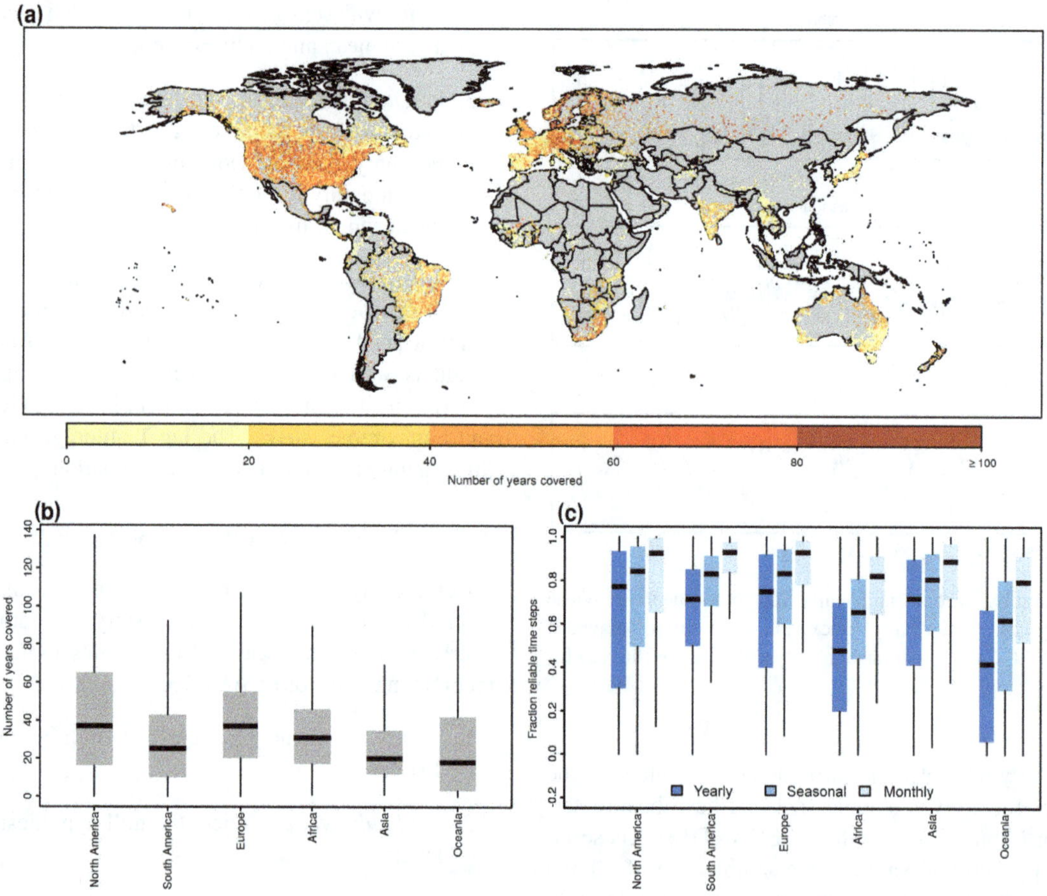

Figure 4. Temporal coverage of streamflow time-series indices. **(a)** Map of the number of years covered by each time series under consideration. **(b)** Distribution of the number of years available per time series for the continental regions of the world. **(c)** Distribution of the fraction of time steps that are classified as reliable using the ECA&D13 data availability criteria. Boxplots show the interquartile range (box) and the median (vertical bar); the whiskers extend to the most extreme point, which is not more than 1.5 times the interquartile range away from the box; outliers are omitted.

time steps that are classified as reliable (Sect. 3.1.3) are considered. This results in a conservative assessment as (i) strict data-availability criteria are applied, and (ii) because inhomogeneities could occur in a time window not relevant to a study. Therefore, the presented results can be used for a general overview of time-series homogeneity, but their suitability should always be re-considered prior to specific applications.

Figure 5 illustrates the results of the homogeneity assessment for the MEAN index for the North Umpqua River in the US. The top panel shows the monthly MEAN index, which displays a sudden jump after the first third of the record. This jump may for example be the result of upstream flow regulation and would be detrimental for climatological investigations. The lower panel shows the time series after the above-mentioned pre-whitening procedure was applied. The seasonal cycle is effectively removed and obtaining the residuals from the lag-1 autoregressive model reduced the magnitude of the sudden jump. Note also the spurious trend, which is an artefact of the de-trending that occurs in the presence of strong, sudden shifts in the mean. Nevertheless, three of the four considered tests identify this inhomogeneity at the 0.01 significance level, and the series is classified as *suspect*.

Global summaries of the number of stations in different homogeneity classes are shown in Fig. 6. Owing to the reduced number of time steps, the homogeneity testing could only be applied for approximately half of the locations at yearly resolution. Nevertheless, the homogeneity assessment highlights that the other half of the yearly indices can be considered "useful" at many locations. Only a small number of the low-flow indices (e.g. MIN, P10, P20, P30) had "constant" values and other issues were rarely detected. For both seasonal and monthly resolution, the number of stations with sufficient data for homogeneity assessment increased significantly, although it is important to recall that the homogeneity tests were in many cases applied to relatively short records (i.e. at least 20 seasons or 20 months respectively). Most of the seasonal and monthly time series with sufficient data are

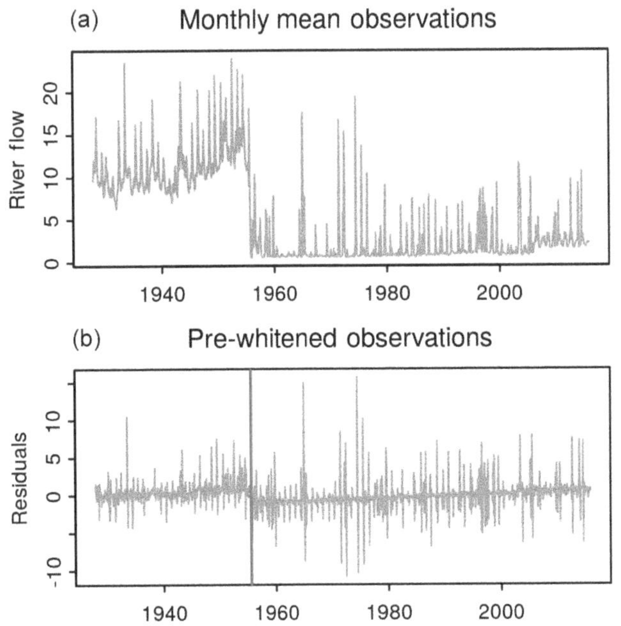

Figure 5. Homogeneity assessment of monthly mean flow of the North Umpqua River, US. **(a)** Monthly mean observations. **(b)** Pre-whitened observations together with the time step at which the standard normal homogeneity test, the Buishand range test and the Pettitt test identified a breakpoint at the 0.01 significance level.

classified as "useful", but a number of "doubtful" and "suspect" values were also detected. At a few locations, low-flow indices had constant values.

Figure 7 shows continental summaries of the homogeneity assessment at yearly, seasonal and monthly timescales and highlights the number of stations at which all indices were classified as useful according to the ECA&D13 criteria. Interestingly, the fraction of time series for which all indices have been classified as "useful" remains approximately constant irrespective of the considered time resolution. Figure 8 illustrates the effect of data availability criteria (Sect. 3.1.3) and the homogeneity assessment of the number of stations for each time step. Regardless of the temporal resolution, the number of stations reduces significantly when the homogeneity criterion is applied. This effect is more prominent at finer temporal resolution (monthly), as adding the "all indices homogenous" criterion removes approximately half of the eligible time series (bottom panel of Fig. 8). Note, however, that the presented summaries can only act as a rough guide on data availability, as criteria for including or excluding specific stations will depend on the objectives of individual future assessments.

5 Data availability and overview of the data product

5.1 Data availability

The data described in this paper are freely available as a compressed zip archive that can be downloaded from https://doi.pangaea.de/10.1594/PANGAEA.887470 (Gudmundsson et al., 2018). The zip archive contains (i) a readme file, (ii) all time-series indices and (iii) the results of all homogeneity tests. Note that the data are accompanied by additional information on the data collection process, catchment boundaries and selected catchment properties (Do et al., 2018a, b).

5.2 Time series of yearly, seasonal and annual indices

The indices derived from daily streamflow time series as described in Sects. 2 and 3 are stored in the INDICES directory. To address the different temporal resolution of the available indices (yearly, seasonal and monthly scales), the GSIM indices were organized into three respective subdirectories where each GSIM station is represented through a text file. For instance, indices at yearly resolution derived from the station with the identifier "AR_0000006" are stored as a text file called "AR_0000006.year" in the "yearly" sub-directory. Indices at seasonal and monthly resolution are stored as "AR_0000006.seas" and "AR_0000006.mon" in the respective ("seasonal", "monthly") sub-directories.

An identical data structure was adopted across all time-series files, with basic metadata (e.g. station identifier, station name, river name) stored in the header, and all index time series written in subsequent lines as a table, where (i) the first column contains the date, which is by convention the last day of the respective yearly, seasonal or monthly time step; (ii) the subsequent columns contain the index values, with column names corresponding to the abbreviations introduced in Table 4; and (iii) the last two columns contain information on the number of (missing) daily values used to compute the index.

5.3 Homogeneity of time-series indices

The results of the homogeneity analysis are stored in three tables, representing indices at yearly, seasonal and monthly resolution which are placed in the HOMOGENEITY directory and contain information on all stations. There is an identical structure for these three text files, with the first 13 columns containing important metadata such as the station identifier, name of the gauging location, and first and last time steps of the index time series. The remaining columns contain the results of four homogeneity tests that are described in the paper, and thus each index is accompanied by four columns (corresponding to the results of the (1) standard normal homogeneity test, (2) the Buishand range test, (3) the Pettitt test and (4) the Neuman ratio test).

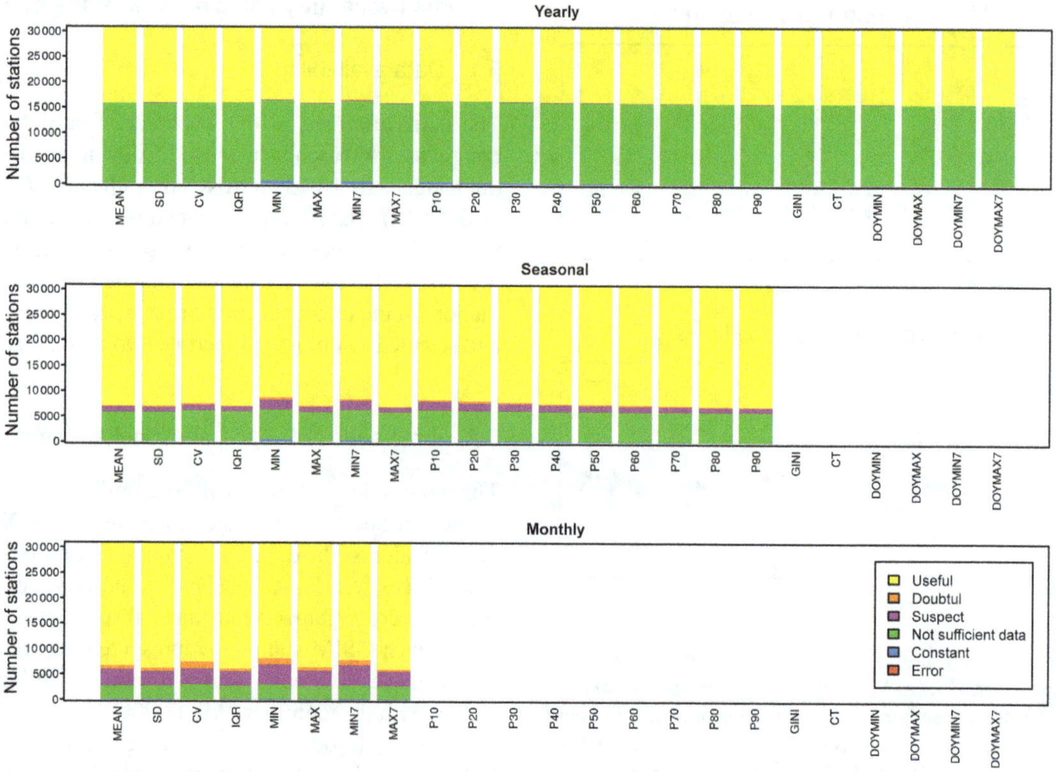

Figure 6. Global summary of the homogeneity analysis for all considered indices at yearly, seasonal and monthly resolution. Shown are the number of stations that are classified as (1) useful, (2) doubtful, (3) suspect, (4) not sufficient data, (5) constant and (6) error according to Sect. 4.1.3. Note that all six categories do occur, although some of them are rare and thus barely visible in the figure.

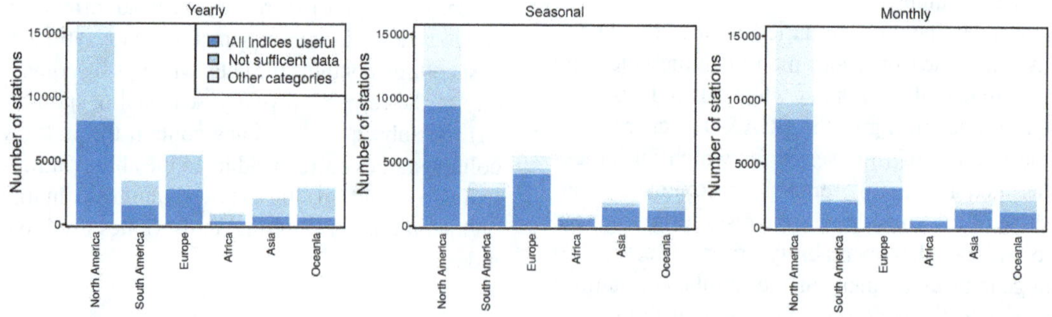

Figure 7. Continental summary of the homogeneity analysis for yearly, seasonal and monthly indices. Shown are the total number of stations at which all indices are classified as useful according to the criteria of ECA&D13, stations that did not have sufficient data for the application of the homogeneity analysis, and all other stations (other categories).

6 Summary and conclusions

Together with Do et al. (2018a) (Part 1), this paper presents the Global Streamflow Indices and Metadata Archive (GSIM), which is a unique collection of streamflow observations at more than 30 000 stations around the globe. In Part 1 (Do et al., 2018a) of the paper series we focussed on the collection and merging of freely available streamflow data worldwide. Part 1 also introduced shapefiles of catchment boundaries together with essential catchment properties such as land cover, topography and mean climatic conditions. As not all data providers allow for a free distribution of unprocessed daily values, we followed in Part 2 an approach that has been established through the ETCCDI in climate research (Klein Tank et al., 2009; Zhang et al., 2011) and introduced a set of time-series indices that can be used to assess the water balance, seasonality, low flows and floods, which are made freely available to serve the scientific community.

While focussing on time-series indices facilitates the redistribution of the data, this approach inevitably comes with

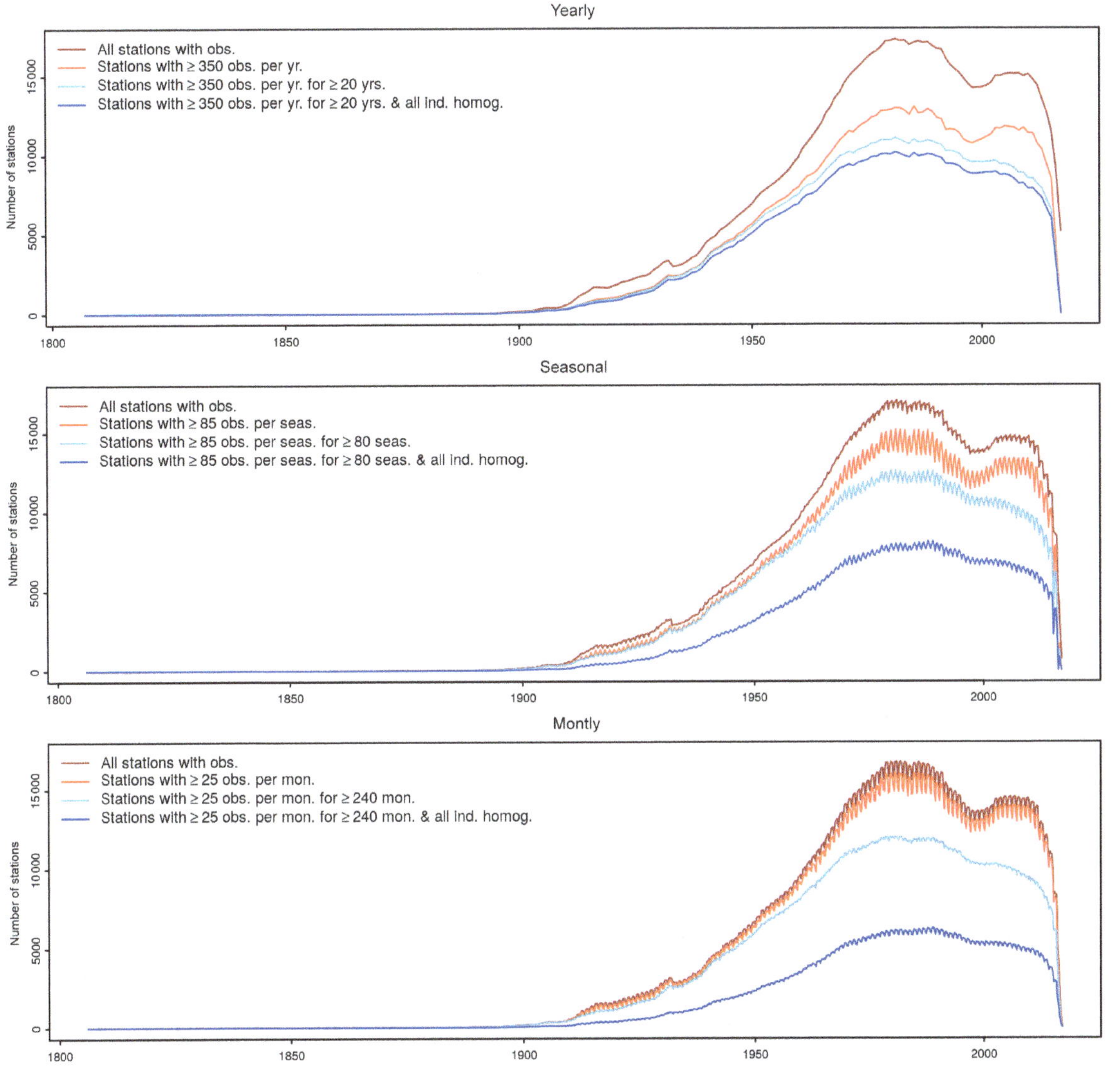

Figure 8. Temporal evolution of global station coverage, conditional on different data-selection criteria for yearly, monthly and seasonal timescales. Successively, the following criteria are applied: (i) all stations that at least one observation for the respective time step (i.e. year, season, month). (ii) Stations that have at least a critical number of observations for each time step (critical values depend on the timescale; see Sect. 3.1.3). (iii) Stations that have at least a critical number of observations for the equivalent of 20 station years (i.e. 20 yearly values, $20 \times 4 = 80$ seasonal values, $20 \times 12 = 240$ monthly values). (iv) Stations where criterion (iii) applied and all indices were considered to be useful in the homogeneity analysis (see Sect. 4.1.3).

inherent limitations. For example, many applications, including hydrological or ecological modelling, may require daily resolution data and other studies may depend on indices not included in the presented collections. Consequently, some users may prefer to seek out the original data sources (see details in Do et al., 2018a) and access the raw daily streamflow values in that manner. Nevertheless, we would like to also highlight the advantages of time-series indices: a benefit of having pre-processed the daily streamflow data into indices is that they can be readily used in studies across large regions with minimal handling of raw data files. In addition, the selected indices foster a wide variety of assessments, including water balance calculations, extreme event analysis and the identifications of trends in the world's freshwater resources.

To ensure the reliability of the published data, we first evaluated the quality of individual daily values through a combination of quality flags developed by the data providers and a transparent numerical screening approach. Subsequently, the homogeneity of yearly, seasonal and monthly indices was assessed using reproducible methods, aiming at aiding potential users to gauge the suitability of individual time series for their research questions. Note, however, that it is not the intent of this project to derive a single "best" dataset, for example, by considering a pre-defined baseline period which gauges must cover, or by derivation of a so-called "high-quality" dataset by applying a rigorous set of quality criteria to available stations. While these approaches are of high value if a dataset is tailored to a specific application, the emphasis of GSIM is to provide a large database of streamflow observations by collating and standardizing many data sources around the world.

Given that data quality requirements can vary substantially, it will remain the work of individual users to establish selection criteria for each study, thereby finding a trade-off between data quantity (number of gauges) and data quality (record length, missing periods). While the criteria used to gauge the usability of the indices are based on the recommendations of ECA&D13, they necessarily rely on subjective decisions on what constitutes a "reliable index". For example, in some climates a gauge may be "reliable" and yet unable to provide measurements for part of the year (e.g. seasonally dry or cold climates). For this reason, attempts have been made to provide flexibility, aiming at facilitating the user to judge upon "reliability" in the context of their applications. Nonetheless, it is our hope that enabling a wide usage of streamflow indices might also lead to greater scrutiny of the data, accumulated knowledge of performance of each site and improved methods for judging the quality of streamflow observations.

There are numerous unsettled scientific questions at the global scale that this dataset has the potential to support. For example, there are unresolved questions around the relationship between trends in rainfall extremes and hydrological extremes (Do et al., 2017; Westra et al., 2013), as well as developing a better understanding of the influence of human activities on the hydrological cycle more broadly (Barnett et al., 2008; Blöschl et al., 2017; Destouni et al., 2013; Gudmundsson et al., 2017; Hegerl et al., 2015; Jaramillo and Destouni, 2015). Expanding upon recent methodological developments (Gudmundsson and Seneviratne, 2015, 2016), the newly assembled data may act as a basis for developing gridded global-scale observation-based data products. There are also likely to be many applications in fields as diverse as hydro-ecology, water quality modelling, environmental assessment and socio-hydrology. We therefore expect the presented data to be a valuable source of information to answer pending questions in global freshwater research, e.g. in the context of the World Climate Research Program Grand Challenge on Water Availability (Trenberth and Asrar, 2014) or the international research efforts on "Change in hydrology and society" (Montanari et al., 2013).

The significant increase in global gauge density and record length through the GSIM archive would not have been possible without the fact that water agencies are increasingly making data accessible online. However, the benefits of this new collection are overshadowed by challenges that are essentially bureaucratic in nature: how to systematically collate, maintain and improve streamflow data globally and who should do it. While agencies such as the GRDC would provide a natural fit for this type of task, they are currently constrained in their capacity to commit to a regular and systematic upkeep of such a global dataset. This paper series represents a one-off initiative of the authors, requiring over a year's worth of checking and evaluation and with little to no capacity for updating or extending the dataset. While it is possible that updates might be achieved through similar future efforts from the community, they are likely to be ad hoc and far from ideal. There are many troubles that can result from patchwork efforts of data collating, including (i) orphaned versions that persist in usage despite updated data being available, (ii) gauges or regions becoming out-of-sync, (iii) repeated needs to identify duplicates in overlapping datasets, (iv) information loss between versions and poor upkeep of documentation, (v) competing or "forked" databases, and many more. To remedy this situation, the hydrological community needs to collectively improve the organization of initiatives for coordinated systems that facilitate updating, storage and documentation of existing data, and to lobby for existing closed databases to be made open and accessible. As part of a global imperative for improved streamflow data, there are a number of additional activities researchers might undertake. These include (i) providing new analyses that improve the quality and understanding of the existing database; (ii) developing new automated methods that can be used systematically to maintain or improve the quality of the instrumental record; (iii) providing additional streamflow observations from missing or currently inaccessible datasets; and (iv) deriving new observational data products though better ground-truthing of remote-sensed variables, reanalysis from hydrological models or upscaling of in situ observations using machine learning.

Competing interests. The authors declare that they have no conflict of interest.

Acknowledgements. We would like to thank Sonia I. Seneviratne for the fruitful discussions on the creation of the GSIM archive. This work would not have been possible without the tremendous efforts of regional, national and international organizations in collecting and archiving river flow observations. Their work is highly appreciated.

Edited by: David Carlson

References

Alexander, L. V., Zhang, X., Peterson, T. C., Caesar, J., Gleason, B., Klein Tank, A. M. G., Haylock, M., Collins, D., Trewin, B., Rahimzadeh, F., Tagipour, A., Rupa Kumar, K., Revadekar, J., Griffiths, G., Vincent, L., Stephenson, D. B., Burn, J., Aguilar, E., Brunet, M., Taylor, M., New, M., Zhai, P., Rusticucci, M., and

Vazquez-Aguirre, J. L.: Global observed changes in daily climate extremes of temperature and precipitation, J. Geophys. Res., 111, D05109, https://doi.org/10.1029/2005JD006290, 2006.

Alexandersson, H.: A homogeneity test applied to precipitation data, J. Climatol., 6, 661–675, 1986.

Alkama, R., Marchand, L., Ribes, A., and Decharme, B.: Detection of global runoff changes: results from observations and CMIP5 experiments, Hydrol. Earth Syst. Sci., 17, 2967–2979, https://doi.org/10.5194/hess-17-2967-2013, 2013.

Barnett, T. P., Pierce, D. W., Hidalgo, H. G., Bonfils, C., Santer, B. D., Das, T., Bala, G., Wood, A. W., Nozawa, T., Mirin, A. A., Cayan, D. R., and Dettinger, M. D.: Human-Induced Changes in the Hydrology of the Western United States, Science, 319, 1080–1083, 2008.

Beck, H. E., de Roo, A., and van Dijk, A. I. J. M.: Global maps of streamflow characteristics based on observations from several thousand catchments, J. Hydrometeorol., 16, 1478–1501, 2015.

Beck, H. E., van Dijk, A. I. J. M., de Roo, A., Dutra, E., Fink, G., Orth, R., and Schellekens, J.: Global evaluation of runoff from 10 state-of-the-art hydrological models, Hydrol. Earth Syst. Sci., 21, 2881–2903, https://doi.org/10.5194/hess-21-2881-2017, 2017.

Becker, A., Finger, P., Meyer-Christoffer, A., Rudolf, B., Schamm, K., Schneider, U., and Ziese, M.: A description of the global land-surface precipitation data products of the Global Precipitation Climatology Centre with sample applications including centennial (trend) analysis from 1901–present, Earth Syst. Sci. Data, 5, 71–99, https://doi.org/10.5194/essd-5-71-2013, 2013.

Blöschl, G., Hall, J., Parajka, J., Perdigão, R. A. P., Merz, B., Arheimer, B., Aronica, G. T., Bilibashi, A., Bonacci, O., Borga, M., v Canjevac, I., Castellarin, A., Chirico, G. B., Claps, P., Fiala, K., Frolova, N., Gorbachova, L., Gül, A., Hannaford, J., Harrigan, S., Kireeva, M., Kiss, A., Kjeldsen, T. R., Kohnová, S., Koskela, J. J., Ledvinka, O., Macdonald, N., Mavrova-Guirguinova, M., Mediero, L., Merz, R., Molnar, P., Montanari, A., Murphy, C., Osuch, M., Ovcharuk, V., Radevski, I., Rogger, M., Salinas, J. L., Sauquet, E., Sraj, M., Szolgay, J., Viglione, A., Volpi, E., Wilson, D., Zaimi, K., and Zivković, N.: Changing climate shifts timing of European floods, Science, 357, 588–590, 2017.

Buishand, T. A.: Some methods for testing the homogeneity of rainfall records, J. Hydrol., 58, 11–27, 1982.

Burn, D. H. and Elnur, M. A. H.: Detection of hydrologic trends and variability, J. Hydrol., 255, 107–122, 2002.

Ceriani, L. and Verme, P.: The origins of the Gini index: extracts from Variabilità e Mutabilità (1912) by Corrado Gini, J. Econ. Inequal., 10, 421–443, 2012.

Chu, M. L., Ghulam, A., Knouft, J. H., and Pan, Z.: A Hydrologic Data Screening Procedure for Exploring Monotonic Trends and Shifts in Rainfall and Runoff Patterns, J. Am. Water Resour. As., 50, 928–942, https://doi.org/10.1111/jawr.12149, 2013.

Cunderlik, J. M. and Ouarda, T. B. M. J.: Trends in the timing and magnitude of floods in Canada, J. Hydrol., 375, 471–480, 2009.

Dee, D. P., Uppala, S. M., Simmons, A. J., Berrisford, P., Poli, P., Kobayashi, S., Andrae, U., Balmaseda, M. A., Balsamo, G., Bauer, P., Bechtold, P., Beljaars, A. C. M., van de Berg, L., Bidlot, J., Bormann, N., Delsol, C., Dragani, R., Fuentes, M., Geer, A. J., Haimberger, L., Healy, S. B., Hersbach, H., Hólm, E. V., Isaksen, L., Kållberg, P., Köhler, M., Matricardi, M., McNally, A. P., Monge-Sanz, B. M., Morcrette, J. J., Park, B. K., Peubey, C., de Rosnay, P., Tavolato, C., Thépaut, J. N., and Vitart, F.: The ERA-Interim reanalysis: configuration and performance of the data assimilation system, Q. J. Roy. Meteor. Soc., 137, 553–597, 2011.

Destouni, G., Jaramillo, F., and Prieto, C.: Hydroclimatic shifts driven by human water use for food and energy production, Nat. Clim. Change, 3, 213–217, 2013.

Do, H. X., Westra, S., and Leonard, M.: A global-scale investigation of trends in annual maximum streamflow, J. Hydrol., 552, 28–43, 2017.

Do, H. X., Gudmundsson, L., Leonard, M., and Westra, S.: The Global Streamflow Indices and Metadata Archive (GSIM) – Part 1: The production of a daily streamflow archive and metadata, Earth Syst. Sci. Data, 10, 765–785, https://doi.org/10.5194/essd-10-765-2018, 2018a.

Do, H. X., Gudmundsson, L., Leonard, M., and Westra, S.: The Global Streamflow Indices and Metadata Archive – Part 1: Station catalog and Catchment boundary, PANGAEA, https://doi.pangaea.de/10.1594/PANGAEA.887477, 2018b.

ECA & D Project Team and Royal Netherlands Meteorological Institute: Algorithm Theoretical Basis Document (ATBD), available at: https://www.ecad.eu/documents/atbd.pdf, 2013.

Ehsanzadeh, E. and Adamowski, K.: Trends in timing of low stream flows in Canada: impact of autocorrelation and long-term persistence, Hydrol. Process., 24, 970–980, 2010.

Fekete, B. M., Looser, U., Pietroniro, A., and Robarts, R. D.: Rationale for Monitoring Discharge on the Ground, J. Hydrometeorol., 13, 1977–1986, 2012.

Fekete, B. M., Robarts, R. D., Kumagai, M., Nachtnebel, H.-P., Odada, E., and Zhulidov, A. V.: Time for in situ renaissance, Science, 349, 685–686, 2015.

Frich, P., Alexander, L. V., Della-Marta, P., Gleason, B., Haylock, M., Tank, A. M. G. K., and Peterson, T.: Observed coherent changes in climatic extremes during the second half of the twentieth century, Clim. Res., 19, 193–212, 2002.

Gudmundsson, L. and Seneviratne, S. I.: Towards observation-based gridded runoff estimates for Europe, Hydrol. Earth Syst. Sci., 19, 2859–2879, https://doi.org/10.5194/hess-19-2859-2015, 2015.

Gudmundsson, L. and Seneviratne, S. I.: Observation-based gridded runoff estimates for Europe (E-RUN version 1.1), Earth Syst. Sci. Data, 8, 279–295, https://doi.org/10.5194/essd-8-279-2016, 2016.

Gudmundsson, L., Tallaksen, L. M., and Stahl, K.: Spatial cross-correlation patterns of European low, mean and high flows, Hydrol. Process., 25, 1034–1045, 2011.

Gudmundsson, L., Tallaksen, L. M., Stahl, K., Clark, D. B., Dumont, E., Hagemann, S., Bertrand, N., Gerten, D., Heinke, J., Hanasaki, N., Voss, F., and Koirala, S.: Comparing Large-Scale Hydrological Model Simulations to Observed Runoff Percentiles in Europe, J. Hydrometeorol., 13, 604–620, 2012a.

Gudmundsson, L., Wagener, T., Tallaksen, L. M., and Engeland, K.: Evaluation of nine large-scale hydrological models with respect to the seasonal runoff climatology in Europe, Water Resour. Res., 48, W11504, https://doi.org/10.1111/jawr.12149, 2012b.

Gudmundsson, L., Seneviratne, S. I., and Zhang, X.: Anthropogenic climate change detected in European renewable freshwater resources, Nat. Clim. Change, 7, 813–816, 2017.

Gudmundsson, L., Do, H. X., Leonard, M., and Westra, S.: The Global Streamflow Indices and Metadata Archive (GSIM) – Part 2: Time Series Indices and Homogeneity Assessment, PANGAEA, https://doi.pangaea.de/10.1594/PANGAEA.887470, 2018.

Haddeland, I., Clark, D. B., Franssen, W., Ludwig, F., Voß, F., Arnell, N. W., Bertrand, N., Best, M., Folwell, S., Gerten, D., Gomes, S., Gosling, S. N., Hagemann, S., Hanasaki, N., Harding, R., Heinke, J., Kabat, P., Koirala, S., Oki, T., Polcher, J., Stacke, T., Viterbo, P., Weedon, G. P., and Yeh, P.: Multimodel Estimate of the Global Terrestrial Water Balance: Setup and First Results, J. Hydrometeorol., 12, 869–884, 2011.

Hall, J., Arheimer, B., Aronica, G. T., Bilibashi, A., Boháč, M., Bonacci, O., Borga, M., Burlando, P., Castellarin, A., Chirico, G. B., Claps, P., Fiala, K., Gaál, L., Gorbachova, L., Gül, A., Hannaford, J., Kiss, A., Kjeldsen, T., Kohnová, S., Koskela, J. J., Macdonald, N., Mavrova-Guirguinova, M., Ledvinka, O., Mediero, L., Merz, B., Merz, R., Molnar, P., Montanari, A., Osuch, M., Parajka, J., Perdigão, R. A. P., Radevski, I., Renard, B., Rogger, M., Salinas, J. L., Sauquet, E., Šraj, M., Szolgay, J., Viglione, A., Volpi, E., Wilson, D., Zaimi, K., and Blöschl, G.: A European Flood Database: facilitating comprehensive flood research beyond administrative boundaries, P. Int. Ass. Hydrol. Sci., 370, 89–95, 2015.

Hannah, D. M., Demuth, S., van Lanen, H. A. J., Looser, U., Prudhomme, C., Rees, G., Stahl, K., and Tallaksen, L. M.: Large-scale river flow archives: importance, current status and future needs, Hydrol. Process., 25, 1191–1200, 2011.

Harris, I., Jones, P. D., Osborn, T. J., and Lister, D. H.: Updated high-resolution grids of monthly climatic observations – the CRU TS3.10 Dataset, Int. J. Climatol., 34, 623–642, 2014.

Haylock, M. R., Hofstra, N., Klein Tank, A. M. G., Klok, E. J., Jones, P. D., and New, M.: A European daily high-resolution gridded data set of surface temperature and precipitation for 1950–2006, J. Geophys. Res., 113, D20119, https://doi.org/10.1029/2008JD010201, 2008.

Hegerl, G. C., Black, E., Allan, R. P., Ingram, W. J., Polson, D., Trenberth, K. E., Chadwick, R. S., Arkin, P. A., Sarojini, B. B., Becker, A., Dai, A., Durack, P. J., Easterling, D., Fowler, H. J., Kendon, E. J., Huffman, G. J., Liu, C., Marsh, R., New, M., Osborn, T. J., Skliris, N., Stott, P. A., Vidale, P.-L., Wijffels, S. E., Wilcox, L. J., Willett, K. M., and Zhang, X.: Challenges in quantifying changes in the global water cycle, B. Am. Meteorol. Soc., 96, 1097–1115, 2015.

Hidalgo, H. G., Das, T., Dettinger, M. D., Cayan, D. R., Pierce, D. W., Barnett, T. P., Bala, G., Mirin, A., Wood, A. W., Bonfils, C., Santer, B. D., and Nozawa, T.: Detection and Attribution of Streamflow Timing Changes to Climate Change in the Western United States, J. Climate, 22, 3838–3855, 2009.

Hisdal, H., Stahl, K., Tallaksen, L. M., and Demuth, S.: Have streamflow droughts in Europe become more severe or frequent?, Int. J. Climatol., 21, 317–333, 2001.

Hodgkins, G. A., Whitfield, P. H., Burn, D. H., Hannaford, J., Renard, B., Stahl, K., Fleig, A. K., Madsen, H., Mediero, L., Korhonen, J., Murphy, C., and Wilson, D.: Climate-driven variability in the occurrence of major floods across North America and Europe, J. Hydrol., 552, 704–717, 2017.

Jaramillo, F. and Destouni, G.: Local flow regulation and irrigation raise global human water consumption and footprint, Science, 350, 1248–1251, 2015.

Katz, R. W., Parlange, M. B., and Naveau, P.: Statistics of extremes in hydrology, Adv. Water Resour., 25, 1287–1304, 2002.

Klein Tank, A. M. G., Zwiers, F. W., and Zhang, X.: Guidelines on Analysis of extremes in a changing climate in support of informed decisions for adaptation, available at: https://www.ecad.eu/documents/WCDMP_72_TD_1500_en_1.pdf, 2009.

Kumar, S., Merwade, V., Kam, J., and Thurner, K.: Streamflow trends in Indiana: Effects of long term persistence, precipitation and subsurface drains, J. Hydrol., 374, 171–183, 2009.

Kundzewicz, Z. W., Graczyk, D., Maurer, T., Piskwar, I., Radziejewski, M., Svensson, C., and Szwed, M. G.: Trend detection in river flow series: 1. Annual maximum flow, Hydrolog. Sci. J., 50, 797–810, 2005.

Lettenmaier, D. P., Wood, E. F., and Wallis, J. R.: Hydro-Climatological Trends in the Continental United States, 1948–88, J. Climate, 7, 586–607, 1994.

Lins, H. F. and Slack, J. R.: Streamflow trends in the United States, Geophys. Res. Lett., 26, 227–230, 1999.

Loon, V. and Anne, F.: Hydrological drought explained, Wiley Interdisciplinary Reviews: Water, 2, 359–392, 2015.

Masaki, Y., Hanasaki, N., Takahashi, K., and Hijioka, Y.: Global-scale analysis on future changes in flow regimes using Gini and Lorenz asymmetry coefficients, Water Resour. Res., 50, 4054–4078, 2014.

McCabe, G. J. and Wolock, D. M.: A step increase in streamflow in the conterminous United States, Geophys. Res. Lett., 29, 2185–2185, 2002.

McKee, T. B., Doesken, N. J., and Kleist, J.: The relationship of drought frequency and duration to time scales, 8th Conference on Applied Climatology, 179–184, 1993.

Milly, P. C. D., Dunne, K. A., and Vecchia, A. V.: Global pattern of trends in streamflow and water availability in a changing climate, Nature, 438, 347–350, 2005.

Montanari, A., Young, G., Savenije, H. H. G., Hughes, D., Wagener, T., Ren, L. L., Koutsoyiannis, D., Cudennec, C., Toth, E., Grimaldi, S., Blöschl, G., Sivapalan, M., Beven, K., Gupta, H., Hipsey, M., Schaefli, B., Arheimer, B., Boegh, E., Schymanski, S. J., Di Baldassarre, G., Yu, B., Hubert, P., Huang, Y., Schumann, A., Post, D. A., Srinivasan, V., Harman, C., Thompson, S., Rogger, M., Viglione, A., McMillan, H., Characklis, G., Pang, Z., and Belyaev, V.: "Panta Rhei-Everything Flows": Change in hydrology and society – The IAHS Scientific Decade 2013–2022, Hydrolog. Sci. J., 58, 1256–1275, 2013.

Moore, J. N., Harper, J. T., and Greenwood, M. C.: Significance of trends toward earlier snowmelt runoff, Columbia and Missouri Basin headwaters, western United States, Geophys. Res. Lett., 34, L16402, https://doi.org/10.1029/2007GL031022, 2007.

Oki, T. and Kanae, S.: Global Hydrological Cycles and World Water Resources, Science, 313, 1068–1072, 2006.

Olden, J. D. and Poff, N. L.: Redundancy and the choice of hydrologic indices for characterizing streamflow regimes, River Res. Appl., 19, 101–121, 2003.

Oliveira, P. J. C., Davin, E. L., Levis, S., and Seneviratne, S. I.: Vegetation-mediated impacts of trends in global radiation on land hydrology: a global sensitivity study, Glob. Change Biol., 17, 3453–3467, 2011.

Pettitt, A. N.: A Non-Parametric Approach to the Change-Point Problem, J. Roy. Stat. Soc. C-App., 28, 126–135, 1979.

Poli, P., Hersbach, H., Dee, D. P., Berrisford, P., Simmons, A. J., Vitart, F., Laloyaux, P., Tan, D. G. H., Peubey, C., Thépaut, J.-

Pettitt, A. N.: A Non-Parametric Approach to the Change-Point Problem, J. Roy. Stat. Soc. C-App., 28, 126–135, 1979.

Poli, P., Hersbach, H., Dee, D. P., Berrisford, P., Simmons, A. J., Vitart, F., Laloyaux, P., Tan, D. G. H., Peubey, C., Thépaut, J.-N., Trémolet, Y., Hólm, E. V., Bonavita, M., Isaksen, L., and Fisher, M.: ERA-20C: An Atmospheric Reanalysis of the Twentieth Century, J. Climate, 29, 4083–4097, 2016.

Rajah, K., OLeary, T., Turner, A., Petrakis, G., Leonard, M., and Westra, S.: Changes to the temporal distribution of daily precipitation, Geophys. Res. Lett., 41, 8887–8894, 2014.

Rauscher, S. A., Pal, J. S., Diffenbaugh, N. S., and Benedetti, M. M.: Future changes in snowmelt-driven runoff timing over the western US, Geophys. Res. Lett., 35, L16703, https://doi.org/10.1029/2008GL034424, 2008.

Reek, T., Doty, S. R., and Owen, T. W.: A Deterministic Approach to the Validation of Historical Daily Temperature and Precipitation Data from the Cooperative Network, B. Am. Meteorol. Soc., 73, 753–762, 1992.

Regonda, S. K., Rajagopalan, B., Clark, M., and Pitlick, J.: Seasonal Cycle Shifts in Hydroclimatology over the Western United States, J. Climate, 18, 372–384, 2005.

Sawicz, K., Wagener, T., Sivapalan, M., Troch, P. A., and Carrillo, G.: Catchment classification: empirical analysis of hydrologic similarity based on catchment function in the eastern USA, Hydrol. Earth Syst. Sci., 15, 2895–2911, https://doi.org/10.5194/hess-15-2895-2011, 2011.

Sawicz, K. A., Kelleher, C., Wagener, T., Troch, P., Sivapalan, M., and Carrillo, G.: Characterizing hydrologic change through catchment classification, Hydrol. Earth Syst. Sci., 18, 273–285, https://doi.org/10.5194/hess-18-273-2014, 2014.

Shiklomanov, I. A., Babkin, V. I., Penkova, N. V., Georgievsry, V. Y., Zaretskaya, I. P., Izmailova, A. V., Balonishnikova, J. A., Grigorkina, T. E., Grube, T. V., Skoryatina, K. V., Tsytsenko, and Yunitsyna, V. P.: World Water Resources at the Beginning of the Twenty-First Century, Cambridge University Press, 2004.

Shukla, S. and Wood, A. W.: Use of a standardized runoff index for characterizing hydrologic drought, Geophys. Res. Lett., 35, L02405, https://doi.org/10.1029/2007GL032487, 2008.

Sippel, S., Zscheischler, J., Heimann, M., Otto, F. E. L., Peters, J., and Mahecha, M. D.: Quantifying changes in climate variability and extremes: Pitfalls and their overcoming, Geophys. Res. Lety., 42, 9990–9998, 2015.

Small, D., Islam, S., and Vogel, R. M.: Trends in precipitation and streamflow in the eastern US: Paradox or perception?, Geophys. Res. Lett., 33, L03403, https://doi.org/10.1029/2005GL024995, 2006.

Stahl, K., Hisdal, H., Hannaford, J., Tallaksen, L. M., van Lanen, H. A. J., Sauquet, E., Demuth, S., Fendekova, M., and Jódar, J.: Streamflow trends in Europe: evidence from a dataset of near-natural catchments, Hydrol. Earth Syst. Sci., 14, 2367–2382, https://doi.org/10.5194/hess-14-2367-2010, 2010.

Stahl, K., Tallaksen, L. M., Hannaford, J., and van Lanen, H. A. J.: Filling the white space on maps of European runoff trends: estimates from a multi-model ensemble, Hydrol. Earth Syst. Sci., 16, 2035–2047, https://doi.org/10.5194/hess-16-2035-2012, 2012.

Stewart, I. T., Cayan, D. R., and Dettinger, M. D.: Changes toward Earlier Streamflow Timing across Western North America, J. Climate, 18, 1136–1155, 2005.

Svensson, C., Kundzewicz, W. Z., and Maurer, T.: Trend detection in river flow series: 2. Flood and low-flow index series, Hydrolog. Sci. J., 50, 811–824, 2005.

Tallaksen, L. M. and van Lanen, H. A. J.: Hydrological Drought: Processes and Estimation Methods for Streamflow and Groundwater, Elsevier, 2004.

Tallaksen, L. M., Madsen, H., and Clausen, B.: On the definition and modelling of streamflow drought duration and deficit volume, Hydrolog. Sci. J., 42, 15–33, 1997.

Trenberth, K. E. and Asrar, G. R.: Challenges and Opportunities in Water Cycle Research: WCRP Contributions, Surv. Geophys., 35, 515–532, 2014.

Vogel, R. M. and Fennessey, N. M.: Flow-Duration Curves. I: New Interpretation and Confidence Intervals, J. Water Res. Plan. Man., 120, 485–504, 1994.

von Neumann, J.: Distribution of the Ratio of the Mean Square Successive Difference to the Variance, Ann. Math. Stat., 12, 367–395, 1941.

Vörösmarty, C. J., Green, P., Salisbury, J., and Lammers, R. B.: Global Water Resources: Vulnerability from Climate Change and Population Growth, Science, 289, 284–288, 2000.

Westerberg, I. K., Wagener, T., Coxon, G., McMillan, H. K., Castellarin, A., Montanari, A., and Freer, J.: Uncertainty in hydrological signatures for gauged and ungauged catchments, Water Resour. Res., 52, 1847–1865, 2016.

Westra, S., Alexander, L. V., and Zwiers, F. W.: Global Increasing Trends in Annual Maximum Daily Precipitation, J. Climate, 26, 3904–3918, 2013.

Wijngaard, J. B., Klein Tank, A. M. G., and Können, G. P.: Homogeneity of 20th century European daily temperature and precipitation series, Int. J. Climatol., 23, 679–692, 2003.

Zaitchik, B. F., Rodell, M., and Olivera, F.: Evaluation of the Global Land Data Assimilation System using global river discharge data and a source-to-sink routing scheme, Water Resour. Res., 46, W06507, https://doi.org/10.1029/2009WR007811, 2010.

Zhang, X., Harvey, K. D., Hogg, W. D., and Yuzyk, T. R.: Trends in Canadian streamflow, Water Resour. Res., 37, 987–998, 2001.

Zhang, X., Hegerl, G., Zwiers, F. W., and Kenyon, J.: Avoiding Inhomogeneity in Percentile-Based Indices of Temperature Extremes, J. Climate, 18, 1641–1651, 2005.

Zhang, X., Alexander, L., Hegerl, G. C., Jones, P., Tank, A. K., Peterson, T. C., Trewin, B., and Zwiers, F. W.: Indices for monitoring changes in extremes based on daily temperature and precipitation data, Wires. Clim. Change, 2, 851–870, 2011.

11

Basin-scale water-balance dataset (BSWB)

Martin Hirschi and Sonia I. Seneviratne

Institute for Atmospheric and Climate Science, ETH Zurich, Universitätstrasse 16, 8092 Zürich, Switzerland

Correspondence to: Martin Hirschi (martin.hirschi@env.ethz.ch)

Abstract. This paper presents an update of a basin-scale diagnostic dataset of monthly variations in terrestrial water storage for large river basins worldwide (BSWB v2016, doi:10.5905/ethz-1007-82). Terrestrial water storage comprises all forms of water storage on land surfaces, and its seasonal and inter-annual variations are mostly determined by soil moisture, groundwater, snow cover, and surface water. The dataset presented is derived using a combined atmospheric and terrestrial water-balance approach with conventional streamflow measurements and reanalysis data of atmospheric moisture flux convergence. It extends a previous, existing version of the dataset (Mueller et al., 2011) temporally and spatially.

1 Introduction

Terrestrial water storage (TWS) plays a key role in the hydrological cycle. It encompasses all water stored on land surfaces, and its seasonal and inter-annual variations are determined by soil moisture, groundwater, snow cover, and surface water. Soil moisture, especially, contributes to land–atmosphere coupling in an essential way (e.g. Koster et al., 2006; Seneviratne et al., 2006, 2010; Hirschi et al., 2011; Miralles et al., 2014; Guillod et al., 2015). In particular, it is important for numerical weather prediction (e.g. Beljaars et al., 1996; Drusch, 2007; Orth et al., 2016) and seasonal forecasting (e.g. Koster et al., 2000; van den Hurk et al., 2010), as well as for simulations of present and future climate (e.g. Shukla and Mintz, 1982; Milly and Dunne, 1994; Koster et al., 2009; Seneviratne et al., 2013).

Despite recent activities in assembling in situ soil moisture observations (Dorigo et al., 2013), global coverage remains limited. This is even more the case for in situ observations of the other components of TWS. Remote sensing can help to increase the spatial coverage with observations. For soil moisture, the European Space Agency (ESA) Climate Change Initiative (CCI, http://www.esa-soilmoisture-cci.org) provides a long-term global soil moisture product from merging data from active and passive microwave sensors (Liu et al., 2012). However, penetration depth is limited to the top few centimetres of the soil (Owe and Van de Griend, 1998). On the other hand, remote-sensing-based measurements of TWS from the Gravity Recovery and Climate Experiment (GRACE; Tapley et al., 2004; Ramillien et al., 2008; Humphrey et al., 2016) only go back to 2002. Thus, for further retrospective evaluation of TWS, alternative approaches are required.

Here, we rely on a combination of streamflow measurements (relatively broadly available) and an observation-assimilating atmospheric reanalysis system to diagnose TWS variations on a basin scale (e.g. Seneviratne et al., 2004; Hirschi et al., 2006a). This allows for an evaluation of TWS variations in gauged river basins worldwide and over a longer time period (mostly limited by the availability of streamflow data). The basin-scale water-balance dataset of monthly TWS variations (BSWB v2016) presented here extends a previous version of the dataset (Mueller et al., 2011; hereafter referred to as BSWB v2011; data available at www.iac.ethz.ch/url/bswb) temporally and spatially.

2 Methodology

The method used to derive the BSWB v2016 dataset is based on publications describing previous versions of the data (Seneviratne et al., 2004; Hirschi et al., 2006a; Mueller et al., 2011). For a given river basin, the terrestrial water balance can be expressed as

$$\left\{\frac{\partial S}{\partial t}\right\} = \{\overline{P} - \overline{E}\} - \{\overline{R}\}\,, \tag{1}$$

where S represents the TWS of the given basin, P the precipitation, E the evapotranspiration, and R the measured streamflow, which is assumed to include both the surface and the groundwater run-off of the area. The overbar denotes a temporal average (i.e. monthly means) and {} a space average over the basin.

Neglecting the contribution of the liquid and solid water in clouds (Peixoto and Oort, 1992), the atmospheric water balance for the same area can be expressed as

$$\left\{\frac{\partial W}{\partial t}\right\} = -\{\overline{\nabla_H \cdot Q}\} - \{\overline{P} - \overline{E}\}\,, \tag{2}$$

where W represents the column storage of water vapour and Q the vertically integrated two-dimensional water vapour flux. The operator ($\nabla_H \cdot$) denotes the horizontal divergence. Eliminating the term $\{\overline{P} - \overline{E}\}$ in Eqs. (1) and (2) results in the combined atmospheric and terrestrial water-balance equation:

$$\left\{\frac{\partial S}{\partial t}\right\} = -\left\{\frac{\partial W}{\partial t}\right\} - \{\overline{\nabla_H \cdot Q}\} - \{\overline{R}\}\,. \tag{3}$$

The monthly variations in TWS of the studied basin can thus be expressed as the sum of three terms only: the change in atmospheric water vapour content, the water vapour flux convergence, and the measured river streamflow. The term $\{\overline{\partial W/\partial t}\}$ is usually negligible for annual means but not for monthly means, particularly during the spring and fall seasons (Rasmusson, 1968; Seneviratne et al., 2004).

3 Data sources

3.1 Reanalysis data

The vertically integrated atmospheric moisture flux divergence and water vapour content are taken from the ERA-Interim reanalysis product (Dee et al., 2011) of the European Centre for Medium-Range Weather Forecasts (ECMWF).

ERA-Interim is produced with a 2006 version of the IFS (Integrated Forecasting System, Cy31r2). It has a T255 spherical harmonic representation of the atmospheric dynamic and thermodynamic fields, corresponding to grid spacings of about 80 km, on 60 vertical levels from the surface up to 0.1 hPa. Here, we use the interpolated $0.5 \times 0.5°$ product. The reanalysis covers the period from 1979 to the near present and uses a 12 h 4D-Var assimilation technique. The two ERA-Interim fields used in the water-balance computations (i.e. atmospheric moisture flux divergence and water vapour content) contain assimilated humidity and wind observations from radiosondes. Additional documentation on ERA-Interim can be found at http://www.ecmwf.int/en/research/climate-reanalysis/era-interim.

Table 1. Correlations between BSWB v2011 and BSWB v2016 time series (drift-corrected data) of absolute TWS variations as well as of the anomalies in the river basins covered by both datasets and with at least 4 overlapping years of data.

River	Absolute	Anomalies
Lena at Stolb	1.00	0.96
Selenga at Mostovoy	1.00	0.91
Yenisey at Igarka	1.00	0.98
Irtysh at Omsk	0.99	0.89
Ob' at Salekhard	0.99	0.91
Yukon River at Pilot Station, AK	0.98	0.85
Columbia River at The Dalles, OR	0.99	0.95
Mackenzie River at Arctic Red River	0.99	0.96
Rhône at Beaucaire	1.00	0.96
Rhine River at Rees	0.94	0.87
Weser at Intschede	1.00	0.98
Elbe River at Neu Darchau	0.97	0.93
Volga at Volgograd power plant	1.00	0.96
Mean	0.99	0.93

3.2 Streamflow data and catchment boundaries

The monthly streamflow data have been obtained from the Global Runoff Data Center (GRDC). We use the GRDC reference dataset (http://www.bafg.de/GRDC/EN/04_spcldtbss/43_GRfN/refDataset_node.html), which compiles time series of river discharge data of 718 stations of the GRDC database longer than 20 years, each capturing a basin area greater than 10 000 km^2. The GRDC reference dataset time series are updated regularly.

Catchment boundaries are provided by the GRDC as shape polygons. These are used to average the ERA-Interim fields over the basin area (see below).

4 BSWB v2016 dataset

4.1 Processing

The ERA-Interim atmospheric moisture flux divergence and water vapour content are processed to monthly averages first. Then these fields are averaged over the basin area using the fractional coverage of the catchments in each ERA-Interim grid cell as a weighting factor. For the basin averaging, we use the R-package "raster" (http://cran.r-project.org/web/packages/raster/). Note that basin masks with the fraction of the grid cell inside the catchment are provided for different spatial resolutions as part of the BSWB v2016 dataset (see Sect. 7).

From the GRDC reference dataset, the monthly data are used. The flags provided are applied in the following way:

- Flag 99 (use not recommended by the provider) is set to missing data.

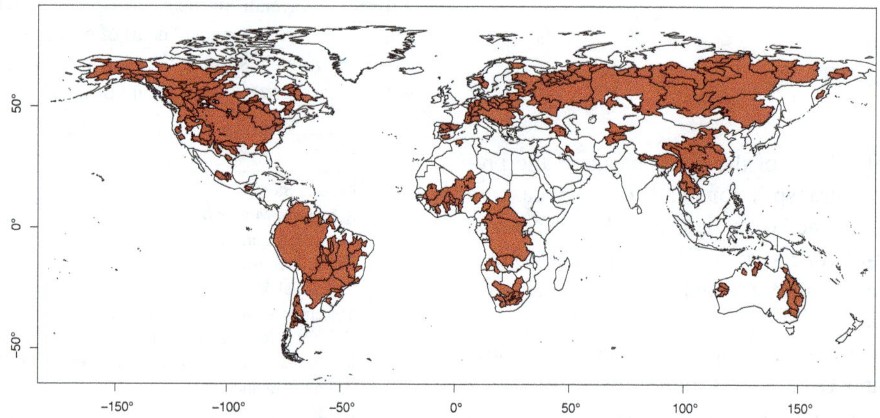

Figure 1. Global coverage of river basins of the BSWB v2016 dataset.

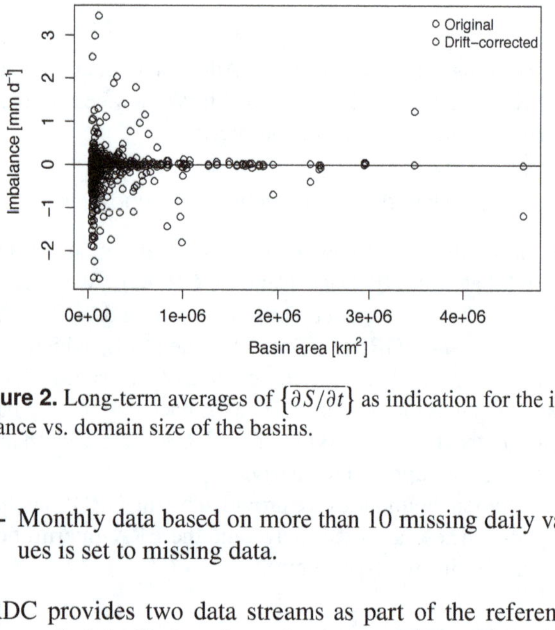

Figure 2. Long-term averages of $\{\overline{\partial S/\partial t}\}$ as indication for the imbalance vs. domain size of the basins.

- Monthly data based on more than 10 missing daily values is set to missing data.

GRDC provides two data streams as part of the reference dataset, "original" data from the data provider and "calculated" data that was modified by GRDC. As suggested by GRDC, we use the calculated values for times when they are available, otherwise the original values are used (GRDC, personal communication). Finally, the monthly variations in TWS are calculated using Eq. (3).

The critical domain size for water-balance computations using high-resolution reanalysis data is assumed to be of the order of 10^5 km^2 (Yeh et al., 1998; Berbery and Rasmusson, 1999; Seneviratne et al., 2004). Smaller basins often suffer from an imbalance between moisture convergence and streamflow (see below, and e.g. Hirschi et al., 2006a; Mueller et al., 2011). For consistency with BSWB v2011, we consider basins greater than 35 000 km^2.

Moreover, only basins covering more than 6 overlapping years of streamflow and reanalysis data are presented. Consequently, the BSWB v2016 dataset covers the time period 1979–2015, but is often limited by availability of streamflow data. The resulting global coverage with basins is displayed in Fig. 1. Currently, the BSWB v2016 dataset encompasses 341 river basins.

4.2 Imbalance and drift correction

In the long term, the water input to a basin should be balanced by the water output. As the contribution of the changes in column storage of water vapour are negligible for annual to long-term means, column integrated moisture flux convergence should be balanced by streamflow. This assumption is generally correct for multi-year means, although some regions may show persistent trends, e.g. due to groundwater withdrawal (e.g. Zektser et al., 2005; Rodell et al., 2009). Other neglected factors are inter-basin groundwater flow as well as direct groundwater discharge to the ocean. This latter term is not included in the measured streamflow and represents approximatively 6 % of the total annual global water gain by the oceans (Zektser and Loaiciga, 1993).

In particular smaller river basins preferentially suffer from an imbalance between long-term means of the vertically integrated water vapour flux convergence and the streamflow of the basin. Figure 2 displays this imbalance as the long-term average of $\{\overline{\partial S/\partial t}\}$ vs. the domain size of the river basins. Our results roughly confirm the threshold for the critical domain size for water-balance computations (10^5 km^2, see Sect. 4.1) as the imbalance decreases above this basin size.

As a consequence of the imbalance, the temporal integration of the diagnosed monthly variations in TWS, i.e

$$S(t) = S_0 + \int_{t_0}^{t} \left\{ \frac{\partial S}{\partial t} \right\} \mathrm{d}t \quad \text{with} \quad S_0 = 0\,\mathrm{mm}, \tag{4}$$

shows a drift in TWS. The likely reason for this drift are biases in the atmospheric moisture convergence data (Seneviratne et al., 2004), though actual drifts in TWS can be important in some regions and could contribute to part of the

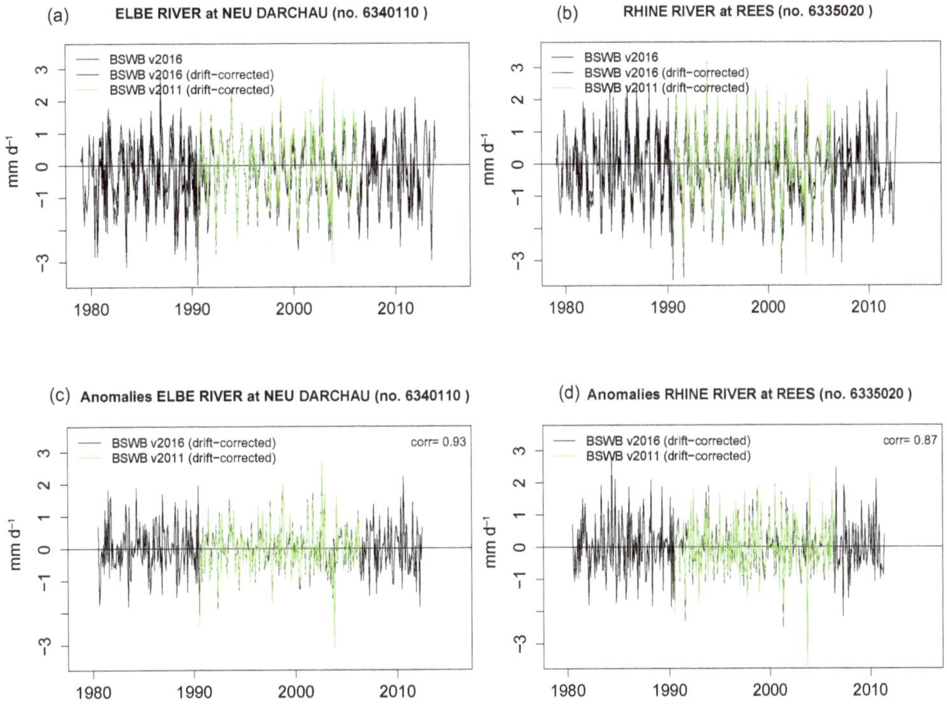

Figure 3. Comparison of BSWB v2011 and BSWB v2016 data of the Elbe (at Neu Darchau) and the Rhine (at Rees) river basins (with numbers in brackets denoting the GRDC station number). Top panels show the absolute time series and bottom panels the anomalies with respect to the mean seasonal cycle (with the correlation between the anomaly time series noted as well). For BSWB v2016, both the original and the drift-corrected data are displayed for the absolute time series.

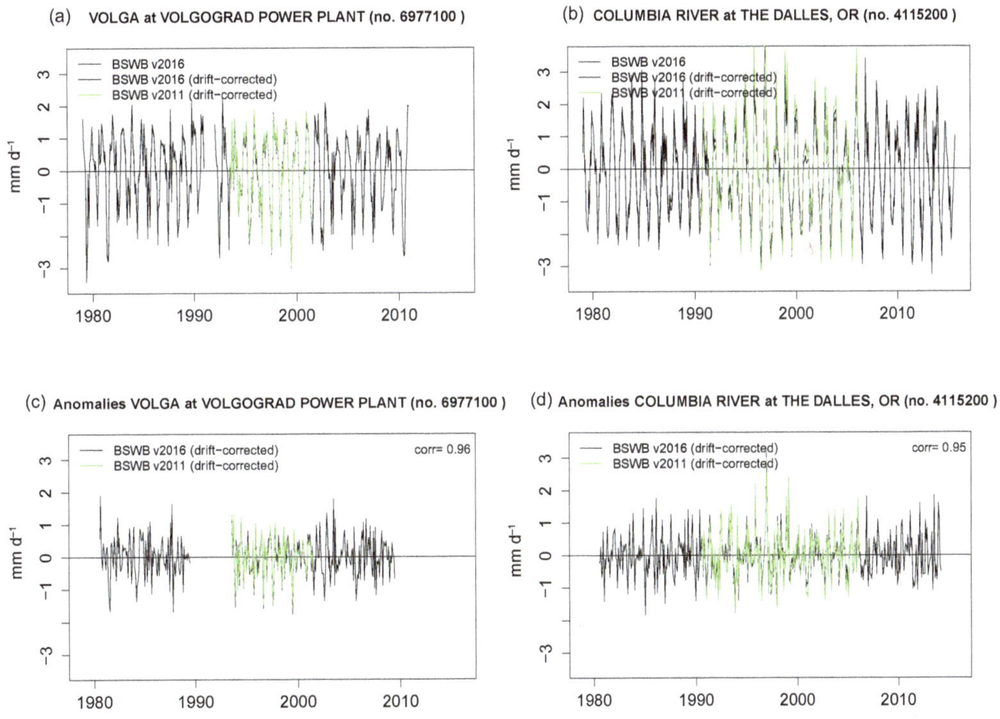

Figure 4. As Fig. 3, but for the Volga (at Volgograd power plant) and the Columbia (at The Dalles) river basins.

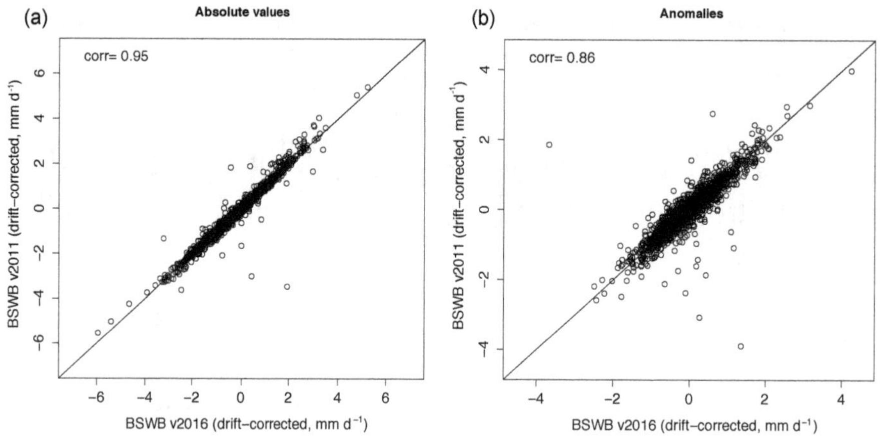

Figure 5. Scatter plots of BSWB v2011 vs. BSWB v2016; all basins covered by both datasets. **(a)** Absolute values of monthly TWS variations; **(b)** their anomalies with respect to the mean seasonal cycle.

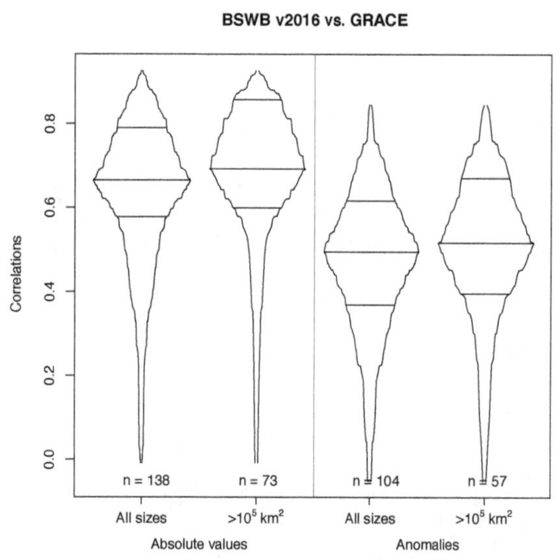

Figure 6. Box–percentile plot showing the distributions of the correlations between BSWB v2016 and GRACE time series for absolute values and for their anomalies with respect to the mean seasonal cycle. Basins with at least 4 overlapping years of data and with at least 9 years for the calculation of the mean seasonal cycle are considered, with the resulting number of basins denoted in the plot. The distributions are displayed for all basin sizes as well as for basins larger than 10^5 km^2. The width of the box at any given height is proportional to the percent of observations that are more extreme in that direction. The median and the 25th and 75th percentiles are marked by lines across the box.

signal (see above). The errors in the run-off measurements are expected to be small, i.e. around 5 % (Winter, 1981) for longer-term averages. As widespread information on natural sources of drifts in TWS is not available, the most appropriate procedure is to assume that the observed drifts are purely artificial and to remove them by a high-pass filter.

The drift correction is achieved by subtracting a running mean with a 3-year window from the original estimates of TWS variations. This forces the long-term average of $\overline{\{\partial S/\partial t\}}$ to zero (cf. red dots in Fig. 2) and allows the removal of the artificial drift without losing the short-term variability. Note that a remaining imbalance might persist on Fig. 2 due to non-complete years of data.

We provide both the original and the drift-corrected estimates as part of the BSWB v2016 dataset. In this way, the application of alternative filters (e.g. locally weighted regression LOESS) to the original estimates is still possible for the user. Note that due to the reasons presented, the BSWB data are not applicable in trend analyses.

5 Comparison with previous BSWB v2011

Due to the availability of ERA-Interim and streamflow data, BSWB v2011 was restricted to the 1989–2008 time period. ERA-Interim has been updated and temporally extended since then (see http://www.ecmwf.int/en/about/media-centre/news/2011/extension-era-interim-reanalysis-1979). In addition, the BSWB v2016 datasets relies on a consistent run-off database from GRDC (i.e. the GRDC reference dataset), while BSWB v2011 was based on various heterogeneous data sources for run-off (apart from GRDC also the US Geological Survey and local sources with varying data formatting and quality checks). By using the GRDC reference dataset, enhanced consistency between the basins can be ensured, and future updates of the BSWB dataset are more easily feasible by relying on one source for run-off only.

To check the consistency of the updated BSWB v2016 dataset with the previous BSWB v2011, we compare time series of both versions for some river basins. Note that despite the varying data sources (see above), the differences between the two dataset versions should only be minor. Figures 3 and

4 show time series of the BSWB v2011 and BSWB v2016 datasets for selected basins. The datasets agree very well, both for the absolute TWS variations as well as for their anomalies (i.e. anomalies with respect to the mean seasonal cycle). Correlations amount to 0.99 on the average for absolute TWS variations and are mostly higher than 0.85 for the anomalies (for basins with at least 4 overlapping years of data, see Table 1). This good agreement is also visible in the scatter plots based on all basins covered by both datasets (Fig. 5), which again show high correlations between BSWB v2011 and BSWB v2016 both for absolute values and the anomalies. Despite this close agreement, differences between BSWB v2011 and BSWB v2016 exist and are likely related to changes in ERA-Interim. For instance the previously existing negative outlier in Europe in 2003 was caused by a strong moisture divergence which has been alleviated in the extended ERA-Interim version (see Fig. 3).

6 Comparison with GRACE

The BSWB v2016 dataset is also compared with independent remote-sensing-based estimates of TWS from GRACE (Tapley et al., 2004). We use the Jet Propulsion Laboratory (JPL)-RL05M GRACE mascon (mass concentration blocks) solution, providing equivalent water thickness with a spatial sampling of $0.5 \times 0.5°$ (doi:10.5067/TEMSC-OCL05; Watkins et al., 2015). The BSWB v2016 drift-corrected monthly variations in TWS have been temporally integrated for the comparison with GRACE (see also Sect. 4.2).

Figure 6 displays the distributions of the time series correlations between BSWB v2016 and GRACE for the absolute values as well as for their anomalies (i.e. anomalies with respect to the mean seasonal cycle) as box–percentile plots (Esty and Banfield, 2003). Basins with at least 4 overlapping years of data and with at least 9 years for the calculation of the mean seasonal cycle are considered. The analysis is done separately for all basin sizes as well as for basins larger than 10^5 km^2. The two datasets mostly agree well, with slightly higher correlations for the larger basins. Mean correlations amount to 0.7 for the absolute values and 0.51 for the anomalies.

7 Data availability

The basin-scale diagnostic dataset of monthly variations in terrestrial water storage (BSWB v2016) presented here is available for download at the ETH data archive: doi:10.5905/ethz-1007-82. The data are provided in individual ASCII files for each of the river basins and contain time series of the uncorrected monthly variations in terrestrial water storage (in units of mm d^{-1}) as well as the drift-corrected data (see Sect. 4.2). Moreover, gridded masks of the river basins are provided to facilitate comparison with other gridded data (e.g. climate model output). The masks are made available in NetCDF format for three spatial resolutions (1×1, 0.5×0.5, and $0.25 \times 0.25°$) and contain the fraction of the grid cell inside the respective catchment. Older versions of the BSWB dataset are available at the following website: www.iac.ethz.ch/url/bswb.

8 Conclusions

We present an update of a basin-scale diagnostic dataset of monthly variations in terrestrial water storage for large river basins worldwide (BSWB v2016). The dataset is derived using a combined atmospheric and terrestrial water-balance approach with conventional streamflow measurements and atmospheric reanalysis data from ECMWF ERA-Interim. It extends the existing version of the dataset (Mueller et al., 2011) temporally and spatially (i.e. 1979–2015 vs. 1989–2008, 341 vs. 36 river basins). Overall, the update shows very good agreement with the previous version of the dataset. It also compares well with independent remote-sensing-based estimates of TWS. BSWB data proved to be valuable for climate and land-surface model evaluation (e.g. van den Hurk et al., 2005; Hirschi et al., 2007; Balsamo et al., 2009; Dutra et al., 2010) and the investigation of land-surface processes (e.g. Fischer et al., 2007), as well as for the evaluation of other large-scale estimates of TWS (e.g. Hirschi et al., 2006b; Troch et al., 2007).

Competing interests. The authors declare that they have no conflict of interest.

Acknowledgements. We thank ECMWF for providing the atmospheric reanalysis data and GRDC for the run-off data. GRACE land mass grids were processed by the Jet Propulsion Laboratory (JPL) GRACE Team and are available at http://grace.jpl.nasa.gov, supported by the NASA MEaSUREs Program. We thank Vincent Humphrey for help with the GRACE data. Partial support from the ERC DROUGHT-HEAT project (grant agreement FP7-IDEAS-ERC-617518) is acknowledged.

Edited by: D. Carlson

References

Balsamo, G., Viterbo, P., Beljaars, A., van den Hurk, B., Hirschi, M., Betts, A. K., and Scipal, K.: A revised hydrology for the ECMWF model: Verification from field site to terrestrial water storage and impact in the Integrated Forecast System, J. Hydrometeorol., 10, 623–643, doi:10.1175/2008JHM1068.1, 2009.

Beljaars, A. C. M., Viterbo, P., Miller, M. J., and Betts, A. K.: The anomalous rainfall over the United States during July 1993: Sensitivity to land surface parameterization and soil moisture anomalies, Mon. Weather Rev., 124, 362–383, 1996.

Berbery, E. H. and Rasmusson, E. M.: Mississippi moisture budgets on regional scale, Mon. Weather Rev., 127, 2654–2673, 1999.

Dee, D. P., Uppala, S. M., Simmons, A. J., Berrisford, P., Poli, P., Kobayashi, S., Andrae, U., Balmaseda, M. A., Balsamo, G., Bauer, P., Bechtold, P., Beljaars, A. C. M., van de Berg, L., Bidlot, J., Bormann, N., Delsol, C., Dragani, R., Fuentes, M., Geer, A. J., Haimberger, L., Healy, S. B., Hersbach, H., Holm, E. V., Isaksen, L., Kallberg, P., Kohler, M., Matricardi, M., McNally, A. P., Monge-Sanz, B. M., Morcrette, J. J., Park, B. K., Peubey, C., de Rosnay, P., Tavolato, C., Thepaut, J. N., and Vitart, F.: The ERA-Interim reanalysis: configuration and performance of the data assimilation system, Q. J. Roy. Meteor. Soc., 137, 553–597, doi:10.1002/qj.828, 2011.

Dorigo, W. A., Xaver, A., Vreugdenhil, M., Gruber, A., Hegyiová, A., Sanchis-Dufau, A. D., Zamojski, D., Cordes, C., Wagner, W., and Drusch, M.: Global Automated Quality Control of In Situ Soil Moisture Data from the International Soil Moisture Network, Vadose Zone J., 12, doi:10.2136/vzj2012.0097, 2013.

Drusch, M.: Initializing numerical weather prediction models with satellite-derived surface soil moisture: Data assimilation experiments with ECMWF's Integrated Forecast System and the TMI soil moisture data set, J. Geophys. Res., 112, D03102, doi:10.1029/2006JD007478, 2007.

Dutra, E., Balsamo, G., Viterbo, P., Miranda, P. M. A., Beljaars, A., Schär, C., and Elder, K.: An Improved Snow Scheme for the ECMWF Land Surface Model: Description and Offline Validation, J. Hydrometeorol., 11, 899–916, doi:10.1175/2010JHM1249.1, 2010.

Esty, W. and Banfield, J.: The Box-Percentile Plot, J. Stat. Softw., 8, 1–14, doi:10.18637/jss.v008.i17, 2003.

Fischer, E. M., Seneviratne, S. I., Vidale, P. L., Lüthi, D., and Schär, C.: Soil moisture–atmosphere interactions during the 2003 European summer heatwave, J. Climate, 20, 5081–5099, doi:10.1175/JCLI4288.1, 2007.

Guillod, B. P., Orlowsky, B., Miralles, D. G., Teuling, A. J., and Seneviratne, S. I.: Reconciling spatial and temporal soil moisture effects on afternoon rainfall, Nat. Commun., 6, 6443, doi:10.1038/ncomms7443, 2015.

Hirschi, M., Seneviratne, S. I., and Schär, C.: Seasonal variations in terrestrial water storage for major midlatitude river basins, J. Hydrometeorol., 7, 39–60, 2006a.

Hirschi, M., Viterbo, P., and Seneviratne, S. I.: Basin-scale water-balance estimates of terrestrial water storage variations from ECMWF operational forecast analysis, Geophys. Res. Lett., 33, L21401, doi:10.1029/2006GL027659, 2006b.

Hirschi, M., Seneviratne, S. I., Hagemann, S., and Schär, C.: Analysis of seasonal terrestrial water storage variations in regional climate simulations over Europe, J. Geophys. Res., 112, D22109, doi:10.1029/2006JD008338, 2007.

Hirschi, M., Seneviratne, S. I., Alexandrov, V., Boberg, F., Boroneant, C., Christensen, O. B., Formayer, H., Orlowsky, B., and Stepanek, P.: Observational evidence for soil-moisture impact on hot extremes in southeastern Europe, Nat. Geosci., 4, 17–21, doi:10.1038/ngeo1032, 2011.

Humphrey, V., Gudmundsson, L., and Seneviratne, S. I.: Assessing Global Water Storage Variability from GRACE: Trends, Seasonal Cycle, Subseasonal Anomalies and Extremes, Surv. Geophys., 37, 357–395, doi:10.1007/s10712-016-9367-1, 2016.

Koster, R. D., Suarez, M. J., and Heiser, M.: Variance and predictability of precipitation at seasonal-to-interannual timescales, J. Hydrometeorol., 1, 26–46, 2000.

Koster, R. D., Guo, Z., Dirmeyer, P. A., Bonan, G., Chan, E., Cox, P., Davies, H., Gordon, C. T., Kanae, S., Kowalczyk, E., Lawrence, D., Liu, P., Lu, C. H., Malyshev, S., McAvaney, B., Mitchell, K., Mocko, D., Oki, T., Oleson, K. W., Pitman, A., Sud, Y. C., Taylor, C. M., Verseghy, D., Vasic, R., Xue, Y., and Yamada, T.: GLACE: The Global Land-Atmosphere Coupling Experiment. Part I: Overview, J. Hydrometeorol., 7, 590–610, 2006.

Koster, R. D., Guo, Z., Yang, R., Dirmeyer, P. A., Mitchell, K., and Puma, M. J.: On the Nature of Soil Moisture in Land Surface Models, J. Climate, 22, 4322–4335, 2009.

Liu, Y., Dorigo, W., Parinussa, R., de Jeu, R., Wagner, W., McCabe, M., Evans, J., and van Dijk, A.: Trend-preserving blending of passive and active microwave soil moisture retrievals, Remote Sens. Environ., 123, 280–297, doi:10.1016/j.rse.2012.03.014, 2012.

Milly, P. C. D. and Dunne, K. A.: Sensitivity of the global water cycle to the water-holding capacity of the land, J. Climate, 7, 506–526, 1994.

Miralles, D. G., Teuling, A. J., van Heerwaarden, C. C., and Vila-Guerau de Arellano, J.: Mega-heatwave temperatures due to combined soil desiccation and atmospheric heat accumulation, Nat. Geosci., 7, 345–349, doi:10.1038/ngeo2141, 2014.

Mueller, B., Hirschi, M., and Seneviratne, S. I.: New diagnostic estimates of variations in terrestrial water storage based on ERA-Interim data, Hydrol. Process., 25, 996–1008, doi:10.1002/hyp.7652, 2011.

Orth, R., Dutra, E., and Pappenberger, F.: Improving Weather Predictability by Including Land Surface Model Parameter Uncertainty, Mon. Weather Rev., 144, 1551–1569, doi:10.1175/MWR-D-15-0283.1, 2016.

Owe, M. and Van de Griend, A. A.: Comparison of soil moisture penetration depths for several bare soils at two microwave frequencies and implications for remote sensing, Water Resour. Res., 34, 2319–2327, doi:10.1029/98WR01469, 1998.

Peixoto, J. P. and Oort, A. H.: Physics of climate, AIP Press, New York, 1992.

Ramillien, G., Famiglietti, J., and Wahr, J.: Detection of Continental Hydrology and Glaciology Signals from GRACE: A Review, Surv. Geophys., 29, 361–374, 2008.

Rasmusson, E. M.: Atmospheric Water Vapor Transport and the Water Balance of North America. II. Large-scale water balance investigations, Mon. Weather Rev., 96, 720–734, 1968.

Rodell, M., Velicogna, I., and Famiglietti, J. S.: Satellite-based estimates of groundwater depletion in India, Nature, 460, 999–1002, doi:10.1038/nature08238, 2009.

Seneviratne, S. I., Viterbo, P., Lüthi, D., and Schär, C.: Inferring changes in terrestrial water storage using ERA-40 reanalysis data: The Mississippi river basin, J. Climate, 17, 2039–2057, 2004.

Seneviratne, S. I., Lüthi, D., Litschi, M., and Schär, C.: Land-atmosphere coupling and climate change in Europe, Nature, 443, 205–209, doi:10.1038/nature05095, 2006.

Seneviratne, S. I., Corti, T., Davin, E. L., Hirschi, M., Jaeger, E. B., Lehner, I., Orlowsky, B., and Teuling, A. J.: Investigating soil moisture-climate interactions in a changing climate: A review, Earth-Sci. Rev., 99, 125–161, doi:10.1016/j.earscirev.2010.02.004, 2010.

Seneviratne, S. I., Wilhelm, M., Stanelle, T., van den Hurk, B., Hagemann, S., Berg, A., Cheruy, F., Higgins, M. E., Meier, A., Brovkin, V., Claussen, M., Ducharne, A., Dufresne, J.-L., Findell, K. L., Ghattas, J., Lawrence, D. M., Malyshev, S., Rummukainen, M., and Smith, B.: Impact of soil moisture-climate feedbacks on CMIP5 projections: First results from the GLACE-CMIP5 experiment, Geophys. Res. Lett., 40, 5212–5217, doi:10.1002/grl.50956, 2013.

Shukla, J. and Mintz, Y.: Influence of land-surface evaporation on the earth's climate, Science, 215, 1498–1501, 1982.

Tapley, B. D., Bettadpur, S., Ries, J. C., Thompson, P. F., and Watkins, M. M.: GRACE measurements of mass variability in the Earth system, Science, 305, 503–505, 2004.

Troch, P., Durcik, M., Seneviratne, S., Hirschi, M., Teuling, A., Hurkmans, R., and Hasan, S.: New Data Sets to Estimate Terrestrial Water Storage Change, EOS, 88, 469–470, 2007.

van den Hurk, B., Hirschi, M., Schär, C., Lenderink, G., van Meijgaard, E., van Ulden, A., Rockel, B., Hagemann, S., Graham, P., Kjellström, E., and Jones, R.: Soil control on runoff response to climate change in regional climate model simulations, J. Climate, 18, 3536–3551, 2005.

van den Hurk, B., Doblas-Reyes, F., Balsamo, G., Koster, R., Seneviratne, S., and Camargo, H.: Soil moisture effects on seasonal temperature and precipitation forecast scores in Europe, Clim. Dynam., 38, 349–362, doi:10.1007/s00382-010-0956-2, 2010.

Watkins, M. M., Wiese, D. N., Yuan, D.-N., Boening, C., and Landerer, F. W.: Improved methods for observing Earth's time variable mass distribution with GRACE using spherical cap mascons, J. Geophys. Res., 120, 2648–2671, doi:10.1002/2014JB011547, 2015.

Winter, T. C.: Uncertainties in estimating the water balance of lakes, Water Resour. Bull., 17, 82–115, 1981.

Yeh, P. J.-F., Irizarry, M., and Eltahir, E. A. B.: Hydroclimatology of Illinois: A comparison of monthly evaporation estimates based on atmospheric water balance and soil water balance, J. Geophys. Res., 103, 19823–19837, 1998.

Zektser, I. S. and Loaiciga, H. A.: Groundwater fluxes in the global hydrologic cycle: past, present and future, J. Hydrol., 144, 405–427, 1993.

Zektser, I. S., Loaiciga, H. A., and Wolf, J. T.: Environmental impacts of groundwater overdraft: selected case studies in the southwestern United States, Environ. Geol., 47, 396–404, 2005.

12

A database of water and heat observations over grassland in the north-east of Japan

Wenchao Ma[1], **Jun Asanuma**[1,2], **Jianqing Xu**[3], **and Yuichi Onda**[1,2]

[1]Center for Research in Isotopes and Environmental Dynamics, University of Tsukuba, Tsukuba, 305-8577, Japan

[2]Faculty of Life and Environmental Sciences, University of Tsukuba, Tsukuba, 305-8577, Japan

[3]College of Business Administration, Kanto Gakuin University, Yokohama, 236-0037, Japan

Correspondence: Wenchao Ma (wma@ied.tsukuba.ac.jp)

Abstract. A highly valuable database of long-term hydrometeorological measurements is presented, containing in situ observations for a period of 37 years from a well-maintained grassland in the north-east of Japan. The observations include shortwave radiation, net radiation, air and dew point temperatures at three elevations, soil temperature at four depths, sensible heat flux, soil heat flux, wind speed, relative humidity, air pressure and precipitation. The heights of measurements are 1.6, 12.5 and 29.5 m above ground, with the soil-layer observations at depths of 0.02, 0.1, 0.5 and 1 m. This high-quality database includes four temporal resolutions of 10 s, 0.5 h, 1 h and 24 h, with the hourly data presented here. Monthly and annual statistics are presented at the database web page of the Center for Research in Isotopes and Environmental Dynamics and Prediction of the University of Tsukuba, http://doi.org/10.24575/0001.198108. We validated the data by comparing them with published data from the local meteorological agency in Tateno operated by the Japan Metrological Agency, including the average, maximum and minimum values of air temperature, shortwave radiation, wind speed, relative humidity and precipitation. We have generated a daily downward longwave radiation time series with a method developed by Kondo and Xu (1997) based on the observations from the database. This constructed time series agrees well with observations collected between 2002 and 2006, as evaluated based on the values of the Nash–Sutcliffe efficiency (= 0.947) and percent bias (= 1.486). For the whole database, annually averaged values show a positive trend in precipitation, air temperature, shortwave radiation, net radiation and sensible heat flux over the past 37 years, with a negative trend detected for wind speed, soil heat flux and soil temperature.

1 Introduction

In situ observational databases play a dominant role in all research disciplines. In hydrometeorological studies, water and temperature observations from field work are essential for validating differently sourced data (Jackson et al., 2009; Liang et al., 2011; Guillevic et al., 2012), supporting theoretical modeling (Grayson and Blöschl, 2001), estimating the energy budget (Hirschi and Seneviratne, 2017; Makarieva et al., 2018) and recording historical climatic variation (Godsey et al., 2018; Kormos et al., 2018) for different subjects (Qu et al., 2016; Godsey et al., 2018). A full set of in situ hydrometeorological observations is introduced here from a well-maintained grassland located in the north-east of Japan, which has been continuously operated since 1981 by the Environmental Dynamics & Prediction (EDP) department of the Center for Research in Isotopes and Environmental Dynamics of the University of Tsukuba, Japan. This observational site has a long history of providing important contributions to multidisciplinary investigations in many aspects of geological, hydrological, biological and meteorological studies, including micrometeorology both within and outside of the plant community as well as investigations into the transportation of turbulence, evapotranspiration, the soil energy budget and groundwater movement (Kawamura, 1991).

For decades, the EDP department has provided a high-quality database for many valuable studies assessing the energy balance and the degree of evapotranspiration. For example, Nakagawa (1983) tested the possibility for estimating actual evapotranspiration using an equilibrium evaporation model. Sugita et al. (1985) developed an apparatus for measuring the heat pulse velocity for estimation of the transpiration flux by employing the energy-budget method, while investigating the effect of stemflow and vegetation storage on the evapotranspiration. Sugita and Kotoda (1985) tested the effects of the soil-water deficits on forest evapotranspiration based on the observations from the EDP database, from which the effective rate of soil-moisture consumption was estimated and shown to be more than half the amount controlling the suppression of the evapotranspiration. In that same year, another advanced technology was employed (Sugita and Kotoda, 1985), which combined using the remotely sensed, land-surface temperature as well as Priestley–Taylor-type equations, for estimating the regional evapotranspiration. By evaluating the observations in 2001, Yubasaki et al. (2005) tested whether a reduction factor for pasture can be used for the turf site by comparing the evapotranspiration estimated by Penman, energy-budget eddy covariance and energy-balance Bowen ratio methods. Saito and Yamanaka (2005) carried out a quality control of the data and analyzed the evapotranspiration data observed with a weighing lysimeter between 1981 and 2002, with the results of the data quality summarized. In 1983, a model was developed by Tase and Majima (1983) for estimating precipitation under the influence of interception. This model showed good adaptability with the canopy, stem and evapotranspiration components. In addition, latent heat flux was assessed by Hiyama et al. (1993), with flux behavior compared before and after the precipitation event as well as an investigation into temporal variation and an assessment of measurement accuracy.

Investigations into the water and energy transfer, not only above ground, but also within the soil layers, have also been extensively carried out. For example, Sakura (1979) tested the infiltration effect for soil-water movement and soil temperature. Taniguchi (1990) specified the distribution of heat transport within soil layers by using observations of soil temperature. Sakura and Taniguchi (1983) conducted experiments at the EDP site to test the characteristics of soil-water movement during infiltration. Measurement errors were investigated by Iwata and Sugita (2006), who demonstrated the reliability of the observed heat fluxes.

A variety of studies on ecology and vegetation were conducted because of the unique land surface covered by grassland. For example, Kotada and Hayashi (1980) investigated the micrometeorology of the vegetation, with specifying empirical parameters describing the vegetation growth state based on physical modeling. While Nasuno et al. (1989) estimated the turbulent fluxes with the eddy-covariance method to investigate the effect of forest and vegetation on the exchange of energy and water vapor. Another study was carried out by Hayashi et al. (1989) to investigate the water vapor and temperature profile on this grassland. Being a well-maintained observation site, the grassland provides good conditions for fundamental ecological research. For example, Yasui and Oikawa (1993) investigated the CO_2 flux resulting from soil respiration, showing a high correlation between the soil temperature, moisture and respiration. Based on the unique land-surface characters, Yamanaka et al. (2005) conducted experiments to investigate the biomass and root characteristics at the site.

As a newly developed and reliable approach, the isotope method has also been applied extensively in the investigations of the water and temperature observations within the EDP database. Generally, branches of isotopic research in hydrological applications include the fields of hydrometeorology, eco-hydrology, groundwater hydrology, watershed hydrology and the use of isoscapes (Yamanaka, 2012), with isotopic techniques being well accepted as a powerful and stable method for partitioning the transpiration from the evapotranspiration. In particular, Shimizu and Yamanaka (2005) tested the spatial structure of the isotopic composition of the atmospheric water vapor at the micrometeorological scale over this observational site and analyzed the mixing processes of water vapor with different sources. Wang and Yamanaka (2014) developed a new method based on a two-source model for the growing season and demonstrated the behavior of vegetation affected by physiological responses. Since then, numerical models of isotopic tracers have been developed (Wang et al., 2015), from which the Iso-SPAC model has evolved, with a steady-state assumption for the transpiration flux successfully reproducing seasonal variations of all the components of the surface-energy balance.

In recent decades, more attention has been paid to climate change, with the need for more comprehensive and diverse data sources. The advantage of a single-site database is the ability to provide refined, high-quality data, since a large-scale database can easily overlook specific procedures occurring within certain events involving both water and energy transport. The purpose of this paper is to present the available data and the processes involved in their collection for providing high-quality observations over multiple frequency scales for assisting climate studies and energy-balance estimations in multiple ways, such as (1) providing a complete data set for water and temperature observations for modeling studies or theoretical investigations; (2) helping to compare forecasted and estimated values of the climate trend; (3) showing a unique data set representing the meteorological and hydrological variations located in the north-east of Japan; (4) providing a valuable and unique database for biological and ecological studies in a long-term, continually well-maintained grassland.

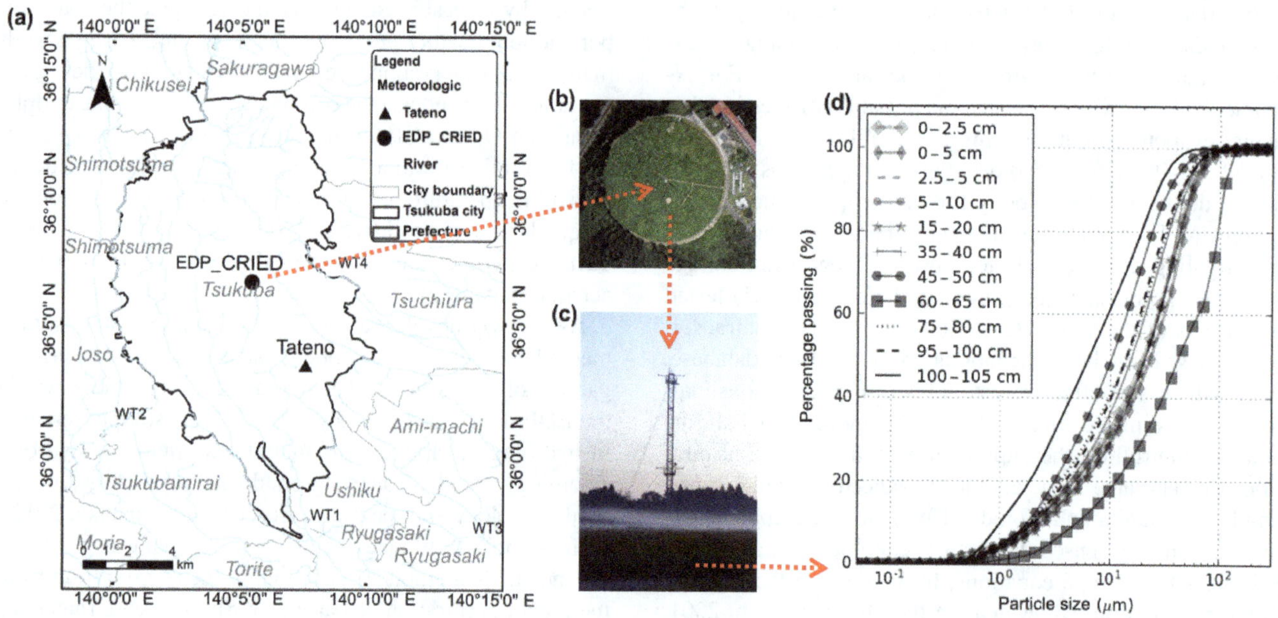

Figure 1. EDP observational site: **(a)** location, **(b)** projected view of the EDP grassland, **(c)** the observation tower and **(d)** results of the soil content analysis.

2 Data description

2.1 Observational site, data and instrumentation

The data were collected from an observational site located at the department of Environmental Dynamics and Prediction at the Center for Research in Isotopes (36°06′35″ N, 140°06′00″ E) in the grounds of the University of Tsukuba (Fig. 1a). This region has a semihumid marine climate with a long-term-average annual precipitation of 1200–1600 mm (Hamada et al., 1998). The observational site consists of a grass-covered circular field that is 160 m in diameter at an altitude of 27 m above sea level (a.s.l.), a meteorological observation tower at a height of 30 m equipped with sensors at heights of 1.6, 10 and 29.5 m, and underground sensors at depths of 0.02, 0.1, 0.5 and 1 m, which monitor the soil temperature and heat flux (Fig. 1b and Table 1). All observed time series have been aggregated at frequencies of 30 min, 1 h and 24 h since 1981 and at a frequency of 10 s since 2003. Being a continuous, long-term observational site, the maintenance and modernization of instrumentation has taken place to ensure high-quality observations, with the date on which each instrument was introduced found in Table 1.

The observational site was artificially filled with loam and volcanic ash soil in the top of $1 \sim 2$ m and a clay layer with thickness of $4 \sim 5$ m, underneath (Sakura, 1977). Each soil layer consists of brown to dark sand, clay and loam in different ratios. A soil-particle-size analysis was carried out using a Shimadsu SALD-3100 particle analyzer, with the results shown in Fig. 1d. During the growing season, the lowest soil layer encounters groundwater at a depth of around 2 m. The observational site is covered by multiple species of vegetation, with a similar percentage of species each year since 1977 (Imasu et al., 2002) providing consistent meteorological and hydrological research conditions throughout this period.

The vegetation is naturally grown C_3 and C_4 vegetation, such as *Imperata cylindrica*, *Andropogon virginicus*, *Miscanthus sinensis* as C_4 and *Solidago altissima*, *Artemisia princeps*, *Lespedeza cuneata*, *Lespedeza pilosa*, *Equisetum arvense*, *Festuca arundinacea*, *Potentilla freyniana*, *Lysimachia clethroides* as C_3. The similarities in grass species, depth and leaf-area index (LAI) were confirmed annually by two different surveys. Tanaka and Oikawa (1998) conducted a survey of the seasonal dynamics of the LAI for both C_3 and C_4 vegetation types, with results showing that the percentage of the species and the height of the vegetation had not changed dramatically since 1977. Another survey was carried out between 2000 and 2002 to directly measure the LAI and height, with similar results found as before, such as the value of the LAI, which ranged from less than 0.1 in winter to nearly 5 during September, while the height of the grass is about 1 cm during winter and around 1.8 m in September and October (Imasu et al., 2002). Since 2006, the grass has been mown twice each year (summer and winter); dead plants and grass clippings were redistributed.

Most observations we present were observed directly, except for the dew point temperature T_d, which, according to the historical record, has been collected by different instrumentation. While the initial instrumentation for the dew point temperature was a lithium chloride dew point thermometer, since 15 December 2006 this has been changed to a hygrom-

Table 1. Instrumentation information of the EDP observational system.

Item	Elevation (m)	Period	Instrument	Maker (model)
Shortwave radiation	1.5 1.5	Dec 2015 ~ Aug 1981 ~[a]	Four component radiation balance meter Thermocouple type total solar pyrometer	Hukseflux (CHF-NR01) EKO (MS-402F) EKO (MS-43F) EKO (A75510(MS43F)) EKO (A84517)
Net radiation	1.5	Mar 2014 ~ Aug 1981 ~	Four component radiation balance meter Draft-type thermocouple type radiation budget meter	Hukseflux (CHF-NR01) EKO (MF-11, CN-11)
Air temperature[b]	1.6, 12.3, 29.5	Aug 1981 ~	Ventilated platinum resistance thermometer Humidity temperature probe	Vaisala (CVS-HMP155D) Vaisala (CVS-HMP45D) Nakaasa (E-731)
Dew point temperature (relative humidity)	1.6, 12.3, 29.5	Aug 1981 ~ Jan 2006 2006 ~	Calculated from relative humidity Lithium chloride dew point thermometer	Vaisala (CVS-HMP) Nakaasa (E-771)
Sensible heat flux	1.6 12.3 29.5NW 29.5SE	Aug 1981 ~ Aug 1981 ~ 1987 Aug 1981 ~ Aug 1981 ~ Mar 2015 Jul 2015 ~	Ultrasonic wind speed thermometer	KAIJO SONIC (DA650, TA-61A) KAIJO (DA600) Kaijyo Denki
Soil temperature	−0.02, −0.1, −0.5, −1.0	Aug 1981 ~	Waterproof type platinum resistance thermometer	C-PTG-10 Nakaasa (E-751)
Soil heat flux	−0.02 −0.1	Aug 1981–~ May 2013~	Thermocouple type ground heat flow plate Heat flow plate	 EKO (CN-81) Hukseflux (HFP01-10)
Wind speed	1.6 12.3 29.5NW[c] 29.5SE[c] 30.35	Aug 1981 ~ Aug 1981 ~ 1987 Aug 1981 ~ Aug 1981 ~ Mar 2015 Jul 2015 ~ Feb 2014 ~	Ultrasonic wind speed thermometer Propeller type wind direction anemometer	KAIJO SONIC (DA650, TA-61A) KAIJO SONIC (DA600) Kaijyo Denki YOUNG (CYG-5103LM)
Precipitation	0.3	Aug 1981 ~	Tipping falls separated type self-recorded rain gauge	Yokokawa (WB0013-05, B-011-00)
Atmospheric pressure	1.5	Aug 1981 ~	Barometer	Vaisala (PTB210) Nakaasa (F-401)

[a] Frequency for all observation periods: 0.5, 1 and 24 h. A frequency of 10 s was used starting in 2003. [b] Air temperature observations included average, maximum and minimum values.
[c] NW is northwest; SE is southeast.

eter (CVS-HMMP45D, Climatec), with the value of the dew point temperature calculated from the observed specific humidity by Watarai and Yamanaka (2007).

$$T_\mathrm{d} = \frac{b \times \log_{10}\left(\frac{e}{6.11}\right)}{a - \log_{10}\left(\frac{e}{6.11}\right)}, \quad (1)$$

$$e = e_\mathrm{sat} \times \frac{\mathrm{RH}}{100}, \quad (2)$$

where $a = 7.5$, $b = 237.3$, RH is the relative humidity expressed as a percentage, and the saturated vapor pressure e_sat is obtained from

$$e_\mathrm{sat} = 6.11 \times 10^{\left(\frac{aT}{b+T}\right)}, \quad (3)$$

where the temperature T is given in °C. Data collected from supersonic anemometer–thermometers were used to obtain heat and momentum fluxes using the eddy-correlation method. The data are freely available to download from the Center for Research in Isotopes and Environment Dynamics (CRiED) website (http://www.ied.tsukuba.ac.jp/en/edps/database-doi/, last access: 22 October 2018) (formerly known as TERC) as hourly, monthly and annual summaries. Since 2003 the temporal resolution has been at 10 s, 30 min, 60 min and 24 h intervals.

When calculating averaged data (Asanuma et al., 2004) at least 24 records at 30 min were required. Readings with less than 20 records were discarded (marked with "***" in the supplement data file) and data with between 20 and 24 records were annotated as incomplete (Ohba and Yamanaka, 2008; marked with "*" in the supplement data file). In addition to the missing data, the dates of equipment maintenance, construction and mowing information are recorded in the maintenance log (http://www.ied.tsukuba.ac.jp/yosoku/kansoku/hojyo_log/, last access: 22 October 2018).

2.2 Estimation of downward longwave radiation

Using a method developed by Kondo and Xu (1997), the downward longwave radiation flux is estimated from the routine meteorological observations, including the solar radiation, air temperature, relative humidity, air pressure, wind speed and precipitation (see Table 1 for instrumentation information), with the temporal resolution of the calculation chosen based on both the observed forcing data and the estimated downward longwave radiation. This method was successfully applied to the Tibetan Plateau (Xu and Haginoya, 2001; Xu et al., 2005). In this study, the results of the estimation have been validated against the downward longwave radiation as directly observed by the EDP department between 2002 and 2007. To assess the model performance, the coefficient of determination (R^2), Nash–Sutcliffe efficiency (NSE) and the percent bias (PBIAS) are considered.

2.2.1 Estimation of downward longwave radiation

The downward longwave radiation at the ground is estimated following Kondo et al. (1994) as

$$L^{\downarrow} = \sigma T^4 \left[1 - \left(1 - \frac{L_{\mathrm{f}}^{\downarrow}}{\sigma T^4} \right) C \right], \tag{4}$$

$$L_{\mathrm{f}}^{\downarrow} = \left(0.74 + 0.19x + 0.07x^2 \right) \sigma T^4, \tag{5}$$

$$x \equiv \log_{10} w^*, \tag{6}$$

where $L^{\downarrow}$ is the downward longwave radiation for a cloudy day, $L_{\mathrm{f}}^{\downarrow}$ is the downward longwave radiation for a fine day, T (K) is the 1 h air temperature, $\sigma = 5.670 \times 10^{-8}\ \mathrm{W\,m^{-2}\,K^{-4}}$ is the Stefan–Boltzmann constant, and w^* is the effective precipitable water expressed as

$$w^* = \frac{1}{g} \int_0^{P_s} q \frac{p}{p_0} \mathrm{d}p. \tag{7}$$

Here, $p_0 = 1013\ \mathrm{hPa}$ is the standard atmospheric pressure, and p_s is the surface pressure (hPa). The precipitable water (cm) is estimated as

$$\log_{10} w \approx \log_{10} w^* + 0.10, \tag{8}$$

and C is the coefficient expressing the effect of clouds by

$$\begin{aligned} C &= 0.03B^3 - 0.03B^2 + 1.25B - 0.04, \quad B \geq 0.0323 \\ &= 0, \quad B < 0.0323, \end{aligned} \tag{9}$$

where $B \equiv S_{\mathrm{obs}}^{\downarrow} / S_{\mathrm{top}}^{\downarrow}$, Here, $S_{\mathrm{obs}}^{\downarrow}$ is the observed solar radiation flux, $S_{\mathrm{top}}^{\downarrow}$ is the mean downward solar radiation at the top of the atmosphere,

$$S_{\mathrm{top}}^{\downarrow} = \frac{S_{00}}{\pi} d \left(\zeta \sin\phi \sin\delta + \cos\phi \cos\delta \sin\zeta \right), \tag{10}$$

where

$$\zeta = \cos^{-1} \left(\frac{\sin\alpha - \sin\phi \sin\delta}{\cos\phi \cos\delta} \right), \tag{11}$$

$$\begin{aligned} d = {} & 1.00011 + 0.034221 \cos\eta + 0.00128 \sin\eta \\ & + 0.000719 \cos 2\eta + 0.000077 \sin 2\eta, \end{aligned} \tag{12}$$

$$\delta = \sin^{-1} \left[0.398 \times \sin(4.871 + \eta + 0.033 \sin\eta) \right], \tag{13}$$

$$\eta = \frac{2\pi}{365} \mathrm{Day}, \tag{14}$$

S_{00} is the solar constant, ζ is the half-day angle, ϕ is the latitude, δ is the solar declination, and Day is the total number of days from 1 January to the day of observation. According to Kondo et al. (1994), the estimated downward longwave radiation is accurate to within $\pm 10\ \mathrm{W\,m^{-2}}$, with differences resulting from the effects of snow in winter.

2.2.2 Model evaluation

Evaluating the agreement between observed and simulated data is a basic requirement for the data quality control (Moriasi et al., 2007; Singh et al., 2005) for which the following indices are chosen: the coefficient of determination (R^2) for describing the ratio of variance in the observations, the Nash–Sutcliffe efficiency (NSE) for indicating the error in the units of the constituent between observed and simulated data, and the percentage bias (PBIAS) for measuring the average tendency of the simulated data to be larger or smaller than the corresponding observed data (Gupta et al., 1999). Here,

$$\mathrm{NSE} = 1 - \left[\frac{\sum_{i=1}^{n} \left(Y_i^{\mathrm{obs}} - Y_i^{\mathrm{sim}} \right)^2}{\sum_{i=1}^{n} \left(Y_i^{\mathrm{obs}} - Y^{\mathrm{mean}} \right)^2} \right] \tag{15}$$

$$\mathrm{PBIAS} = \left[\frac{\sum_{i=1}^{n} \left(Y_i^{\mathrm{obs}} - Y_i^{\mathrm{sim}} \right) \cdot (100)}{\sum_{i=1}^{n} \left(Y_i^{\mathrm{obs}} \right)} \right], \tag{16}$$

where n is the total number of observations, Y_i^{obs} is the observed value for the ith step, Y_i^{sim} is the simulated value, Y^{mean} is the mean value for all the observed value (Moriasi et al., 2007). Here, a NSE value of unity represents the optimal value, while the optimal value is zero for the PBIAS, with low-magnitude values indicating an accurate simulation. These modeling indices are evaluated for checking the accuracy of the time series of the estimated downward longwave radiation.

3 Data quality control

The data quality was checked for the entire observation period, and errors were deleted that resulted from maintenance activities or abnormal system behavior and marked as missing values in the EDP database, with Table 2 showing the

Table 2. Parameter information within the EDP database. Locations A, B and C are located within the observation site of the EDP department as shown in Fig. 1.

Item	Unit	Height (m)	Period	Data coverage (%)	Average	Maximum	Minimum
Shortwave downward radiation[a]	$W m^{-2}$	1.5	1981 ~[a]	94.64	146.63	356.50	3.50
Net radiation	$W m^{-2}$	1.5	1981 ~	93.84	67.39	237.20	−15.00
Longwave downward radiation[b]	$W m^{-2}$	1.5	2002–2007[b]	84.75	346.51	450.75	227.37
Air temperature_Layer1_average[a]	°C	1.6	1981 ~	95.58	13.93	31.10	−5.30
Air temperature_Layer1_maximum[a]	°C	1.6	1981 ~	98.26	19.06	37.40	−0.60
Air temperature_Layer1_minimum[a]	°C	1.6	1981 ~	97.30	9.03	26.70	−11.30
Air temperature_Layer2	°C	12.5	1981 ~	97.79	14.23	31.10	−3.80
Air temperature_Layer3	°C	29.5	1981 ~	96.11	14.51	31.20	−3.00
Dew point temperature_Layer1	°C	1.6	1981 ~	95.51	8.86	26.30	−14.70
Dew point temperature_Layer2	°C	12.5	1981 ~	89.46	8.68	27.50	−19.50
Dew point temperature_Layer3	°C	29.5	1981 ~	89.60	8.53	27.30	−15.80
Soil temperature_Layer1	°C	−0.02	1981 ~	97.45	14.79	32.30	−1.50
Soil temperature_Layer2	°C	−0.10	1981 ~	97.73	14.72	30.30	0.60
Soil temperature_Layer3	°C	−0.50	1981 ~	92.50	14.95	25.90	3.10
Soil temperature_Layer4	°C	−1.00	1981 ~	97.63	14.78	25.20	5.00
Sensible heat flux	$°C m s^{-1}$	1.6	1981 ~	90.91	0.01	0.07	−0.03
Soil heat flux	$W m^{-2}$	−0.02	1981 ~	85.13	−0.34	20.00	−20.00
Wind speed[a]	$m s^{-1}$	1.6	1981 ~	99.44	0.97	3.80	0.16
Relative humidity[a, c]	%	1.6	2003 ~	97.17[b]	75.25	100.10	28.11
Precipitation[a]	$mm day^{-1}$	0.5	1981 ~	96.51	3.20	181.50	0.00
Atmosphere pressure	hPa	1.6	1983 ~	90.67	1010.41	1033.60	977.30

[a] Data items are compared with data from the Japan Meteorological Agency (JMA_Tateno). [b] The data coverage for relative humidity was estimated starting in 2003. [c] The observation of longwave downward radiation was between 2002 and 2007. Data coverage, average, maximum and minimum were determined using estimated longwave radiation based on other observation items. The estimated longwave radiation data comprise dates since 1983.

data availability within the database after error deletion. To preserve only original observational data from the EDP site, we do not conduct any gap filling. For most of the parameters, the daily values are average values from hourly data, except the daily precipitation value is the accumulation from hourly data.

This database includes three versions: ver. 1.0, 1.1 and 2.0. For ver. 1.0, the data were collected in integer data format following a former system standard, which applied to the observed data until April 2003. Then, the new system was started from May 2003 and the data set was updated as ver. 1.1. The data quality is guaranteed by the consistent quality control of all raw observation data. The quality control includes removing error data due to instrumental problems and missing data caused by observed values out of the specified range (http://www.ied.tsukuba.ac.jp/yosoku/terc/, last access: 22 October 2018). The data format in ver. 1.1 were established in accordance with Asanuma et al. (2004). Ver. 2.0 is the newest version, which is a comprehensive version containing both ver. 1.0 and ver. 1.1 for the purpose of improving data reliability by performing quality evaluation and quality control. Ver. 2.0 has two main sections. The first section is composed of the hourly, monthly and annual average values with highly consistent quality control from August 1981 to December 2005. The other one is composed of the raw data, which include data in time frequencies of 30 min, 60 min, 24 h and 10 s from 2003 to the present (http://www.ied.tsukuba.ac.jp/yosoku/kansoku/rawdata/, last access: 22 October 2018).

3.1 Comparison with the Japan Meteorological Agency

After the deletion of error values, the reliability of the observations from the EDP database is assessed with respect to the reference database from the nearest local meteorological agency in Tateno, which belongs to the Japan Meteorological Agency. The Tateno observations are at a height of 2 m above ground level, with only a few items available for data-quality checking, including shortwave radiation, air temperature (average, maximum and minimum), wind speed, relative humidity and precipitation, at a daily resolution between 1981 and 2017 (see Table 2, Fig. 2). For the EDP database, observations from the 1.6 m height are compared with the Tateno data, with the data at other heights compared with the adjacent measurement heights. Both of these two data sets are compared in Fig. 2 along with the corresponding values of R^2.

As the value of $R^2 > 0.99$ for the average, maximum and minimum values of the air temperature in correlations between the Tateno and EDP databases, a high similarity ex-

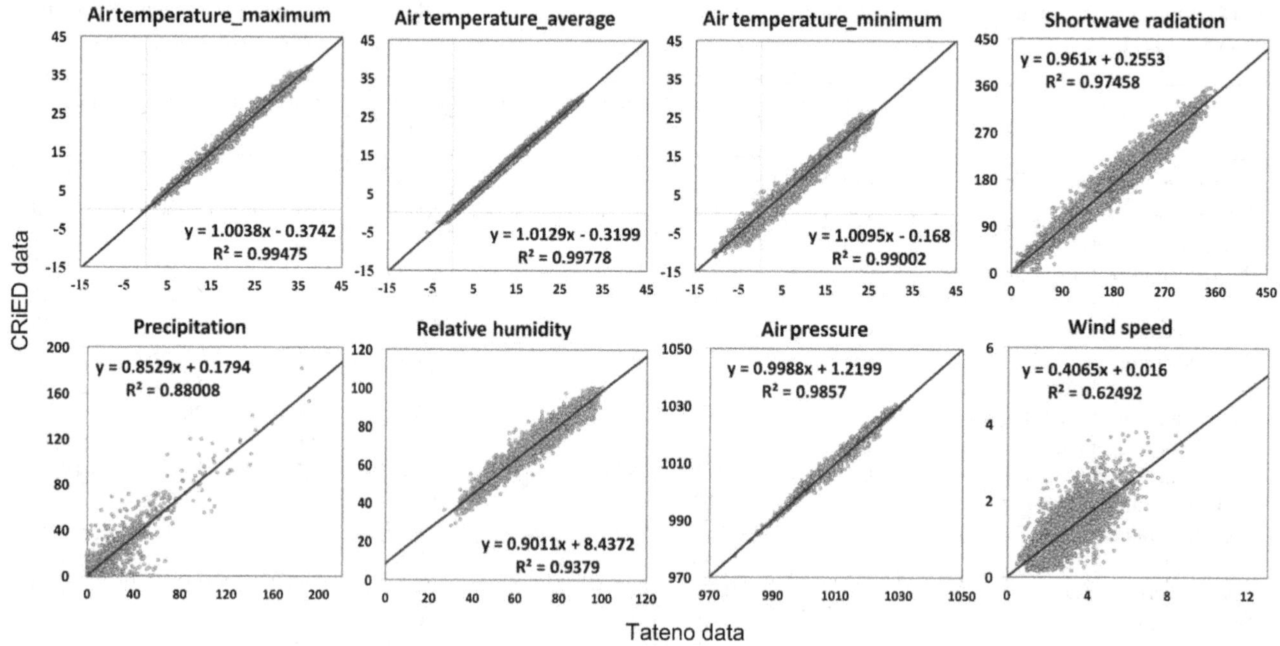

Figure 2. Comparisons of the EDP data with the Tateno data from the Japan Meteorological Agency from 1981 to 2017. Further parameter information is found in Table 2.

ists between the two closely located stations separated by < 10 km and an elevation difference of 1.8 m (the Tateno and EPD sites are 25.2 m and 27 m a.s.l., respectively). However, slight differences between these two databases were found, which is caused by the different land surface, such as the presence of different vegetation and artificial structures. Because of the diverse land-surface cover and varied moisture distribution, the value of the relative humidity RH differs slightly, but is still well correlated. Differences in shortwave radiation between the EDP and Tateno are mainly governed by solar radiation, whereas absorption and reflection may be caused by atmospheric conditions (clouds). Overall, most of the variables are highly correlated, except wind speed and precipitation.

According to the regression analysis, a lower correlation is found for the precipitation and wind speed between these two stations (Fig. 2), because of the highly local characteristics of these two variables. In the case of the wind speed, the sensor of the EDP site located at a height of 1.6 m is surrounded by grass during the growing season and prior to mowing when the height of the grass reaches 1.8 m, which generates enhanced local turbulence, resulting in a reduced wind speed. The precipitation amounts are significantly affected by the local conditions depending on the humidity, temperature, cloud distribution and wind speed (Shuttleworth, 2011). In general, after error deletion, the EDP database compares reasonably well to the Tateno database.

3.2 Downward longwave radiation

The downward longwave radiation was calculated by using the method introduced in Sect. 2.2.1 as estimated from routine meteorological observations from 1983 to 2017 and compared with a 5-year data set collected by the EDP department from 2002 to 2006 (Fig. 3), giving $R^2 = 0.974$, $\text{NSE} = 0.947$ and $\text{PBIAS} = 1.486$ and indicating good agreement of the estimated values with the observed values according to Moriasi et al. (2007). Although the estimated values are slightly higher than the observed values, this slight bias of estimation is consistent with the statement by Kondo et al. (1994) regarding the effect of winter resulting from the presence of snow. However, most of the estimated values correspond well with the observed values.

3.3 Statistical analysis

Plotted in Fig. 4 are daily observed values of the air and soil temperatures for all layers, the precipitation, air pressure, humidity, longwave radiation, solar radiation, net radiation (R_n), the soil heat flux (G) and sensible heat flux (H) after data quality control. For each soil layer, the temperature varies between the deep and shallow layers, with the long-term-average values varying from 14.7 to 15 °C, the maximum values from 25.2 to 32.3 °C and the minimum values from -1.5 to 5 °C. A regular annual variation is found for the four soil layers, with the response of the deeper layers being slower than the shallow layers. The larger amplitude variation is evident in the shallowest soil layer, with the amplitude gradually decreasing for deeper soil layers. Some ex-

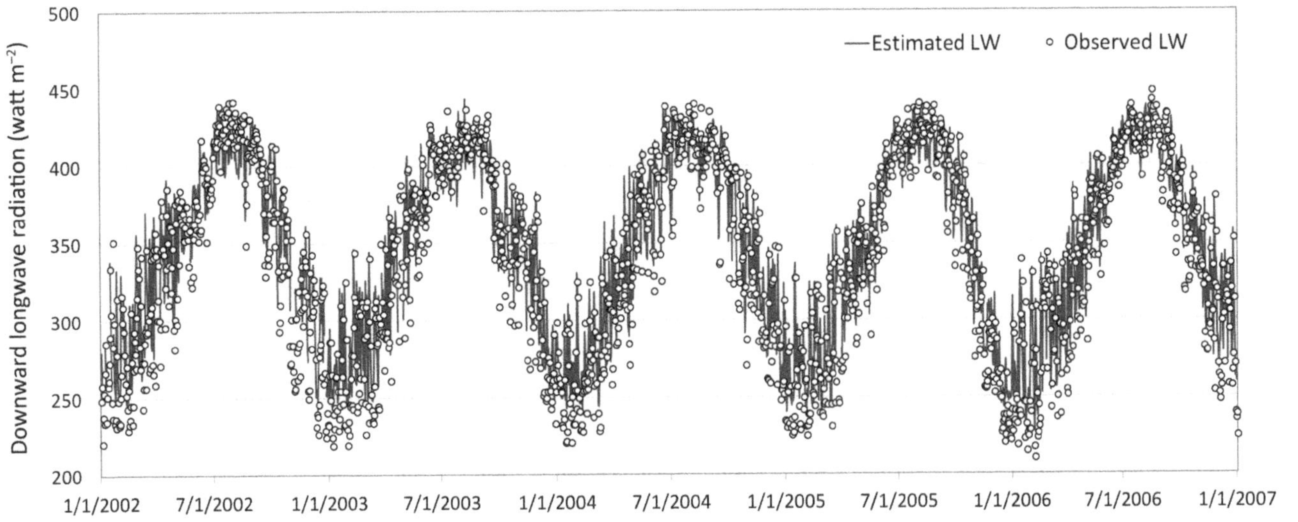

Figure 3. Comparison of estimated and observed downward longwave radiation at the EDP site from 1 January 2002 to 31st December 2006.

tremely high values occurred in 1990, 1991 and 2005, and extremely low values were found from 1983 to 1985 and 2011 to 2013. In contrast to the temperature of the different soil layers, the three air-temperature measurements show similar patterns of variation, with the 37-year average values for the heights of 1.6, 12.5 and 29.5 m corresponding to 13.9, 14.2 and 14.5 °C, respectively. The maximum value is 31.1 °C for all three layers, but the minimum values show obvious differences of −5.3 °C at 1.6 m, −3.8 °C at 12.9 m and −3.0 °C at 29.5 m. For the dew point temperature, the three layers show very similar ranges of observation values. From 1.6 to 29.5 m, the long-term average varies from 8.5 to 8.9 °C, the maximum value from 26.3 to 27.9 °C and the minimum value from −14.7 to −19.5 °C. The minimum values for the air and dew point temperature at 1.6 m are more different than the other layers, since the 1.6 m measurement height is more easily affected by any dramatic temperature or moisture exchanges with the ground.

The 37-year average value of precipitation is 1183.8 mm year^{-1}, and the maximum precipitation event occurred in 1986 with an amount of 181.5 mm day^{-1}. The relative humidity takes an average value of 75.2 % for the 37-year period. The long-term-average value of the air pressure is 1010.4 hPa.

The long-term-average value of the net radiation is 66.9 W m^{-2}, while the maximum value is 237.2 W m^{-2}, and the minimum is −15 W m^{-2}. The interannual variation indicates an increasing tendency for both the summer and winter from 1981 to 2004, but which has reduced since 2005. The average value of the shortwave radiation is about 146.6 W m^{-2}, while the maximum value is 356.5 W m^{-2}, and the minimum is −3.5 W m^{-2}. The average value of the longwave radiation is 346.5 W m^{-2}, while the maximum value is 450.8 W m^{-2}, and the minimum is 227.4 W m^{-2}. The soil heat flux and sensible heat flux show less regular variations compared with the values of the temperatures presented above, because of the difficulties in the accurate monitoring of the soil heat flux resulting from the high spatial variation in the soil properties. Furthermore, the soil heat flux is easily influenced by the soil temperature, air temperature and heat capacity as well as the thermal peculiarities of heterogeneous soil layers. However, since the observation of these thermal components is rare, they are highly valuable for hydrometeorological investigations. Further analysis of the database is highly encouraged for specific research topics.

4 Results

The complete long-term hydrometeorological data set at our grassland site represents a valuable resource for climate studies. For example, Duan et al. (2015) demonstrated the uniqueness of this area, which indicates that changes in the precipitation amounts in Japan are in the opposite direction to the extremes seen in the surrounding countries. Moreover, the database is capable of providing data for assessing both the energy and water budgets. Here, the main tendencies in the last 37 years are summarized in Fig. 5, where the annual values represent the years with data availability > 90 % (328 days for 1 year), because gap filling has not been performed.

Based on the annually averaged values, most variables show a positive trend, except the soil temperature, soil heat flux, wind speed and precipitation, which show a slightly negative trend. First, the shortwave, longwave and net radiation increased between 1981 and 2017 (Fig. 5a–c). The shortwave radiation was compared with data from the Tateno database above, from which a similar positive tendency was

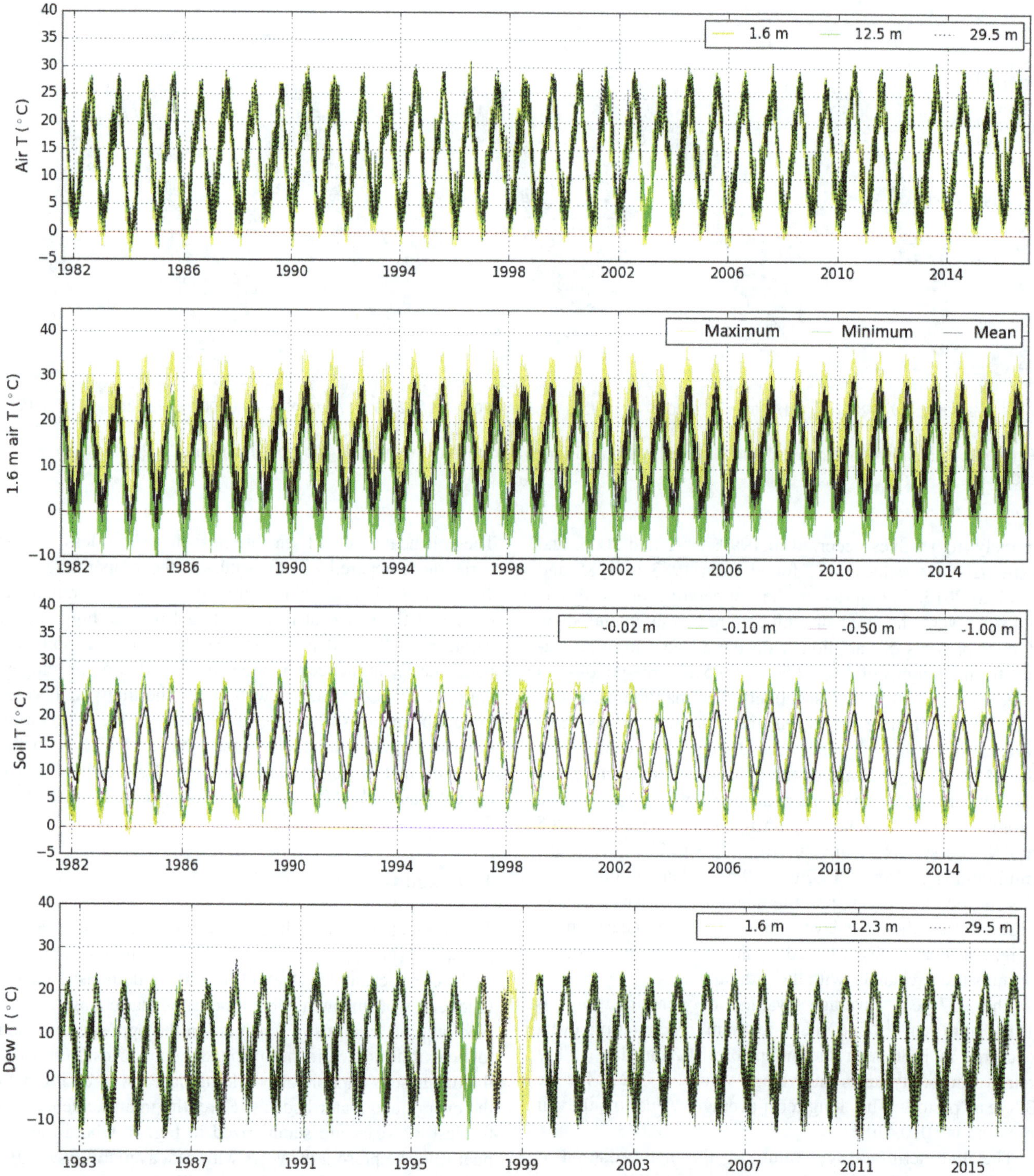

Figure 4.

found. That the estimated longwave radiation has a positive trend is consistent with the air-temperature trend with reference to the Stefan–Boltzmann law (Shuttleworth, 2011). Since the shortwave radiation has increased, it is reasonable to expect the net radiation to have increased as well, because the balance of net radiation is governed by the balance of the shortwave and longwave radiation. An observed positive air-temperature trend is evident since 1981 (Fig. 5j), which is consistent with the global trend related to climate change and an increasing temperature. The average, maximum and min-

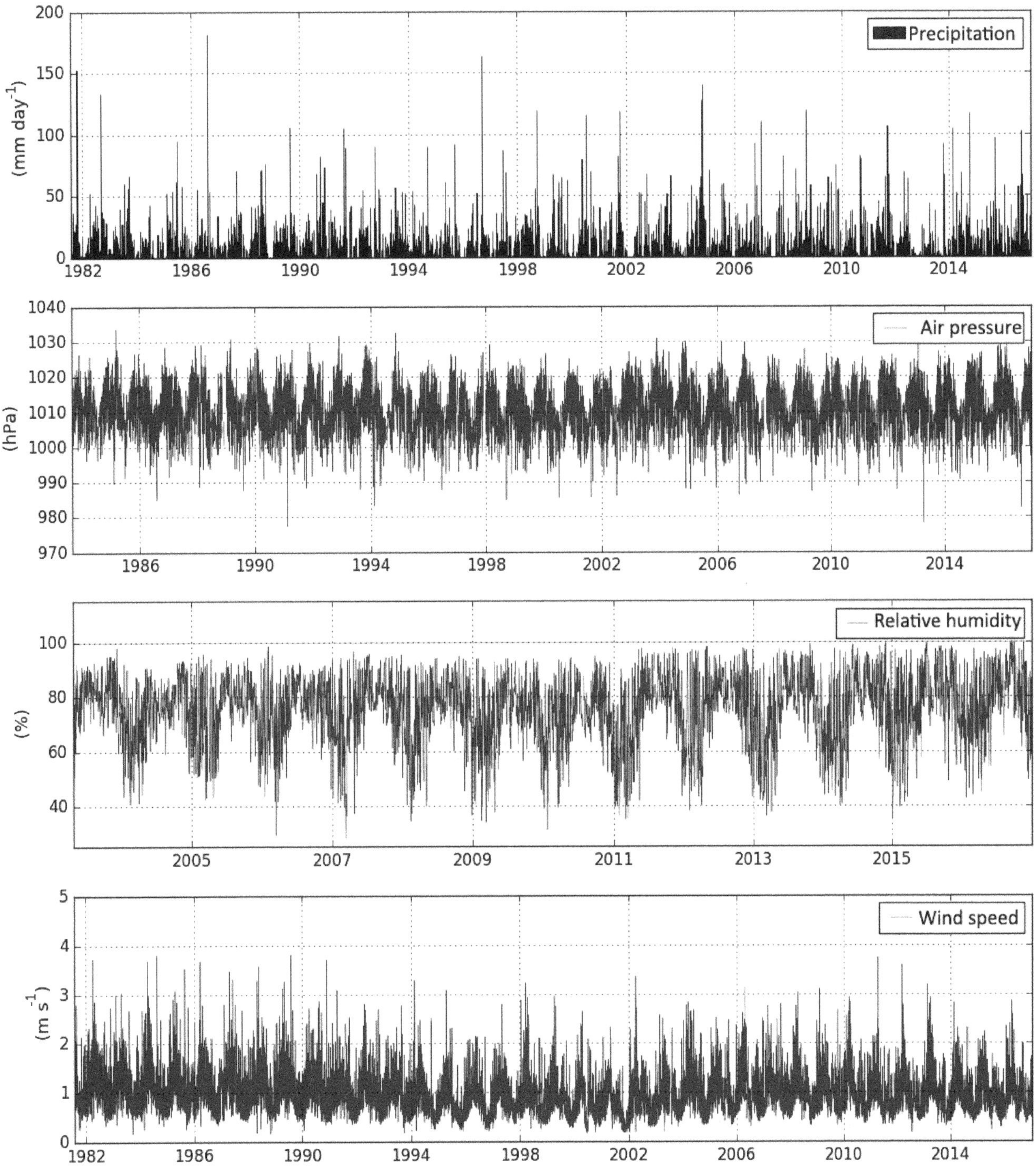

Figure 4.

imum values of the air temperature all show a positive trend with an increase of 0.04 °C $year^{-1}$. Here, the minimum value increased more slowly than the other two statistics, because the minimum air temperature is highly affected by urbanization (Kondo et al., 1994), where the main construction period of Tsukuba occurred within the 1980s and slowed down thereafter, so that the minimum air temperature shows a relatively smaller positive trend compared with the maximum and average values. The dew point temperature mainly followed a similar positive trend as the air temperature, with similarly increasing values at all three heights, including the first height, which is often surrounded by grass (see

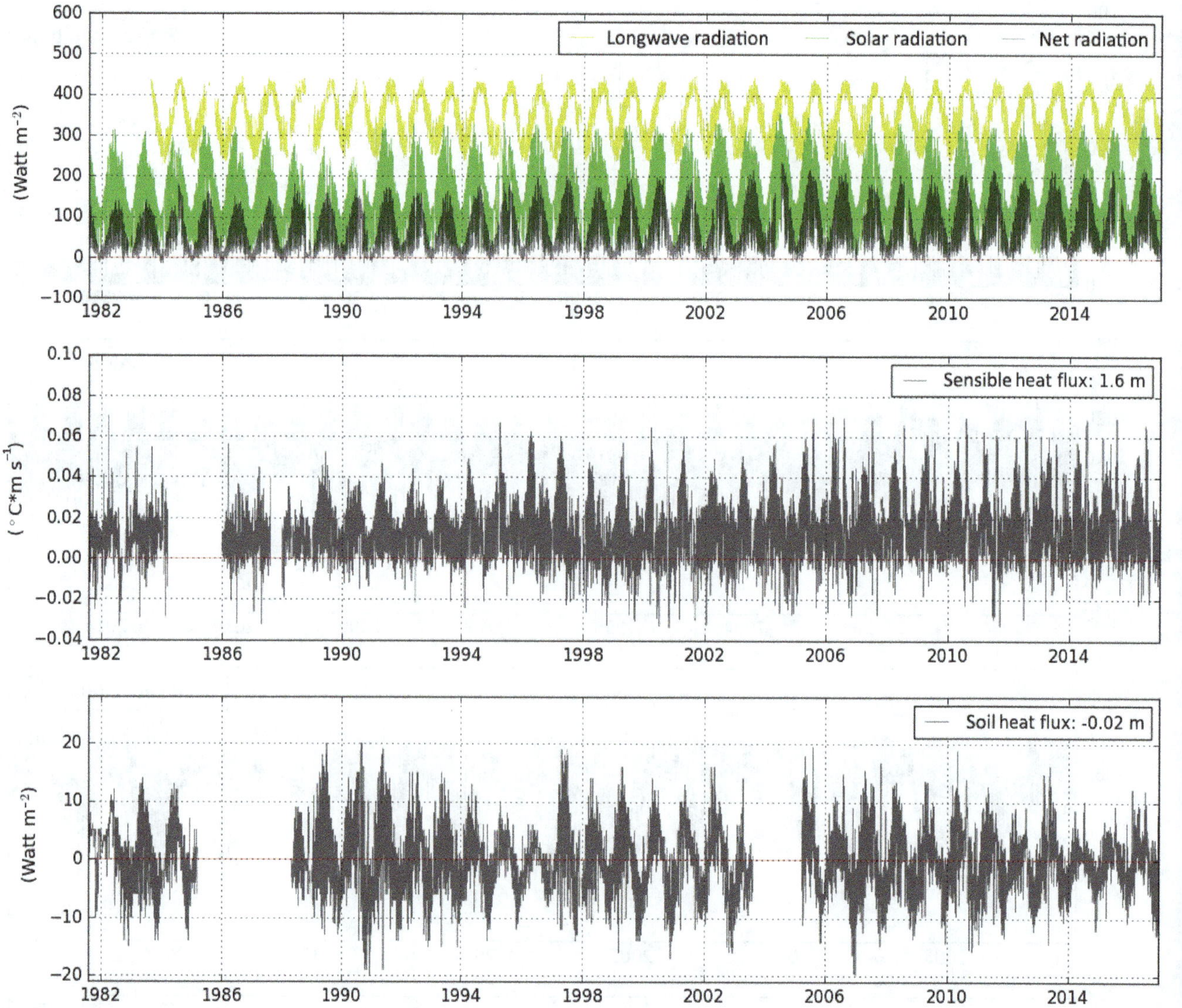

Figure 4. Daily observed values of the air temperatures for all layers, maximum, minimum and mean air temperature at a height of 1.6 m, soil and dew temperatures for all layers, the precipitation, air pressure, humidity, wind speed, longwave radiation, solar radiation, net radiation, sensible heat flux and the soil heat flux at the EDP site from 1981 to 2017.

Figs. 5k, 1). The variation in the surface soil layer (−0.02 m) shows a higher amplitude than the lower soil layers, which are more exposed to the energy exchange occurring in the lower atmosphere, and since the EDP site is covered by bushy grass, the energy transport received has a strong influence. Therefore, the temperature of the soil layer does not show any simple deviation from the annual average values (Fig. 5l). The explanation of the negative tendency of the soil heat flux (Fig. 5e) relates to the energy balance as pointed out by Brutsaert (1982), where $R_n = L_e E + H + G$, L_e is the latent heat of vaporization, and L_p is the thermal conversion factor for the fixation of carbon dioxide, so that the variation in the soil heat flux ($G = R_n - L_e E - H$) is directly related to the variation in the net radiation, and the latent and sensible heat fluxes. The tendency of evapotranspiration is possibly positive since the temperature, which is the dominant factor, shows a significant positive trend. Furthermore, although there is a slightly positive trend of net radiation as well as a negative trend of the sensible heat flux, the soil heat flux shows a negative trend. The corresponding increase in potential evapotranspiration is very possibly caused by the air temperature rising, which is in accordance with global climate change. As the temperature is the most important parameter for evapotranspiration, the increase in temperature probably leads to an increase in evapotranspiration, meaning the decrease in the soil heat flux is reasonable, although no data exist for the observations of the latent heat flux. The sensible heat flux and the relative humidity show positive

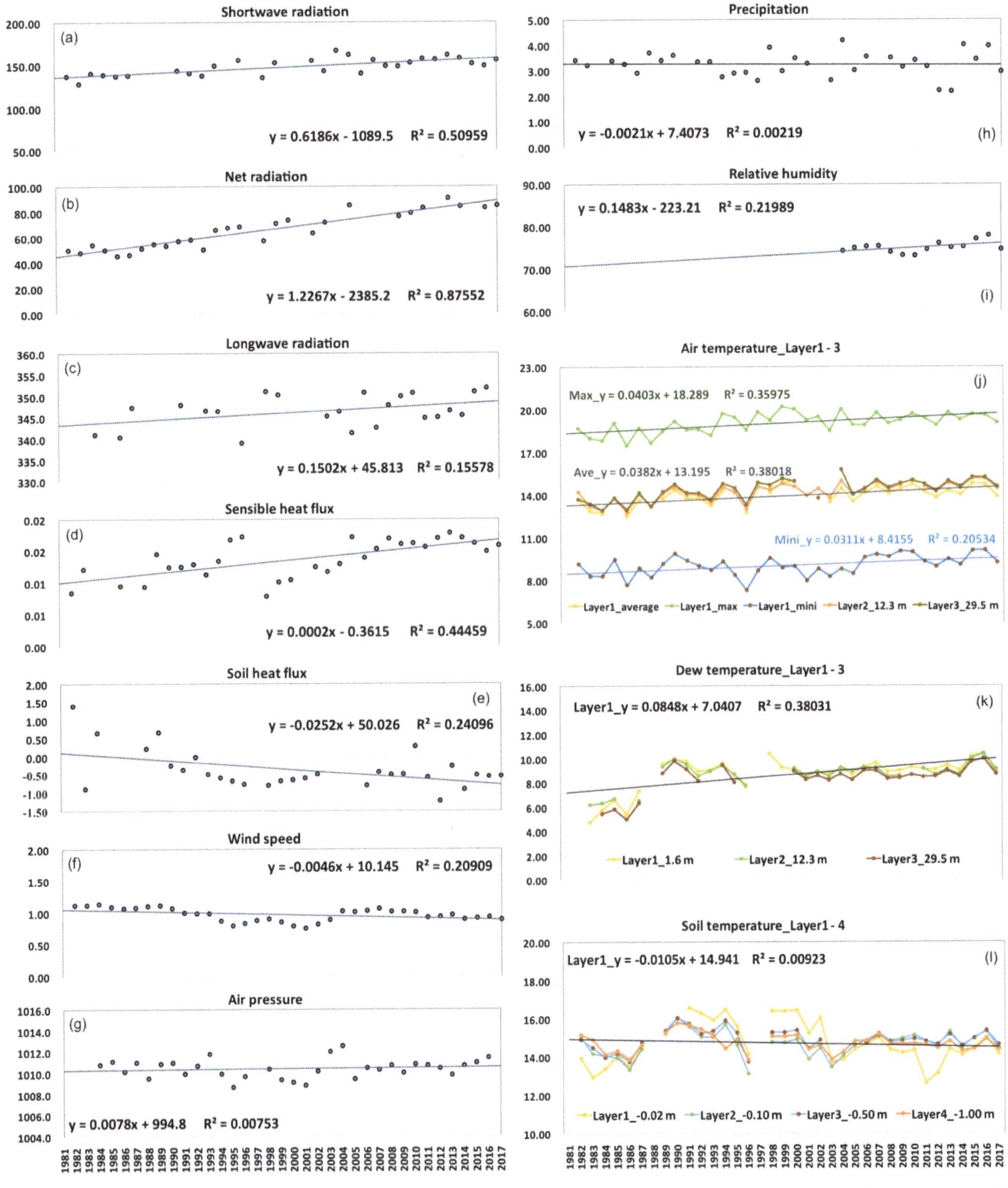

Figure 5. Annual average values of the observational database showing the **(a)** shortwave radiation ($W m^{-2}$), **(b)** net radiation ($W m^{-2}$), **(c)** longwave radiation ($W m^{-2}$), **(d)** sensible heat flux (°C $m s^{-1}$), **(e)** soil heat flux ($W m^{-2}$), **(f)** wind speed ($m s^{-1}$), **(g)** air pressure (hPa), **(h)** precipitation ($mm day^{-1}$), **(i)** relative humidity (%), **(j)** average, maximum and minimum air temperature (°C) at a height of 1.6 m, with the average values at the heights of 12.3 and 29.5 m, **(k)** dew point temperature (°C) at heights of 1.6, 12.3 and 29.5 m, and **(l)** soil temperature (°C) at the depths of 0.02, 0.1, 0.5 and 1 m.

trends (Fig. 5d, i). The main variation in the trend of precipitation is relatively moderate because of the more extreme low and high values recorded in recent decades, which is consistent with the more extreme precipitation observed globally (Donat et al., 2016, Fig. 5h). Compared with precipitation, a similar trend may be found in the relative humidity, although the analysis period is relatively short, beginning in 2004 (Fig. 5i). For the wind speed (Fig. 5f), there is negative tendency shown for the long-term observations, which is easily affected by the variation in the surrounding land cover. During these decades, two minor decreases in the signal are found between 1993 and 2002, and a decrease is also found in the value of the air pressure (Fig. 5g).

5 Data availability

The database described here has a digital object identifier (https://doi.org/10.24575/0001.198108, Asanuma and Ma, 2017) and is freely available at the home page of the EDP data center (https://doi.org/10.24575/0001.198108). The data must be fully referenced for every use as introduced at http://www.ied.tsukuba.ac.jp/en/edps/database-doi/ (last access: 22 October 2018). Supplemental materials may be found by checking the observational data website at http://www.ied.tsukuba.ac.jp/yosoku/kansoku/ (last access: 22 October 2018). Maintenance information is found from the database log at http://www.ied.tsukuba.ac.jp/yosoku/kansoku/hojyo_log/ (last access: 22 October 2018).

6 Conclusions

A high-quality database covering 37 years is presented from the north-east of Japan, encompassing four observational frequencies, three measurement levels above ground and four layers below the ground. The daily and annual average values were presented, including the daily values of shortwave radiation, average, maximum and minimum air temperatures, wind speed, relative humidity and precipitation, which were compared with data from the local meteorological agency at Tateno for validation of the data quality. Highly valuable data are also presented, such as for the net radiation, soil temperature at four depths, the air and dew point temperatures at three heights, the net radiation, soil heat flux and sensible heat flux, especially because the soil heat flux and sensible heat flux are relatively rarely observed or available. Furthermore, a time series of longwave radiation has been generated for this site based on the reliable observations.

The annual average values were analyzed for the years with data availability > 90 %. The annual average values show a positive tendency for the shortwave radiation, net radiation, longwave radiation, sensible heat flux, air pressure, relative humidity, the air and dew point temperatures at the three heights and a negative trend of the soil heat flux, wind speed, precipitation and soil temperature. These trends provide an important database and evidence for understanding the variations occurring within the study area. At the same time, the specific characteristics of the database may be found for the grassland with respect to the values of the wind speed and precipitation based on the comparisons between the daily values from the EDP database and Tateno as well as the regression of annual average values. As the in situ observational site and database have contributed to many previous studies, we shall continue to maintain it, and we wish to make the data available for further research and analysis.

Author contributions. JA created the observation system and data collection system, and controls the measurement. WM prepared the database and wrote this manuscript. JX provided the theoretical support of downward longwave radiation. WM and JX estimated the downward longwave radiation. JA and YO designed this work. All authors provided critical feedback and helped to shape the research, analysis and manuscript.

Competing interests. The authors declare that they have no conflict of interest.

Acknowledgements. This study was supported by "Interdisciplinary Project on Environmental Transfer of Radionuclides". The authors would like to acknowledge Prof. Tsutomu Yamanaka for his valuable suggestions. We thank Tomohiro Sekiguchi and Kentaro Aida for their assistance in the system maintenance. We thank Hideo Iijima, Shiraishi Izumi and Naomi Nakajima for providing digital and paper-based bulletins for this review. We thank two anonymous reviewers for their thoughtful and constructive comments, which helped to improve the quality of this work.

Edited by: David Carlson

References

Asanuma, J. and Ma, W.: Heat and Water Balance Measurement System, Center for Research in Isotopes and Environmental Dynamics at University of Tsukuba, https://doi.org/10.24575/0001.198108, 2017.

Asanuma, J., Nohara, D., Hara, M., and Yorisaki, T.: The Third Generation of the Data Collecting System for Meteorological and Hydrological Measurements, Bulletin of the TERC, the University of Tsukuba, 5, 157–174, https://doi.org/10.15068/00146972, 2004.

Brutsaert, W.: Evaporation into the Atmosphere, in: Theory, History and Applications, Kluwer Academic, Dordrecht, 1, 2–3, 1982.

Donat, M. G., Lowry, A. L., Alexander, L. V., O'Gorman, P. A., and Maher, N.: More extreme precipitation in the worlds dry and wet regions, Nat. Clim. Change, 6, 508–513, https://doi.org/10.1038/nclimate2941, 2016.

Duan, W., He, B., Takara, K., Luo, P., Hu, M., Alias, N. E., and Nover, D.: Changes of precipitation amounts and extremes over Japan between 1901 and 2012 and their connection to climate indices, Clim. Dynam., 45, 2273–2292, https://doi.org/10.1007/s00382-015-2778-8, 2015.

Godsey, S. E., Marks, D., Kormos, P. R., Seyfried, M. S., Enslin, C. L., Winstral, A. H., McNamara, J. P., and Link, T. E.: Eleven years of mountain weather, snow, soil moisture and streamflow data from the rain-snow transition zone – the Johnston Draw catchment, Reynolds Creek Experimental Watershed and Critical Zone Observatory, USA, Earth Syst. Sci. Data, 10, 1207–1216, https://doi.org/10.5194/essd-10-1207-2018, 2018.

Grayson, R. and Blöschl, G. (Eds.): Spatial Patterns in Catchment Hydrology: Observations and Modelling, Cambridge University Press, Cambridge, 2001.

Guillevic, P. C., Privette, J. L., Coudert, B., Palecki, M. A., Demarty, J., Ottle, C., and Augustine, J. A.: Land Surface Temperature product validation using NOAA's surface climate observation networks – Scaling methodology for the Visible Infrared Imager Radiometer Suite (VIIRS), Remote Sens. Environ., 124, 282–298, https://doi.org/10.1016/j.rse.2012.05.004, 2012.

Gupta, H. V., Sorooshian, S., and Yapo. P. O.: Status of automatic calibration for hydrologic models: Comparison with multilevel expert calibration, J. Hydrol. Eng., 4, 135–143, 1999.

Hamada, Y., Fernando, A. R. I., and Tanaka, T.: Physical Properties of Soils of Red Pine Forest and Heat Balance and Water Balance Experiment Field in Environmental Research Center, University of Tsukuba, Bulletin of the TERC, the University of Tsukuba, 23, 1–10, 1998.

Hayashi, Y., Harazono, Y., Nasuno, M., and Toritani, H.: Dynamic response of a Red Pine Canopy to changes of the Outside Environment. 1. Characteristics of temperature field within and above a Red pine canopy, Bulletin of the ERC, the University of Tsukuba, 13, 97–103, 1989.

Hirschi, M. and Seneviratne, S. I.: Basin-scale water-balance dataset (BSWB): an update, Earth Syst. Sci. Data, 9, 251–258, https://doi.org/10.5194/essd-9-251-2017, 2017.

Hiyama, T., Sugita, M., and Mikami, M.: Comparisons of the Latent Heat Fluxes Evaluated by a Weighing Lysimeter and an Energy Balance Method, Bulletin of the TERC, the University of Tsukuba, 18, 41–53, 1993.

Imasu, F., Mo, W., Kato, M., and Oikawa, T.: Seasonal changes in LAI and aboveground biomass of a C_3/C_4 mixed grassland in 2001 in the Terrestrial Environment Research Center, Bulletin of the TERC, the University of Tsukuba, 3, 17–25, 2002.

Iwata, H. and Sugita, M.: Correction of sensible heat flux measurement errors using a universal function of standard deviation of vertical wind velocity, Bulletin of the TERC, the University of Tsukuba, 6, 87–97, https://doi.org/10.15068/00147136, 2006.

Jackson, T. J., Vindlish, R., and Cosh, M.: Validation of AMSR-E Soil Moisture Products Using In Situ Observations, Journal of The Remote Sensing Society of Japan, 29, 263–270, https://doi.org/10.11440/rssj.29.263, 2009.

Kawamura, R.: Studies on Global Climate System, Bulletin of the ERC, the University of Tsukuba, 15, 109–115, 1991.

Kondo, J.: Meteorology of the water environment, Asakura Press, 86–91, 1994 (in Japanese).

Kondo, J. and Xu, J.: Seasonal Variations in the Heat and Water Balances for Nonvegetated Surfaces, J. Appl. Meteorol., 36, 1676–1695, 1997.

Kormos, P. R., Marks, D. G., Seyfried, M. S., Havens, S. C., Hedrick, A., Lohse, K. A., Sandusky, M., Kahl, A., and Garen, D.: 31 years of hourly spatially distributed air temperature, humidity, and precipitation amount and phase from Reynolds Critical Zone Observatory, Earth Syst. Sci. Data, 10, 1197–1205, https://doi.org/10.5194/essd-10-1197-2018, 2018.

Kotada, K. and Hayashi, Y.: Numerical modelling on wind speed profile of vegetation community, Bulletin of the ERC, the University of Tsukuba, 4, p. 102, 1980.

Liang, L., Schwartz, M. D., and Fei, S.: Validating satellite phenology through intensive ground observation and landscape scaling in a mixed seasonal forest, Remote Sens. Environ., 115, 143–157, https://doi.org/10.1016/j.rse.2010.08.013, 2011.

Makarieva, O., Nesterova, N., Lebedeva, L., and Sushansky, S.: Water balance and hydrology research in a mountainous permafrost watershed in upland streams of the Kolyma River, Russia: a database from the Kolyma Water-Balance Station, 1948–1997, Earth Syst. Sci. Data, 10, 689–710, https://doi.org/10.5194/essd-10-689-2018, 2018.

Moriasi, D. H., Arnold, J. G., Van Liew, M. W., Bingner, R. L., Harmel, R. D., and Veith, T. L.: Model evaluation guidelines for systematic quantification of accuracy in watershed simulations, American Society of Agricultural and Biological Engineers, 50, 885–900, 2007.

Nakagawa, S.: Possibility of Equilibrium Evaporation Model for Estimating Actual Evapotranspiration, Bulletin of the ERC, the University of Tsukuba, 7, 17–26, 1983.

Nasuno, M., Hayashi, Y., Harazono, Y., Nishizawa, T.: Dynamic response of a Red Pine Canopy to changes of the Outside Environment. 3. Turbulence Transport over and within a Red Pine Canopy, Bulletin of the ERC, the University of Tsukuba, 13, 113–121, 1989.

Ohba, M. and Yamanaka, T.: Observational data of heat balance and water balance: 2007, Bulletin of the TERC, the University of Tsukuba, 9, 49–74, https://doi.org/10.15068/00147251, 2008.

Qu, W., Bogena, H. R., Huisman, J. A., Schmidt, M., Kunkel, R., Weuthen, A., Schiedung, H., Schilling, B., Sorg, J., and Vereecken, H.: The integrated water balance and soil data set of the Rollesbroich hydrological observatory, Earth Syst. Sci. Data, 8, 517–529, https://doi.org/10.5194/essd-8-517-2016, 2016.

Saito, M. and Yamanaka, T.: Analysis of Long-term Evapotraspiration Data Observed by Weighing Lysimeter and Its Quality Control, Bulletin of the TERC, the University of Tsukuba, 6, 53–62, https://doi.org/10.15068/00147122, 2005.

Sakura, Y.: Miscellaneous, IV Water balance observation facility, Bulletin of the ERC, the University of Tsukuba, 1, 87–90, 1977.

Sakura, Y.: Soil Water Movement Estimated from Soil Temperature Changes during Infiltration, Bulletin of the ERC, the University of Tsukuba, 3, 33–38, 1979.

Sakura, Y. and Taniguchi, M.: Experiments of rain infiltration on characteristics of soil water movement using a soil column, Geographical Review of Japan, 56, 81–93, https://doi.org/10.4157/grj.56.81, 1983.

Shimizu, R. and Yamanaka, T.: The Spatial Structure of Isotopic Composition of Atmospheric Water Vapor at Micrometeorological Scale, Bulletin of the TERC, the University of Tsukuba, 6, 3–9, https://doi.org/10.15068/00147114, 2005.

Shuttleworth, W. J.: Terrestrial Hydrometeolology, Wiley-Black Well, 59–63, 155–174, 2011.

Singh, J., Knapp, H. V., and Demissie, M.: Hydrologic modeling of the Iroquois River watershed using HSPF and SWAT, ISWS CR 2004-08, Champaign, Ill: Illinois State Water Survey, available at: https://pdfs.semanticscholar.org/8e94/d50b49c6f1b8ad03d86e8605890f30f25847.pdf (last access: 7 December 2018), 2005.

Sugita, M. and Kotoda, K.: Effects of Soil Water Deficits on Forest Evapotranspiration, Bulletin of the ERC, the University of Tsukuba, 9, 83–88, 1985.

Sugita, M., Degutschi, K., and Kotoda, K.: An Apparatus for Measuring Heat Pulse Velocity and Its Application for Transpiration Flux Estimation – Examples of Measurements in Coniferous and Deciduous Tress, Bulletin of the ERC, the University of Tsukuba, 9, 83–88, 1985.

Tanaka, K. and Oikawa, T.: Seasonal dynamics of LAI in a C_3/C_4 co-occurring grassland at the Environmental Research Center (1996, 1997), Bulletin of the ERC, the University of Tsukuba, 23, 87–89, 1998.

Taniguchi, M.: Groundwater Recharge Rates and Soil Temperature in Pasture Land and Pine Forest, Bulletin of the ERC, the University of Tsukuba, 14, 69–74, 1990.

Tase, N. and Majima, M.: Rainfall Simulation in a Red Pine Forest, Bulletin of the ERC, the University of Tsukuba, 7, 9–15, 1983.

Wang, P. and Yamamaka, T.: Application of a two-source model for partitioning evapotranspiration and assessing its controls in temperate grasslands in central Japan, Ecohydrology, 7, 345–353, https://doi.org/10.1002/eco.1352, 2014.

Wang, P., Yamanaka, T., Li, X. Y., and Wei, Z. W.: Partitioning evapotranspiration in a temperate grassland ecosystem: Numerical modeling with isotopic tracers, Agr. Forest Meteorol., 208, 16–31, https://doi.org/10.1016/j.agrformet.2015.04.006, 2015.

Watarai, Y. and Yamanaka, T.: Observational Data of Heat Balance and Water Balance – 2006, Bulletin of the TERC, the University of Tsukuba, 8, 55–80, https://doi.org/10.15068/00147239, 2007.

Xu, J. and Haginoya, S.: An Estimation of Heat and Water Balances in the Tibetan Plateau, Journal of the Meteorological Society of Japan, 79, 485–504, https://doi.org/10.2151/jmsj.79.485, 2011.

Xu, J., Haginoya, S., Masuda, K., and Suzuki, R.: Heat and Water Balance Estimates over the Tibetan Plateau in 1997-1998, Journal of the Meteorological Society of Japan, 83, 577–593, https://doi.org/10.2151/jmsj.83.577, 2005.

Yamanaka, T.: Summary of Studies Using Environmental Isotope Tracers, Bulletin of the TERC, the University of Tsukuba, 13, 45–58, https://doi.org/10.15068/00147045, 2012.

Yamanaka, T., Matsuo, D., Yano, M., Kakubari, J., Iida, S., Wakui, H., Shimizu, R., and Tanaka, T.: Difference in Root System among Co-occurring Plants in a Secondary Pine Forest Undergoing Succession, Bulletin of the TERC, the University of Tsukuba, 6, 39–44, https://doi.org/10.15068/00147120, 2005.

Yasui, Y. and Oikawa, T.: Daily and Seasonal changes of Soil Respiration Rates and Micrometeorological Conditions of a Red Forest Floor in University of Tsukuba, Bulletin of the TERC, the University of Tsukuba, 18, 77–91, 1993.

Yubasaki, S., Tase, N., and Haginoya, S.: Consideration for the Estimation Methods of Evapotranspiration at Terrestrial Environment Research Center, Bulletin of the TERC, the University of Tsukuba, 6, 45–51, https://doi.org/10.15068/00147121, 2005.

13

Historical gridded reconstruction of potential evapotranspiration for the UK

Maliko Tanguy[1], Christel Prudhomme[2,3,1], Katie Smith[1], and Jamie Hannaford[1]

[1]NERC Centre for Ecology & Hydrology, Maclean Building, Benson Lane, Crowmarsh Gifford, Wallingford, Oxon, OX10 8BB, UK
[2]European Centre for Medium-Range Weather Forecasts, Shinfield Road, Reading, RG2 9AX, UK
[3]Department of Geography, Loughborough University, Loughborough, LE11 3TU, UK

Correspondence: Maliko Tanguy (malngu@ceh.ac.uk)

Abstract. Potential evapotranspiration (PET) is a necessary input data for most hydrological models and is often needed at a daily time step. An accurate estimation of PET requires many input climate variables which are, in most cases, not available prior to the 1960s for the UK, nor indeed most parts of the world. Therefore, when applying hydrological models to earlier periods, modellers have to rely on PET estimations derived from simplified methods. Given that only monthly observed temperature data is readily available for the late 19th and early 20th century at a national scale for the UK, the objective of this work was to derive the best possible UK-wide gridded PET dataset from the limited data available.

To that end, firstly, a combination of (i) seven temperature-based PET equations, (ii) four different calibration approaches and (iii) seven input temperature data were evaluated. For this evaluation, a gridded daily PET product based on the physically based Penman–Monteith equation (the CHESS PET dataset) was used, the rationale being that this provides a reliable "ground truth" PET dataset for evaluation purposes, given that no directly observed, distributed PET datasets exist. The performance of the models was also compared to a "naïve method", which is defined as the simplest possible estimation of PET in the absence of any available climate data. The "naïve method" used in this study is the CHESS PET daily long-term average (the period from 1961 to 1990 was chosen), or CHESS-PET daily climatology.

The analysis revealed that the type of calibration and the input temperature dataset had only a minor effect on the accuracy of the PET estimations at catchment scale. From the seven equations tested, only the calibrated version of the McGuinness–Bordne equation was able to outperform the "naïve method" and was therefore used to derive the gridded, reconstructed dataset. The equation was calibrated using 43 catchments across Great Britain.

The dataset produced is a 5 km gridded PET dataset for the period 1891 to 2015, using the Met Office 5 km monthly gridded temperature data available for that time period as input data for the PET equation. The dataset includes daily and monthly PET grids and is complemented with a suite of mapped performance metrics to help users assess the quality of the data spatially.

This dataset is expected to be particularly valuable as input to hydrological models for any catchment in the UK.

1 Introduction

Potential evapotranspiration is a conceptual variable which measures the atmospheric demand for moisture from an open surface water. Reference crop evapotranspiration (also referred to as potential evapotranspiration PET) is the rate of evapotranspiration of an idealised short grass actively growing and not short of water (Shuttleworth, 1993), providing an upper limit of the evaporative losses of grass. As evapotranspiration is a major factor in the catchment water balance (Beven, 2012), PET is used as input data for rainfall–runoff models.

Different approaches have been proposed for estimating PET. The most complex, combination methods, are based on physical processes accounting for the energy available to a plant to evaporate during photosynthesis, and the amount of water that can be dissipated in the atmosphere (Penman, 1948; Monteith, 1965). They are referred to as combination methods as they combine the energy balance with the mass transfer method. The simplest methods aim to capture the dominant climatic factors in the plant evapotranspiration processes. The simplified methods can be broadly divided between the radiation-based methods (e.g. Doorenbos and Pruitt, 1984; Hargreaves and Samani, 1983; Jensen and Haise, 1963), which use measured data such as net solar radiation, sunshine hours or cloudiness factors; and the temperature-based methods (e.g. Blaney and Criddle, 1950; Thornthwaite, 1948; Oudin et al., 2005; McGuinness and Bordne, 1972), which use temperature as a proxy for the radiative energy available, along with extraterrestrial radiation estimated from the date of the year and latitude. Both radiation and temperature methods are used instead of combination methods when the full set of climatic variables necessary for the latter is not readily available. However, there is no general agreement on the best performing method, and the final choice of the equation often depends on the application and data availability, as well as the particular environmental setting (e.g. Donohue et al., 2010; Federer et al., 1996; Oudin et al., 2005; Prudhomme and Williamson, 2013; Xu and Singh, 2000, 2001). When little or no climatic variables are available, an alternative is to use remote-sensing data in simplified empirical PET methods (Barik, 2014; Barik et al., 2016; Knipper, 2017; Mu, 2013), however this can only be applied to the satellite era (from the 1970s or the 1980s).

In the UK, the Met Office Rainfall and Evaporation Calculation System (MORECS) (Thompson et al., 1982) is one of the main sources of PET estimates, available as an approximately 40×40 km monthly gridded product, with time series from 1961. MORECS is based on the Penman–Monteith formulation (Monteith, 1965), but includes evaporation from rainfall intercepted by the canopy and considers 14 different vegetation types and three different soils (Hough, 2003). Recently, the Centre for Ecology and Hydrology (CEH) published the Climate, Hydrology and Ecology research Support System (CHESS), a 1 km gridded daily meteorological and land state dataset for Great Britain (Robinson et al., 2016a, b, 2017) spanning the period 1961–2015. It includes PET data calculated from the meteorological variables using the Penman–Monteith equation for a well-watered grass surface, following the Food and Agriculture Organisation of the United Nations (FAO) guidelines for computing reference crop PET (Allen, 1998), both with and without corrections for water intercepted by the canopy (Robinson et al., 2016b).

In the UK, PET data is widely used for hydrological modelling, where streamflow time series are generated from rainfall and PET inputs. This is particularly useful in providing information where streamflow observations do not exist, i.e. in reconstructing flows for pre-observational periods, or to explore the response of a changing climate on hydrology. Currently, there is no readily available source of PET time series for studying long-term variability and change in hydrological regimes before the 1960s, including water resources availability and drought patterns. This is a major obstacle, because historical drought periods are used in water resources and drought planning (Watts et al., 2012), as well as for providing a baseline of past hydrological variability for future change assessments. In practice, however, limited availability of atmospheric variables makes it difficult to account for the majority of evapotranspiration processes for the pre-1960 period using the Penman–Monteith equation. Simpler methods therefore need to be used as an alternative, but this requires a thorough evaluation of the differences they bring when compared with established datasets. This study focuses on temperature-based PET equations as temperature (together with precipitation) are among the climate variables that have observed, spatially distributed records for the longest period for the UK. High-resolution gridded (5 km) temperature and precipitation data from the Met Office are available from 1910, and have recently been extended back in time as part of historical data rescuing effort by the Met Office funded by the Historic Droughts project. Monthly temperature and precipitation data were available to project partners from 1891 and 1862 respectively. Detailed climatic variables are available in the UK only from 1961 onwards. Some other variables such as sea level pressure data are also available from late 19th century (Met Office HADSLP2 product), however the spatial resolution is much coarser (5°).

While the focus was on applying PET methods for historical reconstruction, this undertaking will provide useful information for other applications where only temperature data are available; for example, hydrological forecasting or long-term climate change impact studies.

This paper describes the derivation of a 5 km gridded daily and monthly PET dataset for the UK from 1891 to 2015, with hydrological modelling being the main targeted application. First, the data used for calibration, validation and production of the gridded dataset are presented. This is followed by the methods, where the temperature-based PET equations, the calibration strategies and the evaluation approach used are described. Thirdly, the results of the evaluation of the PET equations, and the assessment of the final PET grids are presented. Lastly, uncertainties and limitations of the product are discussed, and recommendations for users are listed.

2 Data

This study has used various gridded temperature and PET datasets, which are described in this section. Table 1 provides a summary of all datasets used.

2.1 Temperature data

Three main sources of high-resolution gridded national-scale temperature data exist for the UK:

- CHESS-met high-resolution mean daily temperature (1 km grids, daily time series) for 1961–2015 for Great Britain (henceforth CHESS-temp daily). This dataset is part of a larger dataset developed by CEH for environment modelling applications; its derivation is fully described in Robinson et al. (2016a).
- UKCP09 mean monthly temperature (5 km grids, monthly time series) for 1910–2015 for the UK, including Northern Ireland (henceforth UKCP09-temp monthly). This is part of a larger dataset developed by UK Met Office, its derivation is fully described in Perry and Hollis (2005). The monthly mean temperature is derived from the average of daily maximum and minimum temperature averaged across the month at each contributing station. Stations with no more than two missing days within a calendar month are used to create the gridded product.
- Historic droughts mean monthly temperature (5 km grids, monthly time series) for 1891–1909 for the UK, including Northern Ireland (henceforth HD-temp monthly). This was derived using the same methodology as UKCP09, using historic weather station data rescued by the Met Office in the Historic Droughts project (NERC grant number: NE/L01016X/1).

The two latter were combined in this study and treated as a single dataset to provide a single, continuous record of temperature from 1891 to 2015, which is used to derive the long-term potential evapotranspiration dataset that is the focus of this study. The combined dataset will be referred to as UKCP09 in the rest of the paper. The uncertainty associated with the temperature dataset is discussed in Sect. 5.1. The shorter (1961–2015) CHESS dataset is used for calibration and sensitivity testing.

Prior to 1961, temperature data is only available at a 5 km spatial resolution and monthly time step. Because of this coarser temporal and spatial resolution of temperature data in the earlier period, alternative datasets were generated and used in the analysis to quantify the sensitivity of PET derivation to temperature input, and are summarised in Table 1a and b:

- CHESS daily mean temperature climatology (1 km grids) (henceforth CHESS-temp clim): long-term average (1961–1990) of daily mean temperature, derived from CHESS-temp daily. This provides a default option that could be used even if no temperature data were available in the past (or future). This gives a day-to-day variability pattern of temperature throughout the year, which is then repeated every year.
- CHESS daily mean temperature derived from monthly averages (1 km grids). Different methods to disaggregate monthly temperature into daily data were tested:
    i. Constant temperature during the month (henceforth CHESS-temp monthly I). This means there are step changes in temperature between consecutive months.
    ii. Interpolated using pchip method for a smooth transition between months (henceforth CHESS-temp monthly II). Pchip stands for piecewise cubic hermite interpolating polynomial, which is an interpolation method in which a cubic polynomial approximation is assumed over each subinterval. Aràndiga et al. (2016) describe this interpolation scheme in detail together with its advantages, mainly that it is both accurate (preserves values at the nodes) and preserves monotonicity. Pchip was selected for the present study because (i) the fitted curve passes through observed values at inflexion points unlike spline or quadratic methods, for example, and (ii) it does not require re-fitting when the period of application is extended as each subinterval is treated separately.
    iii. Disaggregated to daily using CHESS daily mean temperature climatology pattern (henceforth CHESS-temp monthly III). The daily relative variation in temperature follows the climatology, but for each month, the daily values are adjusted so that monthly mean temperatures are correct. In other words, CHESS-temp clim data is shifted uniformly so the monthly mean temperature matches the CHESS monthly temperature data.
- UKCP09 daily mean temperature (5 km grids) derived from monthly averages. Two different methods to disaggregate monthly temperature into daily data were tested:
    i. Constant during the month (henceforth UKCP09-temp monthly I).
    ii. Interpolated using pchip method (henceforth UKCP09-temp monthly II).

Table 1. Summary of temperature and PET datasets used in this study. **(a)** Temperature data used to investigate the effect of temporal distribution of data on the output PET estimation, **(b)** temperature data to investigate effect of spatial resolution of the data on the output PET estimation, and **(c)** PET data used to calibrate and evaluate the equation and final PET output.

(a) Temperature datasets used as input data for temperature-based PET equations. Multiple versions were used to investigate the effect of temporal distribution of the data on output PET estimation			
Dataset short name	Resolution	Description	Comment
CHESS-temp daily	1 km × 1 km Daily	CHESS-met high-resolution mean daily temperature: part of a larger dataset developed by CEH for environment modelling applications, available for 1961–2015	"Best" available gridded daily temperature data for Great Britain
CHESS-temp clim	1 km × 1 km Daily	CHESS daily mean temperature climatology: long-term average (1961–1990) of daily mean temperature, derived from CHESS-temp daily	Default option that could be used even if no temperature data were available
CHESS-temp monthly I	1 km × 1 km Monthly disaggregated to daily	CHESS daily mean temperature derived from monthly averages, constant during the month. Step changes in temperature between consecutive months	To investigate whether temporal disaggregation method (from monthly to daily) has an effect on output PET estimation
CHESS-temp monthly II	1 km × 1 km Monthly disaggregated to daily	CHESS daily mean temperature derived from monthly averages, interpolated using pchip	
CHESS-temp monthly III	1 km × 1 km Monthly disaggregated to daily	CHESS daily mean temperature derived from monthly averages, disaggregated to daily using CHESS-temp clim pattern	
(b) Temperature datasets used to assess spatial resolution for the best performing PET method			
Dataset short name	Resolution	Description	Comment
UKCP09-temp monthly I	5 km × 5 km Monthly disaggregated to daily	UKCP09 daily mean temperature derived from monthly averages, constant during the month	Two temporal disaggregation methods tested.
UKCP09-temp monthly II	5 km × 5 km Monthly disaggregated to daily	UKCP09 daily mean temperature derived from monthly averages, interpolated using pchip	
(c) PET datasets used to calibrate the equations and assess the output PET			
Dataset short name	Resolution	Description	Use
CHESS-PM	1 km × 1 km Daily and monthly	CHESS-PET 1 km grids, daily (and monthly) time series available for 1961–2015, calculated using the Penman–Monteith (PM) equation for FAO-defined well-watered grass	(1) calibration of the temperature-based PET equations (1961–1990) (2) Evaluation of the equations (1991–2012) (3) Evaluation of the final gridded product (1991–2012)
CHESS-PM climatology	1 km × 1 km Daily and monthly	Daily (and monthly) PET long-term average, calculated from CHESS-PM for 1961 to 1990	used as a "naïve method" against which the PET reconstruction methodology can be tested to assess performance

Figure S1 in the Supplement illustrates what these different temperature time series look like for an example catchment. Figure S2 in the Supplement shows the spatial coverage of the different datasets used in this study (Fig. S2a and b), whereas Fig. S2c shows the geographical extend of Great Britain, Northern Ireland and the UK.

In summary, seven different daily temperature datasets were used as input data to the temperature-based PET equations: CHESS-temp daily, CHESS-temp clim, CHESS-temp monthly I, CHESS-temp monthly II, CHESS-temp monthly III, UKCP09-temp monthly I and UKCP09-temp monthly II. CHESS-temp daily is an existing dataset (https://catalogue.ceh.ac.uk/documents/b745e7b1-626c-4ccc-ac27-56582e77b900, last access: 30 April 2018), whereas the other six daily datasets are manipulated versions of existing datasets (CHESS or UKCP09).

2.2 PET

One main source of national-scale mean daily PET time series was available: CHESS-PET 1 km grids, daily time series (Robinson et al., 2016b) available for 1961–2015, calculated using the Penman–Monteith (PM) equation (Monteith, 1965) for FAO-defined well-watered grass (Allen, 1998). Because the PM equation is a physically based model which combines the energy balance with the mass transfer method, and is recommended by the FAO to calculate PET, CHESS-PET daily is here considered as "ground truth" and proxy for observations (hereafter referred to as CHESS-PM).

CHESS-PM daily climatology (hereafter CHESS-PM climatology) was also calculated from 1961 to 1990, and is used as a "naïve method" against which the PET reconstruction methodology can be tested to assess performance. "Naïve method" here refers to the simplest way PET could be estimated in the absence of any climate data.

3 Methods

To produce the PET gridded reconstruction product, a sequence of assessments and tests were first undertaken. In the first stage, a set of seven temperature-based equations (presented in Sect. 3.1) were tested with seven different input temperature datasets (Sect. 2.1) and four calibration strategies (Sect. 3.2) (in addition to the non-calibrated equations). These combinations of equation/calibration strategy/temperature input data were evaluated in the second stage (Sect. 3.3.1), leading to the selection of the best combination. In the third stage, the effect of the spatial resolution was investigated (Sect. 3.2.2.), followed by a study of the effect of averaging over the catchments in the fourth stage (Sect. 3.2.2.). Finally, in the fifth stage, the final PET gridded product was evaluated with the calculation of performance metrics both at catchment scale and grid scale (Sect. 3.2.2.). Figure 1 is a workflow diagram summarising the different stages of the work.

3.1 Temperature-based equations

Seven temperature-based equations were evaluated (see Table 2). Four of them (Eqs. 1–4 in Table 2) were calibrated testing different calibration strategies (Sect. 3.2): Hamon (Hamon, 1961), McGuinness–Bordne (McGuinness and Bordne, 1972), Blaney–Criddle (Blaney and Criddle, 1950) and Kharrufa (Kharrufa, 1985). Each contains a number of parameters representative of the climatic region where the equation was originally developed, which can be calibrated to match the climatic regime of the UK. The other three equations were not suitable for simple calibration techniques, or had set calibrations: Oudin (Oudin et al., 2005), MOHYSE (Fortin and Turcotte, 2006) and Thornthwaite (Thornthwaite, 1948) (Eqs. 5 to 7 in Table 2).

The physical basis for estimating evaporation using temperature alone is that both terms of the combination equation (the energy required to sustain evaporation and the energy removed from the surface as water vapour) are generally related to temperature (Shuttleworth, 1993).

The main difference between the different temperature-based formulations, lies in the way temperature is linked to PET to simulate the effect of the full set of variables normally required in the combination equations. Most temperature-based equations use day length or related variables (Hamon, 1961; Blaney and Criddle, 1950; Kharrufa, 1985; Fortin and Turcotte, 2006; Thornthwaite, 1948), except McGuinness and Bordne (1972), and the derived Oudin et al. (2005)'s equation which use extraterrestrial radiation instead. Blaney–Criddle equation has also an additional parameter k, which depends on crop type. Most of these equations were developed for the USA, except MOHYSE (which was developed in Quebec), Kharrufa (developed for arid regions) and Oudin (developed in Australia, USA and France).

Note that equations requiring minimum and maximum temperature (Droogers and Allen, 2002; Hargreaves and Samani, 1983; Heydari and Heydari, 2014) were not considered here as only low-data demanding methods that could be easily reproduced and extended in cases of minimal data availability were selected.

3.2 Calibration strategies

To compare the different calibration methods (stage 1 in Fig. 1), the calibration and testing was done at catchment scale using two independent sets of catchments representative of typical hydroclimatic conditions prevailing in the UK and with good spatial coverage: 43 were used for calibration and assessment of the equations, and an additional 263 (making a total of 306 catchments) used for evaluation of the final PET grids (Fig. 2). Table S1 (Excel spreadsheet) in the Supplement shows the catchments with some of their

Table 2. Temperature-based equations evaluated in this study.

Equation number and name	Formula	Abbreviations	Reference	Comments	Calibrated in this study
Eq. (1): Hamon	Version 1: $PE\ [\text{mm day}^{-1}] = 13.97\left(\frac{N}{12}\right)^2 \rho_s$ Version 2: $PE\ [\text{mm day}^{-1}] = \left(\frac{N}{12}\right)^2 \exp\left(\frac{T}{16}\right)$	– N maximum possible daylight hours [h] – ρ_s saturated vapour density [g m^{-3}] – T average temperature [°C]	Version1: Hamon (1961) Version 2: Oudin et al. (2005)	Originally developed for the USA. Version 2 was used in this study.	YES
Eq. (2): McGuinness–Bordne	$PE\ [\text{mm day}^{-1}] = \frac{1}{\lambda} S_0 \left(\frac{T+5}{68}\right)$	– λ latent heat of vaporisation [MJ kg^{-1}] – T temperature [°C] – S_0 extraterrestrial radiation [MJ m^{-2} day^{-1}]	McGuinness and Bordne (1972) Oudin et al. (2005)	Originally developed for the USA. Oudin et al. (2005) version used in this study.	
Eq. (3): Blaney–Criddle	$PE\ [\text{mm day}^{-1}] = kT\,p_d$ with $p_d = 100\frac{N_d}{\sum_{i=1}^{365} N_i}$	– p_d mean daily percent of annual daytime hours for day d – T mean air temperature – k monthly consumptive use coefficient. – The coefficients k depends on crop, location and season	Blaney and Criddle (1950)	Originally developed to estimate the irrigation requirements of crops in western USA.	
Eq. (4): Kharrufa	$PET\ [\text{mm day}^{-1}] = 0.34\left(100 \cdot \frac{DL}{365\times 12}\right) \cdot T_a^{1.3}$	– DL day length [h day^{-1}] – T_a air temperature [°C]	Kharrufa (1985)	Originally developed for arid regions.	
Eq. (5): Oudin	$\begin{cases} PE\ [\text{mm day}^{-1}] = \frac{1}{\lambda} S_0 \left(\frac{T+5}{100}\right) & \text{if } T > -5° \\ PE\ [\text{mm day}^{-1}] = 0 & \text{if } T \leq -5° \end{cases}$	– λ latent heat of vaporisation [MJ kg^{-1}] – T temperature [°C] – S_0 extraterrestrial radiation [MJ m^{-2} day^{-1}]	Oudin et al. (2005)	Version of McGuinness–Bordne equation calibrated for catchments in Australia, USA and France, and which was assessed as the best temperature-based PE equation following a review of various PE methods for use as input to hydrological models (Oudin et al., 2005).	NO
Eq. (6): MOHYSE	$PET = \frac{4.088}{\pi} \cdot \omega \cdot \exp\left(\frac{17.3T_a}{238+T_a}\right)$	– ω the sunset hour angle [rad] – T_a air temperature [°C]	Fortin and Turcotte (2006)	Developed in Quebec.	
Eq. (7): Thornthwaite	$PE'\ [\text{mm month}^{-1}] = 16\left(\frac{DL}{360}\right)\left(\frac{10T_a}{I}\right)^K$ $a = 0.49239 + 0.01792I - 7.71 \times 10^{-5} I^2$ with $I = \sum_{m}^{12}\left(\frac{T_m}{5}\right)^{1.514}$ annual heat index	– T_a air temperature [°C] – T_m mean temperature of month m [°C]	Thornthwaite (1948)	Thornthwaite correlated mean monthly air temperature with PE as determined by water balance studies in valleys of east-central USA.	

Figure 1. Work flow diagram of the evaluation procedure of the PET equations and final PET gridded product. This process was made in five stages: in stage 1, the equations were calibrated using different calibration strategies and different input temperature data; in stage 2, the multiple combinations of PET equation/calibration approach/temperature input data were evaluated; in stage 3, the effect of spatial resolution of the input temperature data was assessed. These first three stages led to the selection of PET equation, calibration strategy and input dataset used to produce the final gridded PET product. In the fourth stage, the effect of calibrating the equations at the catchment scale was investigated; and finally, in stage 5, a final evaluation of the new gridded PET product was carried out both at the catchment scale and at the grid scale. Stages 1, 2 and 3 used the set of 43 catchments shown in Fig. 2a, whereas stages 4 and 5 used the full set of 306 evaluation catchments shown in Fig. 2b.

catchment characteristics. The spatial averaging of temperature and PET time series to conduct the analysis has the advantage of smoothing out any discontinuity that could exist at the grid-scale level due to different interpolation algorithms and recording stations and which could consequently impact on local performance of the PET generation technique. In addition, for many practical hydrological modelling applications, PET is required at the catchment scale. The impact of catchment-scale vs. grid-scale calibration is assessed in Sect. 4.2 (stage 4 in Fig. 1).

As previously mentioned, temperature-based PET equations use parameters to link temperature to PET as a simpli-

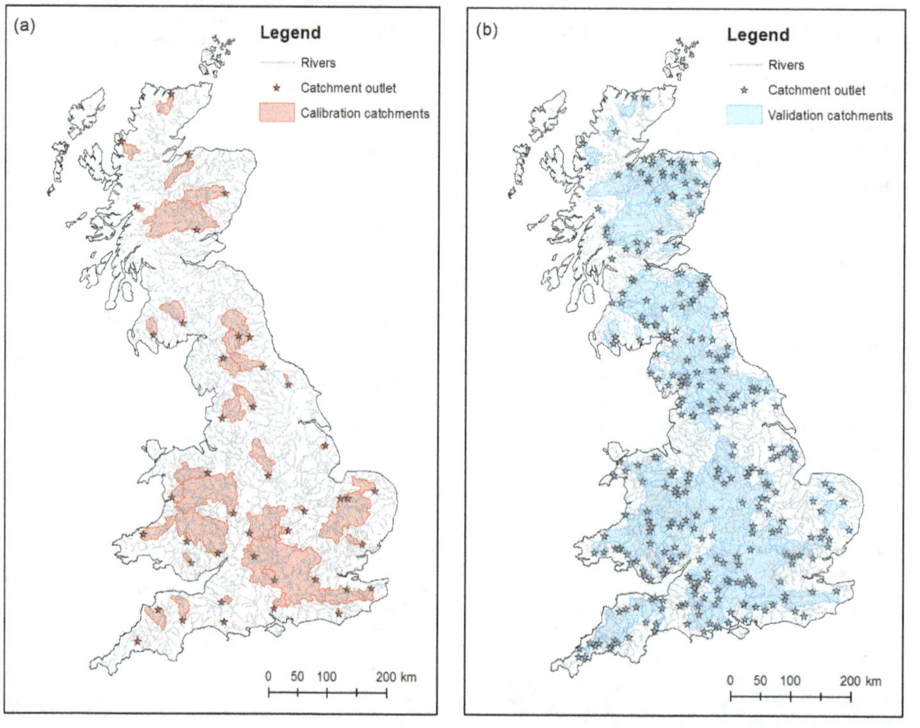

Figure 2. Maps of the boundaries and outlets of **(a)** catchments that were used to calibrate the PET equations and to calculate the performance metrics of the PET equations, described in Sect. 3.3.1 (stages 1, 2 and 3 in Fig. 1); and **(b)** catchments that were used to carry out the assessment of the final PET grids using the performance metrics described in Sect. 3.3.2 (stages 4 and 5 in Fig. 1).

fication of full evaporation dynamics. Because of important climatic variation across space, and in time across the year, it might be possible that optimal parameterisation could be achieved by letting the parameters vary in time and space. Therefore, four calibration strategies, which are graphically represented in Fig. 3, were considered. The simplest one consists in a global parameterisation (GB) leading to a single equation (1P) for all 43 catchments (1P-GB). In the most complex approach, a local and monthly parametrisation leads to 12 equations for each of the 43 catchments (12P-ind). The trade-off between a simplified method (global parameterisation), which is much easier to implement, and a local method, which requires a long calibration procedure and a parameter transfer methodology for Northern Ireland (where no daily PET dataset is available), is discussed in the results section.

Two independent time periods were selected for the calibration (1961–1990) and evaluation (1991–2012) procedures. The equations' parameters were calibrated using the ordinary least squares (OLS) method against CHESS-PM. The data showed some heteroscedasticity and a moderate degree of autocorrelation, which violates the assumption of OLS. However the effect of these violations has been investigated and does not affect parameter estimations in our particular case. More detail on this can be found in the Supplement (Sect. S1).

3.3 Evaluation

3.3.1 Catchment-scale performance metrics for evaluating the combinations of PET equations/calibration strategy/input temperature data

This evaluation corresponds to stage 2 in Fig. 1, and was done on the 43 calibration catchments shown in Fig. 2a for the period 1991–2012.

Two metrics were used to evaluate the best combination of temperature data, PET equation, and calibration strategy: the mean absolute percentage error (MAPE) and Nash–Sutcliffe efficiency (NSE) coefficient, using CHESS-PM daily as ground truth.

MAPE is widely used in the forecasting community to evaluate accuracy of output from models (Danladi et al., 2017; Lefebvre and Bensalma, 2015). Applied to PET, it is calculated as follows:

$$\mathrm{MAPE} = \frac{100}{n}\sum_{t=1}^{n}\left|\frac{\mathrm{PET}_{\mathrm{O}}^{t} - \mathrm{PET}_{\mathrm{m}}^{t}}{\mathrm{PET}_{\mathrm{O}}^{t}}\right|$$

where t is the time step, n is the number of time steps, $\mathrm{PET}_{\mathrm{O}}^{t}$ is the actual value of PET at time t and $\mathrm{PET}_{\mathrm{m}}^{t}$ is the modelled value of PET at time t.

In order to be able to apply MAPE, values of observed PET of 0 were replaced by 0.1. Smaller values of MAPE

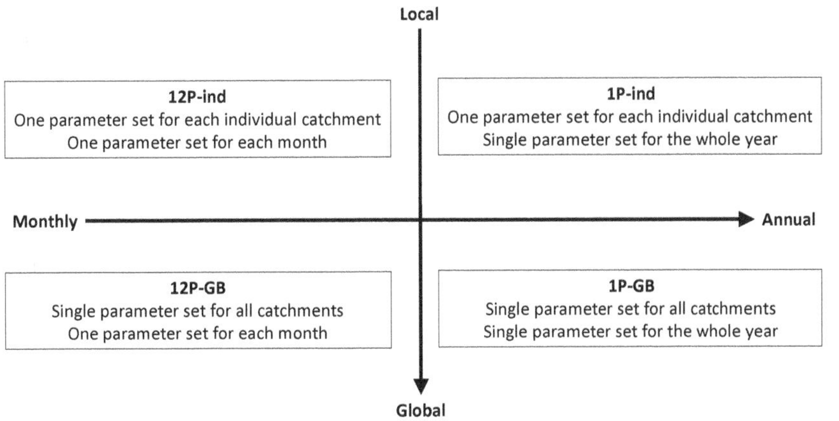

Figure 3. Schematic of the calibration strategies. Four calibration approaches were considered to calibrate the PET equations: from local and monthly parametrisation leading to 12 equations for each of the 43 catchments (12P-ind), to a global parameterisation leading to a single equation for all 43 catchments (1P-GB).

indicate greater accuracy of the model prediction. Observed PET was found to be equal to 0 about 3 % of the time, which is not frequent enough to significantly skew the MAPE score.

The Nash–Sutcliffe efficiency (NSE) coefficient was initially developed to assess hydrological models (Nash and Sutcliffe, 1970), but has since then also been widely used to evaluate PET models (Ershadi et al., 2014; Guerschman et al., 2009; Liu et al., 2005; Schneider et al., 2007; Spies et al., 2014; Srivastava Prashant et al., 2013). NSE, which is also referred to as mean square error skill score (MSESS) in the forecasting community, looks at how much superior a given model is in predicting a variable (here: PET) compared to the long-term average (climatology). It is calculated as follows:

$$\mathrm{NSE} = 1 - \frac{\sum_{t=1}^{n}\left(\mathrm{PET}_{\mathrm{m}}^{t} - \mathrm{PET}_{\mathrm{O}}^{t}\right)^2}{\sum_{t=1}^{n}\left(\mathrm{PET}_{\mathrm{O}}^{t} - \overline{\mathrm{PET}_{\mathrm{O}}}\right)^2},$$

where $\overline{\mathrm{PET}_{\mathrm{O}}}$ is the mean of observed PET, $\mathrm{PET}_{\mathrm{m}}^{t}$ is modelled PET at time t, and $\mathrm{PET}_{\mathrm{O}}^{t}$ is observed PET at time t.

Nash–Sutcliffe efficiency can range from $-\infty$ to 1. An efficiency of one (NSE = 1) corresponds to a perfect match of modelled PET to the observed data. An efficiency of zero (NSE = 0) indicates that the model predictions are as accurate as the mean of the observed data, whereas an efficiency less than zero (NSE < 0) occurs when the observed mean is a better predictor than the model or, in other words, when the residual variance (described by the numerator in the expression above), is larger than the data variance (described by the denominator).

The performance of each of the different combinations (PET equations/calibration approaches/input temperature data) was compared against an independent benchmark (reference) for comparison – CHESS-PM clim, used as an alternative way to estimate daily PET locally when no data is available (e.g. for the past or the future). It is worth noting that $\overline{\mathrm{PET}_{\mathrm{O}}}$ used in the calculation of NSE is different from CHESS-PM clim, in that (i) the latter has a daily value for each day of year (which is repeated for every year), whereas $\overline{\mathrm{PET}_{\mathrm{O}}}$ is just a single value (CHESS-PM averaged over time); and (ii) CHESS-PM clim is the daily average PET calculated for the period 1961–1990, whereas $\overline{\mathrm{PET}_{\mathrm{O}}}$ is the average CHESS-PM value for the evaluation period which is 1991–2012.

The same two metrics (MAPE and NSE) are also used to assess the effect of the input temperature data's spatial resolution on the estimated PET (stage 3 in Fig. 1).

3.3.2 National-scale performance metrics for final PET grid quality assessment

One of the possible issues with a catchment-scale calibration such as implemented here is its applicability at a finer spatial scale. To test the validity of catchment-scale calibration (stage 4 in Fig. 1), catchment-average daily PET time series extracted from the final 5 km daily PET gridded product were compared with daily PET series based on catchment average temperature, derived using the same equation. The correlation coefficient (r) was calculated to measure the goodness of fit.

To assess the quality of the final gridded PET product (stage 5 in Fig. 1), a series of performance metrics were calculated at national scale, and provided together with the final product. Once again, CHESS-PM daily was used as ground truth.

In addition to MAPE, NSE and the correlation coefficient (r) described previously, the following three metrics were also calculated:

- Bias ratio (β), calculated as $\beta = \frac{\overline{\mathrm{PET}_{\mathrm{m}}}}{\overline{\mathrm{PET}_{\mathrm{O}}}}$ where $\overline{\mathrm{PET}_{\mathrm{m}}}$ and $\overline{\mathrm{PET}_{\mathrm{O}}}$ are the mean modelled and observed PET.

- Variability ratio (VR), calculated as $\mathrm{VR} = \frac{\mathrm{CV_m}}{\mathrm{CV_O}} = \frac{\sigma_\mathrm{m}/\overline{\mathrm{PET_m}}}{\sigma_\mathrm{O}/\overline{\mathrm{PET_O}}}$, where CV is the coefficient of variation and σ is the standard deviation of PET.
- Kling–Gupta efficiency (KGE) which is a combination of r, β and VR (Gupta et al., 2009; Kling et al., 2012) and is calculated as $\mathrm{KGE} = 1 - \sqrt{(r-1)^2 + (\beta-1)^2 + (\mathrm{VR}-1)^2}$.

KGE, VR, r and β coefficient all have their optimum at unity (1). All the metrics were calculated both at catchment scale (using 306 validation catchments shown in Fig. 2b) and at grid scale. MAPE was also calculated for each month as the error varies seasonally. Monthly MAPE can be used as a measure of uncertainty in the data.

These six metrics were chosen as they assess different aspects of the modelled data. NSE looks as how much better our model is in predicting PET compared to the long-term average (climatology), MAPE gives an indication of the uncertainty, r informs about how well the modelled PET fits the observed values (or "proxy" to observed in our case), β tells us whether the estimations are biased, VR whether the spread of the estimated values matches the observed spread, and finally KGE informs on the combined effect of r, β and VR.

4 Results

4.1 Assessment of the temperature-based PET equations

In this section, results from the evaluation represented in stage 2 and 3 of Fig. 1 are presented.

Figure 4 is a summary graphic showing the average MAPE and NSE for all combinations of forcing data, PET equation and calibration strategy tested. For simplicity, Fig. 4 does not show the results from the following:

- models that were not calibrated in this study, i.e. Oudin, MOHYSE and Thornthwaite (Eqs. 5 to 7 in Table 2) (as these were performing worse than the calibrated models);
- using CHESS-temp monthly III forcing (similar results to those for CHESS-temp monthly II);
- using UKCP09-temp monthly I and II, as they were only used with the final selected equation as an additional test to check the effect of spatial resolution on the results (stage 3 in Fig. 1).

The full list of performance metrics is given in Tables S1 and S2 in the Supplement.

Figure 4 displays the following:

i. Calibration yields substantial improvement in performance, except for Hamon (Eq. 1 in Table 2) which performed well before calibration.

ii. Calibration strategy has very little effect on the performance. Both annual and global calibrations show a similar performance to the locally calibrated, monthly models. The simplest calibration approach was hence adopted: national-scale application was conducted using the 1P-GB strategy (see Fig. 3).

iii. Daily temperature data only performs marginally better than forcing based on monthly temperature time series. This might be explained by the small day-to-day variability in temperature fields (and hence, in any resulting PET field) compared with other climate variables such as wind speed, humidity or radiation, which provide a much larger contribution to the daily variability of PET than temperature. The artificial daily pattern introduced by temporal disaggregation of monthly temperature is in fact small compared with the error introduced by using temperature-only forcing to estimate PET. This is illustrated in Fig. S1 (Supplement). Also, the temperature seasonal variability is a main component to the PET, and is well captured by monthly values, with sub-monthly values only adding some noise. This is why the choice of temperature data only has a marginal effect, because the daily variance is of secondary importance in comparison to an accurate representation of the seasonality.

iv. CHESS-PM climatology is only outperformed by the calibrated version of McGuinness–Bordne equation (Eq. 2 in Table 2). This suggests a small inter-annual variability of PET, with a daily climatology being a good alternative when no other time series is available. Note however that the evaluation period (1991–2012) is too short for investigating the possible impact of trends (e.g. temperature trends, interdecadal variability, climate change signal) in the PET signal, which might reduce the overall ability of a climatology average to represent PET correctly. A surprising result is that, in the absence of any climate data available, calibrating the McGuinness–Bordne equation with CHESS-temp clim (long-term daily temperature climatology) outperforms using CHESS-PM climatology. NSE scores are equivalent for both approaches but MAPE is worse for the latter. The two approaches give similar results, but running the McGuinness–Bordne equation using CHESS-temp clim produces smoother time series than directly using CHESS-PM climatology. The latter displays random noise which explains the larger values of MAPE compared to the smoother version. This is illustrated in Fig. S4 in the Supplement.

A single annual McGuinness–Bordne PET equation calibrated over all catchments simultaneously was selected as the most appropriate method to derive daily PET time series from monthly mean temperature data.

To investigate the effect of coarser spatial resolution in the forcing temperature data, McGuinness–Bordne 1P-GB was

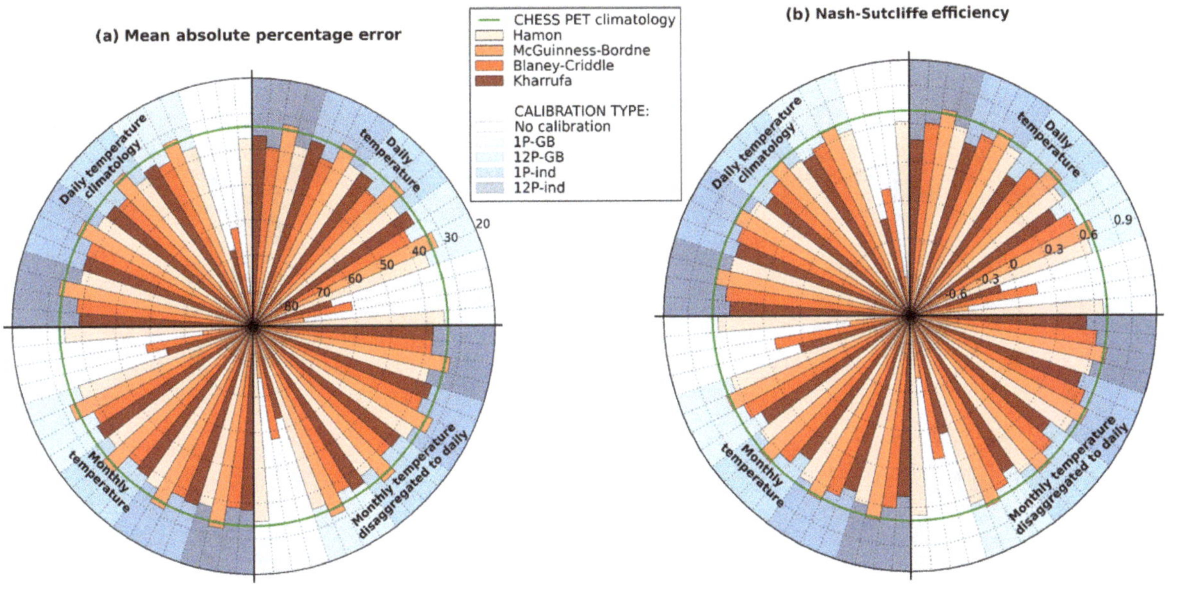

Figure 4. Performance of the different combinations of PET equations (shown in different shades of red), calibration approaches (shown in different shades of blue) and input temperature data (one in each quadrant). The green line on the plots shows the reference CHESS PET climatology for comparison. **(a)** Mean absolute percentage error (MAPE) – note that the y axis is inverted so that lower MAPE values (which indicate better performance) are shown towards the outside of the radial plot. **(b)** Nash–Sutcliffe efficiency (NSE); a higher NSE indicates better performance.

applied using UKCP09-temp monthly I and II (5 km gridded data) as forcing data (stage 3 in Fig. 1). Results show (Table S1 in the Supplement) that at the catchment scale, the spatial resolution of the forcing temperature data has virtually no effect on the performance, with MAPE and NSE values almost identical when using 1 km gridded CHESS-temp monthly I data (MAPE = 32.02, NSE = 0.72) or 5 km gridded UKCP09-temp monthly I data (MAPE = 31.65, NSE = 0.72), and when using CHESS-temp monthly II data (MAPE = 32.13, NSE = 0.72) or UKCP09-temp monthly II data (MAPE = 32.06, NSE = 0.72). This suggests that for the reconstruction prior to 1961, when only 5 km monthly temperature time series are available across the UK, performance is expected to be equivalent to if finer spatial and temporal resolution of mean temperature data existed.

The relationship between performance and catchment area was also tested, but no clear relationship was found.

4.2 Assessment of the final PET grids

This section presents the results of the assessment described in stages 4 and 5 in Fig. 1.

Based on the results in Sect. 4.1, the McGuinness–Bordne 1P-GB equation calibrated on 43 catchments was selected to generate a 5 km PET dataset covering the period 1891 to 2015, using UKCP09-temp monthly II data. A monthly version of the dataset (monthly aggregation of the daily PET for consistency) was also produced for applications requiring a coarse temporal resolution such as groundwater modelling, which has the advantage of a smaller data volume. The final gridded PET data produced here is hereafter referred to as "historic PET dataset".

4.2.1 Catchment-scale calibration (stage 4, Fig. 1)

At catchment scale, there is virtually no difference in deriving PET time series from the historic PET dataset or from PET calculated with the same equation using catchment-average temperature. The correlation coefficient is close to 1 for the 306 catchments. This validates our assumption that the selected equation calibrated at catchment scale is applicable at grid scale.

4.2.2 Final evaluation (stage 5, Fig. 1)

PET extracted from the historic PET dataset was compared with CHESS-PM (ground truth), both at daily and monthly timescale, to evaluate the performance of the final reconstructed product, at catchment scale and grid scale.

At catchment scale, the results are more varied. Spatial differences can be observed between the different metrics and are represented in detail in Figs. S5 and S6 in the Supplement. The results are not discussed here as they are very similar to the grid-to-grid comparison described in the following.

At the grid scale, the following observations can be made (Fig. 5 for daily PET and Fig. 6 for monthly PET; note differences in the legend colour scale):

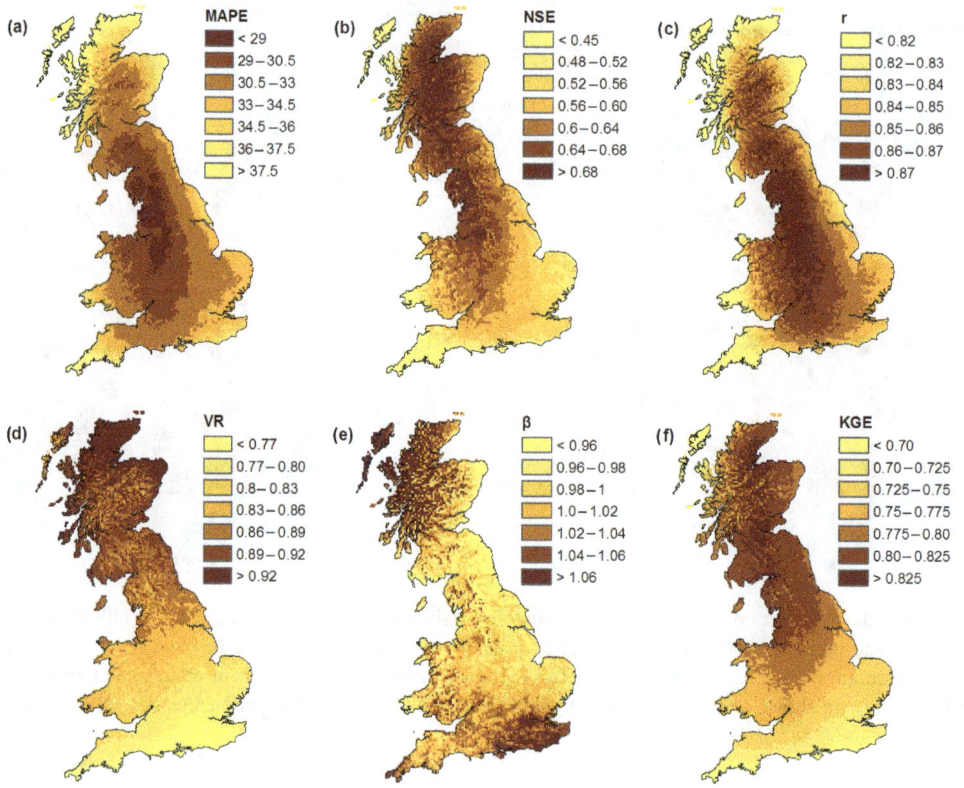

Figure 5. Grids of evaluation metrics for the new daily gridded PET dataset. The darker the colour, the better the performance for all metrics represented, except for the bias ratio (β) **(e)** where the middle-range colour is optimal (see Sect. 3.3).

i. Performance is greater for monthly (Fig. 6) than daily (Fig. 5) PET, except for the bias ratio β which is very similar for both. This suggests that the error is greater at daily than at monthly resolution, likely to be due to the smoothing of the day-to-day variability absent in a temperature-based method. The PET equation shows very good performance at a monthly scale with values of NSE > 0.9, $r > 0.97$ and KGE > 0.8 for the whole country. For daily values, the performance is more moderate (NSE > 0.4, $r > 0.8$ and KGE > 0.7).

ii. Performance varies spatially, but this variability depends on the metrics chosen and is different for monthly and daily PET. For daily PET, MAPE (Fig. 5a), NSE (Fig. 5b) and r (Fig. 5c) show lower performance near the coasts. This is probably because daily variation of wind and humidity are higher near the coast, which is not captured in temperature-based PET equations and hence results in larger errors. VR (Fig. 5d) displays a north–south gradient in performance, the north being better. This is because the coefficient of variability in observed PET is smaller in the north than it is in the south, with less daily noise (see Fig. S3 in the Supplement, grey line). The bias ratio β (Fig. 5e) is close to one everywhere across Great Britain, which indicates that the calibrated McGuinness–Bordne equation shows very little bias. KGE (Fig. 5f) which is a combination of r, VR and β, shows a north–south gradient as the strongest influence comes from VR. For monthly PET, the daily noise in the climate variables is absent, which explains smaller differences in performance scores for most metrics.

The metrics grids are provided as part of the dataset, and can inform the users on the quality of the PET estimation for a given location.

5 Data availability

The new PET dataset is called "Historic Gridded Potential Evapotranspiration (PET)" based on the temperature-based equation McGuinness–Bordne calibrated for the UK (1891–2015)" and is available from https://doi.org/10.5285/17b9c4f7-1c30-4b6f-b2fe-f7780159939c (Tanguy et al., 2017). The dataset is stored in NetCDF4 format.

For the monthly grids, the dataset is structured as three-dimensional grids covering the UK, with twelve time steps (monthly grids) in the time dimension in each yearly file, and a spatial resolution of 5 km.

For the daily grids, the dataset is structured as three-dimensional grids covering the UK, with 365 or 366 (leap

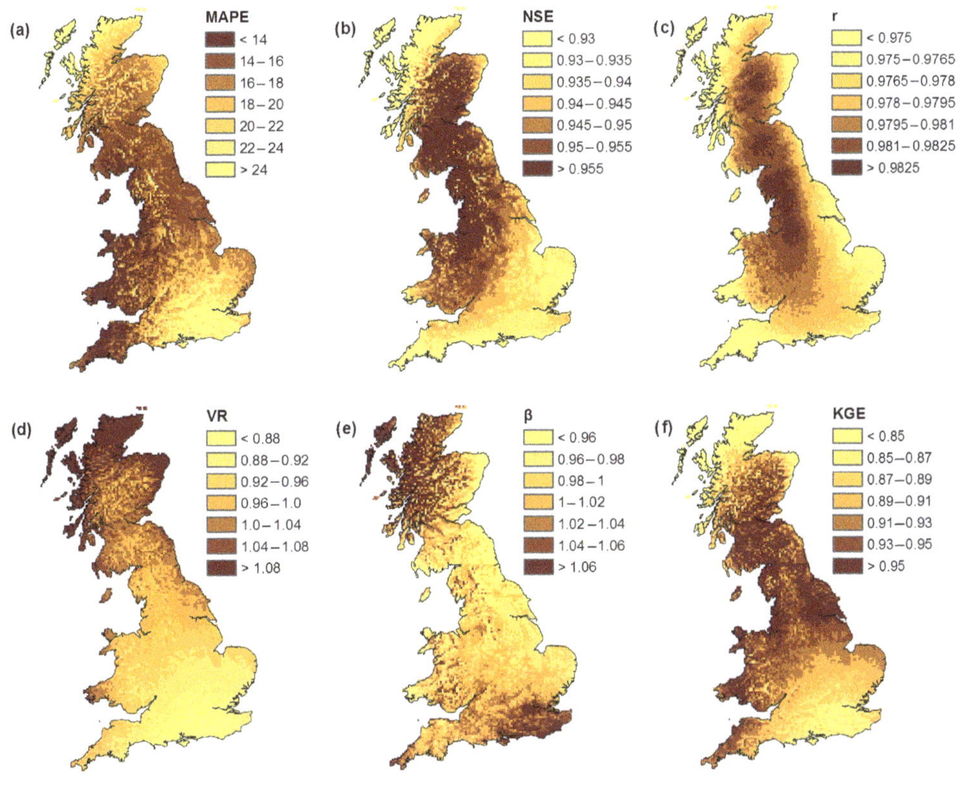

Figure 6. Grids of evaluation metrics for the new monthly gridded PET dataset. The darker the colour, the better the performance for all metrics represented, except for the variability ratio (VR) **(d)** and bias ratio (β) **(e)** where the middle-range colour is optimal.

year) time steps (daily grids) in each yearly file, and a spatial resolution of 5 km.

In addition, four metric files, also in NetCDF format, accompany the PET files (two for daily grids and two for monthly grids), also at a spatial resolution of 5 km.

The data are projected using the British National Grid coordinate system.

The following citation should be used for every application of the data: Tanguy, M., Prudhomme, C., Smith, K., and Hannaford, J.: Historic Gridded Potential Evapotranspiration (PET) based on temperature-based equation McGuinness–Bordne calibrated for the UK (1891–2015), NERC Environmental Information Data Centre, 2017.

The dataset is available for download from the CEH Environmental Information Data Centre (EIDC).

The temperature and PET datasets used in this study are available to download from the following links:

- CHESS temperature data (Robinson et al., 2016a) can be downloaded from the EIDC catalogue: https://catalogue.ceh.ac.uk/documents/b745e7b1-626c-4ccc-ac27-56582e77b900.
- CHESS PET data (Robinson et al., 2016b) can be downloaded from the EIDC catalogue: https://catalogue.ceh.ac.uk/documents/8baf805d-39ce-4dac-b224-c926ada353b7.
- UKCP09 temperature data (Perry and Hollis, 2005) can be downloaded from CEDA catalogue: http://catalogue.ceda.ac.uk/uuid/87f43af9d02e42f483351d79b3d6162a.

6 Discussion, limitations and recommendations

In this section, the uncertainties linked to the temperature dataset and the PET method are discussed. Subsequently, recommendations to users depending on the intended application of the data are listed. Lastly, a summary of findings and potential future work are presented.

6.1 Uncertainties

Firstly, the uncertainties linked to the underpinning temperature data should be considered. The data rescuing work that the UK Met Office has undertaken to extend the temperature data back to the late 19th century raises some questions about how the change in network density might affect the accuracy of the spatial data.

According to information provided by the Met Office, the station density gradually increased from 74 stations across the country in 1891 to a peak of 672 in the mid-1990s, after which it decreased again to reach a total of 355 stations in 2015. Legg (2015) has extensively investigated the effect of network density on the error in gridded dataset in the UK, and his results suggest that the change in density observed here would only lead to a minor increase in error in temperature. An increase in the root mean square error of less than 0.2 °C is observed for most cases when the network density changes from 570 to 75 stations across the UK. This reflects the spatial coherence in the temperature data.

A sensitivity analysis of McGuinness–Bordne PET on errors in input temperature was conducted. It was found that a ±0.2 °C in input temperature translates into a 0.5 to 2 % difference (with an average of 0.8 %) in PET estimation. We consider these differences negligible in comparison to the uncertainties arising from the PET method itself (MAPE ranging from 14 to 24 % for monthly PET estimation, see Fig. 6a).

Some additional considerations regarding the joining of the two temperature datasets can be found in the Supplement (Sect. S2).

The main limitation of the historic PET dataset comes from the method used to derive it, which only takes temperature into account. This is the case particularly for the daily version. The PM evapotranspiration equation has radiative and convective components. In simplified temperature-based equations, temperature is used as a proxy for radiation but does not account for the convective aspect. Therefore, temperature-based equations are not able to reproduce the full daily fluctuation of PET, and are only a smoothed version of reality. This has to be kept in mind for applications where the daily variability of PET is important, such as the estimation of daily water balance, flood peaks and crop water demand, among others. Users are strongly advised to look at performance metrics associated with the dataset in their study area, such as monthly MAPE for example, which provide information on the uncertainty in the estimates. Note that because of the absence of the reference dataset CHESS-PM in Northern Ireland, no quality metrics are available in that region. Datasets based on physically based equations such as CHESS-Penman–Monteith (CHESS-PM) are a better option when and where they are available, which is not the case in Great Britain before 1961, and in Northern Ireland. When and where such high-resolution physically based PET datasets are not available, temperature-based PET datasets such as the historical PET dataset reconstructed here provide a valuable substitute.

At a monthly timescale, the magnitude in the seasonal cycle is well captured, which is reflected in better performance metrics for the monthly PET data compared to the daily PET (Figs. 5 and 6). This makes this dataset particularly suitable for deriving monthly or seasonal river discharges or run-offs, as its accuracy is adequate at this coarser timescale, and its daily temporal resolution is sufficient for most hydrological modelling applications.

6.2 Applications

While uncertainties in the PET dataset are quite large, especially in the daily version, the impact it might have will depend on the intended purpose of the data.

For hydrological applications, the choice of PET equation was shown to affect the estimated streamflow when using hydrological models (Seiller and Anctil, 2016), in particular at high and low flows (Samadi, 2016). However, several studies show that hydrological models are much more sensitive to errors in rainfall than to errors in PET, especially in temperate climates such as the UK (Bastola et al., 2011; Guo et al., 2017; Paturel et al., 1995). Furthermore, other studies (Bai et al., 2016; Seiller and Anctil, 2016) show that hydrological model parameter calibration can eliminate the influences of different PET inputs on runoff simulations. Oudin et al. (2005) have also demonstrated that temperature-based methods are suitable for conceptual hydrological modelling, and when available at a fine spatial scale, are also suitable for distributed hydrological modelling. Therefore, the historic PET dataset is considered particularly suitable for use in hydrological models, especially if these are being calibrated using this dataset, as the impact of PET uncertainties will be small compared to those of rainfall. It's also worth mentioning that the McGuinness–Bordne equation used to derive the historic PET dataset was calibrated against CHESS-PM. There is no systematic bias (bias ratio ≈ 1, see Figs. 5 and 6) between the two datasets. The use of the historic PET data would therefore be adequate in hydrological models that have been calibrated using CHESS-PM, but recalibration would be recommended if any other PET source was used in the original calibration.

For macroecology and biogeography studies, Fisher Joshua et al. (2010) have produced a global "guide to choosing an ET model for geographical ecology", according to the climate zone of the study area. For temperate climates such as the UK, their conclusion is that any PET model type (temperature-based, radiation-based or combination) is equally adequate for its use in biodiversity modelling. Therefore, the historic PET dataset would be appropriate for this type of application. However, for crop modelling, greater caution is required as modelled crop yield is highly sensitive to the choice of PET model (Balkovič et al., 2013; Liu et al., 2016; Luo et al., 2009).

Regarding the derivation of drought indices which use PET, some seem insensitive to the choice of PET model, such as the Reconnaissance Drought Index (Tsakiris et al., 2007) as demonstrated by Vangelis et al. (2013); whereas for others such as the Standardized Precipitation-Evapotranspiration Index (Vicente-Serrano et al., 2009) or the Palmer Drought Severity Index (Palmer, 1965), different formulations of PET have a significant impact on the result (Beguería et al., 2013; Sheffield et al., 2012; Stagge et al., 2014). However, this is less important in humid areas such as the UK (Beguería et al., 2013). Therefore, the impact of uncertainties in PET for deriving drought indices will depend on the choice of drought index.

In general, for the use of the historic PET dataset to derive drought indices, or any other application not mentioned above, we would recommend that the user compares the results over the more recent period (1961–2015) using (i) CHESS-PM and (ii) the historic PET dataset, to estimate the impact of PET uncertainties in their study. This way, the user can truly assess the sensitivity of their specific application to the errors in PET, investigate how the uncertainties propagate in their model and make an informed decision on whether the historic PET dataset is suitable for their needs or not.

6.3 Further findings and future work

Beyond generating a new 125 years gridded daily PET dataset for the UK, this research has highlighted valuable insights for PET calculation in the UK:

i. calibration is essential for realistic results, but the choice of calibration method (global/annual or local/monthly) has a minimal effect, and therefore the easiest, most cost-effective calibration method is recommended (global/annual);

ii. the temporal resolution of the input temperature data and the temporal disaggregation method when using monthly data has little influence on the results;

iii. temperature-based equations perform better at a monthly scale than at a daily scale, as the full daily fluctuation of PET due to other climate variables (wind speed, humidity, radiation) are not being accounted for, but these are smoothed out at the monthly scale;

iv. the temperature-based PET equation (from the seven equations tested) that produces the best results for the UK is the calibrated version of the McGuinness–Bordne equation;

v. for this equation, the spatial resolution (1 or 5 km) of the input temperature data has virtually no effect in the results at catchment scale;

vi. CHESS-PM daily climatology is the second best of the tested options, and is therefore a possible alternative source of PET if no climate variables are available. (Whilst mean seasonal PET or climatology can be used in hydrological modelling (Burnash, 1995; Calder et al., 1983; Fowler, 2002), McGuinness–Bordne derived PET time series are preferable as they are able to reproduce the inter-annual variability existing in PET, absent from any climatology); and finally,

vii. performance of the McGuinness–Bordne equation across the UK is variable in space, and the gridded metrics provided within the dataset can inform future work on the adequacy of using this approach for estimating PET in particular areas.

Future research could explore the use of reanalysis data as an alternative or complementary source of data to derive past spatio-temporal PET data. The use of reanalysis data would enable the calculation of PET through the more accurate combined methods (such as PM). However, the uncertainties associated with reanalysis data should be carefully examined, as some of the modelled variables can display large errors (Reichler and Kim, 2008), and PM has also shown sensitivity to input data inaccuracy (Oudin et al., 2005; Debnath et al., 2015; Estévez et al., 2009; Gong et al., 2006).

Competing interests. The authors declare that they have no conflict of interest.

Acknowledgements. This is an outcome of the IMPETUS (grant number: NE/L010267/1) and Historic Droughts (grant number: NE/L01016X/1) projects, funded by the Natural Environment Research Council.

The authors would like to thank their CEH colleague Cath Sefton for offering valuable feedback on the manuscript, and the Met Office, in particular Mark McCarthy and Tim Legg for providing the historic monthly temperature data. Finally, we are grateful to Mark McCarthy and a second anonymous referee who both provided constructive comments during the peer review process, which contributed to improving the paper.

Edited by: David Carlson

References

Allen, R. G., Pereira, L. S., Raes, D., and Smith, M.: Fao irrigation and drainage paper 56 – crop evapotranspiration – guidelines for computing crop water requirements, Rome, 1998.

Aràndiga, F., Donat, R., and Santágueda, M.: The PCHIP subdivision scheme, Appl. Math. Comput., 272, 28–40, https://doi.org/10.1016/j.amc.2015.07.071, 2016.

Bai, P., Liu, X., Yang, T., Li, F., Liang, K., Hu, S., and Liu, C.: Assessment of the Influences of Different Potential Evapotranspiration Inputs on the Performance of Monthly Hydrological Models under Different Climatic Conditions, J. Hydrometeorol., 17, 2259–2274, doi10.1175/JHM-D-15-0202.1, 2016.

Balkovič, J., van der Velde, M., Schmid, E., Skalský, R., Khabarov, N., Obersteiner, M., Stürmer, B., and Xiong, W.: Pan-European crop modelling with EPIC: Implementation, up-scaling and regional crop yield validation, Agr. Syst., 120, 61–75, https://doi.org/10.1016/j.agsy.2013.05.008, 2013.

Barik, M. G.: Remote Sensing-based Estimates of Potential Evapotranspiration for Hydrologic Modeling in the Upper Colorado River Basin Region, PhD, Civil Engineering 0300 UCLA, University of California, Los Angeles, 146 pp., 2014.

Barik, M. G., Hogue, T. S., Franz, K. J., and Kinoshita, A. M.: Assessing Satellite and Ground-Based Potential Evapotranspiration for Hydrologic Applications in the Colorado River Basin, J. Am. Water Resour. As., 52, 48–66, https://doi.org/10.1111/1752-1688.12370, 2016.

Bastola, S., Murphy, C., and Sweeney, J.: The sensitivity of fluvial flood risk in Irish catchments to the range of IPCC AR4 climate change scenarios, Sci. Total Environ., 409, 5403–5415, https://doi.org/10.1016/j.scitotenv.2011.08.042, 2011.

Beguería, S., Vicente-Serrano Sergio, M., Reig, F., and Latorre, B.: Standardized precipitation evapotranspiration index (SPEI) revisited: parameter fitting, evapotranspiration models, tools, datasets and drought monitoring, Int. J. Climatol., 34, 3001–3023, https://doi.org/10.1002/joc.3887, 2013.

Beven, K.: Rainfall-runoff modelling. The primer, 2nd Edn., John Wiley & Sons Ltd, Oxford (UK), 457 pp., 2012.

Blaney, H. F. and Criddle, W. D.: Determining water requirements in irrigated areas from climatological and irrigation data, Washington, USA, 48 pp., 1950.

Burnash, R. J. C.: The NWS River Forecast System – Catchment Modeling, in: Computer models of watershed hydrology, edited by: Singh, V., Water Resources Publications, Highlands Ranch, Colorado, USA, 1995.

Calder, I. R., Harding, R. J., and Rosier, P. T. W.: An objective assessment of soil-moisture deficit models, J. Hydrol., 60, 329–355, https://doi.org/10.1016/0022-1694(83)90030-6, 1983.

Danladi, A., Stephen, M., Aliyu, B. M., Gaya, G. K., Silikwa, N. W., and Machael, Y.: Assessing the influence of weather parameters on rainfall to forecast river discharge based on short-term, Alexandria Engineering Journal, https://doi.org/10.1016/j.aej.2017.03.004, online first, 2017.

Debnath, S., Adamala, S., and Raghuwanshi, N. S.: Sensitivity Analysis of FAO-56 Penman-Monteith Method for Different Agro-ecological Regions of India, Environmental Processes, 2, 689–704, https://doi.org/10.1007/s40710-015-0107-1, 2015.

Donohue, R. J., McVicar, T. R., and Roderick, M. L.: Assessing the ability of potential evaporation formulations to capture the dynamics in evaporative demand within a changing climate, J. Hydrol., 386, 186–197, 2010.

Doorenbos, J. and Pruitt, W. O.: Guidelines for predicting crop water requirements, FAO irrigation and drainage paper 24, Food and Agriculture organization of the United Nations, Rome, 144 pp., 1984.

Droogers, P. and Allen, R. G.: Estimating Reference Evapotranspiration Under Inaccurate Data Conditions, Irrigation and Drainage Systems, 16, 33–45, https://doi.org/10.1023/A:1015508322413, 2002.

Ershadi, A., McCabe, M. F., Evans, J. P., Chaney, N. W., and Wood, E. F.: Multi-site evaluation of terrestrial evaporation models using FLUXNET data, Agr. Forest Meteorol., 187, 46–61, https://doi.org/10.1016/j.agrformet.2013.11.008, 2014.

Estévez, J., Gavilán, P., and Berengena, J.: Sensitivity analysis of a Penman–Monteith type equation to estimate reference evapotranspiration in southern Spain, Hydrol. Process., 23, 3342–3353, https://doi.org/10.1002/hyp.7439, 2009.

Federer, C. A., Vorosmarty, C. J., and Fekete, B.: Intercomparison of methods for calculating potential evaporation in regional and global water balance models, Wat. Resour. Res., 32, 2315–2321, 1996.

Fisher Joshua, B., Whittaker Robert, J., and Malhi, Y.: ET come home: potential evapotranspiration in geographical ecology, Global Ecol. Biogeogr., 20, 1–18, https://doi.org/10.1111/j.1466-8238.2010.00578.x, 2010.

Fortin, V. and Turcotte, R.: Le modèle hydrologique MOHYSE, Note de cours pour SCA7420., Département des sciences de la terre et de l'atmosphère, Montréal: Université du Québec à Montréal., 2006.

Fowler, A.: Assessment of the validity of using mean potential evaporation in computations of the long-term soil water balance, J. Hydrol., 256, 248–263, doi10.1016/S0022-1694(01)00542-X, 2002.

Gong, L., Xu, C.-Y., Chen, D., Halldin, S., and Chen, Y. D.: Sensitivity of the Penman–Monteith reference evapotranspiration to key climatic variables in the Changjiang (Yangtze River) basin, J. Hydrol., 329, 620–629, https://doi.org/10.1016/j.jhydrol.2006.03.027, 2006.

Guerschman, J. P., Van Dijk, A. I. J. M., Mattersdorf, G., Beringer, J., Hutley, L. B., Leuning, R., Pipunic, R. C., and Sherman, B. S.: Scaling of potential evapotranspiration with MODIS data reproduces flux observations and catchment water balance observations across Australia, J. Hydrol., 369, 107–119, https://doi.org/10.1016/j.jhydrol.2009.02.013, 2009.

Guo, D., Westra, S., and Maier, H. R.: Use of a scenario-neutral approach to identify the key hydro-meteorological attributes that impact runoff from a natural catchment, J. Hydrol., 554, 317–330, https://doi.org/10.1016/j.jhydrol.2017.09.021, 2017.

Gupta, H. V., Kling, H., Yilmaz, K. K., and Martinez, G. F.: Decomposition of the mean squared error and NSE performance criteria: Implications for improving hydrological modelling, J. Hydrol., 377, 80–91, https://doi.org/10.1016/j.jhydrol.2009.08.003, 2009.

Hamon, W. R.: Estimating potential evapotranspiration, J. Hydr. Eng. Div.-ASCE, 87, 107–120, 1961.

Hargreaves, G. H. and Samani, Z. A.: Estimating potential evapotranspiration, Journal of the Irrigation and Drainage Division, Proceedings of the American Society of Civils Engineers, 108, 225–230, 1983.

Heydari, M. M. and Heydari, M.: Evaluation of pan coefficient equations for estimating reference crop evapotranspiration in the arid region, Arch. Agron. Soil Sci., 60, 715–731, https://doi.org/10.1080/03650340.2013.830286, 2014.

Hough, M.: An historical comparison between the Met Office Surface Exchange Scheme-Probability Distributed Model (MOSES-PDM) and the Met Office Rainfall and Evaporation Calculation System (MORECS), Met Office, Bracnkell, Report for the Environment Agency and the Met. Office, 38 pp., 2003.

Jensen, M. E. and Haise, H. R.: Estimating evapotranspiration from solar radiation, J. Irrig. Drain. Div.-ASCE, 89, 15–41, 1963.

Kharrufa, N. S.: Simplified equation for evapotranspiration in arid regions, Beiträge Zur Hydrologie Sonderheft, 5, 39–47, 1985.

Kling, H., Fuchs, M., and Paulin, M.: Runoff conditions in the upper Danube basin under an ensemble of climate change scenarios, J. Hydrol., 424, 264–277, https://doi.org/10.1016/j.jhydrol.2012.01.011, 2012.

Knipper, K., Hogue, T., Scott, R., and Franz, K.: Evapotranspiration Estimates Derived Using Multi-Platform Remote Sensing in a Semiarid Region, Remote Sensing, 9, 184, https://doi.org/10.3390/rs9030184, 2017.

Lefebvre, M. and Bensalma, F.: Modeling and forecasting river flows by means of filtered Poisson processes, Appl. Math. Model., 39, 230–243, https://doi.org/10.1016/j.apm.2014.05.027, 2015.

Legg, T.: Uncertainties in gridded area-average monthly temperature, precipitation and sunshine for the United Kingdom, Int. J. Climatol., 35, 1367–1378, https://doi.org/10.1002/joc.4062, 2015.

Liu, S., Graham, W. D., and Jacobs, J. M.: Daily potential evapotranspiration and diurnal climate forcings: influence on the numerical modelling of soil water dynamics and evapotranspiration, J. Hydrol., 309, 39–52, https://doi.org/10.1016/j.jhydrol.2004.11.009, 2005.

Liu, W., Yang, H., Folberth, C., Wang, X., Luo, Q., and Schulin, R.: Global investigation of impacts of PET methods on simulating crop-water relations for maize, Agr. Forest Meteorol., 221, 164–175, https://doi.org/10.1016/j.agrformet.2016.02.017, 2016.

Luo, W., Jing, W., Jia, Z., Li, J., and Pan, Y.: The effect of PET calculations in DRAINMOD on drainage and crop yields predictions in a subhumid vertisol soil district, Sci. China Ser. E, 52, 3315, https://doi.org/10.1007/s11431-009-0349-0, 2009.

McGuinness, J. L. and Bordne, E. F.: A Comparison of Lysimeter-Derived Potential Evapotranspiration With Computed Values, United States Department of Agriculture, Economic Research Service, 1972.

Monteith, J. L.: Evaporation and environment, Symposia of the society for experimental biology, Cambridge University Press, Cambridge, 1965.

Mu, Q., Zhao, M., and Running, S. W.: MODIS Global Terrestrial Evapotranspiration (ET), Product (NASA MOD16A2/A3), Algorithm Theoretical Basis Document, Collection 5, NASA Headquarters, 2013.

Nash, J. E. and Sutcliffe, J. V.: River flow forecasting through conceptual models part I – A discussion of principles, J. Hydrol., 10, 282–290, https://doi.org/10.1016/0022-1694(70)90255-6, 1970.

Oudin, L., Hervieu, F., Michel, C., Perrin, C., Andréassian, V., Anctil, F., and Loumagne, C.: Which potential evapotranspiration input for a lumped rainfall–runoff model?: Part 2 – Towards a simple and efficient potential evapotranspiration model for rainfall–runoff modelling, J. Hydrol., 303, 290–306, https://doi.org/10.1016/j.jhydrol.2004.08.026, 2005.

Palmer, W. C.: Meteorological droughts, U.S. Department of Commerce Weather Bureau, Research Paper 45, 58 pp., 1965.

Paturel, J. E., Servat, E., and Vassiliadis, A.: Sensitivity of conceptual rainfall-runoff algorithms to errors in input data – case of the GR2M model, J. Hydrol., 168, 111–125, https://doi.org/10.1016/0022-1694(94)02654-T, 1995.

Penman, H. L.: Natural evaporation from open water, bare soil and grass, P. Roy. Soc. Lond. A Mat., 1032, 120–145, 1948.

Perry, M. and Hollis, D.: The generation of monthly gridded datasets for a range of climatic variables over the UK, Int. J. Climatol., 25, 1041–1054, https://doi.org/10.1002/joc.1161, 2005.

Prudhomme, C. and Williamson, J.: Derivation of RCM-driven potential evapotranspiration for hydrological climate change impact analysis in Great Britain: a comparison of methods and associated uncertainty in future projections, Hydrol. Earth Syst. Sci., 17, 1365–1377, https://doi.org/10.5194/hess-17-1365-2013, 2013.

Reichler, T. and Kim, J.: Uncertainties in the climate mean state of global observations, reanalyses, and the GFDL climate model, J. Geophys. Res.-Atmos., 113, D05106, https://doi.org/10.1029/2007JD009278, 2008.

Robinson, E. L., Blyth, E. M., Clark, D. B., Comyn-Platt, E., Finch, J., and Rudd, A. C.: Climate hydrology and ecology research support system meteorology dataset for Great Britain (1961–2015) [CHESS-met], NERC Environmental Information Data Centre, https://doi.org/10.5285/10874370-bc58-4d23-a118-ea07df8a07f2, 2016a.

Robinson, E. L., Blyth, E. M., Clark, D. B., Finch, J., and Rudd, A. C.: Climate hydrology and ecology research support system potential evapotranspiration dataset for Great Britain (1961–2015) [CHESS-PE] NERC Environmental Information Data Centre, https://doi.org/10.5285/8baf805d-39ce-4dac-b224-c926ada353b7, 2016b.

Robinson, E. L., Blyth, E. M., Clark, D. B., Finch, J., and Rudd, A. C.: Trends in atmospheric evaporative demand in Great Britain using high-resolution meteorological data, Hydrol. Earth Syst. Sci., 21, 1189–1224, https://doi.org/10.5194/hess-21-1189-2017, 2017.

Samadi, S. Z.: Assessing the sensitivity of SWAT physical parameters to potential evapotranspiration estimation methods over a coastal plain watershed in the southeastern United States, Hydrol. Res., 48, 395–415, 2016.

Schneider, K., Ketzer, B., Breuer, L., Vaché, K. B., Bernhofer, C., and Frede, H.-G.: Evaluation of evapotranspiration methods for model validation in a semi-arid watershed in northern China, Adv. Geosci., 11, 37–42, https://doi.org/10.5194/adgeo-11-37-2007, 2007.

Seiller, G. and Anctil, F.: How do potential evapotranspiration formulas influence hydrological projections?, Hydrolog. Sci. J., 61, 2249–2266, https://doi.org/10.1080/02626667.2015.1100302, 2016.

Sheffield, J., Wood, E. F., and Roderick, M. L.: Little change in global drought over the past 60 years, Nature, 491, 435, https://doi.org/10.1038/nature11575, 2012.

Shuttleworth, W. J.: Evaporation, in: Handbook of hydrology, edited by: Maidment, D. R., McGraw-Hill, 4.1–4.53, New York, NY, USA, 1993.

Spies, R. R., Franz, K. J., Hogue, T. S., and Bowman, A. L.: Distributed Hydrologic Modeling Using Satellite-Derived Potential Evapotranspiration, J. Hydrometeorol., 16, 129–146, https://doi.org/10.1175/JHM-D-14-0047.1, 2014.

Srivastava Prashant, K., Han, D., Rico Ramirez Miguel, A., and Islam, T.: Comparative assessment of evapotranspiration derived from NCEP and ECMWF global datasets through Weather Research and Forecasting model, Atmos. Sci. Lett., 14, 118–125, https://doi.org/10.1002/asl2.427, 2013.

Stagge, J., Tallaksen, L., Xu, C.-Y., and Van Lanen, H.: Standardized precipitation-evapotranspiration index (SPEI): Sensitivity to potential evapotranspiration model and parameters, Hydrology in a changing world: Environmental and Human Dimensions. Proceedings of FRIEND-Water 2014, Montpellier, France, IAHS Publ., 363, 367–373, 2014.

Tanguy, M., Prudhomme, C., Smith, K., and Hannaford, J.: Historic Gridded Potential Evapotranspiration (PET) based on temperature-based equation McGuinness-Bordne calibrated for the UK (1891–2015), NERC Environmental Information Data Centre, https://doi.org/10.5285/17b9c4f7-1c30-4b6f-b2fe-f7780159939c, 2017.

Thompson, N., Barrie, I. A., and Ayles, M.: The Meteorological Office Rainfall and Evaporation Calculation System: MORECS (July 1981), Met. Office, Bracknell, UK, 1982.

Thornthwaite, C. W.: An approach toward a rational classification of climate, Geographical Review, 38, 55–94, 1948.

Tsakiris, G., Pangalou, D., and Vangelis, H.: Regional Drought Assessment Based on the Reconnaissance Drought Index (RDI), Water Resour. Manage., 21, 821–833, https://doi.org/10.1007/s11269-006-9105-4, 2007.

Vangelis, H., Tigkas, D., and Tsakiris, G.: The effect of PET method on Reconnaissance Drought Index (RDI) calculation, J. Arid Environ., 88, 130–140, https://doi.org/10.1016/j.jaridenv.2012.07.020, 2013.

Vicente-Serrano, S. M., Beguería, S., and López-Moreno, J. I.: A Multiscalar Drought Index Sensitive to Global Warming: The Standardized Precipitation Evapotranspiration Index, J. Climate, 23, 1696–1718, https://doi.org/10.1175/2009JCLI2909.1, 2009.

Watts, G., Christierson, B. V., Hannaford, J., and Lonsdale, K.: Testing the resilience of water supply systems to long droughts, J. Hydrol., 414–415, 255–267, https://doi.org/10.1016/j.jhydrol.2011.10.038, 2012.

Xu, C.-Y. and Singh, V. P.: Evaluation and generalization of radiation-based methods for calculating evaporation, Hydrol. Process., 14, 339–349, 2000.

Xu, C. Y. and Singh, V. P.: Evaluation and generalization of temperature-based methods for calculating evaporation, Hydrol. Process., 15, 305–319, 2001.

14

High-resolution global grids of revised Priestley–Taylor and Hargreaves–Samani coefficients for assessing ASCE-standardized reference crop evapotranspiration and solar radiation

Vassilis G. Aschonitis[1], Dimitris Papamichail[2], Kleoniki Demertzi[2], Nicolo Colombani[1], Micol Mastrocicco[3], Andrea Ghirardini[1], Giuseppe Castaldelli[1], and Elisa-Anna Fano[1]

[1]Department of Life Sciences and Biotechnology, University of Ferrara, Ferrara, Italy
[2]Department of Hydraulics, Soil Science and Agricultural Engineering, Aristotle University of Thessaloniki, Thessaloniki, Greece
[3]Department of Environmental, Biological and Pharmaceutical Sciences and Technologies, University of Campania Luigi Vanvitelli, Caserta, Italy

Correspondence to: Vassilis G. Aschonitis (schvls@unife.it)

Abstract. The objective of the study is to provide global grids (0.5°) of revised annual coefficients for the Priestley–Taylor (P-T) and Hargreaves–Samani (H-S) evapotranspiration methods after calibration based on the ASCE (American Society of Civil Engineers)-standardized Penman–Monteith method (the ASCE method includes two reference crops: short-clipped grass and tall alfalfa). The analysis also includes the development of a global grid of revised annual coefficients for solar radiation (R_s) estimations using the respective R_s formula of H-S. The analysis was based on global gridded climatic data of the period 1950–2000. The method for deriving annual coefficients of the P-T and H-S methods was based on partial weighted averages (PWAs) of their mean monthly values. This method estimates the annual values considering the amplitude of the parameter under investigation (ET_o and R_s) giving more weight to the monthly coefficients of the months with higher ET_o values (or R_s values for the case of the H-S radiation formula). The method also eliminates the effect of unreasonably high or low monthly coefficients that may occur during periods where ET_o and R_s fall below a specific threshold. The new coefficients were validated based on data from 140 stations located in various climatic zones of the USA and Australia with expanded observations up to 2016. The validation procedure for ET_o estimations of the short reference crop showed that the P-T and H-S methods with the new revised coefficients outperformed the standard methods reducing the estimated root mean square error (RMSE) in ET_o values by 40 and 25 %, respectively. The estimations of R_s using the H-S formula with revised coefficients reduced the RMSE by 28 % in comparison to the standard H-S formula. Finally, a raster database was built consisting of (a) global maps for the mean monthly ET_o values estimated by ASCE-standardized method for both reference crops, (b) global maps for the revised annual coefficients of the P-T and H-S evapotranspiration methods for both reference crops and a global map for the revised annual coefficient of the H-S radiation formula and (c) global maps that indicate the optimum locations for using the standard P-T and H-S methods and their possible annual errors based on reference values. The database can support estimations of ET_o and solar radiation for locations where climatic data are limited and it can support studies which require such estimations on larger scales (e.g. country, continent, world). The datasets produced in this study are archived in the PANGAEA database (https://doi.org/10.1594/PANGAEA.868808) and in the ESRN database (http://www.esrn-database.org or http://esrn-database.weebly.com).

1 Introduction

The reference crop evapotranspiration ET_o is defined as the maximum value of water losses by evaporation and transpiration above a reference crop (e.g. grass), which can be achieved under no water restrictions. It is also one of the most important parameters for water balance estimations and irrigation planning of crops (Allen et al., 1998). Several methods have been proposed for ET_o estimations (Itenfisu et al., 2003; Allen et al., 2005; Wang and Dickinson, 2012; McMahon et al., 2013; Valipour, 2017; Valipour and Gholami Sefidkouhi, 2017; Valipour et al., 2017) with the most popular being the FAO-56 Penman–Monteith (Allen et al., 1998), the Priestley–Taylor (P-T; Priestley and Taylor, 1972) and the Hargreaves–Samani (Hargreaves and Samani, 1982, 1985) methods. The FAO-56 has been updated to the ASCE (American Society of Civil Engineers)-standardized method (Allen et al., 2005), which reflects the current state of the art, providing ET_o estimations for two reference crops (a short and a tall reference crop, which correspond to clipped grass and alfalfa, respectively). The ASCE-standardized method has been proposed by the ASCE-EWRI (American Society of Civil Engineers – Environmental & Water Resources Institute) Task Committee as the most precise method and requires a wide range of climatic parameters, which in many cases are not available. The problem of data availability can be solved by other methods, such as the Priestley–Taylor and Hargreaves–Samani, which require less information for their determination. In fact, they are considered the most precise among the simplified methods with reduced parameters (Xu and Singh, 2002; Sumner and Jacobs, 2005; Valipour, 2012, 2014).

The Priestley–Taylor method requires net solar radiation and temperature data. The P-T formula includes an empirical factor known as advection coefficient a_{pt}, which is usually set equal to 1.26 (Priestley and Taylor, 1972) and generally ranges between 1.08 and 1.34 (Tateishi and Ahn, 1996; Xu et al., 2013). Other studies for various climatic conditions have shown that a_{pt} presents significant spatial and seasonal variability (Castellvi et al., 2001; Moges et al., 2003; Pereira, 2004; Tabari and Talaee, 2011; Aschonitis et al., 2015). Weiß and Menzel (2008) used the value 1.26 for wet and the value 1.75 for dry climatic conditions, as suggested by Maidment (1992). The value $a_{pt} = 1.26$ has been verified experimentally for bare irrigated soil (Eichinger et al., 1996). Theoretical simulations for the case of the reference crop in saturated soil have also verified $a_{pt} = 1.26$ for the case of no or restricted advection effects (Lhomme, 1997; McMahon et al., 2013). Lower values of the advection coefficient have been reported by Singh and Irmak (2011) ($a_{pt} = 1.14$) for Nebraska (USA), by Abtew (1996) ($a_{pt} = 1.18$) for Florida (USA), by Kellner (2001) ($a_{pt} = 0.8$) for central Sweden, and by Xu and Singh (2002) ($a_{pt} = 0.9$) for Switzerland. Values of $a_{pt} < 1$ have been reported for forested steep areas (Shuttleworth and Calder, 1979; Giles et al., 1984; Flint and Childs, 1991). On the other hand, high values ranging between 1.82 and 2.14 have been reported for cold dry areas of Iran (Tabari and Talaee, 2011). Aschonitis et al. (2015) analysed the monthly variation in a_{pt} for Italian regions and observed through regression analysis that more than 90 % of the spatial variability in the seasonal a_{pt} was explained by the spatial variability in vapour pressure deficit (DE; positive correlation). The rate of a_{pt} variation per unit DE was found significantly different between seasons, and it was negatively correlated to net solar radiation and/or temperature. The general trends in a_{pt} led to the conclusion that colder drier conditions due to low net radiation and high vapour pressure deficit tend to increase its values.

The Hargreaves–Samani (H-S) method requires only temperature data, including four empirical factors (or three depending on the formula). A part of the equation empirically describes the incident solar radiation R_s. A basic problem of the Hargreaves–Samani method is that it tends to underestimate ET_o under high-wind conditions ($u_2 > 3\,\mathrm{m\,s^{-1}}$) and to overestimate ET_o under conditions of high relative humidity (Allen et al., 1998). In the last years, many scientists have performed analysis and recalibration of the Hargreaves–Samani method for various climates (Trajkovic, 2007; Tabari, 2010; Tabari and Talaee, 2011; Azhar and Perera, 2011; Aschonitis et al., 2012; Mohawesh and Talozi, 2012; Rahimikhoob et al., 2012; Ravazzani et al., 2012; Bachour et al., 2013; Long et al., 2013; Mendicino and Senatore, 2013; Ngongondo et al., 2013; Berti et al., 2014; Heydari and Heydari, 2014), which indicates a global interest in simplified methods, mainly driven by the lack of data.

The analysis of ET_o on a global scale is of special interest since it provides a general view of the spatiotemporal variation in this parameter, while (together with rainfall) it provides significant information about the aridity of terrestrial systems. A basic limitation of global analysis is the lack of homogeneously distributed meteorological stations around the globe and especially in mountainous regions. In the last years, climatic models, advanced interpolation and other methods have succeeded in generating datasets of various climatic parameters (Hijmans et al., 2005; Sheffield et al., 2006; Osborn and Jones, 2014; Brinckmann et al., 2016), facilitating the attempts to develop ET_o maps. Significant work relating to global ET_o estimations has been carried out by various scientists. Mintz and Walker (1993) used the Thorthwaite (1948) method and provided isoline maps of ET_o. Tateishi and Ahn (1996) used the Priestley–Taylor method and provided ET_o maps at 0.5° resolution. Droogers and Allen (2002) used the FAO-56 Penman–Monteith method, providing ET_o maps at 10 arcmin resolution and a modified Hargreaves–Samani method, which considers rainfall. Weiß and Menzel (2008) compared four different methods (Priestley–Taylor, Kimberly Penman, FAO-

56 Penman–Monteith and Hargreaves–Samani) and provided ET_0 maps at 0.5° resolution. Zomer et al. (2008) used Hargreaves–Samani method and provided the highest resolution (30 arcsec) available for ET_0 maps.

The objectives of the study are

a. to develop mean monthly maps of ET_0 for the period 1950–2000 on a global scale using the most precise ASCE-standardized method for both reference crops (short-clipped grass and tall alfalfa)

b. to develop global maps that provide the possible annual error in ET_0 estimations using the standard P-T and H-S evapotranspiration methods in comparison to the ASCE method for the short reference crop and the possible annual error in solar radiation estimations using the temperature-based H-S radiation formula (this attempt will allow us to identify the optimum locations for the application of the standard H-S and P-T evapotranspiration formulas based on their proximity to the results of ASCE for the short reference crop)

c. to develop global maps of readjusted annual coefficients for the H-S and P-T evapotranspiration methods for both short and tall reference crop based on a new method that estimates partial weighted averages of the monthly coefficients (the same procedure was also followed for the coefficients of the H-S radiation formula)

d. to validate the results of the readjusted P-T and H-S coefficients using data from meteorological stations from different locations with different climatic conditions and

e. to compare the predictive ability of the readjusted P-T and H-S coefficients for the short reference crop evapotranspiration with the respective predictions obtained from other models that have low data requirements.

The analysis and the produced datasets of this study were based on mean monthly climatic data of 0.5° resolution for the period 1950–2000. The final datasets of revised H-S and P-T coefficients will provide a global overview of the variation in their values and a common base for comparing the values of different regions since they are calibrated using common datasets and using the same technique. The produced global datasets of this study can support estimations of ET_0 and solar radiation for locations where climatic data are limited, and it can support studies which require such estimations on larger scales (e.g. country, continent, world).

2 Data and methods

2.1 Global climatic data

The analysis presented in this study was based on global climatic data obtained from the following databases:

- The database of Hijmans et al. (2005) provides mean monthly values for the parameters of precipitation and maximum, minimum and mean temperature at 30 arcsec spatial resolution. The data are provided as grids of mean monthly values of the period 1950–2000 (http://www.worldclim.org/). The database also includes a revised digital elevation model (DEM) of GTOPO30 (global topographic at 30 arcsec) based on SRTM (Shuttle Radar Topographic Mission), which was used for the estimation of atmospheric pressure. The DEM was also used as a base to calculate the distance from the coastlines in raster format at 30 arcsec spatial resolution based on the Euclidean distance.

- The database of Sheffield et al. (2006) provides monthly values of parameters such as wind speed at the height of 10 m above the ground surface, solar radiation, specific humidity, precipitation and temperature for the period 1948–2006 at 0.5° spatial resolution. The data are available in the form of netcdf files of monthly values of each year for the period 1948–2006 (http://hydrology.princeton.edu/data.pgf.php).

- The database of Peel et al. (2007) provides the revised global Köppen–Geiger climate map. The data are provided in raster form with 0.1 degree spatial resolution. The climate map was developed using the GHCN (Global Historical Climatology Network) version 2.0 dataset (Peterson and Vose, 1997), which includes precipitation data from 12 396 stations and temperature data from 4844 station data for the periods 1909–1991 and 1923–1993, respectively. The Köppen–Geiger map was used to obtain the climatic type of the meteorological stations used in the validation dataset.

In this study, the ET_0 is estimated combining the databases of Hijmans et al. (2005) and Sheffield et al. (2006), as follows: (a) mean monthly values of maximum, minimum and mean temperature and precipitation were obtained from Hijmans et al. (2005), while (b) wind speed, specific humidity and incident solar radiation were obtained from the Sheffield et al. (2006) database. The specific humidity was converted to actual vapour pressure using the equation given by Peixoto and Oort (1996). The final results and analysis presented in this study are based on the coarser 0.5° resolution.

All the calculations presented in the next sections were performed in an ArcGIS 9.3 ESRI environment in a WGS84 ellipsoid coordinate system. For area coverage calculations or for estimations of mean global values of various parameters, coordinate system conversions were performed from WGS84 to projected Cylindrical Equal Area system (any percentage global coverage calculations and the derivation of mean global values of various parameters are performed without considering Antarctica).

2.2 Methods

2.2.1 The ASCE standardized reference evapotranspiration method

The estimation of ET_o using the ASCE method is performed by the following equation (Allen et al., 2005):

$$ET_o = \frac{0.408\Delta(R_n - G) + \frac{\gamma u_2 (e_s - e_a) C_n}{(T_{mean} + 273.16)}}{\Delta + \gamma(1 + C_d u_2)}, \quad (1)$$

where ET_o is the reference crop evapotranspiration (mm d^{-1}), R_n is the net solar radiation at the crop surface (MJ m^{-2} d^{-1}), u_2 is the wind speed at 2 m height above the soil surface (m s^{-1}), T_{mean} is the mean daily air temperature (°C), G is the soil heat flux density at the soil surface (MJ m^{-2} d^{-1}), e_s is the saturation vapour pressure (kPa), e_a is the actual vapour pressure (kPa), Δ is the slope of the saturation vapour pressure–temperature curve (kPa °C^{-1}), γ is the psychrometric constant (kPa °C^{-1}) and C_n and C_d are constants, which vary according to the time step and the reference crop type and describe the bulk surface resistance and aerodynamic roughness. The short reference crop (ASCE-short) corresponds to clipped grass of 12 cm height and surface resistance of 70 s m^{-1}, where the constants C_n and C_d have the values 900 and 0.34, respectively. The tall reference crop (ASCE-tall) corresponds to full-cover alfalfa of 50 cm height and surface resistance of 45 s m^{-1}, where the constants C_n and C_d have the values 1600 and 0.38, respectively (Allen et al., 2005). The use of Eq. (1) at a daily or monthly step for the short reference crop is equivalent to the FAO-56 method (Allen et al., 1998).

2.2.2 The Priestley–Taylor method

The calculation of Priestley–Taylor (P-T) method is performed by the following equation (Priestley and Taylor, 1972):

$$ET_o = a_{pt} \frac{\Delta}{\lambda(\Delta + \gamma)} (R_n - G), \quad (2)$$

where ET_o is the potential evapotranspiration (mm d^{-1}), R_n is the net solar radiation (MJ m^{-2} d^{-1}), G is the soil heat flux density at the soil surface (MJ m^{-2} d^{-1}), Δ is the slope of the saturation vapour pressure–temperature curve (kPa °C^{-1}), γ is the psychrometric constant (kPa °C^{-1}), λ is the latent heat of vaporization (MJ kg^{-1}) and a_{pt} is the P-T advection coefficient. The value of λ was considered equal to 2.45 MJ kg^{-1} (Allen et al., 1998) (this value is also constant in Eq. (1) and appears as $1/\lambda = 0.408$). Equation (1) strictly refers to the reference crop evapotranspiration (i.e. short or tall crop), whereas Eq. (2) has been used for the calculation of evapotranspiration under non-limiting water conditions of the short reference crop, bare soil or open-water surface and for this reason is also called potential evapotranspiration, which is a more general term. Equation (2) is applied in this study as a reference crop evapotranspiration method and for this reason is compared with Eq. (1) for the short reference crop using the standard mean global value 1.26 for the factor a_{pt}.

2.2.3 The Hargreaves–Samani method

The Hargreaves–Samani (H-S) method (Hargreaves and Samani, 1982, 1985) for ET_o includes an internal function, which estimates the incoming shortwave solar radiation R_s (MJ m^{-2} day^{-1}) as follows:

$$R_s = K_{RS} \cdot R_a \cdot (TD)^{0.5}, \quad (3)$$

where K_{RS} is the adjustment coefficient for the H-S radiation formula (°$C^{-0.5}$), R_a is the extraterrestrial radiation (MJ m^{-2} day^{-1}), and TD is the temperature difference between maximum and minimum daily temperature (°C). According to Allen et al. (1998), the empirical K_{RS} coefficient differs for "interior" or "coastal" regions: (a) $K_{RS} = 0.16$ for interior locations, where land mass dominates and air masses are not strongly influenced by a large water body, and (b) $K_{RS} = 0.19$ for coastal locations, situated on or adjacent to the coast of a large land mass and where air masses are influenced by a nearby water body. For general use of Eq. (3), a mean global value of $K_{RS} = 0.17$ has been adopted in this study. R_a and R_s divided by λ change units from MJ m^{-2} day^{-1} to mm day^{-1} as it is required in the next equation of ET_o. The formula for estimating ET_o by the H-S method is given by the following equation (Hargreaves and Samani, 1982, 1985):

$$ET_o = 0.0135(T_{mean} + 17.8)\frac{R_s}{\lambda} = 0.0135(T_{mean} + 17.8) K_{RS} \cdot \frac{R_a}{\lambda} \cdot (TD)^{0.5}. \quad (4a)$$

Considering Eq. (4a), the K_{RS} and the exponent 0.5 are adjustment factors of the radiation formula (Eq. 3), while 0.0135 and 17.8 are adjustment factors of the ET_o formula, leading to a total amount of four empirical coefficients. Using the mean global value of $K_{RS} = 0.17$, Eq. (4a) is simplified according to the following (Allen et al., 1998):

$$ET_o = c_{hs2} \cdot (T_{mean} + 17.8) \cdot \frac{R_a}{\lambda} \cdot (TD)^{0.5} = 0.0023(T_{mean} + 17.8) \cdot \frac{R_a}{\lambda} \cdot (TD)^{0.5}, \quad (4b)$$

where in both Eqs. (4a) and (4b), the ET_o is the potential evapotranspiration (mm d^{-1}), R_a is the extraterrestrial radiation (MJ m^{-2} d^{-1}), λ is the latent heat of vaporization (MJ kg^{-1}) and T_{mean} is the mean daily temperature (°C). Eq. (4b) is applied in this study as a reference crop evapotranspiration method and for this reason is compared with Eq. (1) for the short reference crop. For the case of $T_{mean} < -17.8$ °C, the term of $(T_{mean} + 17.8)$ was set to 0, which is necessary for a global application (Weiß and Menzel, 2008).

In further steps of analysis, the coefficient 0.0135 (Eq. 4a) is represented as c_{hs1}, while the coefficient 0.0023 is represented as c_{hs2}, which is equal to $c_{hs2} = c_{hs1} \cdot K_{RS}$.

In order to reduce the errors of the aforementioned methods in the high latitudes and altitudes (polar and alpine environments) where negative temperatures exist, a filter was used in all methods to set mean monthly $ET_o = 0$ when mean monthly T_{max} is ≤ 0 (conditions of extreme frost).

2.2.4 Steps of analysis

Step 1: comparative analysis between the standard ET_o formulas of ASCE, P-T and H-S and error analysis of H-S radiation formula

The first step of the analysis includes the estimation of mean monthly and mean annual ET_o using the ASCE method (Eq. 1) for the two reference crops (short and tall), the standard P-T method (Eq. 2) with $a_{pt} = 1.26$ and the standard H-S method according to Eq. (4b) with $c_{hs2} = 0.0023$. The difference between the ET_o methods will be captured using as a base the mean annual and the mean monthly ET_o values of ASCE-short.

In the case of mean annual ET_o, the analysis is based on the percentage of mean annual difference (MAD %) of each method M versus the mean annual ET_o of ASCE-short, which is given by

$$\mathrm{MAD}\%_{(M)} = 100\left[\mathrm{YET}_{\mathrm{o}(M)} - \mathrm{YET}_{\mathrm{os}}\right]/(\mathrm{YET}_{\mathrm{os}}), \quad (5)$$

where YET_{os} is the mean annual ET_o of ASCE-short method, $YET_{o(M)}$ is the mean annual ET_o of the M method (M can be either the ASCE-tall method or the standard P-T method or the standard H-S method). The MAD % for ASCE-tall was estimated in order to assess the effects of reference crop type at different climatic environments on the annual estimations of ET_o. The MAD % of the P-T and H-S methods was used to investigate the strength of the two standard methods to approximate the annual ET_o of ASCE-short. Positive values of MAD % indicate overestimation of the mean annual ET_o values using the M method in comparison to the ASCE-short method, while negative values indicate underestimation. Furthermore, the difference between the absolute MAD % values (DMAD) of the standard P-T (with $a_{pt} = 1.26$) and H-S (with $c_{hs2} = 0.0023$) methods was estimated in order to assess which of the two methods is more appropriate to be used locally, based on its proximity to the ASCE-short method. The DMAD is estimated as follows:

$$\mathrm{DMAD} = |\mathrm{MAD}\,\%_{(\mathrm{H\text{-}S})}| - |\mathrm{MAD}\,\%_{(\mathrm{P\text{-}T})}|, \quad (6)$$

where positive DMAD values indicate better performance of standard P-T, while negative values indicate better performance of the standard H-S method in a region. Regions that showed DMAD values between −1 and +1 were considered transitional zones where both methods showed approximately the same proximity to the annual ASCE-short estimations.

In the case of mean monthly ET_o, the coefficient of determination R^2 and the root mean square difference (RMSD; equivalent to the root mean square error, RMSE) (Droogers and Allen, 2002) were used to compare the mean monthly values of ASCE-tall, standard P-T ($a_{pt} = 1.26$) and standard H-S ($c_{hs2} = 0.0023$) methods with the respective values of the ASCE-short method.

The procedures of MAD %, R^2 and RMSD were also used to analyse the mean annual and mean monthly estimations of R_s by the standard solar radiation formula of H-S (Eq. 3 with $K_{RS} = 0.17$) versus the given R_s values of Sheffield et al. (2006).

Step 2: readjustment of annual P-T and H-S coefficients for both reference crops

For the case of P-T, the readjustment of the mean monthly a_{pt} coefficient was performed directly for each location by solving for a_{pt} after equating Eq. (1) with Eq. (2) of each month. A filter was used in order to set a_{pt} equal to 0 when Eq. (1) and/or Eq. (2) without a_{pt} were equal to 0. In this case, the a_{pt} changes its physical meaning in order to indicate that mean monthly ET_o approximates to 0. By carrying out the above procedures for both short and tall reference crop, 12 images of mean monthly readjusted a_{pt} coefficients were produced for each reference crop.

For the case of the H-S method, the readjustment of the coefficients was performed in two stages. In the first stage, the readjustment was performed in the radiation formula (Eq. 3) only for the K_{RS} coefficient, while the exponent 0.5 (square root) of the DT remained the same. The mean monthly K_{RS} was estimated using the values of solar radiation R_s given by Sheffield et al. (2006). In the second stage, the readjustment was performed in the evapotranspiration formula (Eq. 4b) for the coefficient of c_{hs2} using as a base the ASCE method for both reference crops, while the coefficients of 17.8 and 0.5 remained the same. In this way the readjusted values of c_{hs2} and K_{RS} would also provide readjusted values of the c_{hs1} since $c_{hs1} = c_{hs2}/K_{RS}$. A similar filter to set $c_{hs2} = 0$ as in the case of a_{pt} was used, when Eq. (1) and/or Eq. (4b) without c_{hs2} were equal to 0. Following the above procedures, 12 images of mean monthly readjusted c_{hs2} coefficients for each reference crop (short and tall) and 12 K_{RS} images were produced.

The new mean monthly a_{pt}, c_{hs2} and K_{RS} coefficients were used to build respective mean annual coefficients. The robustness of mean annual coefficients are strongly related to their ability to capture the values of the dependent variable better (i.e. ET_o and R_s), especially in the months that present larger values. For this reason, weighted annual averages of mean monthly a_{pt}, c_{hs2} and K_{RS} coefficients were estimated. Under cold conditions, the monthly coefficients may present unrealistic values that significantly affect the weighted averages. In order to solve this problem, threshold values for the mean monthly dependent variables (i.e. ET_o and R_s) were set before their inclusion in the weighted average estimations. Preliminary analysis for the readjustment of a_{pt} and c_{hs2} coefficients (based on the values of ASCE-short) showed that when the mean monthly ET_o values of ASCE-short, H-S and P-T were below $45\,mm\,month^{-1}$ ($\sim 1.5\,mm\,d^{-1}$), unrealistic mean monthly values of a_{pt} and c_{hs2} coefficients appeared. The values considered unrealistic were those that were at least 1 order of magnitude larger or smaller than the standard values of $a_{pt} = 1.26$ and $c_{hs2} = 0.0023$. Taking into account the above, the following procedure was performed in order to obtain partial weighted annual averages (after excluding months with $ET_o \leq 45\,mm\,month^{-1}$) of mean monthly a_{pt} and c_{hs2} coefficients for the short reference crop based on the ASCE-short method:

$$\text{when } ET_{os\,i} > 45\,mm\,month^{-1},\ Fr_i = 1 \text{ else} = 0, \tag{7a}$$

and

$$\text{when } ET_{oi(M)} > 45\,mm\,month^{-1},\ Fm_i = 1 \text{ else} = 0. \tag{7b}$$

$$ET_{os\ adj.i} = ET_{os\ i} \cdot Fr_i \cdot Fm_i \tag{7c}$$

$$YET_{os\ adj.} = \sum_{i=1}^{12} \left(ET_{os\ adj.i}\right) \tag{7d}$$

$$C = \sum_{i=1}^{12} \left(\frac{ET_{os\ adj.i}}{YET_{os\ adj.}} \cdot C_i\right), \tag{7e}$$

where $ET_{os\,i}$ is the mean monthly value of ET_o ($mm\,month^{-1}$) obtained from the ASCE-short method, $ET_{oi(M)}$ is the mean monthly value of ET_o ($mm\,month^{-1}$) obtained from the M method (M is either P-T or H-S), Fr_i is the filter function of the reference method with values 0 or 1, Fm_i is the filter function of the M method with values 0 or 1, $ET_{os\,adj.\,i}$ is the adjusted mean monthly value of ET_o from the ASCE-short method, which becomes 0 when Fr_i or Fm_i is 0, $YET_{os\,adj.}$ is the sum of the monthly adjusted $ET_{os\,adj.i}$ values, C_i is the mean monthly coefficient of the M method (i.e. a_{pt} or c_{hs2}) calibrated based on the ASCE-short method (results from the previous step of the analysis), C is the partial weighted average (PWA) of the mean monthly coefficients of the M method (i.e. a_{pt} or c_{hs2}) for the short reference crop, and i is month.

For estimating the PWA of mean monthly a_{pt} and c_{hs2} for the tall reference crop, the same procedure as in Eq. (7) was followed using the mean monthly values of ET_o from ASCE-tall to estimate the Fr_i values in Eq. (7a), while the adjusted values of ET_o ASCE-tall were also used in Eqs. (7c), (7d) and (7e). For C_i values in Eq. (7e), the estimated mean monthly values of a_{pt} or c_{hs2} based on the ASCE-tall method were used. Even though the mean monthly values of ASCE-tall are generally higher than those from ASCE-short, the threshold of $45\,mm\,month^{-1}$ in Eqs. (7a) and (7b) remained the same since it was observed that the difference between ASCE-short and ASCE-tall is very small when $ET_{os\,i}$ falls below $\sim 50\,mm\,month^{-1}$.

A similar procedure (using the set of Eqs. 7) was also followed to obtain the PWA of mean monthly K_{RS} of the H-S method for R_s estimations. The Fr_i values in Eq. (7a) were estimated using as reference the mean monthly R_s values of Sheffield et al. (2006), which were also used after adjustment in Eqs. (7c)–(7e). The Fm_i values in Eq. (7b) were estimated using the respective R_s values of the standard H-S with $K_{RS} = 0.17$. For C_i values in Eq. (7e), the mean monthly values of K_{RS} calibrated based on the R_s values of Sheffield et al. (2006) were used. The threshold used for adjusting R_s values in Eqs. (7a) and (7b) was set a equal to $3.61\,MJ\,m^{-2}\,d^{-1}$ ($\sim 110\,MJ\,m^2\,month^{-1}$), which is equivalent to $45\,mm\,month^{-1}$ (conversion of $mm\,month^{-1}$ to $MJ\,m^{-2}\,month^{-1}$ was performed after multiplying with $\lambda = 2.45\,MJ\,kg^{-1}$). The threshold for R_s adjustment was tested before its use, and it was found to work satisfactorily excluding unrealistic monthly values of K_{RS}. The values that were considered unrealistic were those that were at least 1 order of magnitude larger or smaller than the standard value of $K_{RS} = 0.17$.

Step 3: use of stations for the validation of the PWA coefficients of the P-T and H-S methods and comparisons with other models of reduced parameters

Stations from two databases (the California Irrigation Management System CIMIS database, http://www.cimis.water.ca.gov, and the Australian Government Bureau of Meteorology AGBM database, http://www.bom.gov.au/) were used in this study in order to validate the PWA coefficients of the P-T and H-S methods. The first database includes stations from California (USA), and it was selected for the following reasons: (a) it has been used as a basis for the development of the Hargreaves–Samani method (Hargreaves and Samani, 1985; Hargreaves and Allen, 2002) and the CIMIS method (Snyder and Pruitt, 1985, 1992), and (b) it provides a dense and descriptive network of stations for a specific region that combines coastal, plain, mountain and desert environments (Table 1, Fig. 1a). The second database includes stations from Australia, and it was selected because the stations network covers a large territory with a large variety of climate classes (Table 1, Fig. 1b) but also because the Priestley–Taylor method has been calibrated for locations of eastern Australia (Priestley and Taylor, 1972). The selection of stations from the AGBM database was performed in such way as to cover all the possible existing Köppen climatic types and elevation ranges of the Australian continent (Table 1). In total, 140 stations were used; 60 stations were selected from CIMIS and 80 stations from the AGBM that have at least 15 years of observations (some stations that do not follow this rule were selected due to their special climate Köppen class or the high elevation of their location). Observations from the years after 2000 up to 2016 were included (if they were available) in the station data, in order to show that the new revised coefficients are applicable to recent periods.

Table 1. Meteorological stations from the USA (California; USA-CA) (CIMIS database) and Australia (AGBM database).

No.	Code	Station	Country	Elevation (m)	Lat (Dec. deg.)	Long (Dec. deg.)	Period	Köppen class[a]
CA-1	006	Davis	USA-CA	18	38.54	−121.78	Sep 1982–Aug 2016	Csa
CA-2	002	FivePoints	USA-CA	87	36.34	−120.11	Jun 1982–Aug 2016	BWk
CA-3	005	Shafter	USA-CA	110	35.53	−119.28	Jun 1982–Aug 2016	BSk
CA-4	007	Firebaugh/Telles	USA-CA	56	36.85	−120.59	Sep 1982–Aug 2016	BWk
CA-5	012	Durham	USA-CA	40	39.61	−121.82	Oct 1982–Aug 2016	Csa
CA-6	008	Gerber	USA-CA	76	40.04	−122.17	Sep 1982–Aug 2014	Csa
CA-7	015	Stratford	USA-CA	59	36.16	−119.85	Nov 1982–Aug 2016	BSk
CA-8	019	Castroville	USA-CA	3	36.77	−121.77	Nov 1982–Aug 2016	Csb
CA-9	021	Kettleman	USA-CA	104	35.87	−119.89	Nov 1982–Aug 2016	BWk
CA-10	027	Zamora	USA-CA	15	38.81	−121.91	Dec 1982–May 2006	Csa
CA-11	030	Nicolaus	USA-CA	10	38.87	−121.55	Jan 1983–Dec 2011	Csa
CA-12	032	Colusa	USA-CA	17	39.23	−122.02	Jan 1983–Aug 2016	Csa
CA-13	033	Visalia	USA-CA	107	36.30	−119.22	Jan 1983–Feb 2007	BSk
CA-14	035	Bishop	USA-CA	1271	37.36	−118.41	Feb 1983–Aug 2016	BSk
CA-15	039	Parlier	USA-CA	103	36.60	−119.50	May 1983–Aug 2016	BSk
CA-16	041	Calipatria/Mulberry	USA-CA	−34	33.04	−115.42	Jul 1983–Aug 2016	BWh
CA-17	043	McArthur	USA-CA	1009	41.06	−121.46	Dec 1983–Aug 2016	Csb
CA-18	044	U.C.Riverside	USA-CA	311	33.96	−117.34	Jun 1985–Aug 2016	BSk
CA-19	047	Brentwood	USA-CA	14	37.93	−121.66	Nov 1985–Aug 2016	Csb
CA-20	049	Oceanside	USA-CA	15	33.26	−117.32	Mar 1986–Oct 2003	BSk
CA-21	054	Blackwells Corner	USA-CA	215	35.65	−119.96	Oct 1986–Aug 2016	BWk
CA-22	056	Los Banos	USA-CA	29	37.10	−120.75	Jun 1988–Aug 2016	BSk
CA-23	061	Orland	USA-CA	60	39.69	−122.15	May 1987–May 2010	Csa
CA-24	062	Temecula	USA-CA	433	33.49	−117.23	Nov 1986–Aug 2016	BSk
CA-25	064	Santa Ynez	USA-CA	149	34.58	−120.08	Nov 1986–Aug 2016	Csb
CA-26	068	Seeley	USA-CA	12	32.76	−115.73	May 1987–Aug 2016	BWh
CA-27	070	Manteca	USA-CA	10	37.83	−121.22	Nov 1987–Aug 2016	BSk
CA-28	071	Modesto	USA-CA	11	37.65	−121.19	Jul 1987–Aug 2016	BSk
CA-29	077	Oakville	USA-CA	58	38.43	−122.41	Jan 1989–Aug 2016	Csb
CA-30	075	Irvine	USA-CA	125	33.69	−117.72	Oct 1987–Aug 2016	BSk
CA-31	078	Pomona	USA-CA	223	34.06	−117.81	Mar 1989–Aug 2016	Csa
CA-32	080	Fresno State	USA-CA	103	36.82	−119.74	Oct 1988–Aug 2016	BSk
CA-33	083	Santa Rosa	USA-CA	24	38.40	−122.80	Jan 1990–Aug 2016	Csb
CA-34	084	Browns Valley	USA-CA	287	39.25	−121.32	Apr 1989–Aug 2016	Csa
CA-35	085	Hopland F.S.	USA-CA	354	39.01	−123.08	Sep 1989–Apr 2016	Csa
CA-36	086	Lindcove	USA-CA	146	36.36	−119.06	May 1989–Aug 2016	Csa
CA-37	087	Meloland	USA-CA	−15	32.81	−115.45	Dec 1989–Aug 2016	BWh
CA-38	088	Cuyama	USA-CA	698	34.94	−119.67	May 1989–Aug 2016	BSk
CA-39	091	Tulelake F.S.	USA-CA	1230	41.96	−121.47	Mar 1989–Aug 2016	Dsb
CA-40	092	Kesterson	USA-CA	23	37.23	−120.88	Oct 1989–Aug 2016	BSk
CA-41	094	Goletta foothills	USA-CA	195	34.47	−119.87	Jul 1989–Jul 2016	Csb
CA-42	099	Santa Monica	USA-CA	104	34.04	−118.48	Dec 1992–Aug 2016	Csb
CA-43	103	Windsor	USA-CA	26	38.53	−122.83	Dec 1990–Aug 2016	Csb
CA-44	104	De Laveaga	USA-CA	91	37.00	−122.00	Sep 1990–Aug 2016	Csb
CA-45	105	Westlands	USA-CA	58	36.63	−120.38	Apr 1992–Aug 2016	BWk
CA-46	106	Sanel Valley	USA-CA	160	38.98	−123.09	Feb 1991–Aug 2016	Csa
CA-47	57	Buntingville	USA-CA	1221	40.29	−120.43	June 1986–Sep 2016	Dsb
CA-48	90	Alturas	USA-CA	1343	41.44	−120.48	Apr 1989–Sep 2016	Dsb
CA-49	151	Ripley	USA-CA	77	33.53	−114.63	Dec 1998–Sep 2016	BWh
CA-50	183	Owens Lake North	USA-CA	1123	36.49	−117.92	Dec 2002–Sep 2016	BWk
CA-51	147	Otay Lake	USA-CA	177	32.63	−116.94	Apr 1999–Sep 2016	Csb
CA-52	175	Palo Verde II	USA-CA	70	33.38	−114.72	Jan 2001–Sep 2016	BWh
CA-53	135	Blynthe NE	USA-CA	84	33.66	−114.56	Jan 1997–Sep 2016	BWh
CA-54	155	Bryte	USA-CA	12	38.60	−121.54	Dec 1998–Sep 2016	Csa
CA-55	159	Monrovia	USA-CA	181	34.15	−117.99	Oct 1999–Sep 2016	Csa
CA-56	161	Patterson	USA-CA	56	37.44	−121.14	Aug 1999–Sep 2016	BSk

Table 1. Continued.

No.	Code	Station	Country	Elevation (m)	Lat (Dec. deg.)	Long (Dec. deg.)	Period	Köppen class[a]
CA-57	174	Long Beach	USA-CA	5	33.80	−118.09	Sep 2000–Sep 2016	Csb
CA-58	173	Torrey Pines	USA-CA	102	32.90	−117.25	Nov 2000–Sep 2016	Csa
CA-59	150	Miramar	USA-CA	136	32.89	−117.14	Apr 1999–Sep 2016	Csa
CA-60	153	Escondido SPV	USA-CA	119	33.08	−116.98	Feb 1999–Sep 2016	Csb
A-1	32040	Townsville Aero	Australia	4	−19.25	146.77	(1940/1996-2016)[b]	Aw
A-2	33307	Woolshed	Australia	556	−19.42	146.54	(1990/2003–2016)	Aw
A-3	2056	Kununurra Aero	Australia	44	−15.78	128.71	(1971/1990–2016)	BSh
A-4	35264	Emerald	Australia	189	−23.57	148.18	(1990/1998–2016)	BSh
A-5	24024	Loxton R.C.	Australia	30	−34.44	140.6	(1984/1998–2016)	BSk
A-6	74037	Yanco AG.I.	Australia	164	−34.62	146.43	(1957/1999–2016)	BSk
A-7	74258	Deniliquin Airp.AWS	Australia	94	−35.56	144.95	(1990/2003–2016)	BSk
A-8	75041	Griffith Airp.AWS	Australia	134	−34.25	146.07	(1958/1990–2016)	BSk
A-9	76031	Mildura Airp.	Australia	50	−34.24	142.09	(1946/1993–2016)	BSk
A-10	24048	Renmark Apt.1	Australia	32	−34.2	140.68	(1990/2003–2016)	BWk
A-11	40082	University of QLD G.	Australia	89	−27.54	152.34	(1897/1990–2016)	Cfa
A-12	40922	Kingaroy Airp.	Australia	434	−26.57	151.84	(1990/2003–2016)	Cfa
A-13	41359	Oakey Aero	Australia	406	−27.4	151.74	(1970/1996–2016)	Cfa
A-14	41522	Dalby Airp.	Australia	344	−27.16	151.26	(1990/2006–2016)	Cfa
A-15	41525	Warwick	Australia	475	−28.21	152.1	(1990/2000–2016)	Cfa
A-16	41529	Toowoomba Airp.	Australia	641	−27.54	151.91	(1990/1997–2016)	Cfa
A-17	80091	Kyabram	Australia	105	−36.34	145.06	(1964/1990–2016)	Cfa
A-18	81049	Tatura I.S.A.	Australia	114	−36.44	145.27	(1942/1990–2016)	Cfa
A-19	81124	Yarrawonga	Australia	129	−36.03	146.03	(1990/2003–2016)	Cfa
A-20	81125	Shepparton Airp.	Australia	114	−36.43	145.39	(1990/1996–2016)	Cfa
A-21	41175	Applethorpe	Australia	872	−28.62	151.95	(1966/2006–2016)	Cfb
A-22	81123	Bendigo Airp.	Australia	208	−36.74	144.33	(1990/2004–2016)	Cfb
A-23	85072	East Sale Airp.	Australia	5	−38.12	147.13	(1943/1996–2016)	Cfb
A-24	85279	Bairnsdale Airp.	Australia	49	−37.88	147.57	(1942/2003–2016)	Cfb
A-25	85280	Morwell L.V.Airp.	Australia	56	−38.21	146.47	(1984/1999–2016)	Cfb
A-26	85296	Mount Moornapa	Australia	480	−37.75	147.14	(1990/2003–2016)	Cfb
A-27	90035	Colac	Australia	261	−38.23	143.79	(1990/2003–2016)	Cfb
A-28	9538	Dwellingup	Australia	267	−32.71	116.06	(1934/1990–2016)	Csb
A-29	9617	Bridgetown	Australia	179	−33.95	116.13	(1990/2003–2016)	Csb
A-30	23373	Nuriootpa Pirsa	Australia	275	−34.48	139.01	(1990/1996–2016)	Csb
A-31	26021	Mount Gambier Aero	Australia	63	−37.75	140.77	(1942/1994–2016)	Csb
A-32	26091	Coonawarra	Australia	57	−37.29	140.83	(1985/1990–2016)	Csb
A-33	66062	Sydney (Obs.Hill)	Australia	39	−33.86	151.21	(1858/1990–2016)	Cfb
A-34	33002	Ayr DPI Res.St.	Australia	17	−19.62	147.38	(1951/1994–2016)	Cwa
A-35	7176	Newman Aero	Australia	524	−23.42	119.8	(1971/2003–2016)	BWh
A-36	13017	Giles	Australia	598	−25.03	128.3	(1956/1990–2016)	BWh
A-37	11052	Forrest	Australia	159	−30.85	128.11	(1990/2003–2016)	BWh
A-38	11003	Eucla	Australia	93	−31.68	128.9	(1876/1995–2016)	BSk
A-39	12071	Salmon Gums	Australia	249	−32.99	121.62	(1932/2003–2016)	BSk
A-40	7045	Meekatharra Airp.	Australia	517	−26.61	118.54	(1944/1992–2016)	BWh
A-41	1025	Doongan	Australia	385	−15.38	126.31	(1988/1990–2016)	Aw
A-42	2012	Halls Creek Airp.	Australia	422	−18.23	127.66	(1944/1996–2016)	BSh
A-43	13015	Carnegie	Australia	448	−25.8	122.98	(1942/1990–2016)	BWh
A-44	3080	Curtin Aero	Australia	78	−17.58	123.83	(1990/2003–2016)	BSh
A-45	6022	Gascoyne Junction	Australia	144	−25.05	115.21	(1907/1990–2016)	BWh
A-46	9789	Esperance	Australia	25	−33.83	121.89	(1969/1990–2016)	Csb
A-47	91223	Marrawah	Australia	107	−40.91	144.71	(1971/1990–2016)	Cfb
A-48	18106	Nullarbor	Australia	64	−31.45	130.9	(1888/2006–2016)	BWk
A-49	16090	Coober Pedy Airp.	Australia	225	−29.03	134.72	(1990/2004–2016)	BWh
A-50	16085	Marla Police St.	Australia	323	−27.3	133.62	(1985/1990–2016)	BWh
A-51	13011	Warburton Airfield	Australia	459	−26.13	126.58	(1940/2003–2016)	BWh

Table 1. Continued.

No.	Code	Station	Country	Elevation (m)	Lat (Dec. deg.)	Long (Dec. deg.)	Period	Köppen class[a]
A-52	15528	Yuendumu	Australia	667	−22.26	131.8	(1952/1990–2016)	BWh
A-53	15666	Rabbit Flat	Australia	340	−20.18	130.01	(1990/1996–2016)	BWh
A-54	14829	Lajamanu Airp.	Australia	316	−18.33	130.64	(1952/1990–2016)	BSh
A-55	15135	Tennant Creek Airp.	Australia	376	−19.64	134.18	(1969/1992–2016)	BSh
A-56	37010	Camooweal Township	Australia	231	−19.92	138.12	(1891/2003–2016)	BWh
A-57	14707	Wollogorang	Australia	60	−17.21	137.95	(1967/1990–2016)	Aw
A-58	14938	Mango Farm	Australia	15	−13.74	130.68	(1980/1990–2016)	Aw
A-59	69134	Batemans Bay	Australia	11	−35.72	150.19	(1985/1991–2016)	Cfb
A-60	14198	Jabiru Airp.	Australia	27	−12.66	132.89	(1971/1990–2016)	Aw
A-61	28008	Lockhart River Airp.	Australia	19	−12.79	143.3	(1956/2001–2016)	Am
A-62	34084	Charters Towers Airp.	Australia	290	−20.05	146.27	(1990/1992–2016)	BSh
A-63	29038	Kowanyama Airp.	Australia	10	−15.48	141.75	(1912/1999–2016)	Aw
A-64	32078	Ingham Composite	Australia	12	−18.65	146.18	(1968/1990–2016)	Am
A-65	40854	Logan City W.T.P.	Australia	14	−27.68	153.19	(1990/1992–2016)	Cfa
A-66	8095	Mullewa	Australia	268	−28.54	115.51	(1896/1990–2016)	BSh
A-67	8251	Kalbarri	Australia	6	−27.71	114.17	(1970/1990–2016)	BSh
A-68	8225	Eneabba	Australia	100	−29.82	115.27	(1964/1990–2016)	Csa
A-69	7139	Paynes Find	Australia	339	−29.27	117.68	(1919/1990–2016)	BWh
A-70	10007	Bencubbin	Australia	359	−30.81	117.86	(1912/1990–2016)	BSh
A-71	10092	Merredin	Australia	315	−31.48	118.28	(1903/1990–2016)	BSk
A-72	12038	Kalgoorlie-Boulder Airp.	Australia	365	−30.78	121.45	(1939/1994–2016)	BSh
A-73	16098	Tarcoola Aero	Australia	123	−30.71	134.58	(1990/1999–2016)	BWh
A-74	18195	Minnipa Pirsa	Australia	165	−32.84	135.15	(1990/2003–2016)	BSk
A-75	46126	Tibooburra Airp.	Australia	176	−29.44	142.06	(1990/2003–2016)	BWh
A-76	48245	Boorke Airp. AWS	Australia	107	−30.04	145.95	(1990/2002–2016)	BSh
A-77	55325	Tamworth Airp. AWS	Australia	395	−31.07	150.84	(1990/2006–2016)	Cfa
A-78	38026	Birdsville Airp.	Australia	47	−25.9	139.35	(1990/2001–2016)	BWh
A-79	30161	Richmond Airp.	Australia	206	−20.7	143.12	(1990/2003–2016)	BSh
A-80	33013	Collinsville Airp.	Australia	196	−20.55	147.85	(1939/1990–2016)	BSh

[a] Köppen classification obtained from Peel et al. (2007). [b] In the case of Australian stations, the periods of observations vary between different climatic parameters; e.g. for the case (1939/1990–2016), the two dates separated by a slash show the starting date of the oldest and newest record of parameters used in calculations, while 2016 is the end date of the records.

In the case of CIMIS stations, the monthly data for all climatic parameters were obtained, including ET_o estimations using the CIMIS method (Snyder and Pruitt, 1985, 1992), but they required quality control before their use. Quality control signs are provided by the database for all climatic data, indicating extreme values, while possible errors are flagged but they are not automatically excluded. For this reason, the user should consider the signs in order to prepare a robust dataset. For this study, proper control was performed and very extreme or erroneous monthly values were excluded. Excluded values were less than 1 ‰ of the total values of all stations and all parameters. The final clean dataset was subjected to a secondary but indirect quality control through the comparison between the estimated mean monthly values of ET_o of the ASCE-short method (Eq. 1) using the clean climatic data of all USA-CA stations versus the respective mean monthly ET_o values given by the CIMIS database (linear regression result between mean monthly values for n obs. $= 12 \times 60 = 720$: $y = 0.994x - 1.07$ with $R^2 = 0.98$) (see Fig. S1 in the Supplement). Data cleaning was not followed in the case of Australian stations, since the AGBM database provides the mean monthly values of the climatic parameters for the total periods of observation and not for individual years. The general statistics of the mean monthly observed values of climatic parameters obtained from the 140 stations of California (USA) and Australia are given in Table S1 of the Supplement. A comparison of the parameters T_{max}, T_{min}, R_s, R_n, DE and u_2 between the rasters (0.5°) and the station data are provided in Fig. S2a, b, c, d, e and f of the Supplement.

The validation procedure was performed using the data of the stations in Table 1 by comparing the mean monthly values of ET_o derived by the P-T (Eq. 2) and H-S (Eq. 4b, c) methods with the standard a_{pt} and c_{hs2} coefficients and with the readjusted ones versus the ASCE method for the short reference crop (Eq. 1). The same procedure was also performed for the new a_{pt} and c_{hs2} coefficients for the tall reference crop and for the readjusted coefficient K_{RS} in the radiation formula of H-S (Eq. 3). For the case of the ASCE method for the short reference crop, additional models of re-

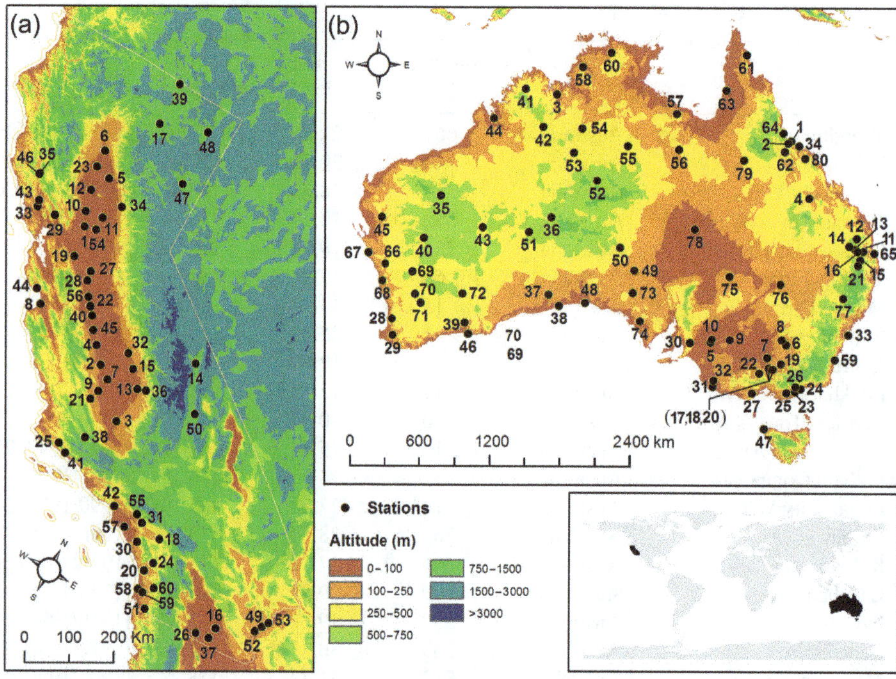

Figure 1. Position of stations **(a)** from California (USA) obtained from the CIMIS database and **(b)** from Australia obtained from AGBM database (the station numbers are those from Table 1 without the abbreviations CA- and A-).

duced parameters were used from the literature in order to perform comparisons with the standard and readjusted P-T and H-S models. The selection of these models was made in such way as to satisfy the following criteria/characteristics:

- The selected models have been calibrated by either using global data or a representative amount of data from California or Australia. Models that have been tested for California and Australia and showed good performance were also included.

- The selected models showed better performance in comparison to other models when tested using the validation datasets of Californian and Australian stations, but they also showed good performance with regard to other regions based on studies from the literature. It has to be mentioned that an extremely large amount of models were examined taking into account the modified H-S and P-T models obtained from literature that has already been cited in the introduction and the large lists of models presented in the work of Valipour (2015a, b, 2017) and Valipour et al. (2017). Strict modifications of the P-T and H-S models with fixed coefficients calibrated for local conditions were not used because they cannot adapt their coefficients to the large climatic variability in the validation dataset.

- The majority of the selected models require additional parameters in comparison to the P-T and H-S models. This criterion was used in order to compare the strength of the readjusted P-T and H-S coefficients versus such models.

Based on the aforementioned criteria, the following eight models were selected for comparisons with the standard and readjusted H-S and P-T models (Table 2):

- Two modified models of H-S by Droogers and Allen (2002), where the second one uses precipitation as an additional parameter. The models were calibrated using global data.

- Three models of reduced parameters given by Valiantzas (2013a, b, 2014), which were calibrated using 535 stations from Europe, Asia and Africa. The first model uses temperature and radiation data, while the other two use temperature, radiation and humidity data. The models have been tested for Californian (Valiantzas, 2013c) and Australian conditions (Ahooghalandari et al., 2017).

- Two models of reduced parameters by Ahooghalandari et al. (2016) calibrated/validated using stations from various locations in Australia. The models use temperature and relative humidity data.

- The Copais model of Alexandris et al. (2006) that uses temperature, radiation and humidity data. The model was calibrated/validated using data from Greece, California and Oregon (USA), while it has shown a very

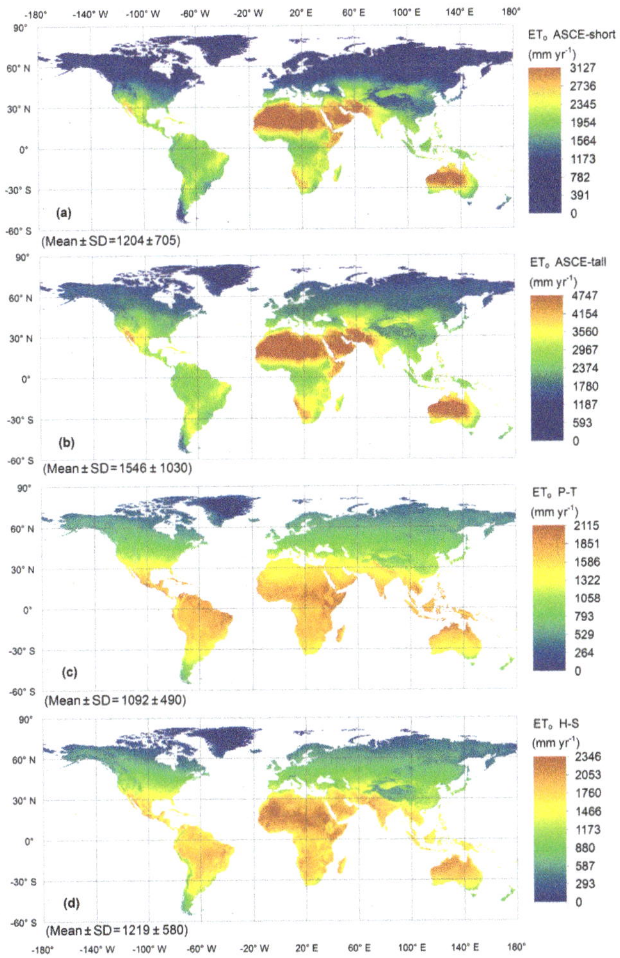

Figure 2. Mean annual values (mm year^{-1}) of ET$_o$ for the period 1950–2000 using **(a)** the ASCE-short method, **(b)** the ASCE-tall method, **(c)** the standard P-T method for $a_{pt} = 1.26$ and **(d)** the standard H-S method for $c_{hs2} = 0.0023$ (0.5° resolution maps; mean ± SD are estimated after conversion from WGS84 to Cylindrical Equal Area coordinate system).

good response to many other regions of the world including Australia (Ahooghalandari et al., 2017).

The following statistical criteria were used in the validation procedure: the coefficient of determination (R^2), the modified coefficient of determination (bR^2) based on $y = bx$ (Krause et al., 2005), mean absolute error (MAE), root mean square error (RMSE), percent bias (PBIAS %), the Nash–Sutcliffe efficiency (NSE; Nash and Sutcliffe, 1970), the index of agreement (d; Willmott, 1981) and the Kling–Gupta efficiency (KGE; Gupta et al., 2009). The criteria were calculated using the package {HydroGOF} in R language (Zambrano-Bigiarini, 2015; see the package manual for formulas).

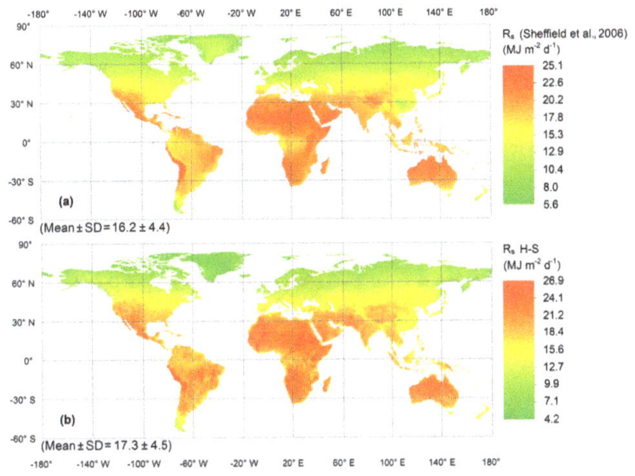

Figure 3. Mean annual values of R_s (MJ m^{-2} d^{-1}) for the period 1950–2000 **(a)** from the database of Sheffield et al. (2006) and **(b)** estimated using the standard H-S radiation formula for $K_{RS} = 0.17$ (Eq. 3) (0.5° resolution maps; mean ± SD are estimated after conversion from WGS84 to Cylindrical Equal Area coordinate system).

3 Results

3.1 Comparative analysis between the standard ET$_o$ formulas of ASCE, P-T and H-S and error analysis of the H-S radiation formula

The global maps of mean monthly ET$_o$ at 0.5° resolution for the period 1950–2000 using (a) the methods of ASCE (Eq. 1) for both reference crops (ASCE-short and ASCE-tall), (b) the standard P-T method (Eq. 2) for $a_{pt} = 1.26$ and (c) the standard H-S method (Eq. 4b) for $c_{hs2} = 0.0023$ were developed. The respective mean annual ET$_o$ maps are given in Fig. 2a, b, c and d. Similarly, the mean annual R_s values provided by Sheffield et al. (2006) and the corresponding R_s values estimated by the standard H-S radiation formula (Eq. 3 with $K_{RS} = 0.17$) are given in Fig. 3a and b, respectively.

The MAD % (Eq. 5) maps of the ASCE-tall, standard P-T and standard H-S methods versus ASCE-short are given in Fig. 4a, b and c, while the MAD % of the standard solar radiation formula of H-S versus the R_s values given by Sheffield et al. (2006) is also given in Fig. 4d. The percentage globe coverage (excluding Antarctica) for different classes of MAD % and the R^2 and RMSD based on respective comparisons of the mean monthly values of ET$_o$ and R_s methods are given in Table 3.

The case of MAD % between the ET$_o$ methods of ASCE-tall and ASCE-short (Fig. 4a and Table 3) indicates that there is 25.2 % map coverage in the MAD % class of ±10 % where the effects of reference crop type are significantly minimized (Table 3). These territories include the regions of tropical rainforests in Latin America, central Africa and Indonesia, regions of large mountain formations and ranges of high ele-

Table 2. Additional models of reduced parameters obtained from the international literature, which provide equivalent results to ET_o for the short reference crop.

Reference	Abbreviation	Formula	Climate data requirements*
Droogers and Allen (2002)	DRAL1 (Eq. 8)	$ET_o = 0.00102 R_a (T_{mean} + 16.8) \cdot (TD)^{0.5}$	T_{max}, T_{min}
Droogers and Allen (2002)	DRAL2 (Eq. 9)	$ET_o = 0.0005304 R_a (T_{mean} + 17.0) \cdot (TD - 0.0123P)^{0.76}$	T_{max}, T_{min}, P
Alexandris et al. (2006)	Copais (Eq. 10)	$ET_o = 0.057 + 0.227C_2 + 0.643C_1 + 0.0124C_1C_2$ $C_1 = 0.6416 - 0.00784RH + 0.372R_s - 0.00264R_sRH$ $C_2 = -0.0033 + 0.00812T_{mean} + 0.101R_s + 0.00584R_sT_{mean}$	T_{mean}, R_s, RH
Valiantzas (2013a, 2014)	VAL1 (Eq. 11)	$ET_o = 0.0393R_s\sqrt{T_{mean} + 9.5} - 0.19R_s^{0.6}\varphi^{0.15} + 0.0061(T_{mean} + 20)(1.12T_{mean} - T_{min} - 2)^{0.7}$	T_{mean}, T_{min}, R_s
Valiantzas (2013a, 2014)	VAL2 (Eq. 12)	$ET_o = 0.0393R_s\sqrt{T_{mean} + 9.5} - 0.19R_s^{0.6}\varphi^{0.15} + 0.078(T_{mean} + 20)\left(1 - \frac{RH}{100}\right)$	T_{mean}, R_s, RH
Valiantzas (2013b)	VAL3 (Eq. 13)	$ET_o = 0.0393R_s\sqrt{T_{mean} + 9.5} - 2.4\left(\frac{R_s}{R_a}\right)^2 + Cu(T_{mean} + 20)\left(1 - \frac{RH}{100}\right)$	T_{mean}, R_s, RH (Cu = 0.054 for RH > 65 % and Cu = 0.083 for RH ≤ 65 %)
Ahooghalandari et al. (2016)	AKJ1 (Eq. 14)	$ET_o = 0.252 \cdot 0.408R_a + 0.221T_{mean}\left(1 - \frac{RH}{100}\right)$	T_{mean}, RH
Ahooghalandari et al. (2016)	AKJ2 (Eq. 15)	$ET_o = 0.29 \cdot 0.408R_a + 0.15T_{max}\left(1 - \frac{RH}{100}\right)$	T_{max}, RH

* $T_{mean, max, min}$: mean, maximum and minimum temperature (°C); TD: difference between maximum and minimum temperature (°C); R_s: incident solar radiation (MJ m^{-2} d^{-1}); R_a: extraterrestrial solar radiation (MJ m^{-2} d^{-1}); RH: relative humidity (%); φ: absolute value of latitude (rads); P: precipitation (mm $month^{-1}$).

Table 3. The percent coverage[a] of MAD % classes based on mean annual values (according to Fig. 4), R^2 and RMSD based on comparisons of the mean monthly values of ET_o and R_s methods (comparisons based on 0.5° resolution maps).

MAD % range	ET_o[b](ASCE-tall) for $C_n = 1600$, $C_d = 0.38$ (Eq. 1)	ET_o[b](P-T) for $a_{pt} = 1.26$ (Eq. 2)	ET_o[b] (H-S) for $c_{hs2} = 0.0023$ (Eq. 4b)	R_s[c] (H-S) for $K_{RS} = 0.17$ (Eq. 3)
≤ −50 %	0.0 %[a]	0.8 %	0.0 %	1.0 %
−50 up to −25 %	0.0 %	14.8 %	5.2 %	2.2 %
−25 up to −10 %	0.0 %	10.8 %	15.4 %	7.1 %
−10 up to 10 %	25.2 %	21.3 %	24.8 %	55.3 %
10 up to 25 %	40.9 %	22.5 %	19.6 %	32.8 %
25 up to 50 %	33.6 %	21.9 %	29.2 %	1.6 %
> 50 %	0.3 %	7.9 %	5.8 %	0.0 %
R^2	0.98	0.77	0.89	0.92
RMSD	39.6[d]	36.0[d]	24.5[d]	2.4[e]

[a] The percent coverage was estimated after conversion from WGS84 ellipsoid to projected Cylindrical Equal Area coordinate system without considering Antarctica. [b] MAD % of the three ET_o methods is estimated versus ASCE-short. [c] MAD % of the standard solar radiation method of H-S is estimated versus the R_s data of Sheffield et al. (2006). [d] The unit of RMSD for ET_o is mm $month^{-1}$. [e] The unit of RMSD for R_s is MJ m^{-2} d^{-1}.

vation, and the regions of taigas and tundras of North America and Asia (Fig. 4a). The low values of the vapour pressure deficit are the main characteristic of these regions. By contrast, the largest differences between the two reference crops appear in arid and semi-arid environments due to the high values of the vapour pressure deficit. The high correlation $R^2 = 0.98$ (Table 3) between the mean monthly ET_o values of ASCE-tall and ASCE-short suggests that it is feasible to develop reliable regional monthly coefficients or regression models, which can convert the ET_o estimations from the short to the tall reference crop especially when the ET_o of the short reference crop is estimated with a method of reduced parameters (e.g. P-T or H-S) (a paradigm has been presented by Aschonitis et al., 2012).

Even though MAD %, R^2 and RMSD for the standard P-T and H-S methods (Fig. 4b and c; Table 3) indicate a better performance of the second one to approximate the ASCE-short on a global scale, both methods seem to be equally valuable because their proximity to ASCE-short is maximized at relatively different climatic regions. This is indi-

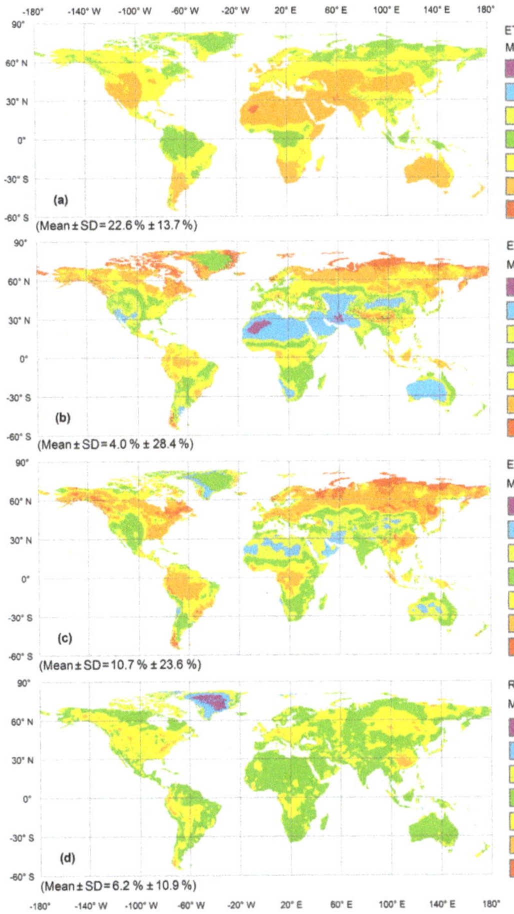

Figure 4. Mean annual difference percentage (MAD %) of ET_o between the ASCE-short and **(a)** the ASCE-tall method, **(b)** the standard P-T method for $a_{pt} = 1.26$, **(c)** the standard H-S method for $c_{hs2} = 0.0023$, and **(d)** MAD % between R_s values of Sheffield et al. (2006) and the standard solar radiation formula of H-S for $K_{RS} = 0.17$ (0.5° resolution maps; mean ± SD are estimated after conversion from WGS84 to Cylindrical Equal Area coordinate system).

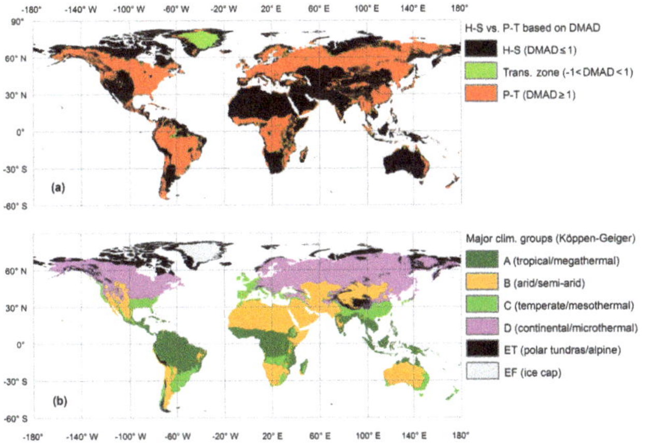

Figure 5. (a) Spatial extent of DMAD classes (P-T versus H-S) (0.5° resolution map) and **(b)** spatial extent of the major climatic groups of the Köppen–Geiger climate classification according to Peel et al. (2007).

cated by the difference between the absolute MAD % values (DMAD) (Eq. 6) of the P-T and H-S methods (Fig. 5a). The interpretation of Fig. 5a was performed using as a base the major climatic groups (CGs) of the Köppen–Geiger climate map in Peel et al. (2007) (Fig. 5b). The spatial extent of the major CGs of the Köppen–Geiger climate classification (without Antarctica) and the respective extent of DMAD classes (P-T versus H-S) in the CGs is given in Table 4. According to Table 4 and Fig. 5a and b, the H-S method prevails in regions of the B group (arid and semi-arid) and the E group (polar/alpine/tundras), while the P-T method prevails in the regions of the A group (tropical/megathermal), the C group (temperate/mesothermal climates) and the D group (continental/microthermal). Even though the P-T method seems to be more powerful in more climatic zones, in reality the H-S method prevails in 49.3 % of the regions, while P-T does so in 46.6 % (the remaining proportion of 4.1 % mainly corresponds to inner Greenland and very high mountain areas with annual $ET_o = 0$ or to regions where both methods gave equal results). The better suitability of the standard H-S method to drier environments and the respective suitability of the standard P-T method in more humid environments can be explained by the fact that the standard coefficient of H-S was calibrated for Californian conditions (semi-arid/arid environments) (Hargreaves and Samani, 1985), while the standard coefficient of P-T was calibrated taking into account more humid environments (Priestley and Taylor, 1972).

The spatial variation in MAD % for the case of R_s estimations using the standard solar radiation formula of H-S for $K_{RS} = 0.17$ (Eq. 3) versus the mean annual R_s values of Sheffield et al. (2006) is given in Fig. 4d. It is indicative that 55.3 % of the territories are included in the MAD % range ±10 %, while 95.2 % is included in the range between ±25 % (Table 3). Significant deviations of R_s estimations using the standard H-S method appear mostly in the region of Greenland (Fig. 4d). The values of R^2 and RMSD (Table 3) indicate a good performance of the method in the case of monthly estimations. The overall results indicate that the use of the standard value $K_{RS} = 0.17$ can provide satisfactory indirect estimations of R_s for the greatest part of the world only by the use of temperature data.

3.2 Partial weighted averages of mean monthly a_{pt}, c_{hs2} and K_{RS}

The PWAs of mean monthly a_{pt} and c_{hs2} for short (PWA-s) and tall (PWA-t) reference crop were derived from the application of Eq. (7), and they are given in Fig. 6, while the respective PWAs of mean monthly K_{RS} values are given in Fig. 7. The global means of PWA of a_{pt} and c_{hs2} for the short reference crop (presented below each map of Fig. 6a

Table 4. Spatial extent of the major climatic groups CGs of the Köppen–Geiger climate map (Peel et al., 2007); spatial extent of DMAD classes (P-T versus H-S) within each CG.

CG of Köppen–Geiger	Percent coverage of CGs[a] based on Peel et al. (2007) map	Percent coverage of DMAD classes[b] in a CG (P-T vs. H-S)		
		H-S ($\text{DMAD} \leq -1$)	Trans. zone $-1 < \text{DMAD} < 1$[c]	P-T ($\text{DMAD} \geq 1$)
A – tropical/megathermal	20.66 %	32.0 %	3.6 %	64.4 %
B – arid/semi-arid	32.90 %	86.4 %	1.3 %	12.3 %
C – temperate/mesothermal	14.58 %	32.8 %	3.2 %	64.1 %
D – continental/microthermal	27.00 %	26.9 %	2.1 %	71.0 %
E – polar/alpine (without Antarctica)	4.86 %	71.1 %	16.3 %[d]	12.5 %

[a] The percent coverage was estimated after conversion from a WGS84 ellipsoid to projected Cylindrical Equal Area coordinate system without considering Antarctica. [b] Percentage coverage of DMAD values was estimated after pixel resampling using the resolution of the Köppen map. [c] DMAD range where the results of both the P-T and H-S methods present almost similar deviation from the results of the ASCE-short method (transitional zone). [d] A large part of this percentage corresponds to regions with an annual ET_o equal to 0 (e.g. inner Greenland). Such cases are included in the transitional zone of Fig. 5a.

and c) and the global mean of PWA of K_{RS} values for R_s (presented below Fig. 7) approximate to the standard values of $a_{pt} = 1.26$, $c_{hs2} = 0.0023$ and $K_{RS} = 0.17$, respectively.

As regards the spatial variation in a_{pt} for the short reference crop (Fig. 6a), the higher values were observed in extremely arid and desert environments exceeding the value of 1.8 (mainly due to extremely high vapour pressure deficit), while the extremely cold and extremely humid environments presented values < 1.0. Interesting cases are the alpine tundra and extreme humid tropical environments, which presented similar values between ~0.8–1.0, due to the low values of the vapour pressure deficit. Values of a_{pt} below 0.8 were observed in subpolar areas. The spatial variation in a_{pt} for the tall reference crop (Fig. 6b) follows similar patterns to the a_{pt} of the short reference crop but with increased values, which can be described by the following relationship $a_{pt(PWA-t)} = 1.73 \times a_{pt(PWA-s)} - 0.58$, $R^2 = 0.996$, $p < 0.0001$. This relationship is valid for $a_{pt(PWA-s)} > 0.8$ for preserving $a_{pt(PWA-t)} \geq a_{pt(PWA-s)}$.

As regards the spatial variation in c_{hs2} for the short reference crop (Fig. 6c), the higher values were observed in extremely arid and desert environments exceeding 0.0026 (mainly due to extremely high vapour pressure deficit) while the extremely cold and extremely humid environments presented values < 0.0018. Similarities appear again in the case of alpine tundra and extreme humid tropical environments, which presented values between ~0.0014 and 0.0018, due to the low values of the vapour pressure deficit. Values of c_{hs2} below 0.0014 were observed in subpolar areas. The spatial variation in c_{hs2} for the tall reference crop (Fig. 6d) follows similar patterns with c_{hs2} of the short reference crop but with increased values, which can be described by the following relationship $c_{hs2(PWA-t)} = 1.793 \times c_{hs2(PWA-s)} - 0.00114$, $R^2 = 0.967$, $p < 0.0001$. This relationship is valid for $c_{hs2(PWA-s)} > 0.0014$ for preserving $c_{hs2(PWA-t)} \geq c_{hs2(PWA-s)}$.

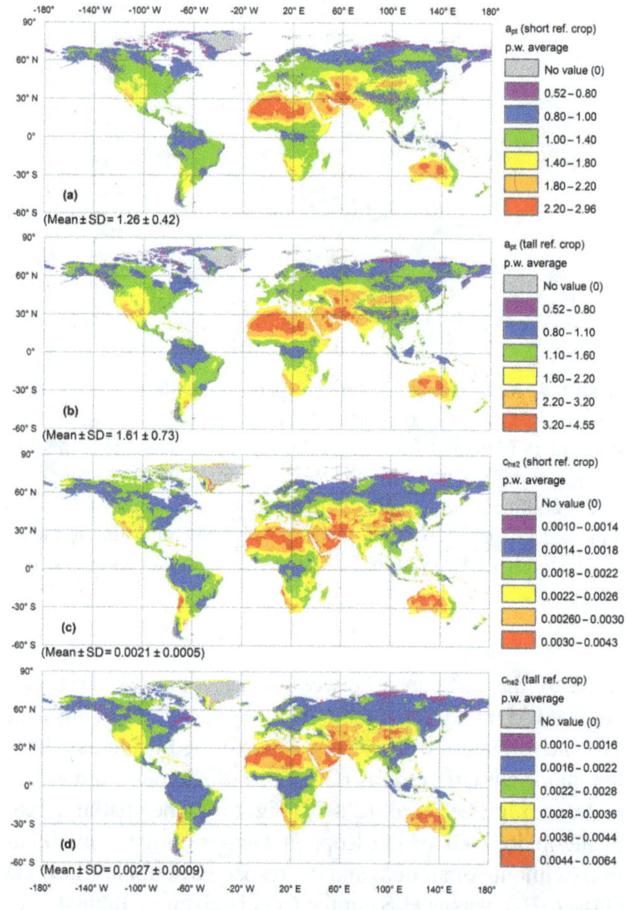

Figure 6. Partial weighted averages of mean monthly **(a)** a_{pt} for the short reference crop, **(b)** a_{pt} for the tall reference crop, **(c)** c_{hs2} for the short reference crop and **(d)** c_{hs2} for the tall reference crop (0.5° resolution maps; mean ± SD are estimated after conversion from WGS84 to Cylindrical Equal Area coordinate system excluding pixels of 0 value).

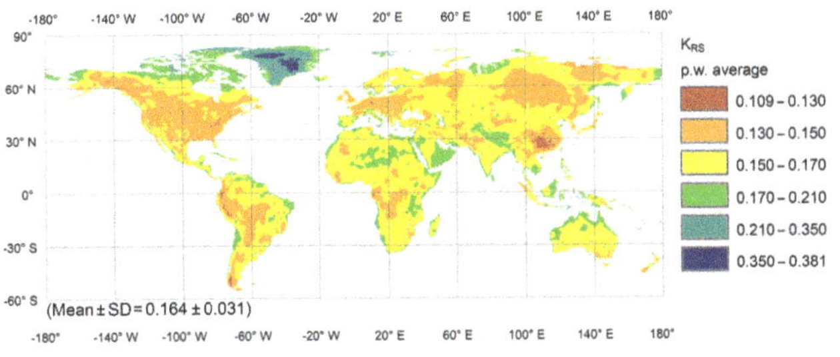

Figure 7. Partial weighted averages of mean monthly K_{RS} (0.5° resolution maps; mean ± SD are estimated after conversion from WGS84 to Cylindrical Equal Area coordinate system).

In the case of K_{RS} (Fig. 7), extreme deviations from the value of 0.17 were observed in Greenland with values above 0.21 and in south-east China with values below 0.13 (regions of Chongqing, Guizhou, Hunan, Jiangxi and Guangxi). The spatial variation in K_{RS} does not follow a specific pattern in relation to climate zones, while in many cases, an increasing trend in its values closer to the coastlines was observed (Fig. 7). Additional observations about the effect of distance from the coastline (Dc) on K_{RS} are given in the Discussion section.

3.3 Validation of the readjusted a_{pt}, c_{hs2} and K_{RS} coefficients

The validation of the readjusted a_{pt} and c_{hs2} coefficients for ET_o estimations (for both reference crops) and theK_{RS} coefficient for R_s was performed taking into account the mean monthly values of the climatic parameters of all stations from Table 1. The readjusted coefficients for each station obtained from the 0.5° resolution maps are given in Table S2 of the Supplement, while the comparison between ET_o estimations (for both reference crops) and rasters and stations is provided in Fig. S2g and h, respectively. The comparison of different methods is described in the next paragraphs, while the overall results of the statistical criteria for all the examined cases are given in Table 5.

Table 5a and Fig. 8 show the ET_o (mm month^{-1}) comparisons between the ASCE-short values versus the values of the P-T and H-S methods with the standard and the readjusted (PWA-s) coefficients and versus the values of the additional models given in Table 2. The following observations were derived from the results of Fig. 8 together with the results of the statistical criteria (Table 5a):

- The P-T(PWA-s) and H-S(PWA-s) models (Fig. 8b, d) outperformed the respective standard P-T(1.26) and H-S(0.0023) models (Fig. 8a, c) with regard to all the statistical criteria (Table 5a), reducing the RMSE values at 40 and 25 %, respectively.

- The comparison of statistical criteria between H-S(0.0023), H-S(PWA-s), DRAL1 and DRAL2, which follow the general formula of the H-S method and are based on calibrations with global data, showed the following order of accuracy H-S(PWA-s) > DRAL1 > DRAL2 > H-S(0.0023).

- The standard P-T(1.26) showed the worst results for all criteria (Table 4a), while the use of P-T(PWA-s) succeeded in improving the predictions, giving better results for the H-S(0.0023), DRAL2, VAL1 and AKJ2 models.

- The H-S(PWA-s) provided better results for DRAL1, DRAL2, VAL1, AKJ1 and AKJ2 where the latter four require data for more climatic parameters.

- The order of accuracy of the models was the following: VAL3 > VAL2 > Copais > H-S(PWA-s) > AKJ1 > P-T(PWA-s) > DRAL1 > DRAL2 > H-S(0.0023) > AKJ2 > VAL1 > P-T(1.26) (the order was based on absolute comparisons of the accuracy rankings for each criterion; see Table S3 in the Supplement). The RMSE difference between H-S(PWA-s) and the best VAL3 model was 6.8 mm month^{-1} (or 0.23 mm d^{-1}), while the respective difference between P-T(PWA-s) and VAL3 was 13.5 mm month^{-1} (or 0.45 mm d^{-1}). These differences are satisfactory, especially for the case of H-S(PWA-s), which uses fewer climatic data from VAL3. Of course, these differences are even smaller when compared to VAL2 and Copais, which also use more climatic parameters. Explanations of the less satisfactory performance of P-T(PWA-s) are given in the Discussion section.

Table 5b and Fig. 9a and b show the ET_o (mm month^{-1}) comparisons between the ASCE-tall values versus the values of the P-T and H-S methods using the readjusted a_{pt} and c_{hs2} coefficients for tall reference crop (PWA-t), respectively. Since there are currently no other methods of reduced parameters calibrated based on ASCE ET_o for the tall reference crop, the comparison between the two methods is restricted. The results of Fig. 9a and b together with the results of the statistical criteria (Table 5b) indicate a better performance of the H-S (with c_{hs2} = PWA-t). The higher errors observed in H-S(PWA-t) and P-T(PWA-t) in comparison to the respective errors of H-S(PWA-s) and P-T(PWA-s) for the short reference crop are explained by the fact that ASCE-tall is significantly higher than ASCE-short, especially in drier environments (ASCE-tall was found to be ~ 28 % higher than ASCE-short on a global scale based on the mean values given in Fig. 2a and b, and ~ 38 % higher based on the comparison of the total mean values estimated by the Californian (USA) and Australian station data).

Table 5c and Fig. 10a and b show the comparisons between the R_s (MJ m^{-2} d^{-1}) of station data versus the respective values of the standard radiation formula of H-S (Eq. 3) with K_{RS} = 0.17 and with K_{RS} = PWA, respectively. The results

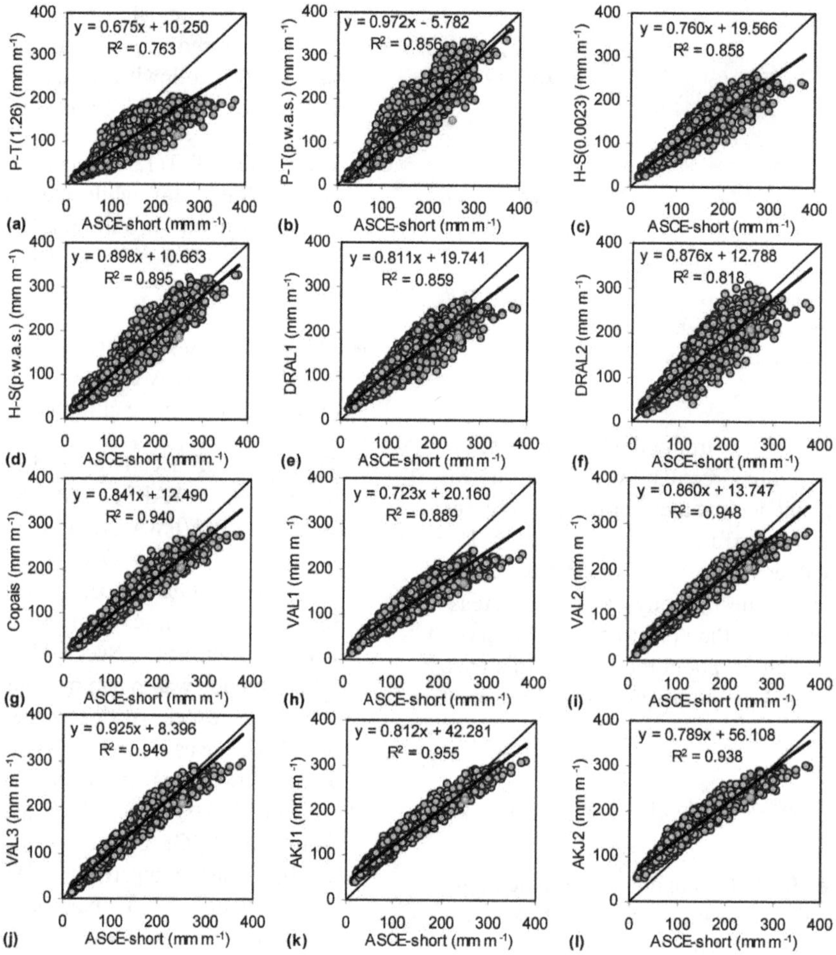

Figure 8. Comparative 1 : 1 plots between the results of ET_o ASCE-short (mm month^{-1}) versus **(a)** the standard P-T method with $a_{pt} = 1.26$, **(b)** the P-T method with a_{pt} = PWA-s (0.5° resolution), **(c)** the standard H-S method with $c_{hs2} = 0.0023$, **(d)** the H-S method with c_{hs2} = PWA-s (0.5° resolution), **(e)** DRAL1 (Eq. 8), **(f)** DRAL2 (Eq. 9), **(g)** Copais (Eq. 10), **(h)** VAL1 (Eq. 11), **(i)** VAL2 (Eq. 12), **(j)** VAL3 (Eq. 13), **(k)** AKJ1 (Eq. 14) and **(l)** AKJ2 (Eq. 15).

of Fig. 10a and b together with the results of the statistical criteria (Table 5c) indicate a better performance of the H-S R_s with K_{RS} = PWA even though the performance of the standard H-S R_s is also satisfactory.

4 Discussion

4.1 Uncertainties in the data used for calibrating and validating the revised coefficients of the P-T and H-S methods

The recalibrated coefficients of the H-S and P-T methods were estimated using raster datasets that cover the period 1950–2000 assuming stationary climate conditions, while the validation datasets of Californian (USA) and Australian stations are expanded up to 2016. Recent studies have shown changes/anomalies after 2000 in temperature (Hansen et al., 2010; Sun et al., 2017), solar radiation (Wild et al., 2013), wind speed (McVicar et al., 2012a, b) and atmospheric humidity (Willet et al., 2014) and such changes could affect the validity of the revised coefficients. The comparisons of T_{max}, T_{min}, R_s, R_n, DE and u_2 values between the raster data and the station data showed a very good correspondence for the case of T_{max}, T_{min}, R_s and R_n (Fig. S2a, b, c, d) and a relatively good correspondence for the case of DE (Fig. S2e). In the case of u_2, a discrepancy was observed between the rasters and station data (Fig. S2f). The separate examination of u_2 for the USA-CA and Australian stations (Fig. S3) showed that the total average of mean monthly u_2 values of USA-CA stations was lower from the raster data of Sheffield et al. (2006) (data extracted from the stations' positions), while the opposite trend was found for the u_2 values of Australian stations. This discrepancy between the u_2 values of rasters and stations can be explained by the following:

Table 5. Statistical criteria from the comparisons **(a)** between ET_o values from ASCE-short and the methods used for estimating short reference crop evapotranspiration (i.e. P-T with standard and readjusted coefficients, H-S with standard and readjusted coefficients and all equations given in Table 2), **(b)** between ET_o values from ASCE-tall and P-T, H-S methods with readjusted coefficients for the tall reference crop, **(c)** R_s values from stations and R_s obtained from the H-S radiation formula with standard and readjusted coefficients.

Case	Criterion	MAE	RMSE	NRMSE %	PBIAS %	R^2	bR^2	NSE	d	KGE
	Optimum value	0	0	0	0	1	1	1	1	1
(a)	P-T (Eq. 2) with $a_{pt} = 1.26$	36.92	48.87	90.9	33.3	0.763	0.591	0.173	0.849	0.539
	P-T (Eq. 2) with a_{pt} = PWA-s	22.71	29.43	40.3	7.5	0.856	0.832	0.837	0.956	0.883
	H-S (Eq. 4b) with $c_{hs2} = 0.0023$	21.19	30.36	53.2	10.8	0.858	0.772	0.717	0.941	0.746
	H-S (Eq. 4b) with c_{hs2} = PWA-s	17.13	22.72	34.4	2.5	0.895	0.878	0.881	0.971	0.921
	DRAL1 (Eq. 8)	19.53	27.05	44.5	4.8	0.859	0.818	0.802	0.955	0.833
	DRAL2 (Eq. 9)	22.92	30.28	45.0	3.2	0.818	0.808	0.798	0.949	0.894
	Copais (Eq. 10)	14.49	20.70	34.3	7.3	0.940	0.870	0.882	0.974	0.829
	VAL1 (Eq. 11)	21.36	31.87	59.8	15.1	0.888	0.763	0.642	0.932	0.657
	VAL2 (Eq. 12)	12.13	17.96	29.3	4.2	0.948	0.900	0.914	0.981	0.859
	VAL3 (Eq. 13)	11.45	15.94	24.1	1.4	0.949	0.934	0.942	0.986	0.940
	AKJ1 (Eq. 14)	21.17	24.24	42.0	−10.6	0.955	0.887	0.824	0.964	0.771
	AKJ2 (Eq. 15)	30.36	33.69	59.5	−16.3	0.938	0.820	0.645	0.931	0.718
(b)	P-T (Eq. 2) with a_{pt} = PWA-t	40.43	52.38	50.6	8.4	0.770	0.754	0.743	0.930	0.845
	H-S (Eq. 4b) with c_{hs2} = PWA-t	31.87	42.34	45.2	3.7	0.823	0.806	0.795	0.950	0.885
(c)	H-S R_s (Eq. 3) with $K_{RS} = 0.17$	1.64	1.99	29.6	−4.5	0.930	0.885	0.912	0.977	0.932
	H-S R_s (Eq. 3) with K_{RS} = PWA	1.05	1.43	22.3	−0.8	0.952	0.944	0.950	0.988	0.972

- Possible changes in wind speeds after 2000, since the majority of wind speed data in the station datasets corresponds to periods after 2000.

- Uncertainties in the Sheffield et al. (2006) wind data due to the scarce existing wind data for calibrating their model on a global scale during the period of 1950–2000 and especially during the years belonging to the first half of the simulation period.

- The effect of the equation $u_2 = 4.87u_z / \ln(67.8z–5.42)$ (Allen et al., 1998, 2005), which was used to adjust the wind rasters of Sheffield et al. (2006) and the wind data of Australian stations from $z = 10$ to 2 m height. The degree of accuracy of the aforementioned equation to convert wind data at 2 m is unknown. This equation is usually not calibrated for meteorological stations with anemometers positioned above 2 m height, while the uncertainty is even larger when it is applied on a global scale and for a pixel of 0.5° resolution, which may contain high topographic variability.

- The bias that may have been introduced after cleaning extreme wind values in the data of USA-CA stations, which may be associated with hurricane events. The region of California is strongly affected by hurricanes and the higher wind speeds in the rasters of the Sheffield et al. (2006) data may partly have occurred because they included such events in their climatic simulations.

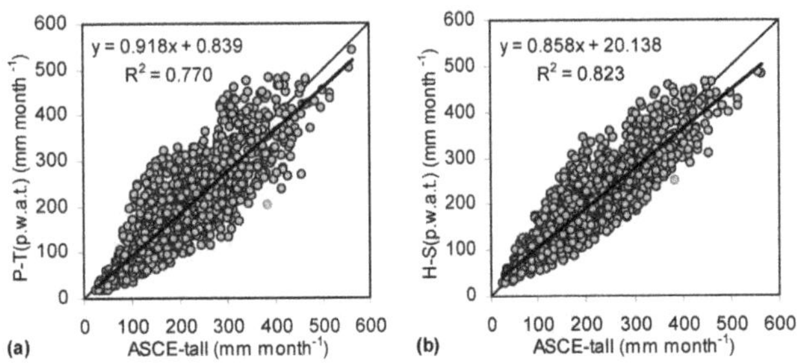

Figure 9. Comparative 1 : 1 plots between the results of ET_o ASCE-tall (mm month^{-1}) versus **(a)** the P-T method with a_{pt} = PWA-t (0.5° resolution) and **(b)** the H-S method with c_{hs2} = PWA-t (0.5° resolution).

- The bias that may have been introduced by the wind data of Australian stations. The AGBM database provides 12 values of mean monthly wind speeds of the total observation periods for 09:00 and another 12 values for 15:00 local time. In order to get the mean monthly wind speeds of the stations, the average value of 09:00 and 15:00 conditions was used for each month.

- Combinations of all the aforementioned cases.

Uncertainties may also exist in the case of DE = e_s–e_a (Fig. S2e), since Sheffield et al. (2006) provide data of specific humidity that were directly converted to actual vapour pressure e_a using the equation of Peixoto and Oort (1996), which uses the additional parameter of atmospheric pres-

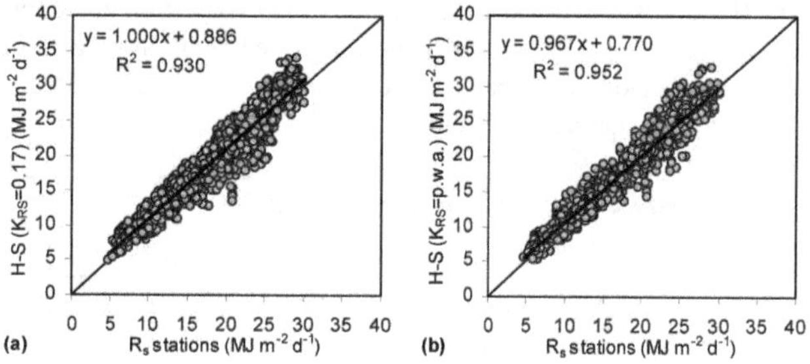

Figure 10. Comparative 1 : 1 plots between the R_S ($\mathrm{MJ\,m^{-2}\,d^{-1}}$) values of USA-CA and Australian stations versus the results of H-S radiation formula (Eq. 3) **(a)** with $K_{RS} = 0.17$ and **(b)** with $K_{RS} = $ PWA (0.5° resolution).

sure as an internal parameter. The atmospheric pressure in the case of rasters was estimated based on elevation data of 1 km resolution (30 arcsec), which were further converted to 0.5° resolution. The use of e_a data from 0.5° resolution pixels may also have added additional error, especially when there is large topographic variability within the 0.5° pixel. On the other hand, the e_a of stations was estimated by relative humidity and temperature data.

Thus, uncertainties exist in both rasters and station data. In future studies, further improvements in the revised coefficients can be made by using global raster data, which incorporate the conditions after 2000, and by solving many of the aforementioned problems related to both station data and raster data produced by climatic models.

4.2 Reasons for using annual PWA coefficients instead of monthly or seasonal ones in the case of H-S and P-T methods

The analysis presented in this study passed through various stages before the selection of the annual PWA form of the coefficients (Eq. 7). Some steps in the preliminary analysis were to analyse (a) the different forms of averages (e.g. mean, mode, median, geometric mean, harmonic mean) for deriving annual coefficients and (b) the strength of the derived mean monthly and seasonal coefficients versus the annual PWA coefficients and versus the coefficients of the standard methods,.

As regards the use of the weighted annual average (w.a.) of the mean monthly coefficients instead of other forms of averages (e.g. mean, mode, median, geometric mean – g.m., harmonic mean – h.m.), preliminary analysis was performed using data extracted by the climatic rasters from many positions in the world. During this analysis, trials to derive annual coefficients were made using an optimization algorithm separately for each position. The results showed that the optimized annual values were always closer to the monthly coefficients of the warmer months since the optimization algorithms try to reduce the total error, which is mainly dominated by the months that show larger ET_o values (or R_s for the case of K_{RS} calibration). The optimized values were also compared to the different types of annual averages (e.g. mean, mode, median, g.m., h.m., w.a.), which were estimated after excluding values of monthly coefficients associated with months with ET_o and R_s values $< 45\,\mathrm{mm\,month^{-1}}$ (for R_s, the equivalent is $3.61\,\mathrm{MJ\,m^{-2}\,d^{-1}}$). The w.a. generally showed better performance in comparison to other types of annual averages because it is the only form that considers the amplitude of the parameter under investigation (ET_o and R_s) (Eq. 7), giving more weight to the monthly coefficients that are related to the warmer months. This attribute of w.a. is extremely significant since it is the only type that considers the seasonal observed differences in monthly ET_o (for a_{pt} and c_{hs2}) and R_s (for K_{RS}) minimizing the possible errors during warmer months.

The case of mean monthly coefficients was also examined (results not shown). The results showed that the assessment of annual ET_o and seasonal ET_o during the warm season using the mean monthly coefficients is more accurate in comparison to the standard coefficients but not as good as the PWA coefficients, especially during the cold season. Similar findings were observed when different time intervals for calculating seasonal averages of the coefficients were used (e.g. 3-month averages or 6-month averages). The basic observed problem with monthly/seasonal coefficients associated with the global-scale application of this study was that many parts of the world presented unreasonably high or low monthly/seasonal values of the coefficients (at least 1 order of magnitude larger or smaller than the standard values) during cold seasons. This problem occurred because P-T and H-S evapotranspiration models do not include the effect of humidity and wind, which becomes greater when temperature is low (in very low temperatures even the ASCE results can be questioned). Such values may lead to significant errors in monthly/seasonal ET_o estimations during cold periods when there are deviations of climatic conditions (seasonal shifts/disturbances or climate changes in general) from those used for calibrating the coefficients. These were the reasons for using the threshold of $45\,\mathrm{mm\,month^{-1}}$ to exclude such values from the PWA of the coefficients. Thus, the PWA annual values were chosen as the best solution for a global application because they counterbalance the errors that could be introduced by intra-annual/intra-seasonal climatic variability or other errors such as those described in the previous section of the Discussion section (errors associated with the data).

It is also important to note that the derivation of annual coefficients is a pure optimization problem when station data are used. For example, Cristea et al. (2013) derived coefficients of the P-T method for 106 stations that represent a range of climates across the contiguous USA. The coefficients were estimated for each station by minimizing the sum of the squared residuals between the benchmark FAO-56 and P-T using data only for the period April–September. The obtained optimized values of the coefficients were interpolated in order to make a map of the a_{pt} coefficient. In our study, the maps of the coefficients were produced based on raster data and not station data, which means that optimization should be performed pixel by pixel ($\sim 62\,000$ pixels globally for the 0.5° resolution excluding Antarctica). This procedure would require special programming since a readily available tool to perform this procedure does not exist in commercial or free GIS software packages. This is the main reason for using Eq. (7) as an alternative method in a GIS environment, since it can be calculated easily in raster calculators incorporated in the GIS packages. A solution could be the development of a tool for GIS purposes using raster data that is able to run using 24 rasters: 12 for the benchmark ET_o and another 12 for the P-T or H-S ET_o formula without the 1.26 and 0.0023 factors, respectively, in order to provide optimized annual values of their coefficients (for a global application, filters to remove unreasonable values are also required).

4.3 Observations derived by the application of the H-S radiation formula

Special attention was also given in the case of K_{RS} coefficient for estimating R_s. Although there were indications that the spatial variation in the PWA a_{pt} and c_{hs2} coefficients on a global scale may be linked to general climatic characteristics (Fig. 5), the respective variation in PWA K_{RS} coefficient could not be clearly linked with a specific climatic or topographic characteristic. The only observed dependence, which showed some relevance to the spatial variation in K_{RS}, was a relatively negative correlation with Dc. This observed dependence can be only used as a general observation and not as a basis for applying the empirical rule of Allen et al. (1998) in general ($K_{RS} = 0.16$ for interior and $K_{RS} = 0.19$ for coastal locations). The large uncertainty in the aforementioned rule was also indicated by Samani (2000), and it is verified by the analysis presented in Fig. S4a of the Supplement. Figure S4a shows a relatively negative correlation between K_{RS} and Dc (for Dc < 500 km) but also shows an extremely high variability in K_{RS} close to the coastlines where K_{RS} values are not necessarily higher in comparison to the values observed in the interior regions. The observed lower variability in K_{RS} in interior regions is probably related to the fact that coastlines are more affected by oceanic climatic phenomena, which present high spatial variability on a global scale anyway. The raster data of K_{RS} (Fig. 7) can be used as indicator to control the validity of the rule but also to control the validity of the given values 0.16–0.19 for a specific region. Samani (2000) also observed that the monthly K_{RS} values may be influenced by the difference between monthly maximum and minimum temperature, TD. This effect was also investigated through correlations between the mean monthly K_{RS} coefficients and the mean monthly TD values of the station data (Fig. S4b, in the Supplement). The results showed that the hypothesis related to the effect of TD on K_{RS} may be stronger locally in comparison to the effect of Dc, but again the variation in K_{RS} is extremely large in the TD range between 8 and 15 °C (Fig. S4b), not allowing secure conclusions for a global-scale application. The result of Fig. S4b is based only on the stations of Table 1, and for this reason the variation on a global scale is expected to be much larger.

4.4 Recommendations for reducing the uncertainties when the readjusted coefficients of the P-T and H-S models are used

The uncertainties, which may be introduced by climate disturbances/changes or other uncertainties related to the data used for calibrating the coefficients, can be reduced by taking into account some of the following observations and recommendations.

A separate analysis using only the stations of California showed that a regional mean value of the coefficients derived by PWA values may present even better performance because it probably counterbalances other uncertainties associated with the spatial climatic variability within a specific region. A factor for such uncertainties may be rainfall, which may not show significant seasonal deviations or deviations from the expected annual values for a large region but may show different spatial patterns every year within the region affecting the accuracy of the coefficients. The aforementioned observation was verified by the application of the H-S method for ET_o of the short reference crop for the stations of California when the average value of $c_{hs2} = 0.0024$ obtained from the respective PWA-s values of the stations (Table S2) was used (this value also approximates the standard value of 0.0023). The average value of 60 PWA-s coefficients of the USA-CA stations gave better results for the individual coefficients (Fig. S5 and Table S4, in the Supplement). The aforementioned observations suggest that a robust territorial segmentation based on general topographic characteristics (e.g. elevation, slope, latitude and longitude, distance from the coastline) and general climatic characteristics (e.g. Köppen class, general precipitation and temperature patterns) can provide a proper zonation of large territories for deriving very robust mean values of a_{pt}, c_{hs2} and K_{RS} coefficients using the respective PWA values of each zone. Robust zonations based on grids of mean monthly precipitation and temperature using the data of Hijmans et al. (2005) or the mean monthly ET_o rasters provided by this study can easily be performed using cluster analysis in a GIS environment (Demertzi et al., 2014; Aschonitis et al., 2016a, b).

The comparison between P-T and H-S evapotranspiration methods with readjusted coefficients but also their comparison with the other models in Table 2 also provided significant information. From the comparison between P-T and H-S with readjusted coefficients, it was observed that H-S provided better results in both the short and the tall reference crop. The better performance of H-S can be attributed to the fact that more than $\sim 80\,\%$ of stations from Table 1 are located in territories with negative DMAD values (Fig. 5a) giving a general advantage to the H-S method for more robust estimations. This observation can explain the better performance of the standard H-S (with $c_{hs2} = 0.0023$) in comparison to the standard P-T (with $a_{pt} = 1.26$) for the short reference crop (Table 5a) and indirectly validates the DMAD map. Considering these observations, it is recommended to take into account both the MAD (Fig. 4,b,c) and DMAD (Fig. 5a) maps before selecting one of the two methods either using the standard or the PWA coefficients. From the comparisons with the other models in Table 2, it was observed that three models, which use temperature, radiation and humidity data (i.e. VAL3, VAL2, Copais and especially VAL3), provided better estimations. These models have shown very good performance using data from other case studies (Pan et al., 2011; Shiri et al., 2014; Kisi, 2014; Gao et al., 2015; Valipour, 2015a, c; Djaman et al., 2015, 2016, 2017; Ahooghalandari et al., 2017), and their use is recommended instead of the P-T and H-S with readjusted coefficients when the only missing climatic parameter is wind speed.

A very interesting observation was also made about the tall reference crop based on the results of the MAD % map (Fig. 4a). In the MAD % class of ±10 % of Fig. 4a, some small negative values were observed, which correspond to ∼2 % of map coverage. These values indicate slightly larger annual values of ASCE-short in comparison to ASCE-tall. This result was observed in regions of extremely small vapour pressure deficit (areas of very high elevation, either of very cold or extremely humid conditions scattered around the world), and it is a peculiarity of Eq. (1) and probably an artefact. This result occurred because the second term of the nominator in Eq. (1) (which includes the vapour pressure deficit term and the C_n coefficient) approximates to 0 when e_s–e_a becomes extremely small, while the denominator of Eq. (1) is always larger in ASCE-tall in comparison to ASCE-short due to the difference in C_d value (0.34 for the short and 0.38 for the tall reference crop). A recommendation for partly solving this problem for the tall reference crop applications is to use the revised coefficients of the P-T and H-S methods derived for the short reference crop in the places were the annual value of ASCE-tall is lower than ASCE-short. This recommendation is based on the fact that annual ASCE-tall is expected to be always larger than the respective value of ASCE-short. This peculiarity was not corrected in the ASCE-tall maps and the respective a_{pt} and c_{hs2} coefficients for the tall reference crop in order to show the absolute estimations of ASCE-tall and the respective coefficients. Finding the negative values in the MAD % map (Fig. 4a), the users can identify the location of these pixels.

5 Data availability

The datasets produced in this study have been archived in the PANGAEA database (https://doi.org/10.1594/PANGAEA.868808) and in the ESRN database, which is currently supported by the University of Ferrara (Italy), Aristotle University of Thessaloniki (Greece) and the University of Campania Luigi Vanvitelli (Italy) (http://www.esrn-database.org/gis-data.html or http://esrn-database.weebly.com/gis-data.html). Apart from the 0.5° resolution raster datasets, the database contains the same datasets at finer resolution (30 arcsec, 2.5 arcmin, 5 arcmin and 10 arcmin). These finer datasets are provided in order to cover the observed resolution range in the initial climatic data (e.g. the temperature data of Hijmans et al. (2005) are provided at 30 arcsec resolution). The finer resolutions were produced using bilinear interpolation on solar radiation, humidity and wind speed data of Sheffield et al. (2006). This interpolation method is not the most appropriate for such purposes. The data of finer resolutions can only be used as a tool to assess uncertainties associated with temperature variation effects within a 0.5° pixel or to estimate average values of the coefficients for larger territories in order to capture a better representation of the coastlines or islands that do not exist at 0.5° resolution (use of values from individual pixels is not recommended). A complete list of the datasets is provided in the Table S5.

6 Conclusions

The study provided global grids of revised annual coefficients for the P-T and H-S methods for estimating ET_o for both short and tall reference crops. The coefficients were calibrated using respective grids of ET_o estimated with the ASCE-standardized method. Respective grids of annual coefficients were also derived for the radiation formula of H-S. The calibration procedures were based on global gridded climatic data of the period 1950–2000. The method for deriving annual coefficients of the P-T and H-S methods was based on partial weighted averages (PWAs) of the respective mean monthly coefficients. This method estimates the annual values considering the amplitude of the parameter under investigation (ET_o and R_s) giving more weight to the monthly coefficients of the months with higher ET_o values (or R_s values for the case of the H-S radiation formula). The method also eliminates the effect of unreasonable monthly coefficients that may occur during periods when ET_o and R_s fall below a specific threshold. The new coefficients were validated based on data from 140 stations located at various climatic zones of the USA and Australia with expanded observations up to 2016. Additional tests were also performed for the case of the short reference crop evapotranspiration using additional models with low requirements for climatic data. The validation procedure for ET_o estimations of the short reference crop showed that the P-T and H-S methods with revised coefficients outperformed the standard methods reducing the estimated RMSE in ET_o values by 40 and 25 %, respectively. The estimations of R_s using the H-S formula with revised coefficients reduced the RMSE by 28 % in comparison to the standard formula. The comparisons with other models of the short reference crop showed that the P-T and H-S methods with revised coefficients can compete with models of additional climatic parameters. In the case where only wind speed is missing from available data, the use of the VAL2, VAL3 and Copais methods (temperature, radiation and humidity data requirements) is recommended. Finally, a raster database of 0.5° resolution was built consisting of (a) global maps for the mean monthly ET_o values estimated by the ASCE-standardized method for both reference crops, (b) global maps for the revised annual coefficients of the P-T and H-S evapotranspiration methods for both reference crops and a global map for the revised annual coefficients of the H-S radiation formula, and (c) global maps that indicate the optimum locations for using the standard P-T and H-S methods and their possible annual errors based on reference values (MAD % and DMAD maps). The online free availability of the database can support estimations of ET_o and solar radiation for locations where climatic data are limited, and it can support studies which require such estimations on larger scales (e.g. country, continent, world).

The methods used in this study, their respective results and the observed uncertainties can be used as a base for future work focusing on

a. the validation of the coefficients for other parts of the world, especially using climatic data obtained after 2000, and the comparison with other models of low data requirements

b. the recalibration of the coefficients using data from climatic models that include observations from more recent years and an analysis of climate change effects on the coefficients

c. the use of the available climatic datasets obtained from climatic models in order to calibrate models of the coefficients for various locations providing values that are not fixed, such as the ones given in this study,

d analysis of alternative methods for deriving annual coefficients that approximate optimized values or the incorporation of optimization algorithms in a GIS environment to capture the optimum solution per pixel

e. the confrontation of uncertainties related to the data used for calibration and validation (e.g. low representativity of interpolated climatic parameters due to the lack of data in many parts of the world; errors associated with commonly used equations, such as the one used for adjusting wind data at 2 m height; uncertainties associated with the observed data).

Competing interests. The authors declare that they have no conflict of interest.

Acknowledgements. This study was performed in the context of two post-doctoral research studies by Vassilis Aschonitis financed by Ferrara University (Italy) and Aristotle University of Thessaloniki (Greece).

Edited by: David Carlson

References

Abtew, W.: Evapotranspiration measurements and methoding for three wetland systems in South Florida, J. Am. Water Resour. As., 32, 465–473, 1996.

Ahooghalandari, M., Khiadani, M., and Jahromi, M. E.: Developing Equations for Estimating Reference Evapotranspiration in Australia, Water Resour. M., 30, 3815–3828, 2016.

Ahooghalandari, M., Khiadani, M., and Jahromi, M. E.: Calibration of Valiantzas' reference evapotranspiration equations for the Pilbara region, Western Australia, Theor. Appl. Climatol., 128, 845–856, 2017.

Allen, R. G., Pereira, L. S., Raes, D., and Smith, M.: Crop Evapotranspiration: Guidelines for computing crop water requirements, Irrigation and Drainage Paper 56, Food and Agriculture Organization of the United Nations, Rome, 1998.

Allen, R. G., Walter, I. A., Elliott, R., Howell, T., Itenfisu, D., and Jensen M.: The ASCE standardized reference evapotranspiration equation. Final Report (ASCE-EWRI) Pr., in: Environmental and Water Resources Institute, 2005, edited by: Allen, R. G., Walter, I. A., Elliott, R., Howell, T., Itenfisu, D., and Jensen, M., Task Committee on Standardization of Reference Evapotranspiration of the Environmental and Water Resources Institute, 2005.

Alexandris, S., Kerkides, P., and Liakatas, A.: Daily reference evapotranspiration estimates by the "Copais" approach, Agr. Water Manage., 82, 371–386, 2006.

Aschonitis, V., Antonopoulos, V. Z., and Papamichail, D. M.: Evaluation of pan coefficient equations in a semi-arid Mediterranean environment using the ASCE standardized Penman-Monteith method, Agr. Sci., 3, 58–65, 2012.

Aschonitis, V., Demertzi, K., Papamichail, D., Colombani, N., and Mastrocicco, M.: Revisiting the Priestley-Taylor method for the assessment of reference crop evapotranspiration in Italy, Ital. J. Agrometeorol., 20, 5–18, 2015.

Aschonitis, V., Miliaresis, G., Demertzi, K., and Papamichail, D.: Terrain segmentation of Greece using the spatial and seasonal variation of reference crop evapotranspiratiom, Adv. Meteorol., 3092671, 1–14, https://doi.org/10.1155/2016/3092671, 2016a.

Aschonitis, V., Awe, G. O., Abegunrin, T. P., Demertzi, K. A., Papamichail, D. M., and Castaldelli, G.: Geographic segmentation, spatial dependencies, and evaluation of the relative position of rain-gauge based on gridded data of mean monthly precipitation: application in Nigeria, Hydrol. Res., nh2016095, https://doi.org/10.2166/nh.2016.095, in press, 2016b.

Azhar, A. H. and Perera, B. J. C.: Evaluation of reference evapotranspiration estimation methods under Southeast Australian conditions, J. Irrig. Drain E.-ASCE, 137, 268–279, 2011.

Bachour, R., Walker, W. R., Torres-Rua, A. F., and McKee, M.: Assessment of reference evapotranspiration by the hargreaves method in the Bekaa Valley, Lebanon, J. Irrig. Drain. E.-ASCE, 139, 933–938, 2013.

Berti, A., Tardivo, G., Chiaudani, A., Rech, F., and Borin, M.: Assessing reference evapotranspiration by the Hargreaves method in north-eastern Italy, Agr. Water Manage., 140, 20–25, 2014.

Brinckmann, S., Krähenmann, S., and Bissolli, P.: High-resolution daily gridded data sets of air temperature and wind speed for Europe, Earth Syst. Sci. Data, 8, 491–516, https://doi.org/10.5194/essd-8-491-2016, 2016.

Castellvi, F., Stockle, C. O., Perez, P. J., and Ibanez, M.: Comparison of methods for applying the Priestley-Taylor equation at a regional scale, Hydrol. Process., 15, 1609–1620, 2001.

Cristea, N. C., Kampf, S. K., and Burges, S. J.: Revised coefficients for Priestley-Taylor and Makkink-Hansen equations for estimating daily reference evapotranspiration, J. Hydrol. Eng., 18, 1289–1300, 2013.

Demertzi, K., Papamichail, D., Aschonitis, V., and Miliaresis, G.: Spatial and seasonal patterns of precipitation in Greece: The terrain segmentation approach, Global Nest J., 16, 988–997, 2014.

Djaman, K., Balde, A. B., Sow, A., Muller, B., Irmak, S., N'Diaye, M. K., Manneh, B., Moukoumbi, Y. D., Futakuchi, K., and Saito, K.: Evaluation of sixteen reference evapotranspiration methods under sahelian conditions in the Senegal River Valley, J. Hydrol. Reg. Stud., 3, 139–159, 2015.

Djaman, K., Irmak, S., Kabenge, I., and Futakuchi, K.: Evaluation of FAO-56 penman-monteith model with limited data and the valiantzas models for estimating grass-reference evapotranspiration in Sahelian conditions, J. Irrig. Drain. E.-ASCE, 142, 04016044, https://doi.org/10.1061/(ASCE)IR.1943-4774.0001070, 2016.

Djaman, K., Irmak, S., and Futakuchi, K.: Daily reference evapotranspiration estimation under limited data in eastern Africa, J. Irrig. Drain E.-ASCE, 143, 06016015, https://doi.org/10.1061/(ASCE)IR.1943-4774.0001154, 2017.

Droogers, P. and Allen, R. G.: Estimating reference evapotranspiration under inaccurate data conditions, Irrig. Drain. Syst., 16, 33–45, 2002.

Eichinger, W. E., Parlange, M. B., and Strickler, H.: On the concept of equilibrium evaporation and the value of the Priestley-Taylor coefficient, Water Resour. Res., 32, 161–164, 1996.

Flint, A. L. and Childs, S. W.: Use of the Priestley-Taylor evaporation equation for soil water limited conditions in a small forest clearcut, Agr. Forest Meteorol., 56, 247–260, 1991.

Gao, X., Peng, S., Xu, J., Yang, S., and Wang, W.: Proper methods and its calibration for estimating reference evapotranspiration using limited climatic data in Southwestern China, Arch. Agron. Soil Sci., 61, 415–426, 2015.

Giles, D. G., Black, T. A., and Spittlehouse, D. L.: Determination of growing season soil water deficits on a forested slope using water balance analysis, Can. J. Forest Res., 15, 107–114, 1984.

Gupta, H. V., Kling, H., Yilmaz, K. K., and Martinez, G. F.: Decomposition of the mean squared error and NSE performance criteria: Implications for improving hydrological modelling, J. Hydrol., 377, 80–91, 2009.

Hansen, J., Ruedy, R., Sato, M., and Lo K.: Global surface temperature change, Rev. Geophys., 48, RG4004, https://doi.org/10.1029/2010RG000345, 2010.

Hargreaves, G. H. and Allen, R. G.: History and evaluation of hargreaves evapotranspiration equation, J. Irrig. Drain E.-ASCE, 129, 53–63, 2002.

Hargreaves, G. H. and Samani, Z. A.: Estimating potential evapotranspiration, J. Irrig. Drain E.-ASCE, 108, 223–230, 1982.

Hargreaves, G. H. and Samani, Z. A.: Reference crop evapotranspiration from ambient air temperature, American Society of Agricultural Engineers, 12 pp., availabe at: http://libcatalog.cimmyt.org/download/reprints/97977.pdf (last access: 1 October 2016), 1985.

Heydari, M. M. and Heydari, M.: Calibration of Hargreaves-Samani equation for estimating reference evapotranspiration in semiarid and arid regions, Arch. Agron. Soil Sci., 60, 695–713, 2014.

Hijmans, R. J., Cameron, S. E., Parra, J. L., Jones, P. G., and Jarvis, A.: Very high resolution interpolated climate surfaces for global land areas, Int. J. Climatol., 25, 1965–1978, 2005.

Itenfisu, D., Elliott, R. L., Allen, R. G., and Walter, I. A.: Comparison of reference evapotranspiration calculations as part of the ASCE standardization effort, J. Irrig. Drain E.-ASCE, 129, 440–448, 2003.

Kellner, E.: Surface energy fluxes and control of evapotranspiration from a Swedish Sphagnum mire, Agr. Forest Meteorol., 110, 101–123, 2001.

Kisi, O.: Comparison of different empirical methods for estimating daily reference evapotranspiration in mediterranean climate, J. Irrig. Drain E.-ASCE, 140, 04013002, https://doi.org/10.1061/(ASCE)IR.1943-4774.0000664, 2014.

Krause, P., Boyle, D. P., and Bäse, F.: Comparison of different efficiency criteria for hydrological model assessment, Adv. Geosci., 5, 89–97, https://doi.org/10.5194/adgeo-5-89-2005, 2005.

Lhomme, J. -P.: A theoretical basis for the Priestley-Taylor coefficient, Bound.-Lay. Meteorol., 82, 179–191, 1997.

Long, H., Shuai, X., Lei, Q., and Zhang, R.: Spatiotemporal distribution of calibration coefficients of hargreaves equation for estimating potential evapotranspiration in Mainland China, J. Irrig. Drain E.-ASCE, 139, 293–299, 2013.

Maidment, D. R.: Handbook of hydrology, McGraw-Hill, New York, 1992.

McMahon, T. A., Peel, M. C., Lowe, L., Srikanthan, R., and McVicar, T. R.: Estimating actual, potential, reference crop and pan evaporation using standard meteorological data: a pragmatic synthesis, Hydrol. Earth Syst. Sci., 17, 1331–1363, https://doi.org/10.5194/hess-17-1331-2013, 2013.

McVicar, T. R., Roderick, M. L., Donohue, R. J., and Van Niel, T. G.: Less bluster ahead? ecohydrological implications of global trends of terrestrial near-surface wind speeds, Ecohydrology, 5, 381–388, 2012a.

McVicar, T. R., Roderick, M. L., Donohue, R. J., Li, L. T., Van Niel, T. G., Thomas, A., Grieser, J., Jhajharia, D., Himri, Y., Mahowald, N. M., Mescherskaya, A. V., Kruger, A. C., Rehman, S., and Dinpashoh, Y.: Global review and synthesis of trends in observed terrestrial near-surface wind speeds: implications for evaporation, J. Hydrol., 416–417, 182–205, 2012b.

Mendicino, G. and Senatore, A.: Regionalization of the hargreaves coefficient for the assessment of distributed reference evapotranspiration in Southern Italy, J. Irrig. Drain. E.-ASCE, 139, 349–362, 2013.

Mintz, Y. and Walker, G. K.: Global fields of soil moisture and land surface evapotranspiration derived from observed precipitation and surface air temperature, J. Appl. Meteorol., 32, 1305–1334, 1993.

Moges, S. A., Katambara, Z., and Bashar, K.: Decision support system for estimation of potential evapotranspiration in Pangani Basin, Phys. Chem. Earth, 28, 927–934, 2003.

Mohawesh, O. E. and Talozi, S. A.: Comparison of Hargreaves and FAO56 equations for estimating monthly evapotranspiration for semi-arid and arid environments, Arch. Agron. Soil Sci., 58, 321–334, 2012.

Nash, J. E. and Sutcliffe, J. V.: River flow forecasting through conceptual models, Part I – A discussion of principles, J Hydrol., 10, 282–290, 1970.

Ngongondo, C., Xu, C. -Y., Tallaksen, L. M., and Alemaw, B.: Evaluation of the FAO Penman-montheith, Priestley-Taylor and Hargreaves models for estimating reference evapotranspiration in southern Malawi, Hydrol. Res., 44, 706–722, 2013.

Osborn, T. J. and Jones, P. D.: The CRUTEM4 land-surface air temperature data set: construction, previous versions and dissemination via Google Earth, Earth Syst. Sci. Data, 6, 61–68, https://doi.org/10.5194/essd-6-61-2014, 2014.

Pan, Y., Gong, H.-L., Li, X.-J., Zhu, L., and Zhang, J.: Application of Valiantzas approach to estimating reference evapotranspiration in China. Shuikexue Jinzhan, Adv. Water Sci., 22, 30–37, 2011.

Peel, M. C., Finlayson, B. L., and McMahon, T. A.: Updated world map of the Köppen-Geiger climate classification, Hydrol. Earth Syst. Sci., 11, 1633–1644, https://doi.org/10.5194/hess-11-1633-2007, 2007.

Peixoto, J. P. and Oort, A. H.: The climatology of relative humidity in the atmosphere, J. Climate, 9, 3443–3463, 1996.

Pereira, A. R.: The Priestley–Taylor parameter and the decoupling factor for estimating reference evapotranspiration, Agr. Forest Meteorol., 125, 305–313, 2004.

Peterson, T. C. and Vose, R. S.: An overview of the global historical climatology network temperature database, B. Am. Meteorol. Soc., 78, 2837–2849, 1997.

Priestley, C. H. B. and Taylor, R. J.: On the assessment of surface heat flux and evaporation using large-scale parameters, Mon. Weather Rev., 100, 81–92, 1972.

Rahimikhoob, A., Behbahani, M. R., and Fakheri, J.: An evaluation of four reference evapotranspiration models in a subtropical climate, Water Resour. Manage., 26, 2867–2881, 2012.

Ravazzani, G., Corbari, C., Morella, S., Gianol, P., and Mancini, M.: Modified Hargreaves-Samani equation for the assessment of reference evapotranspiration in alpine river basins, J. Irrig. Drain. E.-ASCE, 138, 592–599, 2012.

Samani, Z.: Estimating solar radiation and evapotranspiration using minimum climatological data, J. Irrig. Drain. E.-ASCE, 126, 265–267, 2000.

Sheffield, J., Goteti, G., and Wood, E. F.: Development of a 50-yr high-resolution global dataset of meteorological forcings for land surface modeling, J. Climate, 19, 3088–3111, 2006.

Shiri, J., Nazemi, A. H., Sadraddini, A. A., Landeras, G., Kisi, O., Fakheri Fard, A., and Marti, P.: Comparison of heuristic and empirical approaches for estimating reference evapotranspiration from limited inputs in Iran, Comput. Electron. Agr., 108, 230–241, 2014.

Shuttleworth, W. J. and Calder, I. R.: Has the Priestley-Taylor equation any relevance to forest evaporation?, J. Appl. Meteorol., 18, 639–646, 1979.

Singh, R. K. and Irmak, A.: Treatment of anchor pixels in the METRIC model for improved estimation of sensible and latent heat fluxes, Hydrol. Sci. J., 56, 895–906, 2011.

Snyder, R. L. and Pruitt. W. O.: Estimating reference evapotranspiration with hourly data, VII-1-VII-3, edited by: Snyder, R., Henderson, D. W., Pruitt, W. O., and Dong, A.Calif. Irrig. Mgmt. Systems, Final Rep., Univ. Calif., Davis, 1985.

Snyder, R. L. and Pruitt. W. O.: Evapotranspiration data management in California, presented at the Amer. Soc. of Civil Engr. Water Forum "92", 2–6 August 1992, Baltimore, MD, 1992.

Sumner, D. M. and Jacobs, J. M.: Utility of Penman–Monteith, Priestley–Taylor, reference evapotranspiration, and pan evaporation methods to estimate pasture evapotranspiration, J. Hydrol., 308, 81–104, 2005.

Sun, X., Ren, G., Xu, W., Li, Q., and Ren, Y.: Global land-surface air temperature change based on the new CMA GLSAT data set, Sci. Bull., 62, 236–238, 2017.

Tabari, H.: Evaluation of reference crop evapotranspiration equations in various climates, Water Resour. Manage., 24, 2311–2337, 2010.

Tabari, H. and Talaee, P. H.: Local calibration of the Hargreaves and Priestley–Taylor equations for estimating reference evapotranspiration in arid and cold climates of Iran based on the Penman–Monteith model, J. Hydrol. Eng., 16, 837–845, 2011.

Tateishi, R. and Ahn, C. H.: Mapping evapotranspiration and water balance for global land surfaces, ISPRS J. Photogramm., 51, 209–215, 1996.

Thornthwaite, C. W.: An approach toward a rational classification of climate, Geogr. Rev., 38, 55–94, 1948.

Trajkovic, S.: Hargreaves versus Penman–Monteith under humid condition, J. Irrig. Drain. E.-ASCE, 133, 38–42, 2007.

Valiantzas, J. D.: Simple ETo forms of Penman's equation without wind and/or humidity data. I: Theoretical development, J. Irrig. Drain E.-ASCE, 139, 1–8, 2013a.

Valiantzas, J. D.: Simplified reference evapotranspiration formula using an empirical impact factor for penman's aerodynamic term, J. Hydrol. Eng., 18, 108–114, 2013b.

Valiantzas, J. D.: Simple ETo forms of Penman's equation without wind and/or humidity data. II: Comparisons with reduced set-FAO and other methodologies, J. Irrig. Drain. E.-ASCE, 139, 9–19, 2013c.

Valiantzas, J. D.: Closure to "Simple ETo forms of Penman's equation without wind and/or humidity data. I: Theoretical development" by John D. Valiantzas, J. Irrig. Drain E.-ASCE, 140, 07014017, https://doi.org/10.1061/(ASCE)IR.1943-4774.0000750, 2014.

Valipour, M.: Ability of Box-Jenkins models to estimate of reference potential evapotranspiration (A Case Study: Mehrabad synoptic station, Tehran, Iran), IOSR J. Agr. Vet. Sci., 1, 1–11, 2012.

Valipour, M.: Application of new mass transfer formulae for computation of evapotranspiration, J. Appl. Water Eng. Res., 2, 33–46, 2014.

Valipour, M.: Investigation of Valiantzas' evapotranspiration equation in Iran, Theor. Appl. Climatol., 121, 267–278, 2015a.

Valipour, M.: Evaluation of radiation methods to study potential evapotranspiration of 31 provinces, Meteorol. Atmos. Phys., 127, 289–303, 2015b.

Valipour, M.: Importance of solar radiation, temperature, relative humidity, and wind speed for calculation of reference evapotranspiration, Arch. Agron. Soil Sci., 61, 239–255, 2015c.

Valipour, M.: Analysis of potential evapotranspiration using limited weather data, Appl. Water Sci., 7, 187–197, 2017.

Valipour, M. and Gholami Sefidkouhi, M. A.: Temporal analysis of reference evapotranspiration to detect variation factors, Int. J. Global Warm., https://doi.org/10.1504/IJGW.2018.10002058, in press, 2017.

Valipour, M., Gholami Sefidkouhi, M. A., and Raeini-Sarjaz, M.: Selecting the best model to estimate potential evapotranspiration with respect to climate change and magnitudes of extreme events, Agr. Water Manage., 180, 50–60, 2017.

Wang, K. and Dickinson, R. E.: A review of global terrestrial evapotranspiration: Observation, modeling, climatology, and climatic variability, Rev. Geophys., 50, RG2005, https://doi.org/10.1029/2011RG000373, 2012.

Weiß, M. and Menzel, L.: A global comparison of four potential evapotranspiration equations and their relevance to stream flow modelling in semi-arid environments, Adv. Geosci., 18, 15–23, https://doi.org/10.5194/adgeo-18-15-2008, 2008.

Wild, M., Folini, D., Schär, C., Loeb, N., Dutton, E. G., and König-Langlo, G.: The global energy balance from a surface perspective, Clim. Dynam., 40, 3107–3134, 2013.

Willett, K. M., Dunn, R. J. H., Thorne, P. W., Bell, S., de Podesta, M., Parker, D. E., Jones, P. D., and Williams Jr., C. N.: HadISDH land surface multi-variable humidity and temperature record for climate monitoring, Clim. Past, 10, 1983–2006, https://doi.org/10.5194/cp-10-1983-2014, 2014.

Willmot, C. J.: On the validation of models, Phys. Geogr., 2, 184–194, 1981.

Xu, C.-Y. and Singh V. P.: Cross comparison of empirical equations for calculating potential evapotranspiration with data from Switzerland, Water Resour. Manage., 16, 197–219, 2002.

Xu, J., Peng, S., Ding, J., Wei, Q., and Yu, Y.: Evaluation and calibration of simple methods for daily reference evapotranspiration estimation in humid East China, Arch. Agron. Soil Sci., 59, 845–858, 2013.

Zambrano-Bigiarini, M.: Hydrogof: Goodness-of-fit functions for comparison of simulated and observed hydrological time series, R package, version 0.3-8, available at: https://cran.r-project.org/web/packages/hydroGOF/hydroGOF.pdf, last access: 15 December 2015.

Zomer, R. J., Trabucco, A., Bossio, D. A., van Straaten, O., and Verchot, L. V.: Climate change mitigation: A spatial analysis of global land suitability for clean development mechanism afforestation and reforestation, Agr. Ecosyst. Environ., 126, 67–80, 2008.

15

Deriving a dataset for agriculturally relevant soils from the Soil Landscapes of Canada (SLC) database for use in Soil and Water Assessment Tool (SWAT) simulations

Marcos R. C. Cordeiro[1], Glenn Lelyk[2], Roland Kröbel[1], Getahun Legesse[3], Monireh Faramarzi[4], Mohammad Badrul Masud[4], and Tim McAllister[1]

[1]Science and Technology Branch, Agriculture and Agri-Food Canada, Lethbridge, AB, T1J 4B1, Canada
[2]Science and Technology Branch, Agriculture and Agri-Food Canada, Winnipeg, MB, R3T 2N2, Canada
[3]Department of Animal Science, University of Manitoba, Winnipeg, MB, R3T 2N2, Canada
[4]Department of Earth and Atmospheric Sciences, University of Alberta, Edmonton, AB, T6G 2E3, Canada

Correspondence: Tim McAllister (tim.mcallister@agr.gc.ca)

Abstract. The Soil and Water Assessment Tool (SWAT) model has been commonly used in Canada for hydrological and water quality simulations. However, preprocessing of critical data such as soils information can be laborious and time-consuming. The objective of this work was to preprocess the Soil Landscapes of Canada (SLC) database to offer a country-level soils dataset in a format ready to be used in SWAT simulations. A two-level screening process was used to identify critical information required by SWAT and to remove records with information that could not be calculated or estimated. Out of the 14 063 unique soil records in the SLC, 11 838 records with complete information were included in the dataset presented here. Important variables for SWAT simulations that are not reported in the SLC database (e.g., hydrologic soils groups (HSGs) and erodibility factor (K)) were calculated from information contained within the SLC database. These calculations, in fact, represent a major contribution to enabling the present dataset to be used for hydrological simulations in Canada using SWAT and other comparable models. Analysis of those variables indicated that 21.3 %, 24.6 %, 39.0 %, and 15.1 % of the soil records in Canada belong to HSGs 1, 2, 3, and 4, respectively. This suggests that almost two-thirds of the soil records have a high (i.e., HSG 4) or relatively high (i.e., HSG 3) runoff generation potential. A spatial analysis indicated that 20.0 %, 26.8 %, 36.7 %, and 16.5 % of soil records belonged to HSG 1, HSG 2, HSG 3, and HSG 4, respectively. Erosion potential, which is inherently linked to the erodibility factor (K) was associated with runoff potential in important agricultural areas such as southern Ontario and Nova Scotia. However, contrary to initial expectations, low or moderate erosion potential was found in areas with high runoff potential, such as regions in southern Manitoba (e.g., Red River Valley) and British Columbia (e.g., Peace River watershed).

1 Introduction

Integrated environmental modeling is inspired by modern environmental problems and enabled by transdisciplinary science and computer capabilities that allow the environment to be considered in a holistic way (Laniak et al., 2013). In an agricultural context, synthesis and quantification of multidisciplinary knowledge via process-based modeling are essential to manage systems that can be adapted to continual change (Ahuja et al., 2007). The Soil and Water Assessment Tool (SWAT) (Arnold et al., 1998) is an example of such a process-based model. It has been developed over the past 30 years to evaluate the effects of alternative management decisions on water resources and nonpoint-source pollution in large river basins through the simulation of major processes including hydrology, soil temperature and properties, plant growth, nutrient and pesticides dynamics, bacteria and pathogens transport, and land management (Arnold et al., 2012; Douglas-Mankin et al., 2010). Furthermore, a weather generator is included in the model to fill gaps that may exist in meteorological records.

The SWAT model has been extensively tested around the world for a wide range of hydroclimatic conditions, water and land management practices, and timescales (Douglas-Mankin et al., 2010; Arnold et al., 2012; Gassman et al., 2014). The wide adoption of the SWAT model has been prompted by preprocessing and post-processing software tools such as a GIS interface and extensive user documentation (Arnold et al., 2012), as well as several linked databases for crops, soils, fertilizers, tillage, and pesticides (Santhi et al., 2005). Among these, soil properties are especially important as they are needed for the simulation of influential processes such as evapotranspiration, soil water balance, nutrient dynamics, and sediment transport (Neitsch et al., 2005). However, the existing built-in database is only valid for SWAT applications in the USA. Accordingly, studies outside the USA require the development of a soils dataset by preprocessing available soils data into a format readable by SWAT, a time-consuming process as not all data required by SWAT are readily available for countries outside of the USA.

Worldwide, SWAT has emerged as one of the most widely used water quality watershed- and river-basin-scale models for simulation of a broad range of hydrologic and/or environmental problems (Gassman et al., 2014). These applications in different regions are described in the extensive body of peer-reviewed SWAT literature (Arnold et al., 2012). Specifically in Canada, the SWAT model has been used for hydrological simulations in most provinces, including Prince Edward Island (Edwards et al., 2000), New Brunswick (Chambers et al., 2011; Yang et al., 2009), Nova Scotia (Ahmad et al., 2011), Ontario (Asadzadeh et al., 2015; Rahman et al., 2012), Quebec (Lévsque et al., 2008), Manitoba (Yang et al., 2014), Saskatchewan (Mekonnen et al., 2016), Alberta (Mapfumo et al., 2004; Watson and Putz, 2014; Faramarzi et al., 2015), and British Columbia (Zhu et al., 2012). However, preparation of Canadian soils information in a consistent and usable format for SWAT is time-consuming (Rahman et al., 2012), as information has to be collected from soil reports and cross-checked against GIS datasets, missing soil variables have to be calculated from other physical and hydraulic properties, and all parameters have to be attributed to specific soil grids or polygons.

Some of this preprocessing work can be alleviated by using publically available databases that contain most of the information required by SWAT. The Soil Landscapes of Canada (SLC) database published by Agriculture and Agri-Food Canada (Soil Landscapes of Canada Working Group, 2010) is an example, and has been used in SWAT applications in Ontario (Asadzadeh et al., 2015; Rahman et al., 2012), Saskatchewan (Mekonnen et al., 2016), Alberta (Faramarzi et al., 2015), and British Columbia (Zhu et al., 2012). The SLC contains a GIS dataset series that provides information about the country's agricultural soils at the provincial and national levels. It was compiled at a scale of 1 : 1 million, and the information is organized according to a uniform national set of soil and landscape criteria based on permanent natural attributes (Soil Landscapes of Canada Working Group, 2010). The SLC encompasses the southern portions of the provinces of Ontario and Quebec and a larger portion of the Prairies provinces of Manitoba, Saskatchewan, and Alberta as far north as to the boreal shield. Coverage in the maritime provinces of New Brunswick, Nova Scotia, and Prince Edward Island is nearly complete (Fig. 1).

Although there are more detailed soil datasets available at provincial levels (e.g., AGRASID dataset in Alberta), selection of SLC for integration with SWAT was based on the fact that (i) it covers most soils across the agricultural regions of Canada in a single dataset; (ii) it has been used in regional studies in Canada, as described above; and (iii) it is more easily applicable to large-scale national studies as broad-scale datasets require reduced resources to prepare and process data (Moriasi and Starks, 2010). Modeling studies comparing the performance of a single model (calibrated and uncalibrated) but using soil datasets with varying spatial resolution in the USA (i.e., the State Soil Geographic database (STATSGO) compiled at 1 : 250 000 scale, and the Soil Survey Geographic database (SSURGO) with scales ranging from 1 : 12 000 to 1 : 63 360) also revealed that using either dataset produced comparable results (Mednick, 2008).

Besides the American databases (i.e., STASTSGO and SSURGO), the authors are not aware of any other effort to produce a similar dataset from a national soils database for specific use with SWAT, such as the one presented here for Canada. Past efforts in preparing a large-scale soils dataset for modeling include the standardization of the FAO–UNESCO, but this dataset was not optimized for SWAT and is presented at a much coarser spatial resolution (i.e., 1 : 5 000 000; Batjes, 1997). The SOTER (Soil Terrain) database is another initiative to provide a global soils dataset, which was intended to have a global coverage at 1 : 1 mil-

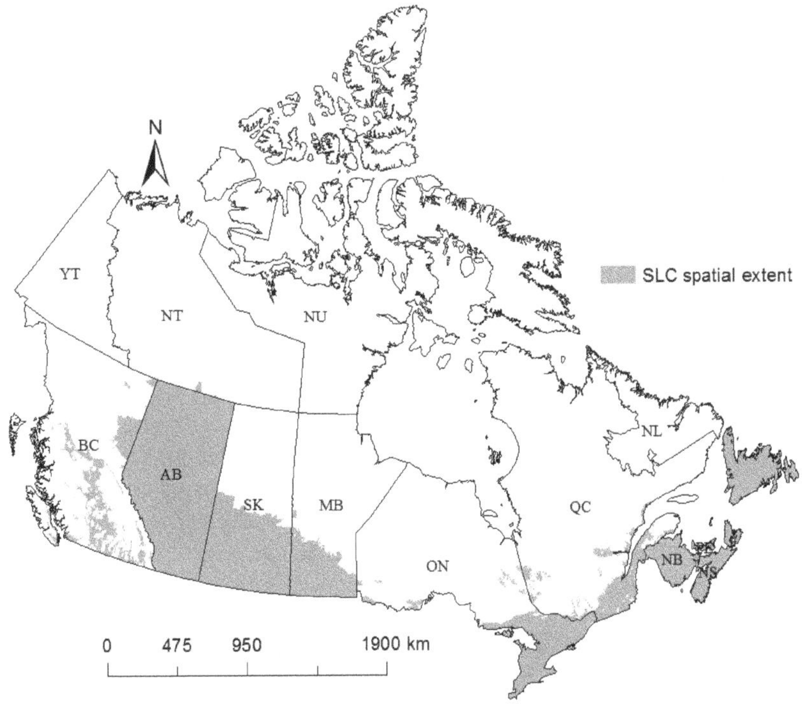

Figure 1. Spatial extent of the Soil Landscapes of Canada (SLC) database showing coverage in the provinces of Newfoundland and Labrador (NL), Prince Edward Island (PE), Nova Scotia (NS), New Brunswick (NB), Quebec (QC), Ontario (ON), Manitoba (MB), Saskatchewan (SK), Alberta (AB), and British Columbia (BC), as well as the Northwest Territories (NT).

lion scale but was later degraded to 1 : 5 million scale due to the lack of means (Dobos et al., 2005). However, SOTER is not optimized for SWAT use and requires some variables to be calculated or estimated to this end (Bossa et al., 2012). Other databases at continental scale, such as the HYPRES in Europe, only cover soil hydrologic properties (Wösten et al., 1999). At national level, only a few countries besides the USA and Canada have a soil electronic database (e.g., Australia, Brazil, and China; Shi et al., 2004; Cooper et al., 2005), while these data are not available in most countries (Cooper et al., 2005). The reduced number of available datasets, coupled with the technicalities involved in translating these datasets into SWAT format and the required variables not reported in them, contribute to the lack of large-scale soil databases fully compatible with SWAT. These limitations emphasize the novelty and importance of the dataset presented in this paper. Besides presenting a soils database ready to use in SWAT simulations in Canada, the present work provides a framework to support similar initiatives in other regions using data from global soil databases.

Due to the importance of the SWAT model for integrated environmental modeling in Canada, and the prominence of the SLC database as a potential input dataset for this model at a national level, the objective of this work was to offer a country-level soils dataset in a format ready to be used in SWAT simulations. The dataset was derived to provide over 20 parameter values for different soil types that are varied for each soil layer. It was prepared in a format suitable for use in the ArcSWAT version of the model, which is attributed to a grid or polygon-based soil map. Such a laborious preprocessing exercise is widely, but inconsistently, adopted in SWAT simulations reported in the literature. Finally, deficiencies in the dataset are also presented and discussed.

2 SLC data structure

The SLC database (http://sis.agr.gc.ca/cansis/nsdb/slc/v3.2/index.html, last access: 20 June 2017) is structured as a component-based GIS layer, whereby a single polygon may contain several soil records. This structure is similar to that of the State Soil Geographic (STATSGO) database in the USA (Srinivasan et al., 2010). Such structure creates a one-to-many relationship, whereby the multiple soil components of a polygon are not spatially defined. The actual soil information in the SLC database is stored in a number of tables linked together through intricate relationships (Soil Landscapes of Canada Working Group, 2010). Among these, four tables are relevant for developing a dataset for SWAT applications:

- The Polygon Attribute Table (PAT) provides the linkage between geographic locations (polygons in the SLC GIS coverage) and soil landscape attributes in the associated database tables (e.g., unique soil ID in the Soil Name

Table 1. Description of variables in SWAT's "usersoil" table.

Variable group	Column number in usersoil table	Variables*
Database indexing	1	OBJECTID
Soil classification	2 through to 6	MUID; SEQN; SNAM; S5ID; CMPPCT
Soil properties		
Profile	7 through to 12	NLAYERS; HYDGRP; SOL_ZMX; ANION_EXCL; SOL_CRK; TEXTURE
Layers	13 through to 132 (12 variables for 10 soil layers)	SOL_Z_x; SOL_BD_x; SOL_AWC_x; SOL_K_x; SOL_CBN_x; $CLAY_x$; $SILT_x$; $SAND_x$; $ROCK_x$; SOL_ALB_x; $USLE\_K_x$; SOL_EC_x
Inactive	133 through to 152	SOL_CAL_x; SOL_PH_x

* Subscript x corresponds to soil layer from 1 to 10.

Table (SNT) and respective number of layers in the Soil Layer Table (SLT)).

- The Component Table (CMP) describes each of the individual soil and landscape features comprising the polygons. That is, it describes which soil records are present in each spatial unit (i.e., polygon) in the GIS layer.

- The Soil Name Table (SNT) describes the general physical and chemical characteristics for all of the soils identified in a geographic region.

- The Soil Layer Table (SLT) contains soil information that varies in the vertical direction (i.e., layered attributes).

The CMP table describes the proportion of each nonspatially defined soil component in a polygon if more than a soil component exists (the soil component(s) refers to the soil(s) element(s) that comprise each polygon). The component numbering follows a sequence of decreasing proportion in a polygon (i.e., first component has the highest proportion; last component has the smallest proportion). This component-based structure of the SLC database does not affect the analysis since all the soil records listed in the SNT table were processed to generate the present dataset. However, it has implications for the SWAT model user, who has to make a decision on how to handle the relationship between the polygon (spatially defined) and each nonspatially defined soil component in multicomponent polygons (e.g., selecting the larger component in a polygon or generating a hybrid soil incorporating properties of each soil component).

3 SWAT soils data structure

The SWAT soils information is stored in the "usersoil" table, located within the SWAT 2012 database in Microsoft Access format (i.e., SWAT2012.mdb). Each soil record is stored as a new record (i.e., row) in the table. Specific soil variables (Table 1) comprise the 152 columns of the usersoil table. The first column is an OBJECTID field assigning a unique identifier for each record. Columns two through six pertain to soil classification. The second column is the map unit identifier (MUID), which is used for mapping a collection of areas grouped by the same soil characteristics. A single MUID may describe different soil types, which are stored with a record counter in the third column (SEQN), while a soil identifying name (SNAM), a soil interpretation record (S5ID), and the percent of each soil component (CMPPCT) are recorded in the fourth, fifth, and sixth columns, respectively (Sheshukov et al., 2009). Columns 7 through 12 describe major soil properties pertaining to the soil record, namely, the number of layers (NLAYERS), the hydrological soil group to which that soil belongs (HYDGRP), the maximum rooting depth of the soil profile (SOL_ZMX), the fraction of soil porosity from which anions are excluded (ANION_EXCL), the potential of maximum crack volume of the soil profile expressed as a fraction of the total soil volume (SOL_CRK), and the texture of the soil layer (TEXTURE).

The next 120 columns starting from column 13 (i.e., columns 13 to 132) describe the information for each layer of the soil record. These columns are arranged in sets of 12 variables each for 10 possible soil layers. The variable NLAYERS indicates how many of these sets should be populated. Variables for any sets beyond NLAYERS should be assigned a value of zero. The variables included in each set of soil layers are the depth from the soil surface to the bottom of the layer (SOL_Z), moist bulk density (SOL_BD), available water capacity of the soil layer (SOL_AWC), saturated hydraulic conductivity (SOL_K), organic carbon (SOL_CBN), clay (CLAY), silt (SILT), sand (SAND), and rock fragment (ROCK) contents, moist soil albedo (SOL_ALB), erodibility factor (USLE_K), and electrical conductivity (SOL_EC). Beyond the columns describing layered soil information, there are 20 columns (i.e., columns 133 to 152) describing two variables (i.e., soil $CaCO_3$ (SOL_CA) and soil pH (SOL_PH)) for 10 soil layers. These variables are not currently active in SWAT and are assigned a value of zero.

4 Merging the two datasets

Despite its usefulness as a source of soil information for hydrological simulations, the SLC dataset is not assembled in a format readable by SWAT or other similar models. For example, SWAT stores all the properties for a specific soil record in a single row in the usersoil table, while this information is stored in the SLC as multiple rows in two different tables (i.e., SNT and SLT). Thus, the information contained in the SLT database has to be processed to satisfy SWAT's format requirements. In addition, all properties in the usersoil table are spatially defined, while those of SLC are often stored in a multi-polygon structure with no unique spatial identification. Variables required by SWAT and contained in the dataset presented here were either extracted from SNT and SLT, or calculated from the information therein. Some other variables were estimated from published values. Extraction or calculation of variables was done through an R code that imported both SNT and SLT, screened the data for missing records and missing SWAT-required information (data screening is described in Sect. 5), and sequentially populated unique soil records in the database. The next section describes how these variables were defined.

5 Data screening

5.1 Screening out incomplete soil information in the SNT

The use of the SNT is necessary as it links the soils information to the GIS coverage containing the PAT. However, a first screening was required to remove soil records from the SNT that are not present in the SLT, as soil layer information is required by SWAT. The mismatch among soil records in both tables can occur for a number of reasons. For example, there are records in both tables that pedologists have identified but whose properties have not yet been characterized. Also, soil records listed in one table may be absent from another table due to changes in soil classification. Finally, soil records listed as unclassified in the SNT (i.e., variable KIND = U) do not have any data associated with them in the SLT and do not occur on any published map.

Out of the 14 063 unique soil records in the SNT, 489 were missing in the SLT and, therefore, removed from the analysis. These 489 soil records correspond to around 3.5 % of the soils listed in the SNT. Most of the missing records were reported as unclassified (305 soils; 62.2 %), suggesting that these soils have been identified, but their properties have not yet been characterized. Mineral soil records corresponded to 29.4 % (144 soils) of the total, likely a reflection of changes in classification. The other two classes comprised non-true soils (e.g., mine tailings, urban land; 33 soils; 6.7 %) and organic soils (8 soils; 1.6 %). Also, only 58 of the 489 missing soil records (11.0 %) could be mapped through linking with the CMP table, making it impossible to do any spatial analysis on the distribution of these soils across the country. However, since the SNT assigns a province for each soil record, it is possible to identify where these missing records occur. Most of the missing soil records were in British Columbia (167 soils; 34.2 %), Manitoba (151 soils; 30.9 %), and Saskatchewan (133 soils; 27.2 %), with smaller proportions in Yukon (13 soils; 2.7 %), Ontario (11 soils; 2.3 %), Nova Scotia (9 soils; 1.8 %), and Newfoundland (5 soils; 1.0 %).

5.2 SWAT requirements

The SWAT data requirements were used as a second level of screening to build the present dataset. The soil input variables in SWAT can be either required or optional (Table 2; Arnold et al., 2013). Required variables that could not be calculated or estimated (e.g., SOL_BD, SOL_K, SOL_CBN, CLAY, SILT, and SAND) were used to separate complete from incomplete records. Soil records in the SLT containing or allowing derivation of all the variables required by SWAT were compiled in a dataset comprising 11 838 unique records that were importable into the model. Soils in the SLT with missing records (i.e., variables entered as −9 in the database) for the required SWAT variables (gray rows in Table 2) were removed from the analysis. These soil records were compiled into a soils list provided as a reference.

As for the nonmatching soil records in the SNT and SLT, only 547 out of 1736 (i.e., 31.5 %) records with missing information could be mapped through linking with the CMP table, which renders any spatial representation of these records nonmeaningful. However, the provinces where these records occur could also be identified. The highest proportions of soil records with incomplete information were in British Columbia (490 records; 28.2 %) and Manitoba (391 records; 22.54 %). Ontario (182 records; 10.5 %) and Alberta (180 records; 10.4 %) had intermediate values, while Newfoundland (123 records; 7.1 %), Saskatchewan (102 records; 5.9 %), New Brunswick (93 records; 5.4 %), the Northwest Territories (80 records; 4.6 %), Nova Scotia (47 records; 2.7 %), Quebec (30 records; 1.7 %), and Yukon (17 records; 1.0 %) had less than 10 % of the soil records missing information.

6 Populating the usersoil table in SWAT

The variables in SWAT's usersoil table refer to record indexing and soil classification, as well as soil properties pertaining to the entire profile or specific layers. The variables in each of these groups are described in the following subsections. The usersoil table starts with a number of columns that define the database and soil classification variables, followed by soil profile and layer information, and inactive soil properties (Table 2).

Table 2. Variables included in the SWAT usersoil table.

Column	Variable[a]	Description	Units	Status
1	OBJECTID	Object identifier	–	Optional
2	MUID	Mapping unit identifier	–	Optional
3	SEQN	Record counter calculated by SWAT	–	Optional
4	SNAM	Soil identifying name	–	Optional
5	S5ID	Soil interpretation record	–	Optional
6	CMPPCT	Soil component percent	–	Optional
7	NLAYERS[b]	Number of layers	–	Required
8	HYDGRP	Hydrologic soil group	–	Required
9	SOL_ZMX	Maximum rooting depth of the soil profile	mm	Required
10	ANION_EXCL	Fraction of soil porosity from which anions are excluded	–	Optional
11	SOL_CRK	Potential of maximum crack volume of the soil profile expressed as a fraction of the total soil volume	$mm^3 mm^{-3}$	Optional
12	TEXTURE	Texture of soil layer	–	Optional
13	SOL_Z_x	Depth from soil surface to bottom of layer	mm	Required
14	SOL_BD_x	Moist bulk density	$Mg\,m^{-3}$ or $g\,cm^{-3}$	Required
15	SOL_AWC_x	Available water capacity of the soil layer	$mm\,mm^{-3}$	Required
16	SOL_K_x	Saturated hydraulic conductivity	$mm\,h^{-1}$	Required
17	SOL_CBN_x	Organic carbon content	% (w/w)	Required
18	$CLAY_x$	Clay content	% (w/w)	Required
19	$SILT_x$	Silt content	% (w/w)	Required
20	$SAND_x$	Sand content	% (w/w)	Required
21	$ROCK_x$	Rock fragment content	% (w/w)	Required
22	SOL_ALB_x	Moist soil albedo	–	Required
23	$USLE\_K_x$	Erodibility factor (K)	0.01 ton ac h ac ft-ton in^{-1}	Required
24	SOL_EC_x	Electrical conductivity	$dS\,m^{-1}$	Optional

Adapted from Arnold et al. (2013) and Sheshukov et al. (2009). [a] Subscript x corresponds to soil layer from 1 to 10. The variables SOL_CAL_x and SOL_PH_x are present in the usersoil table after all the columns listed above for all the 10 preexisting layers. These variables refer to soil $CaCO_3$ and soil pH, respectively, and are not currently active in the model. Thus, their records are entered as zero in the SWAT 2012 database. [b] The number of layers defines how many entries will be required in the layered information, signalled by the subscript x. For example, a soil with NLAYERS = 4 should have subscript x corresponding to soil layer variables from 1 to 4. As a result, the records extend to column 60 in the usersoil table. (i.e., 4 layers × 12 variables + 12 preceding variables = 60).

6.1 Database and soil classification variables

The SWAT soil classification variables include the OBJECTID (general listing number), MUID (map unit identifier), SEQN (sequence number), SNAM (soil name), S5ID (Soils5 ID number for USDA soil series data), and CMPPCT (percentage of the soil component in the MUID). A numbering system used for the OBJECTID variable was chosen to avoid conflicts with existing soil records in the usersoil table. The SWAT model comes with more than 200 soil records in a built-in database that cannot be easily overwritten, and any soil record imported into the database with the same OBJECTID as the existing record will not be imported. Thus, the OBJECTID field was populated sequentially from 1001 to the number of unique soil records in the SLC database plus 1000 (i.e., OBJECTID ends in 12 838 in the case of the COMPLETE dataset, which has 11 838 unique soil records). The map unit ID (MUID) was assigned the SOIL_ID code in the SLC dataset, which is a concatenation of the province code (two digits), a soil code (three digits), a modifier code (five digits), and a profile code (one digit). The sequence number (SEQN) variable was assigned the same value as the OBJECTID variable. This process created a unique SEQN for each recurrence in the SLC dataset.

Similar to the MUID variable, the soil name variable (SNAM) was also assigned the SOIL_ID code in the SLC, despite the soil name being in the database, so as to link the soil information to the GIS layer. The S5ID variable was created as a concatenation between the acronym "SLC" and the province two-digit abbreviation code. For example, all the soil records in the province of Alberta have an S5ID equal to "SLCAB". The CMPPCT variable was assigned a value of 100, meaning that the soil comprises 100 % of this component. As stated in Sect. 2, the user has to make a decision on how to handle multipart polygons in the preprocessing of the SLC GIS dataset since the soil records in multicomponent polygons are not spatially defined.

6.2 Soil profile information

The following six variables in the dataset (i.e., columns 7 to 12) pertain to soil profile information. The number of layer variables (NLAYERS) was defined according to the soil layers in the SLT below the soil surface. The SLT table also contains information for layers above the soil surface, as is the case for litter, which have negative values for upper and lower depths (i.e., the ground surface corresponded to the zero depth, while above-surface and below-surface layers have negative and positive values, respectively). Above-surface layers were removed from the dataset prior to analysis through filtering layers with lower depth above the soil surface (i.e., lower depth less than or equal to zero).

The hydrologic soil group (HSG) variable (HYDGRP) is an influential parameter for estimation of runoff using the SCS curve number method and, consequently, for hydrological simulations in SWAT (Gao et al., 2012; Neitsch et al., 2005). The HSGs were calculated according to the method outlined by USDA-NRCS (1993), which is based on depth to the impermeable layer (e.g., bedrock), depth from soil surface to shallowest water table during the year, hydraulic conductivity of the least conductive layer of the soil profile, and depth range of the hydraulic conductivity. The specific criteria used are provided in tabular form in the Supplement. Soils in the dual HSG classes were assigned to the less restrictive class since most agricultural areas in Canada exhibit some degree of drainage (e.g., municipal drainage network, surface drains, or tile drainage). SWAT translates HSG alphabetical classification into a numeric system, where HSGs A, B, C, and D, are interpreted as 1, 2, 3, and 4, respectively. The runoff potential increases with increasing numeric designations.

The depth to the impermeable layer is not reported in the SLC database and was estimated based on the soil layers available in the SLT. When a bedrock layer or specific soil horizons were present (i.e., fragipan; duripan; petrocalcic; ortstein; petrogypsic; cemented horizon; densic material; placic; bedrock, paralithic; bedrock, lithic; bedrock, densic; or permafrost; USDA-NRCS, 1993), its upper depth was used as the depth to impermeable layer. When a bedrock layer was absent, the lower depth of the deepest mineral soil layer was used as an alternative. The shallowest annual depth to water table is also not reported and was estimated based on drainage class reported in the SNT. Very poorly drained, poorly drained, imperfectly drained, moderately well drained, and well drained (or better) soils were assigned water table depths of 0, 25, 75, 100, and 125 cm, respectively. The variables pertaining to hydraulic conductivity of the least conductive layer of the soil profile and depth range of the hydraulic conductivity were both calculated using information from the SLT.

Out of the 11 838 soil records in the generated dataset, 21.3 %, 24.6 %, 39.0 %, and 15.1 % belonged to HSGs 1, 2, 3, and 4, respectively. These results suggest that more than half of the agricultural soil records in Canada have a relatively high or high runoff generation potential (i.e., HSGs 3 and 4, respectively). A spatial analysis indicated that 20.0 %, 26.8 %, 36.7 %, and 16.5 % of the areal extent of the soil records belonged to HSGs 1, 2, 3, and 4, respectively. Many of the soil records with higher potential for runoff generation are in the humid regions of Ontario, Quebec, and the Maritimes (Fig. 2). Not surprisingly, this region has extensively adopted measures to address excess moisture in agricultural soils, such as tile drainage (Stonehouse, 1995; Rasouli et al., 2014). Excess moisture is also a problem in areas of Canadian Prairies, such as the Red River Valley in Manitoba, where surface drainage (Bower, 2007) and a growing use of tile drainage (Cordeiro and Sri Ranjan, 2012, 2015) have been used to address this problem. Conversely, soil records with low potential for runoff generation are located in Saskatchewan and southeastern Alberta (along the Saskatchewan border), which are among the most arid regions in Canada (Wolfe, 1997).

The maximum rooting depth of the soil profile (SOL_ZMX) was assumed to be the lower depth of the deepest layer in the SLC soil profile. The fraction of soil porosity from which anions are excluded (ANION_EXCL) was not available in the SLC database and was set to the default value of 0.5 in SWAT (Arnold et al., 2013). This variable affects the concentration of nitrate in the mobile water fraction, which is directly related to nitrate leaching. The potential of maximum crack volume of the soil profile expressed as a fraction of the total soil volume (SOL_CRK) can be calculated with the FLOCR model using 30-year weather data (Bronswijk, 1989). However, due to the fact that the model is not readily available for download and the unreasonable time required to run the model for such a large number of soil records, as well as the fact that SOL_CRK is optional in SWAT, its value was set to 0.5. In large-scale studies this value is further adjusted through a spatially explicit calibration scheme (Whittaker et al., 2010). The SOL_CRK variable controls the potential crack volume for the soil profile. This value was selected based on the fact that all of the built-in soil records in the SWAT soils database have the SOL_CRK variable set to 0.5. The TEXTURE variable, although not required for simulations with the SWAT model, was estimated for reference using the "TT.points.in.classes" function from the "soiltexture" R package (Moeys, 2016). The Canadian soil texture classification system was used as a reference.

6.3 Soil layer information

The soil profile variables are followed by 10 sets of 12 variables (i.e., columns 13 to 132) pertaining to layered soil information. The lower depth of each soil layer in the SLT was used as the depth from soil surface to the bottom layer (SOL_Z). The soil bulk density (SOL_BD) was extracted directly from the SLT. The available water capacity of the soil layer (SOL_AWC) was calculated from the water retention of the soil reported in the SLT at different matric potentials. The water moisture content at −33 and −1500 kPa were assumed to represent the soil moisture at field capacity (FC) and per-

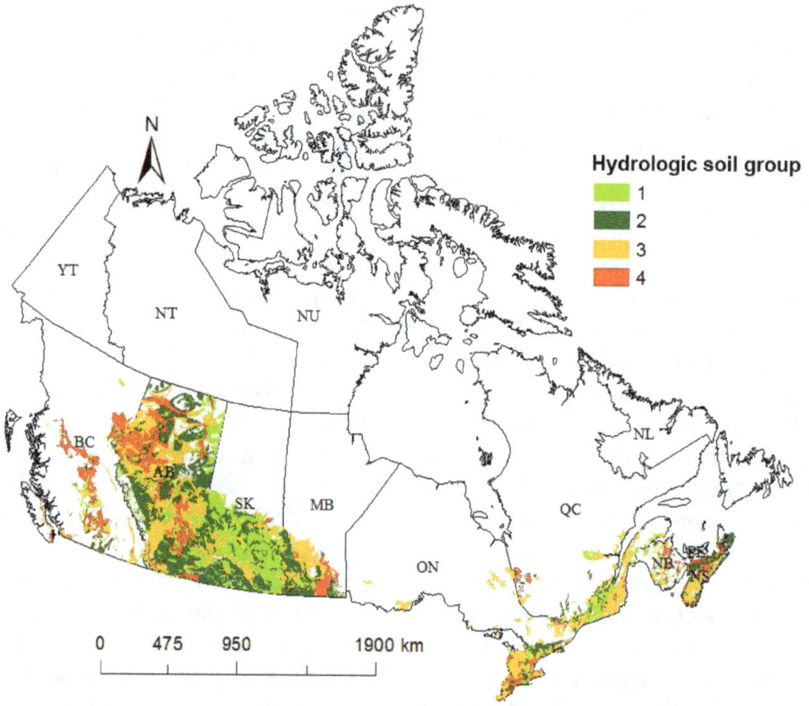

Figure 2. Spatial distribution of the hydrologic soil groups (HYDGRP) variable calculated for the Soil Landscapes of Canada (SLC) database. HSG A = 1, HSG B = 2, HSG C = 3, and HSG D = 4 shown for the provinces of Prince Edward Island (PE), Nova Scotia (NS), New Brunswick (NB), Quebec (QC), Ontario (ON), Manitoba (MB), Saskatchewan (SK), Alberta (AB), and British Columbia (BC). Some HSGs could not be mapped (e.g., province of Newfoundland and Labrador (NL)) due to missing records in the PAT of the GIS layer or being part of the soils with missing data in the SLT.

manent wilting point (PWP), respectively (Givi et al., 2004). The SOL_AWC was calculated as the difference between FC and PWP (Hillel, 1998). Soil moisture content at −33 kPa was not available for 2658 layer records (i.e., 4.3 % of the 61 905 original records in the SLT table), which would result in the variable SOL_AWC not being calculated and the loss of more soil records from the dataset. To avoid this, the moisture content at −10 kPa was used to replace that at −33 kPa. On average, the soil moisture content in the soil profile was around 6 mm larger at −10 kPa than that at −33 kPa (Table 3), indicating an overestimation of SOL_AWC in these records. Larger differences between soil moisture content at −10 and −33 kPa in the top soil layers were likely driven by lower bulk densities, which increase the water-holding capacity of the soil (Table 3).

The variables saturated hydraulic conductivity (SOL_K) and soil organic carbon content (SOL_CBN), as well as the clay (CLAY), silt (SILT), sand (SAND), and rock fragment (ROCK) contents, were extracted directly from the SLT. The moist soil albedo (SOL_ALB) variable was only required for the top layer as subsequent layers were assigned a value of zero. Since this variable is not reported in the SLC database, it was estimated as the average (i.e., 0.10) of the range reported by Maidment (1993) for moist, dark, plowed fields (i.e., 0.05–0.15). Again, this value was selected since the SLC version 3.2 focuses on agricultural areas, which is also the major domain simulated by SWAT.

Another important variable for SWAT is the erodibility factor (USLE_K), used as an input to the Universal Soil Loss Equation (USLE). This equation is used to calculate soil erosion, which is inherently linked to sediment and nutrient transport (Sharpley et al., 1992, 2002; He et al., 1995; Aksoy and Kavvas, 2005; Koiter et al., 2013) and therefore, critical for simulations of non-point sources of pollution. The erodibility factor was calculated using the method presented by Sharpley and Williams (1990), which is based on the sand, silt, clay, and organic carbon content of the soil (Eq. 1):

$$K = \left(0.2 + 0.3 \cdot \exp\left[-0.256 \cdot m_s \cdot \left(1 - \frac{m_{\text{silt}}}{100}\right)\right]\right) \cdot \left(\frac{m_{\text{silt}}}{m_c + m_{\text{silt}}}\right)^{0.3} \cdot \left(1 - \frac{0.25 \cdot \text{orgC}}{\text{orgC} + \exp\left[3.72 - 2.95 \cdot \text{orgC}\right]}\right) \cdot \left(1 - \frac{0.7 \cdot \left(1 - \frac{m_s}{100}\right)}{\left(1 - \frac{m_s}{100}\right) + \exp\left[-5.51 + 22.9 \cdot \left(1 - \frac{m_s}{100}\right)\right]}\right), \quad (1)$$

where K is the erodibility factor (0.01 ton ac h ac ft-ton in^{-1}), m_s is the sand content (%), m_{silt} is the silt content (%), m_c is the clay content (%), and orgC is the organic carbon content (%) of the respective soil layer.

Table 3. Average soil moisture content at matric potentials −10 and −33 kPa and average soil bulk density for discrete layers of the soil profile. The average was calculated for all soils in the dataset. Each layer could have different depths for individual soils used in the average.

Layer	$\overline{\theta}$ at −10 kPa	at −33 kPa	Difference (mm)	Average soil bulk density (g cm^{-3})
1	36.8	29.67	7.13	1.13
2	33.65	26.72	6.93	1.27
3	31.99	25.36	6.63	1.38
4	29.48	23.32	6.16	1.47
5	28.1	22.17	5.93	1.50
6	27.26	21.53	5.73	1.52
7	27.03	21.42	5.61	1.54
8	26.98	21.17	5.81	1.54
9	25.05	18.86	6.19	1.55
Average	29.59	23.36	6.24	1.43

$\overline{\theta}$ is the average soil moisture content (mm).

As for SOL_ALB, USLE_K is only required for the top layer and subsequent layers were also assigned a value of zero. When converted from imperial to SI units (Foster et al., 1981), the range of calculated values (Table 4) generally agrees with the ranges reported for Canada (Wall et al., 2002), taking into consideration that K values may vary, depending on particle size distribution, organic matter, and structure and permeability of individual soils (Wall et al., 2002). However, the units in the dataset presented here were kept in imperial units for consistency with the SWAT input format. The spatial distribution of the erodibility factor (Fig. 3) was anticipated to align with the HSG, which was the case in areas of low erosion potential in Saskatchewan, where sandy soils prevail, and in areas where runoff potential is high such as in southern Ontario. However, the spatial distribution of USLE_K somewhat contrasted to that of the HSG in some areas of Manitoba and British Columbia, where low sediment transport potential was predicted in areas with high runoff potential. This contrast was likely due to other factors reducing the potential for sediment transport, such as soils with high clay to silt ratios or high organic carbon contents (Sharpley and Williams, 1990).

Table 4. Comparison between the average erodibility factor (K) calculated for each soil textural class in the SWAT dataset and values reported in the literature.

Soil textural class	Acronym	Calculated average K	Reported K range[a]
Loam	L	0.14	0.23–0.30
Heavy clay	HCl	0.18	0.05–0.23
Silty clay loam	SiClLo	0.22	0.30–0.38
Clay loam	ClLo	0.14	0.23–0.30
Silt loam	SiLo	0.22	0.30–0.38
Sand	Sa	0.04	< 0.05
Sandy loam	SaLo	0.11	0.05–0.23
Clay	Cl	0.14	0.23–0.30
Silty clay	SiCl	0.22	0.23–0.30
Loamy sand	LoSa	0.07	< 0.05
Sandy clay loam	SaClLo	0.10	0.23–0.30
Silt	Si	0.55	0.30–0.38[b]
Sandy clay	SaCl	0.09	0.05–0.23[c]

[a] Adapted from Wall et al. (2002). [b] Range not reported; value from SiLo used. [c] Range not reported; value from SaLo used.

The soil electrical conductivity (SOL_EC) information was extracted directly from the SLT. The last 20 columns of the dataset (i.e., columns 133 to 152), which correspond to SOL_CAL for the 10 soil layers followed by SOL_PH for the same layers, were all populated with zeros since these variables are not currently active in SWAT. These variables also had values of zero for all the preexisting soil records in the built-in database in the model.

7 Uncertainty

Soil properties are inherently uncertain due to spatial variability and precision of measurement methods (Lacasse and Nadim, 1996). This uncertainty has direct implications for hydrologic simulations and their interpretation (Beven, 2011). The SWAT model simulations, therefore, are subject to the uncertainty of the soil properties used as input. For example, hydraulic conductivity is highly spatially and temporally variable (Hillel, 1998), and these uncertainties are very difficult to be avoided. The processing applied to the original SLC database in the present analysis did not introduce any further uncertainty to the variables reported by SLC (e.g., saturated hydraulic conductivity). There is, however, some uncertainty relating to estimated and calculated parameters. These uncertainties are discussed in this section, although their quantification is beyond the scope of the present work.

An example of introduced uncertainty is the moist soil albedo in the present dataset (0.10), which is the average of a range reported in the literature (Sect. 6.3). However, any value selected would have some uncertainty associated with

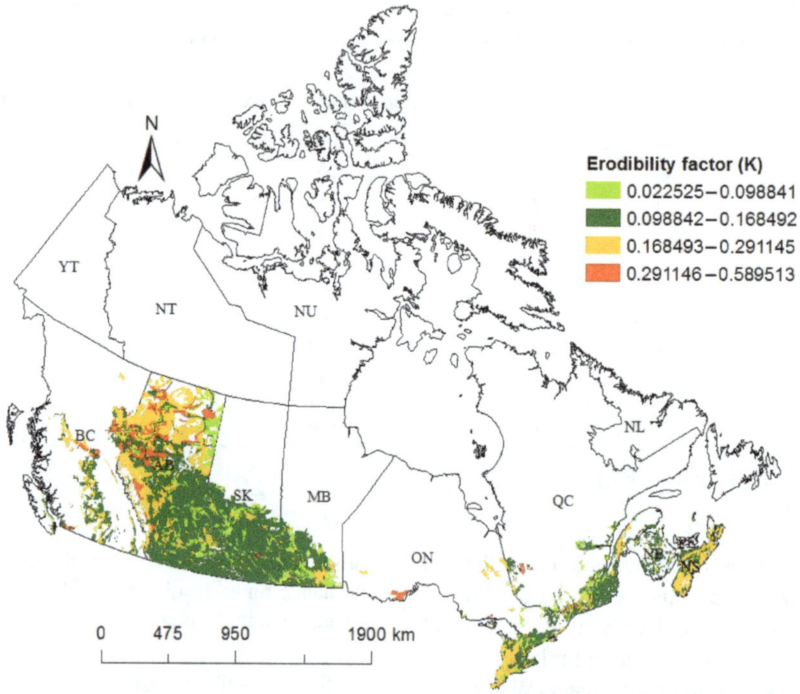

Figure 3. Spatial distribution of the erodibility factor (K) calculated for the Soil Landscapes of Canada (SLC) database (imperial units). The K factor shown for the provinces of Prince Edward Island (PE), Nova Scotia (NS), New Brunswick (NB), Quebec (QC), Ontario (ON), Manitoba (MB), Saskatchewan (SK), Alberta (AB), and British Columbia (BC). Some HSGs could not be mapped (e.g., province of Newfoundland and Labrador (NL)) due to missing records in the PAT of the GIS layer or being part of the soils with missing data in the SLT.

it from a modeling standpoint because a single value cannot represent the variability in moist soil albedo as the soil dries up. This is a recognized problem when trying to represent spatially or temporally variable parameters (e.g., soil moisture) using a point measurement or single value in hydrological models (Beven, 2011).

Another example of uncertainty is the HSG calculations, which required a number of assumptions. For example, the shallowest annual depth to water table was unavailable in the SLC and therefore estimated based on the drainage class of each soil record. Also, the assumption of artificial drainage resulted in assignment of dual-class HSGs to the less restrictive one. An assessment of HSG in the USA indicated a standard error of about one HSG (Stewart et al., 2012), suggesting that classifying soils in the neighboring groups is not uncommon and that there is some uncertainty associated with those estimates.

The estimation of erodibility factor (K; Eq. 1) would also be subject to uncertainty. This is illustrated by the range of erodibility factors reported for a single soil textural class (Table 4). This variability can arise for different reasons. One already mentioned is the precision of the method used to determine the textural classes. A second one is the procedure used to calculate K. For example, Neitsch et al. (2005) present an equation that requires a soil structure code used in soil classification with many types and subtypes. Since this soil structure code is note reported in the SLC, an alternative relationship (Eq. 4; Sharpley and Williams, 1990) that does not require the soil structure code was used. This relationship was selected to avoid the added uncertainty from estimating the soil structure code.

Finally, one last variable worthwhile discussing in term of uncertainty is the available water capacity of the soil layers. This variable was estimated as the difference between field capacity and permanent wilting point. The procedure used here to estimate available water content (i.e., the difference between field capacity and permanent wilting point) follows the same procedure used by SWAT (Neitsch et al., 2005) and is described elsewhere in the soil physics literature (Hillel, 1998). Therefore, it would not introduce any further uncertainty. However, using the soil moisture content at -10 kPa to replace records with missing soil moisture content at -33 kPa (Sect. 6.3) would introduce some uncertainty in available water capacity for the replaced records.

Overall, prediction of uncertainty in regional hydrologic modeling and a careful input data discrimination analysis prior to calibration is unavoidable (Faramarzi et al., 2015). Especially in large transboundary river basins where a consistent soil dataset is not available from a single source, a careful uncertainty assessment provides information on data and model quality. Although the authors are unaware of SWAT hydrologic simulations in binational watersheds that use soil datasets from both the USA and Canada, maybe due to lack of large-scale datasets for Canada, it is expected that the model output is subject to the quality and quantity of both datasets. Some aspects contributing to this uncertainty are (i) possible discontinuity in the mapping units (i.e., polygons) between the GIS layers of both datasets, (ii) the soil record being mapped in multicomponent polygons in the GIS coverage (Soil Survey Staff, 1999; Agriculture and Agri-Food Canada, 1998), (iii) differences in soil taxonomy between the USA system (Soil Survey Staff, 1999) and the Canadian system (Agriculture and Agri-Food Canada, 1998) of soil classification, (iv) the methods used to measure/estimate the physicochemical variables, which may differ in accuracy and precision, and (v) the natural variability in the calculation of some variables that cannot be measured (e.g., HSG; Stewart et al., 2012). Given the number of aspects influencing trans-boundary uncertainty and the large spatial scale of both the USA and the dataset discussed here, an assessment of such uncertainty is beyond the scope of the present study. However, this assessment is suggested to quantify the share of errors from these sources in hydrologic model projection in both upstream and downstream tributaries. These are the subjects of our continuing studies.

8 Importing the SLC dataset into the SWAT database

Although the SWAT database is in a proprietary format (i.e., Microsoft Access), the present soils dataset has been published in a nonproprietary format (i.e., comma-separated values (CSV) file) that can be opened in a variety of software packages. However, the dataset can be easily imported into the SWAT soils database using an automated import routine in Microsoft Access (Fig. 4). This import process consists of opening the SWAT2012 database and using the “Import Text File” tool under the “Import & Link” section of the “External Data” tab to read the CSV file. This action will prompt a window where the user can select the path to where the present dataset is stored and specify how and where the data are stored in the database. The option “Append a copy of the record to the table” should be selected, which activates a drop-down menu from which the usersoil table should be highlighted. Once these options have been processed, an “Import Text Wizard” window will be prompted, where the option “Delimited – Characters such as comma or tab separate each field” should be selected. Processing of this selection will prompt another window where the option “comma” should be automatically selected by the wizard. However, the user should activate the box “First Row Contains Field Names” since the first row of the present dataset contains the variable labels. Confirming the processing of the next windows should finalize the import process, and the data should be ready to be used in SWAT predictions.

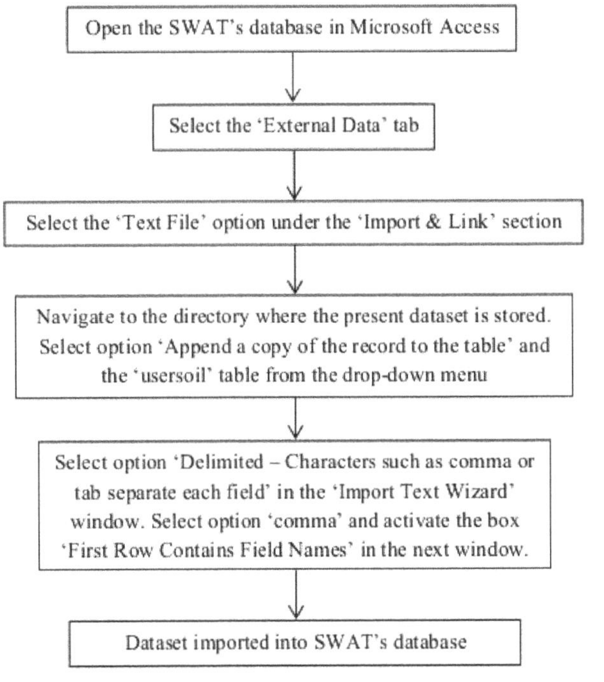

Figure 4. Flowchart showing the steps for importing the present soils dataset into SWAT’s database.

9 Data availability

PANGAEA, an open access library to archive, publish, and distribute georeferenced data, supports database-dependent research. Therefore, the entire dataset (Cordeiro et al., 2017) is published and archived in the PANGAEA database (https://doi.org/10.1594/PANGAEA.877298) under Creative Commons Attribution 3.0 Unported, where the user must give appropriate credit, provide a link to the license, and indicate if changes are made.

10 Conclusions

The soils dataset presented and discussed in this work represents an effort to facilitate hydrological simulations using the SWAT model in Canada. The dataset consists of a compilation of 11 838 different soil records from the SLC database with all the information required by SWAT and is ready to be imported into the model’s soils database. A two-level data screening procedure removed 489 soil records with missing layered information (i.e., not present in the SLT), while 1736 records were removed due to the lack of critical information required by SWAT, such as soil bulk density or saturated hydraulic conductivity. Among the major contributions of this dataset, the calculation and/or estimation of variables not reported in the SLC database are of special importance. The hydrologic soil groups (HSGs) calculated from the SLC database suggest that about half of the soil records in Canada belong to classes with higher potential to generate runoff (i.e., HSG classes 3 and 4). Occurrence of soils in HSG 3 and 4 agree with management practices aimed at addressing excess moisture conditions in agricultural fields, such as subsurface drainage in southern Ontario and Manitoba. The erodibility factor, which is another important variable for SWAT simulations of non-point source pollution, suggests a relationship with runoff potential in portions of southern Ontario and Nova Scotia. However, low erodibility potential, likely driven by high clay to silt ratios or high organic carbon content, was found in areas with higher runoff potential in Manitoba and British Columbia.

Author contributions. MRCC and RK developed the concept for development of the dataset. GL interpreted the soil information contained in the SLC database. MRCC and GL developed the methodology for deriving the soil variables. MRCC developed the code using R programming language to process the SLC dataset and performed data analysis. All the authors revised the dataset and participated in manuscript preparation.

Competing interests. The authors declare that they have no conflict of interest.

Acknowledgements. This research was supported by the Beef Cattle Research Council and Agriculture and Agri-Food Canada through the Beef Cluster, Environmental Footprint of Beef Project, and the Alberta Livestock and Meat Agency (ALMA) of the Alberta Agriculture and Forestry (grant no. 2016E017R).

Edited by: David Carlson

References

Agriculture and Agri-Food Canada: The Canadian System of Soil Classification, 3rd Edn., 93, NRC Research Press, Ottawa, 188 pp., 1998.

Ahmad, H. M. N., Sinclair, A., Jamieson, R., Madani, A., Hebb, D., Havard, P., and Yiridoe, E. K.: Modeling sediment and nitrogen export from a rural watershed in eastern Canada using the Soil and Water assessment Tool, J. Environ. Qual., 40, 1182–1194, https://doi.org/10.2134/jeq2010.0530, 2011.

Ahuja, L. R., Andales, A. A., Ma, L., and Saseendran, S. A.: Whole-System Integration and Modeling Essential to Agricultural Science and Technology for the 21st Century, Journal of Crop Improvement, 19, 73–103, https://doi.org/10.1300/J411v19n01_04, 2007.

Aksoy, H. and Kavvas, M. L.: A review of hillslope and watershed scale erosion and sediment transport models, Catena, 64, 247–271, https://doi.org/10.1016/j.catena.2005.08.008, 2005.

Arnold, J., Kiniry, J., Srinivasan, R., Williams, J., Haney, E., and Neitsch, S.: SWAT 2012 Input/Output Documentation, Texas Water Resources Institute, 2013.

Arnold, J. G., Srinivasan, R., Muttiah, R. S., and Williams, J. R.: Large area hydrologic modeling and assessment part i: Model development, J. Am. Water Resour. As., 34, 73–89, https://doi.org/10.1111/j.1752-1688.1998.tb05961.x, 1998.

Arnold, J. G., Moriasi, D. N., Gassman, P. W., Abbaspour, K. C., White, M. J., Srinivasan, R., Santhi, C., Harmel, R. D., Griensven, A. v., Liew, M. W. V., Kannan, N., and Jha, M. K.: SWAT: Model Use, Calibration, and Validation, T. ASABE, 55, 1491–1508, 2012.

Asadzadeh, M., Leon, L., McCrimmon, C., Yang, W., Liu, Y., Wong, I., Fong, P., and Bowen, G.: Watershed derived nutrients for Lake Ontario inflows: Model calibration considering typical land operations in Southern Ontario, J. Great Lakes Res., 41, 1037–1051, https://doi.org/10.1016/j.jglr.2015.09.002, 2015.

Batjes, N. H.: A world dataset of derived soil properties by FAO–UNESCO soil unit for global modelling, Soil Use Manage., 13, 9–16, https://doi.org/10.1111/j.1475-2743.1997.tb00550.x, 1997.

Beven, K. J.: Rainfall-Runoff Modelling: The Primer, Wiley, West Sussex, UK, 2011.

Bossa, A. Y., Diekkrüger, B., Igué, A. M., and Gaiser, T.: Analyzing the effects of different soil databases on modeling of hydrological processes and sediment yield in Benin (West Africa), Geoderma, 173–174, 61–74, https://doi.org/10.1016/j.geoderma.2012.01.012, 2012.

Bower, S. S.: Watersheds: Conceptualizing Manitoba's drained landscape, 1895–1950, Environ. Hist., 12, 796–819, https://doi.org/10.1093/envhis/12.4.796, 2007.

Bronswijk, J. J. B.: Prediction of actual cracking and subsidence in clay soils, Soil Sci., 148, 87–93, 1989.

Chambers, P. A., Benoy, G. A., Brua, R. B., and Culp, J. M.: Application of nitrogen and phosphorus criteria for streams in agricultural landscapes, Water Sci. Technol., 64, 2185–2191, https://doi.org/10.2166/wst.2011.760, 2011.

Cooper, M., Mendes, L. M. S., Silva, W. L. C., and Sparovek, G.: A National Soil Profile Database for Brazil Available to International Scientists, Soil Sci. Soc. Am. J., 69, 649–652, https://doi.org/10.2136/sssaj2004.0140, 2005.

Cordeiro, M. R. C. and Sri Ranjan, R.: Corn yield response to drainage and subirrigation in the Canadian Prairies, Transactions of the American Society of Agricultural Engineers, 55, 1771–1780, https://doi.org/10.13031/2013.42369, 2012.

Cordeiro, M. R. C. and Sri Ranjan, R.: DRAINMOD simulation of corn yield under different tile drain spacing in the Canadian Prairies, Transactions of the American Society of Agricultural Engineers, 58, 1481–1491, https://doi.org/10.13031/trans.58.10768, 2015.

Cordeiro, M. R. C., Lelyk, G., Kröbel, R., Legesse, G., Faramarzi, M., Masud, M. B., and McAllister, T.: Deriving Canada-wide soils dataset for use in Soil and Water Assessment Tool (SWAT), PANGAEA, https://doi.org/10.1594/PANGAEA.877298, 2017.

Dobos, E., Daroussin, J., and Montanarella, L.: An SRTM-based procedure to delineate SOTER Terrain Units on 1 : 1 and 1 : 5 million scales. EUR 21571 EN, Office for Official Publications of the European Communities, Luxembourg, 2005.

Douglas-Mankin, K. R., Srinivasan, R., and Arnold, J. G.: Soil and Water Assessment Tool (SWAT) Model, Current Developments and Applications, 53, 1423–1431, https://doi.org/10.13031/2013.34915, 2010.

Edwards, L., Burney, J., Brimacombe, M., and Macrae, A.: Nitrogen runoff in a potato-dominated watershed area of Prince Edward

Island, Canada, The Role of Erosion and Sediment Transport in Nutrient and Contaminant Transfer, Waterloo, ON, 93–97, 2000.

Faramarzi, M., Srinivasan, R., Iravani, M., Bladon, K. D., Abbaspour, K. C., Zehnder, A. J. B., and Goss, G. G.: Setting up a hydrological model of Alberta: Data discrimination analyses prior to calibration, Environ. Modell. Softw., 74, 48–65, https://doi.org/10.1016/j.envsoft.2015.09.006, 2015.

Foster, G. R., McCool, D. K., Renard, K. G., and Moldenhauer, W. C.: Conversion of the universal soil loss equation, J. Soil Water Conserv., 36, 355–359, 1981.

Gao, G. Y., Fu, B. J., Lü, Y. H., Liu, Y., Wang, S., and Zhou, J.: Coupling the modified SCS-CN and RUSLE models to simulate hydrological effects of restoring vegetation in the Loess Plateau of China, Hydrol. Earth Syst. Sci., 16, 2347–2364, https://doi.org/10.5194/hess-16-2347-2012, 2012.

Gassman, P. W., Sadeghi, A. M., and Srinivasan, R.: Applications of the SWAT model special section: Overview and insights, J. Environ. Qual., 43, 1–8, https://doi.org/10.2134/jeq2013.11.0466, 2014.

Givi, J., Prasher, S. O., and Patel, R. M.: Evaluation of pedotransfer functions in predicting the soil water contents at field capacity and wilting point, Agr. Water Manage., 70, 83–96, https://doi.org/10.1016/j.agwat.2004.06.009, 2004.

He, Z. L., Wilson, M. J., Campbell, C. O., Edwards, A. C., and Chapman, S. J.: Distribution of phosphorus in soil aggregate fractions and its significance with regard to phosphorus transport in agricultural runoff, Water Air Soil Poll., 83, 69–84, https://doi.org/10.1007/bf00482594, 1995.

Hillel, D.: Environmental Soil Physics: Fundamentals, Applications, and Environmental Considerations, Elsevier Science, London, UK, 1998.

Koiter, A. J., Lobb, D. A., Owens, P. N., Petticrew, E. L., Tiessen, K. H. D., and Li, S.: Investigating the role of connectivity and scale in assessing the sources of sediment in an agricultural watershed in the Canadian prairies using sediment source fingerprinting, J. Soils Sediment., 13, 1676–1691, https://doi.org/10.1007/s11368-013-0762-7, 2013.

Lacasse, S. and Nadim, F.: Uncertainties in characterising soil properties, Uncertainty in the geologic environment: From theory to practice, 49–75, 1996.

Laniak, G. F., Olchin, G., Goodall, J., Voinov, A., Hill, M., Glynn, P., Whelan, G., Geller, G., Quinn, N., Blind, M., Peckham, S., Reaney, S., Gaber, N., Kennedy, R., and Hughes, A.: Integrated environmental modeling: A vision and roadmap for the future, Environ. Model. Softw., 39, 3–23, https://doi.org/10.1016/j.envsoft.2012.09.006, 2013.

Lévsque, É., Anctil, F., Van Griensven, A. V., and Beauchamp, N.: Evaluation of streamflow simulation by SWAT model for two small watersheds under snowmelt and rainfall, Hydrolog. Sci. J., 53, 961–976, https://doi.org/10.1623/hysj.53.5.961, 2008.

Maidment, D. R.: Handbook of hydrology, McGraw-Hill, New York, 1993.

Mapfumo, E., Chanasyk, D. S., and Willms, W. D.: Simulating daily soil water under foothills fescue grazing with the soil and water assessment tool model (Alberta, Canada), Hydrol. Process., 18, 2787–2800, https://doi.org/10.1002/hyp.1493, 2004.

Mednick, A. C.: Comparing the use of STATSGO and SSURGO soils data in water quality modeling: A literature review, Bureau of Science Services, Wisconsin Department of Natural Resources, 2008.

Mekonnen, B. A., Mazurek, K. A., and Putz, G.: Incorporating landscape depression heterogeneity into the Soil and Water Assessment Tool (SWAT) using a probability distribution, Hydrol. Process., 30, 2373–2389, https://doi.org/10.1002/hyp.10800, 2016.

Moeys, J.: soiltexture: Functions for Soil Texture Plot, Classification and Transformation. R package version 1.4.1, available at: https://CRAN.R-project.org/package=soiltexture (last access: 1 June 2017), 2016.

Moriasi, D. N. and Starks, P. J.: Effects of the resolution of soil dataset and precipitation dataset on SWAT2005 streamflow calibration parameters and simulation accuracy, J. Soil Water Conserv., 65, 63–78, https://doi.org/10.2489/jswc.65.2.63, 2010.

Neitsch, S., Arnold, J., Kiniry, J., Williams, J., and King, K.: SWAT2009 Theoretical Documentation, Texas Water Ressources Institute Technical Report, 2005.

Rahman, M., Bolisetti, T., and Balachandar, R.: Hydrologic modelling to assess the climate change impacts in a Southern Ontario watershed, Can. J. Civil Eng., 39, 91–103, https://doi.org/10.1139/l11-112, 2012.

Rasouli, S., Whalen, J. K., and Madramootoo, C. A.: Review: Reducing residual soil nitrogen losses from agroecosystems for surface water protection in Quebec and Ontario, Canada: Best management practices, policies and perspectives, Can. J. Soil Sci., 94, 109–127, https://doi.org/10.4141/cjss2013-015, 2014.

Santhi, C., Muttiah, R. S., Arnold, J. G., and Srinivasan, R.: A gis-based regional planning tool for irrigation demand assessment and savings using SWAT, Transactions of the American Society of Agricultural Engineers, 48, 137–147, https://doi.org/10.13031/2013.17957, 2005.

Sharpley, A. N. and Williams, J. R.: EPIC-erosion/productivity impact calculator: 1. Model documentation, Technical Bulletin-United States Department of Agriculture, Washington DC, 1990.

Sharpley, A. N., Smith, S. J., Jones, O. R., Berg, W. A., and Coleman, G. A.: The transport of bioavailable phosphorus in agricultural runoff, J. Environ. Qual., 21, 30–35, https://doi.org/10.2134/jeq1992.00472425002100010003x, 1992.

Sharpley, A. N., Kleinman, P. J. A., McDowell, R. W., Gitau, M., and Bryant, R. B.: Modeling phosphorus transport in agricultural watersheds: Processes and possibilities, J. Soil Water Conserv., 57, 425–439, 2002.

Sheshukov, A., Daggupati, P., Lee, M.-C., and Douglas-Mankin, K.: ArcMap tool for pre-processing SSURGO soil database for ArcSWAT, 5th SWAT International Conference, Boulder, CO, 2009.

Shi, X. Z., Yu, D. S., Warner, E. D., Pan, X. Z., Petersen, G. W., Gong, Z. G., and Weindorf, D. C.: Soil Database of 1 : 1,000,000 Digital Soil Survey and Reference System of the Chinese Genetic Soil Classification System, Soil Horizons, 45, 129–136, https://doi.org/10.2136/sh2004.4.0129, 2004.

Soil Landscapes of Canada Working Group: Soil landscapes of Canada version 3.2, Agriculture and Agri-Food Canada (digital map and database at 1 : 1 million scale), Ottawa, ON, Canada, 2010.

Soil Survey Staff: Soil Taxonomy: A basic system of soil classification for making and interpreting soil surveys, Agricultural Handbook 436, Soil Use and management, 1, Natural Resources Conservation Service, 57–60, 1999.

Srinivasan, R., Zhang, X., and Arnold, J.: SWAT ungauged: hydrological budget and crop yield predictions in the Upper Mississippi River Basin, Transactions of the American Society of Agricultural Engineers, 53, 1533–1546, 2010.

Stewart, D., Canfield, E., and Hawkins, R.: Curve number determination methods and uncertainty in hydrologic soil groups from semiarid watershed data, J. Hydrol. Eng., 17, 1180–1187, https://doi.org/10.1061/(ASCE)HE.1943-5584.0000452, 2012.

Stonehouse, D. P.: Profitability of soil and water conservation in Canada: A review, J. Soil Water Conserv., 50, 215–219, 1995.

USDA-NRCS: Hydrologic Soil Groups, in: National Engineering Handbook, Part 630, U.S. Department of Agriculture, Soil Conservation Service, 7.1-7.5, 1993.

Wall, G. J., Coote, D. R., Pringle, E. A., and Shelton, I. J.: Revised Universal Soil Loss Equation for Application in Canada: A Handbook for Estimating Soil Loss from Water Erosion in Canada, Agriculture and Agri-Food Canada, Ottawa, Canada, 2002.

Watson, B. M. and Putz, G.: Comparison of temperature-index snowmelt models for use within an operational water quality model, J. Environ. Qual., 43, 199–207, https://doi.org/10.2134/jeq2011.0369, 2014.

Whittaker, G., Confesor, R., Di Luzio, M., and Arnold, J.: Detection of overparameterization and overfitting in an automatic calibration of SWAT, Transactions of the American Society of Agricultural Engineers, 53, 1487–1499, 2010.

Wolfe, S. A.: Impact of increased aridity on sand dune activity in the Canadian Prairies, J. Arid Environ., 36, 421–432, https://doi.org/10.1006/jare.1996.0236, 1997.

Wösten, J. H. M., Lilly, A., Nemes, A., and Le Bas, C.: Development and use of a database of hydraulic properties of European soils, Geoderma, 90, 169–185, https://doi.org/10.1016/S0016-7061(98)00132-3, 1999.

Yang, Q., Meng, F.-R., Zhao, Z., Chow, T. L., Benoy, G., Rees, H. W., and Bourque, C. P. A.: Assessing the impacts of flow diversion terraces on stream water and sediment yields at a watershed level using SWAT model, Agr. Ecosyst. Environ., 132, 23–31, https://doi.org/10.1016/j.agee.2009.02.012, 2009.

Yang, Q., Leon, L. F., Booty, W. G., Wong, I. W., McCrimmon, C., Fong, P., Michiels, P., Vanrobaeys, J., and Benoy, G.: Land use change impacts on water quality in three Lake Winnipeg watersheds, J. Environ. Qual., 43, 1690–1701, https://doi.org/10.2134/jeq2013.06.0234, 2014.

Zhu, Z., Broersma, K., and Mazumder, A.: Impacts of land use, fertilizer and manure application on the stream nutrient loadings in the Salmon River watershed, south-central British Columbia, Canada, J. Environ. Prot., 3, 14, https://doi.org/10.4236/jep.2012.328096, 2012.

16

The National Eutrophication Survey: lake characteristics and historical nutrient concentrations

Joseph Stachelek[1], Chanse Ford[2], Dustin Kincaid[3,6], Katelyn King[1], Heather Miller[4], and Ryan Nagelkirk[5]

[1]Department of Fisheries and Wildlife, Michigan State University, East Lansing, MI, USA
[2]Department of Earth and Environmental Sciences, Michigan State University, East Lansing, MI, USA
[3]Department of Integrative Biology, Michigan State University, East Lansing, MI, USA
[4]Department of Microbiology and Molecular Genetics, Michigan State University, East Lansing, MI, USA
[5]Department of Geography, Environment, and Spatial Sciences, Michigan State University, East Lansing, MI, USA
[6]W.K. Kellogg Biological Station, Michigan State University, Hickory Corners, MI, USA

Correspondence: Joseph Stachelek (stachel2@msu.edu)

Abstract. Historical ecological surveys serve as a baseline and provide context for contemporary research, yet many of these records are not preserved in a way that ensures their long-term usability. The National Eutrophication Survey (NES) database is currently only available as scans of the original reports (PDF files) with no embedded character information. This limits its searchability, machine readability, and the ability of current and future scientists to systematically evaluate its contents. The NES data were collected by the US Environmental Protection Agency between 1972 and 1975 as part of an effort to investigate eutrophication in freshwater lakes and reservoirs. Although several studies have manually transcribed small portions of the database in support of specific studies, there have been no systematic attempts to transcribe and preserve the database in its entirety. Here we use a combination of automated optical character recognition and manual quality assurance procedures to make these data available for analysis. The performance of the optical character recognition protocol was found to be linked to variation in the quality (clarity) of the original documents. For each of the four archival scanned reports, our quality assurance protocol found an error rate between 5.9 and 17 %. The goal of our approach was to strike a balance between efficiency and data quality by combining entry of data by hand with digital transcription technologies. The finished database contains information on the physical characteristics, hydrology, and water quality of about 800 lakes in the contiguous US (Stachelek et al., 2017, https://doi.org/10.5063/F1639MVD). Ultimately, this database could be combined with more recent studies to generate meta-analyses of water quality trends and spatial variation across the continental US.

1 Introduction

Effective management of inland freshwater lakes requires an understanding of the factors that affect water quality and how these factors change over time. One of these factors, termed eutrophication, occurs when excess nutrient inputs from human activities fuels increases in algal growth, which can cause hypoxia and decreases in water clarity. Eutrophication of surface waters from increased phosphorus and nitrogen loading has been observed in connection with altered land use, especially in areas of rapid urbanization and intensive agriculture (Smith et al., 1999, 2014). As human populations and their impacts continue to grow, eutrophication is expected to become more widespread (Bennett et al., 2001; Taranu and Gregory-Eaves, 2008). Historical datasets are needed in order to track, understand, and manage eutroph-

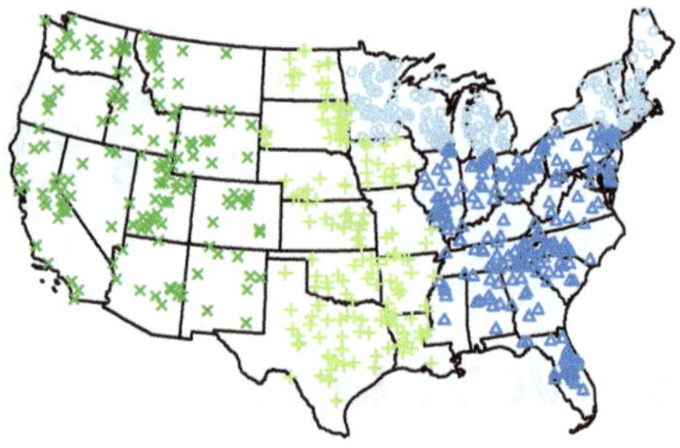

Figure 1. Survey locations colored by sampling year (1972 northeastern: light blue; 1973 southeastern: blue; 1974 central: light green; 1975 western: green).

ication in lakes and reservoirs because they serve as an important baseline for modern studies.

The US Environmental Protection Agency (EPA) designed and implemented the National Eutrophication Survey (NES) in order to investigate the extent of eutrophication in freshwater lakes and reservoirs across the contiguous US. Sampling took place in over 800 lakes and reservoirs from 1972 to 1975 and included a variety of physical, chemical, and biological metrics including data on nutrients and nutrient loading, hydrologic retention time, morphometry, and plankton community diversity. Each lake was sampled on a monthly basis for a period of 1 year. Except for the phytoplankton distribution subset, which we did not transcribe (see Stomp et al., 2011), the NES data are provided as annual averages. Unlike current EPA National Lakes Assessments (NLAs) that select a random sample of lakes across the US, the NES targeted only lakes impacted directly or indirectly by municipal sewage treatment plant discharge (USEPA, 2009, 1975). Until recently, these data were only available in their entirety as four separate scanned reports representing the northeastern and north-central (northeastern), eastern and southeastern (southeastern), central, and western regions of the US (Fig. 1). In the remainder of the present paper we refer to the former two regions as simply the northeastern and southeastern regions.

To our knowledge, there have been no attempts to transcribe the data into a usable, searchable digital database despite its use in previous studies. For example, large portions of the dataset were used to examine large-scale relationships between residence time and phytoplankton abundance (Soballe and Kimmel, 1987). Also, it was used to predict eutrophication incidence in a Bayesian framework (Lamon and Stow 2004). Smaller portions of the data were used to explore drivers of nutrient loading (Stomp et al., 2011; Brett and Benjamin, 2008). However, to our knowledge, the only study to use the NES dataset and provide a publicly available data supplement is that of Stomp et al. (2011), but their data supplement was limited to a small subset of the available variables relating to phytoplankton community diversity.

The present study is the first to leverage digital transcription technologies to unlock the full NES dataset. In this paper, we describe the digital transcription of the full NES dataset with the goal of making the dataset openly accessible to the research community. Specifically, our objective was to exactly reproduce the contents of the original dataset rather than to evaluate its scientific integrity. We introduce and publish the data in an open format that requires no proprietary software. It can be easily downloaded, used for analysis, and amended. The provided summary statistics and figures also allow users to quickly assess the utility of the data. Finally, the code and raw data files are provided to facilitate the extraction of fields not represented in our completed dataset (mostly phytoplankton diversity data).

2 Methods

Data were collected from multiple locations within the water column and included in situ measurements as well as laboratory analyses. Flow estimates and drainage area calculations were provided by the US Geological Survey and were determined from flow gauges when present. More detailed information on sampling methods, units, equipment, and accuracy can be found in the EPA survey methods publication (USEPA, 1975). Due to the historical nature of the dataset, the NES sampling design differs from more modern efforts (USEPA, 2009). For example, the original NES data were collected from four separate regions of the US over the course of 4 years, whereas current assessments complete nationwide sampling in a single summer. As such, NES data values represent the mean of measurements taken in the spring, summer, and fall in either 1972 (northeastern), 1973 (southeastern), 1974 (central), or 1975 (western) rather than summer measurements taken in a single year.

We obtained the NES archival scanned reports from the EPA National Service Center for Environmental Publications (available at: https://www.epa.gov/nscep). The data for each NES region are contained in four separate files. We extracted the data from each file using automated techniques followed by manual quality assurance and checking of each value. To begin, we enhanced (de-noised) each file using the local adaptive filtering algorithm as provided by the ImageMagick program (v6.8.9-9; available at https://www.imagemagick.org/). Next, we processed the enhanced files using the Tesseract optical character recognition (OCR) program (Ooms, 2017; Smith, 2007). The output of these initial extraction steps was recorded in a set of "raw data" files in which each file contains the raw unprocessed text of each document page. The contents of specific fields in the raw data were extracted to a database using the automated rules provided by the nesR software package (Stachelek, 2017). Finally, all values in the database were manually checked for accuracy against the original scanned reports. Inaccurate OCR outputs were corrected by hand in the final database. Because our goal was to reproduce the data from the original reports and not to verify the technical correctness of the original data, we

Table 1. Number of measurements (n) for each variable in each NES region.

Variable	Western	Central	Northeastern	Southeastern
Drainage area	122	138	171	232
Surface area	152	177	200	245
Mean depth	149	174	174	242
Total inflow	124	138	170	232
Retention time	124	140	158	230
Alkalinity	153	177	200	245
Conductivity	153	176	200	245
Secchi depth	153	177	200	245
Total P	153	177	200	245
Total inorg. P	153	177	200	245
Total inorg. N	153	177	200	245
Total N	152	176	1	245
P pt. source mun.	52	83	139	189
P pt. source ind.	7	1	10	24
P pt. source sep.	65	88	111	175
P nonpt. source	122	133	167	231
P total inputs	122	133	167	231
N pt. source mun.	52	84	139	189
N pt. source ind.	7	1	8	22
N pt. source sep.	77	90	111	184
N nonpt. source	122	129	167	231
N total inputs	122	129	167	231
P total exports	119	132	167	227
P retention	99	115	144	201
P load per area	122	133	167	231
N total exports	119	133	166	227
N retention	88	111	122	170
N load per area	122	135	167	231

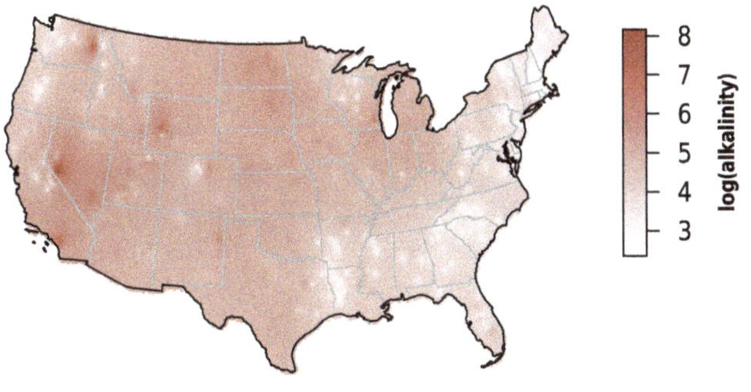

Figure 2. Map of log-scaled alkalinity (mg L^{-1}) interpolated using inverse distance weighting.

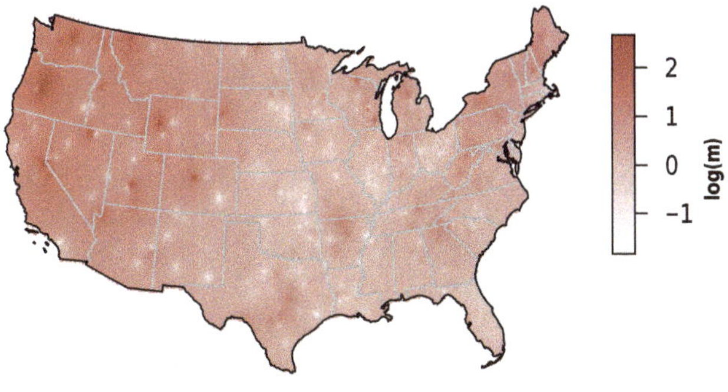

Figure 3. Map of Secchi depth (m) interpolated using inverse distance weighting.

only changed values if they did not match the original data reports. For example, we did not change data from the five NES lakes that had phosphate (PO_4) values exceeding their corresponding total phosphorus (TP) values despite the fact that this is not physically possible (PO^4 is a component of TP).

We provide the final dataset in an open nonproprietary format (comma-delimited, *.csv). In addition, we generated metadata descriptions from the contents of the original scanned reports. All calculations, table construction, and figure generation were performed in R and saved as reproducible R scripts (R Core Team, 2017). Table and figure generation was accomplished with the use of the reshape2, plyr, and sp packages (Wickham, 2016; Pebesma and Bivand, 2017).

3 Results

The final NES dataset contains observations from 775 lakes and the distribution of these lakes was spatially variable. Although there were more lakes measured in the northeastern and southeastern US, the number of locations was close to evenly distributed among the remaining regions (Fig. 1, Table 1). Specifically, the number of lakes sampled in each region were as follows: northeastern – 200 lakes, southeastern – 245 lakes, central – 177 lakes, and western – 153 lakes.

In addition to differences in the total number of lakes measured in each region, there were also differences in the proportion of lakes classified as impoundments rather than as natural lakes. For example, slightly more than half of all the lakes studied (462 of 775) were classified as impoundments yet the northeastern region had only 54 impoundments while the southeastern region had 168 impoundments. Conversely, the number of natural lakes sampled in the northeastern region (146 lakes) was more than double that of any other region (77, 48, and 42 for the southeastern, western, and central US, respectively).

We observed substantial spatial variation in many of the individual lake characteristics. For example, lakes in the eastern subregions were generally smaller and shallower than lakes in the western subregion (Table 2). In addition, lakes in the western subregion generally had higher alkalinity and higher water clarity (Figs. 2 and 3). Lakes with particularly low alkalinity were found in coastal areas, whereas lakes with particularly high alkalinity were found in Nevada, western Washington, and parts of North Dakota. Comparisons among regions was easy for some well-sampled lake chemistry parameters such as TP but more difficult for undersampled lake chemistry parameters. A particularly extreme example of this difficulty was total nitrogen measurements in

Table 2. Mean and standard deviation (SD) for each variable in each NES region.

Region Variable	Western Mean ± SD	Central Mean ± SD	Northeastern Mean ± SD	Southeastern Mean ± SD
Drainage area (km^2)	$2.5 \times 10^4 \pm 7.8 \times 10^4$	$2.1 \times 10^4 \pm 7.5 \times 10^4$	$3.2 \times 10^3 \pm 1.4 \times 10^4$	$5.3 \times 10^3 \pm 1.4 \times 10^4$
Surface area (km^2)	44.57 ± 99.83	$54.38 \pm 1.4 \times 10^2$	27.25 ± 99.01	$42.7 \pm 1.4 \times 10^2$
Mean depth (m)	16.71 ± 27.08	5.97 ± 4.49	7 ± 9.37	6.4 ± 6.07
Total inflow ($m^3 s^{-1}$)	$52.1 \pm 1.1 \times 10^2$	31.82 ± 71.77	23.1 ± 65.26	$82.6 \pm 2.3 \times 10^2$
Retention time (yr)	7.27 ± 43.32	2.78 ± 6.98	2.01 ± 4.77	0.59 ± 1.12
Alkalinity ($mg L^{-1}$)	$1.7 \times 10^2 \pm 3.7 \times 10^2$	$1.5 \times 10^2 \pm 91.51$	$1.2 \times 10^2 \pm 1.6 \times 10^2$	72.18 ± 66.25
Conductivity ($\mu\Omega$)	$4.9 \times 10^2 \pm 1.0 \times 10^3$	$6.4 \times 10^2 \pm 7.6 \times 10^2$	$3.3 \times 10^2 \pm 4.0 \times 10^2$	$2.5 \times 10^2 \pm 2.2 \times 10^2$
Secchi depth (m)	2.86 ± 2.64	1.2 ± 0.91	1.81 ± 1.71	1.22 ± 0.82
Total P ($mg L^{-1}$)	0.07 ± 0.13	0.11 ± 0.16	0.16 ± 0.35	0.12 ± 0.27
Total inorg. P ($mg L^{-1}$)	0.04 ± 0.11	0.04 ± 0.07	0.11 ± 0.3	0.05 ± 0.15
Total inorg. N ($mg L^{-1}$)	0.14 ± 0.23	0.33 ± 0.58	0.47 ± 0.66	0.72 ± 0.91
Total N ($mg L^{-1}$)	0.62 ± 0.65	1.22 ± 1.11	0.12	1.56 ± 1.25
P pt. source mun. ($kg yr^{-1}$)	$2.5 \times 10^4 \pm 8.7 \times 10^4$	$2.3 \times 10^4 \pm 5.6 \times 10^4$	$3.5 \times 10^4 \pm 1.5 \times 10^5$	$4.5 \times 10^4 \pm 1.1 \times 10^5$
P pt. source ind. ($kg yr^{-1}$)	$2.5 \times 10^4 \pm 4.0 \times 10^4$	$1.3 \times 10^4 \pm$ NA	$2.7 \times 10^4 \pm 4.9 \times 10^4$	$1.7 \times 10^4 \pm 4.5 \times 10^4$
P pt. source sep. ($kg yr^{-1}$)	$56.62 \pm 1.4 \times 10^2$	60.62 ± 93.67	$1.6 \times 10^2 \pm 3.4 \times 10^2$	$98.55 \pm 2.3 \times 10^2$
P nonpt. source ($kg yr^{-1}$)	$1.4 \times 10^5 \pm 4.2 \times 10^5$	$1.8 \times 10^5 \pm 6.8 \times 10^5$	$5.6 \times 10^4 \pm 2.1 \times 10^5$	$1.9 \times 10^5 \pm 5.5 \times 10^5$
P total inputs ($kg yr^{-1}$)	$1.5 \times 10^5 \pm 4.7 \times 10^5$	$2.0 \times 10^5 \pm 7.0 \times 10^5$	$8.7 \times 10^4 \pm 3.4 \times 10^5$	$2.3 \times 10^5 \pm 5.8 \times 10^5$
N pt. source mun. ($kg yr^{-1}$)	$7.8 \times 10^4 \pm 2.5 \times 10^5$	$7.3 \times 10^4 \pm 1.7 \times 10^5$	$1.4 \times 10^5 \pm 5.4 \times 10^5$	$1.4 \times 10^5 \pm 3.8 \times 10^5$
N pt. source ind. ($kg yr^{-1}$)	$2.3 \times 10^7 \pm 6.1 \times 10^7$	$4.0 \times 10^3 \pm$ NA	$1.6 \times 10^5 \pm 4.2 \times 10^5$	$1.7 \times 10^5 \pm 5.6 \times 10^5$
N pt. source sep. ($kg yr^{-1}$)	$5.7 \times 10^6 \pm 5.0 \times 10^7$	$2.2 \times 10^3 \pm 3.5 \times 10^3$	$4.3 \times 10^3 \pm 5.5 \times 10^3$	$3.3 \times 10^3 \pm 6.7 \times 10^3$
N nonpt. source ($kg yr^{-1}$)	$1.8 \times 10^6 \pm 4.9 \times 10^6$	$1.8 \times 10^6 \pm 4.4 \times 10^6$	$1.2 \times 10^6 \pm 4.1 \times 10^6$	$3.1 \times 10^6 \pm 8.9 \times 10^6$
N total inputs ($kg yr^{-1}$)	$6.8 \times 10^6 \pm 5.7 \times 10^7$	$1.8 \times 10^6 \pm 4.3 \times 10^6$	$1.3 \times 10^6 \pm 4.6 \times 10^6$	$3.2 \times 10^6 \pm 9.0 \times 10^6$
P total exports ($kg yr^{-1}$)	$6.2 \times 10^4 \pm 1.7 \times 10^5$	$7.4 \times 10^4 \pm 1.9 \times 10^5$	$7.3 \times 10^4 \pm 3.1 \times 10^5$	$1.9 \times 10^5 \pm 6.3 \times 10^5$
P retention (%)	47.77 ± 28.5	57.55 ± 26.01	36.93 ± 25.2	42.7 ± 23.34
P load per area ($g m^{-2} yr^{-1}$)	5.61 ± 21.36	3.3 ± 9.2	28.46 ± 97.49	9.43 ± 17.06
N total exports ($kg yr^{-1}$)	$1.6 \times 10^6 \pm 4.0 \times 10^6$	$1.2 \times 10^6 \pm 2.8 \times 10^6$	$1.2 \times 10^6 \pm 4.9 \times 10^6$	$3.0 \times 10^6 \pm 8.3 \times 10^6$
N retention (%)	39.33 ± 27.13	43.41 ± 23.97	28.41 ± 23.62	26.28 ± 18.85
N load per area ($g m^{-2} yr^{-1}$)	$1.8 \times 10^2 \pm 1.1 \times 10^3$	$42.67 \pm 1.1 \times 10^2$	$2.8 \times 10^2 \pm 9.1 \times 10^2$	$1.3 \times 10^2 \pm 2.4 \times 10^2$

the eastern region, as this parameter was only measured for a single lake (Table 1).

The ability to examine these spatial trends was made possible by our OCR procedure, which had 6–17 % accuracy depending on region and archival report scan quality. In total, we carried out approximately 5000 corrections to the automated data product by hand as part of our manual quality control review. A total of approximately 650 lakes had values for at least 80 % of the total number of variables shown in Table 1. On an individual lake basis, the most common "missing" data were nutrient loading estimates for individual point- and nonpoint-source components. In many cases, these data may not actually be missing but they may not have been a component of the budget for that particular lake. For example, not all lakes have industrial land use so no data are expected in these cases.

4 Code and data availability

Original scanned reports from the EPA are available from the EPA National Service Center for Environmental Publications (https://www.epa.gov/nscep). Our cleaned and useable data are available for download at Stachelek et al. (2017). The data are provided as a zip file, which contains all versions of the data including the raw and quality-checked versions (Stachelek et al., 2017). Moreover, the R package and R code used to scrape and analyze the data are provided by Stachelek (2017) so that the methods may be reproduced and openly available for (re)use. All figures and summary statistics were generated with R scripts available in the data supplement.

5 Discussion

We have demonstrated an approach for rescuing historical data from scanned documents. In particular, our approach involved a two-step process of automated data scraping followed by curation by hand and quality assurance. Overall, we found that OCR was an efficient method for reducing the labor associated with transcribing analog text records (e.g., Drinkwater et al., 2014). Unfortunately, OCR technology does not have absolute accuracy. In our case, transcription was hampered by poor print and scan quality of the source paper documents. We discovered through our manual validation procedure that the OCR computations produced inaccurate values in approximately 6–17 % of the cells in the complete dataset ($n = 4836$). We expect that accuracy could be improved by experimenting with varying the window size of the local adaptive thresholding algorithm relative to the document font size. Our ability to experiment with thresholding window size was limited due to the computationally expensive nature of these extractions.

The end result of our approach was data from every lake and nearly every variable in the NES survey dataset. The only primary subset of the NES data that is not included in our final product is the phytoplankton distribution data, which have already been digitally transcribed by Stomp et al. (2011). The results of the present study could be used to explore anthropogenic and environmental drivers of lake eutrophication as well as to verify previously documented trends. One example is the 2007 National Lakes Assessment Report, which included a reanalysis of some of the NES study lakes (USEPA, 2009). This reanalysis considered population level trends in the NES lakes but did not consider trends in individual lakes or potential environmental drivers contributing to observed trends. On a population basis, the NLA reanalysis found that less than 30 % of the NES lakes had increased chlorophyll and phosphorus concentrations. The results of the present study could be used to verify these claims as well as to compare the NES data with more recent work such as the 2012 National Lakes Assessment. Note that sampling techniques may differ from current techniques; thus, care should be given when making comparisons. In addition to their utility in validating historical trends, this dataset has value because it contains data on a number of hydrographic variables that are difficult to estimate, such as water residence (retention) time. Such data are critical to a variety of hydrological and water quality modeling efforts (Brett and Benjamin, 2008).

Although our goal was to digitally transcribe the full NES dataset to facilitate studies on historical nutrient loading, it is worth noting the similarities between the present study and other scientific record digitization initiatives. Such initiatives are common in the climate and ocean sciences but they are just starting to gain momentum in the biological sciences (Allan et al., 2011; Freeman et al., 2017). To our knowledge, the present study is the first large-scale attempt at digitization of historical limnology records. We hope that by making our analysis open and reproducible we will inspire future efforts to recover important records from the pre-digital era.

Author contributions. All authors contributed to data quality assurance and edited the article text. JS conceived the study and implemented the optical character recognition code. CF, DK, and RN performed the data analysis and made figures. KK, HM, and JS wrote major parts of the paper.

Competing interests. The authors declare that they have no conflict of interest.

Acknowledgements. This work was developed as part of the Reproducible Quantitative Methods course (https://cbahlai.github.io/rqm-template/) led by Christie Bahlai, which was funded by the Mozilla Foundation, the Leona M. and Harry B. Helmsley Charitable Trust, the Michigan State University Program in Ecology and Evolutionary Biology, the BEACON Center for the Study of Evolution in Action, and the Kellogg Biological Station Long-Term Ecological Research site (NSF-DEB no. 1027253). Joseph Stachelek was supported by National Science Foundation grant ICER-1517823.

Edited by: David Carlson

References

Allan, R., Brohan, P., Compo, G. P., Stone, R., Luterbacher, J., and Brönnimann, S.: The international atmospheric circulation reconstructions over the earth (ACRE) initiative, B. Am. Meteorol. Soc., 92, 1421–1425, 2011.

Bennett, E. M., Carpenter, S. R., and Caraco, N. F.: Human impact on erodable phosphorus and eutrophication: a global perspective: increasing accumulation of phosphorus in soil threatens rivers, lakes, and coastal oceans with eutrophication, AIBS Bulletin, 51, 227–234, 2001.

Brett, M. T. and Benjamin, M. M.: A review and reassessment of lake phosphorus retention and the nutrient loading concept, Freshwater Biol., 53, 194–211, https://doi.org/10.1111/j.1365-2427.2007.01862.x, 2008.

Drinkwater, R. E., Cubey, R. W., and Haston, E. M.: The use of Optical Character Recognition (OCR) in the digitisation of herbarium specimen labels, PhytoKeys, 38, 15–30, 2014.

Freeman, E., Woodruff, S. D., Worley, S. J., Lubker, S. J., Kent, E. C., Angel, W. E., Berry, D. I., Brohan, P., Eastman, R., Gates, L., Gloeden, W., Ji, Z., Lawrimore, J., Rayner, N. A., Rosenhagen, G. and Smith, S. R.: ICOADS Release 3.0: a major update to the historical marine climate record, Int. J. Climatol., 37, 2211–2232, https://doi.org/10.1002/joc.4775, 2017.

Ooms, J.: tesseract: Open Source OCR Engine for R, available at: https://CRAN.R-project.org/package=tesseract, R package version 1.6, last access: 2 February 2017.

Pebesma, E. and Bivand, R.: sp: Classes and Methods for Spatial Data, available at: https://CRAN.R-project.org/package=sp, R package version 1.2-5, last access: 2 February 2017.

R Core Team: R: A Language and Environment for Statistical Computing, R Foundation for Statistical Computing, Vienna, Austria, available at: https://www.R-project.org/, last access: 2 February 2017.

Smith, R.: An overview of the Tesseract OCR engine, in: Document Analysis and Recognition, ICDAR 2007, Ninth International Conference on, vol. 2, 629–633, IEEE, 2007.

Smith, V. H., Tilman, G. D., and Nekola, J. C.: Eutrophication: impacts of excess nutrient inputs on freshwater, marine, and terrestrial ecosystems, Environ. Pollut., 100, 179–196, 1999.

Smith, V. H., Dodds, W. K., Havens, K. E., Engstrom, D. R., Paerl, H. W., Moss, B., and Likens, G. E.: Comment: Cultural eutrophication of natural lakes in the United States is real and widespread, Limnol. Oceanogr., 59, 2217–2225, 2014.

Soballe, D. and Kimmel, B.: A large-scale comparison of factors influencing phytoplankton abundance in rivers, lakes, and impoundments, Ecology, 68, 1943–1954, 1987.

Stachelek, J.: nesR: Scrape Data from National Eutrophication Survey archival PDFs, https://doi.org/10.5281/zenodo.1048154, R package version 0.2, 2017.

Stachelek, J., Ford, C., Kincaid, D., King, K., Miller, H., and Nagelkirk, R.: The National Eutrophication Survey: lake characteristics and historical nutrient concentrations. Knowledge Network for Biocomplexity, https://doi.org/10.5063/F1639MVD, 2017.

Stomp, M., Huisman, J., Mittelbach, G. G., Litchman, E., and Klausmeier, C. A.: Large-scale biodiversity patterns in freshwater phytoplankton, Ecology, 92, 2096–2107, 2011.

Taranu, Z. E. and Gregory-Eaves, I.: Quantifying relationships among phosphorus, agriculture, and lake depth at an inter-regional scale, Ecosystems, 11, 715–725, 2008.

USEPA: National Eutrophication Survey Methods 1973–1976 (Working Paper No. 175), Tech. rep., United States Environmental Protection Agency, Office of Research and Development, Corvallis, OR, USA, 1975.

USEPA: National Lakes Assessment: A Collaborative Survey of the Nation's Lakes, Tech. rep., United States Environmental Protection Agency, Office of Research and Development, Washington, D.C., USA, 2009.

Wickham, H.: plyr: Tools for Splitting, Applying and Combining Data, available at: https://CRAN.R-project.org/package=plyr (last access: 2 February 2017), R package version 1.8.4, 2016.

Permissions

All chapters in this book were first published in ESSD, by Copernicus Publications; hereby published with permission under the Creative Commons Attribution License or equivalent. Every chapter published in this book has been scrutinized by our experts. Their significance has been extensively debated. The topics covered herein carry significant findings which will fuel the growth of the discipline. They may even be implemented as practical applications or may be referred to as a beginning point for another development.

The contributors of this book come from diverse backgrounds, making this book a truly international effort. This book will bring forth new frontiers with its revolutionizing research information and detailed analysis of the nascent developments around the world.

We would like to thank all the contributing authors for lending their expertise to make the book truly unique. They have played a crucial role in the development of this book. Without their invaluable contributions this book wouldn't have been possible. They have made vital efforts to compile up to date information on the varied aspects of this subject to make this book a valuable addition to the collection of many professionals and students.

This book was conceptualized with the vision of imparting up-to-date information and advanced data in this field. To ensure the same, a matchless editorial board was set up. Every individual on the board went through rigorous rounds of assessment to prove their worth. After which they invested a large part of their time researching and compiling the most relevant data for our readers.

The editorial board has been involved in producing this book since its inception. They have spent rigorous hours researching and exploring the diverse topics which have resulted in the successful publishing of this book. They have passed on their knowledge of decades through this book. To expedite this challenging task, the publisher supported the team at every step. A small team of assistant editors was also appointed to further simplify the editing procedure and attain best results for the readers.

Apart from the editorial board, the designing team has also invested a significant amount of their time in understanding the subject and creating the most relevant covers. They scrutinized every image to scout for the most suitable representation of the subject and create an appropriate cover for the book.

The publishing team has been an ardent support to the editorial, designing and production team. Their endless efforts to recruit the best for this project, has resulted in the accomplishment of this book. They are a veteran in the field of academics and their pool of knowledge is as vast as their experience in printing. Their expertise and guidance has proved useful at every step. Their uncompromising quality standards have made this book an exceptional effort. Their encouragement from time to time has been an inspiration for everyone.

The publisher and the editorial board hope that this book will prove to be a valuable piece of knowledge for researchers, students, practitioners and scholars across the globe.

List of Contributors

Wei Qu, Heye R. Bogena, Johan A. Huisman, Marius Schmidt, Ralf Kunkel, Ansgar Weuthen, Henning Schiedung, Bernd Schilling, Jürgen Sorg and Harry Vereecken
Agrosphere Institute (IBG-3), Forschungszentrum Jülich GmbH, 52425 Jülich, Germany

Christopher Spence and Newell Hedstrom
Environment and Climate Change Canada, Saskatoon, SK, Canada

Hong Xuan Do, Michael Leonard and Seth Westra
School of Civil, Environmental and Mining Engineering, University of Adelaide, Adelaide, Australia

Lukas Gudmundsson
ETH Zürich, Institute for Atmospheric and Climate Science, Zürich, Switzerland

Daniel Zamrsky
Department of Physical geography, Utrecht University, Utrecht, the Netherlands

Gualbert H. P. Oude Essink and Marc F. P. Bierkens
Department of Physical geography, Utrecht University, Utrecht, the Netherlands Deltares, Utrecht, the Netherlands

Marco A. Hernández-Henríquez, Mark Taylor and Stephen J. Déry
Environmental Science and Engineering Program, University of Northern British Columbia, 3333 University Way, Prince George, British Columbia, V2N 4Z9, Canada

Aseem R. Sharma and Hadleigh D. Thompson
Natural Resources and Environmental Studies Program, University of Northern British Columbia, 3333 University Way, Prince George, British Columbia, V2N 4Z9, Canada

Carsten Montzka, Michael Herbst and Lutz Weihermüller
Forschungszentrum Jülich GmbH, Institute of Bio- and Geosciences: Agrosphere (IBG-3), Jülich, Germany

Anne Verhoef
University of Reading, Department of Geography and Environmental Science, Reading, UK

Harry Vereecken
Forschungszentrum Jülich GmbH, Institute of Bio- and Geosciences: Agrosphere (IBG-3), Jülich, Germany
International Soil Modeling Consortium, Jülich, Germany

Roger Bales, Erin Stacy, Mohammad Safeeq, Xiande Meng and Martha Conklin
Sierra Nevada Research Institute, University of California, Merced, California, USA

Matthew Meadows
Sierra Nevada Research Institute, University of California, Merced, California, USA
Kings River Watershed Association, Fresno, California, USA

Carlos Oroza and Steven Glaser
Department of Civil and Environmental Engineering, University of California, Berkeley, California, USA

Joseph Wagenbrenner
USDA Forest Service, Pacific Southwest Research Station, Arcata, California, USA

Michael J. Bowes, Linda K. Armstrong, Sarah A. Harman, Heather D. Wickham, David J. E. Nicholls, Peter M. Scarlett, Colin Roberts, Helen P. Jarvie, Gareth H. Old, Emma Gozzard, Nuria Bachiller-Jareno and Daniel S. Read
Centre for Ecology and Hydrology, Benson Lane, Crowmarsh Gifford, Wallingford, Oxfordshire, OX10 8BB, UK

Roger C. Bales, Erin M. Stacy, Xiande Meng and Martha H. Conklin
Sierra Nevada Research Institute, University of California, Merced, CA 95343, USA

Peter B. Kirchner
Southwest Alaska Network, Inventory & Monitoring, National Park Service, Anchorage, AK 99501, USA

Zeshi Zheng
Department of Civil and Environmental Engineering, University of California, Berkeley, CA 94720, USA

Lukas Gudmundsson
Institute for Atmospheric and Climate Science, Department of Environmental Systems Science, ETH Zurich, Universitaetstrasse 16, Zurich 8092, Switzerland

Hong Xuan Do, Michael Leonard and Seth Westra
School of Civil, Environmental and Mining Engineering, University of Adelaide, Adelaide, Australia

Martin Hirschi and Sonia I. Seneviratne
Institute for Atmospheric and Climate Science, ETH Zurich, Universitätstrasse 16, 8092 Zürich, Switzerland

Wenchao Ma
Center for Research in Isotopes and Environmental Dynamics, University of Tsukuba, Tsukuba, 305-8577, Japan

Jun Asanuma and Yuichi Onda
Center for Research in Isotopes and Environmental Dynamics, University of Tsukuba, Tsukuba, 305-8577, Japan
Faculty of Life and Environmental Sciences, University of Tsukuba, Tsukuba, 305-8577, Japan

Jianqing Xu
College of Business Administration, Kanto Gakuin University, Yokohama, 236-0037, Japan

Maliko Tanguy, Katie Smith and Jamie Hannaford
NERC Centre for Ecology & Hydrology, Maclean Building, Benson Lane, Crowmarsh Gifford, Wallingford, Oxon, OX10 8BB, UK

Christel Prudhomme
NERC Centre for Ecology & Hydrology, Maclean Building, Benson Lane, Crowmarsh Gifford, Wallingford, Oxon, OX10 8BB, UK
European Centre for Medium-Range Weather Forecasts, Shinfield Road, Reading, RG2 9AX, UK
Department of Geography, Loughborough University, Loughborough, LE11 3TU, UK

Chanse Ford
Department of Earth and Environmental Sciences, Michigan State University, East Lansing, MI, USA

Dustin Kincaid
Department of Integrative Biology, Michigan State University, East Lansing, MI, USA
W.K. Kellogg Biological Station, Michigan State University, Hickory Corners, MI, USA

Vassilis G. Aschonitis, Nicolo Colombani, Andrea Ghirardini, Giuseppe Castaldelli and Elisa-Anna Fano
Department of Life Sciences and Biotechnology, University of Ferrara, Ferrara, Italy

Dimitris Papamichail and Kleoniki Demertzi
Department of Hydraulics, Soil Science and Agricultural Engineering, Aristotle University of Thessaloniki, Thessaloniki, Greece

Micol Mastrocicco
Department of Environmental, Biological and Pharmaceutical Sciences and Technologies, University of Campania Luigi Vanvitelli, Caserta, Italy

Marcos R. C. Cordeiro, Roland Kröbel and Tim McAllister
Science and Technology Branch, Agriculture and Agri-Food Canada, Lethbridge, AB, T1J 4B1, Canada

Glenn Lelyk
Science and Technology Branch, Agriculture and Agri-Food Canada, Winnipeg, MB, R3T 2N2, Canada

Getahun Legesse
Department of Animal Science, University of Manitoba, Winnipeg, MB, R3T 2N2, Canada

Monireh Faramarzi and Mohammad Badrul Masud
Department of Earth and Atmospheric Sciences, University of Alberta, Edmonton, AB, T6G 2E3, Canada

Joseph Stachelek and Katelyn King
Department of Fisheries and Wildlife, Michigan State University, East Lansing, MI, USA

Heather Miller
Department of Microbiology and Molecular Genetics, Michigan State University, East Lansing, MI, USA

Ryan Nagelkirk
Department of Geography, Environment, and Spatial Sciences, Michigan State University, East Lansing, MI, USA

Index

www.ingramcontent.com/pod-product-compliance
Lightning Source LLC
LaVergne TN
LVHW080907210926
848485LV00004B/608
* 9 7 8 1 6 3 9 8 9 0 5 2 1 *